ton

rt

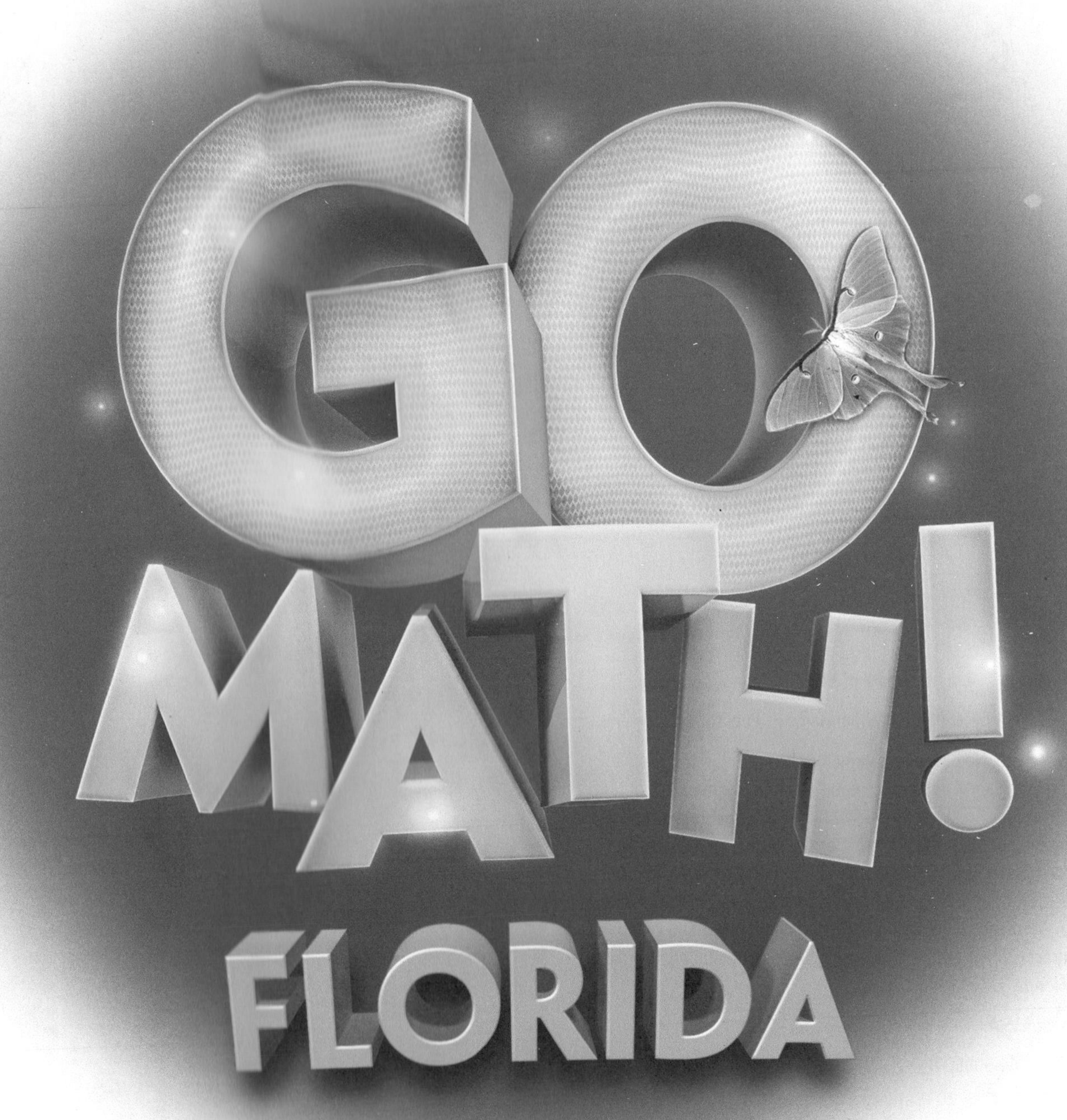

Printed in the U.S.A.

ISBN 978-0-544-50080-8

7 8 9 10 1468 23 22 21 20

4500764866 ^ B C D E F G

Dear Students and Families,

Welcome to **Go Math!**, Grade 3! In this exciting mathematics program, there are hands-on activities to do and real-world problems to solve. Best of all, you will write your ideas and answers right in your book. In **Go Math!**, writing and drawing on the pages helps you think deeply about what you are learning, and you will really understand math!

By the way, all of the pages in your **Go Math!** book are made using recycled paper. We wanted you to know that you can Go Green with **Go Math!**

Sincerely,

The Authors

Made in the United States
Text printed on 100% recycled paper

Authors

Juli K. Dixon
Professor of Mathematics Education
University of Central Florida
Orlando, Florida

Miriam A. Leiva
Founding President, TODOS:
Mathematics for All
Distinguished Professor
of Mathematics Emerita
University of North Carolina Charlotte
Charlotte, North Carolina

Matt Larson
Curriculum Specialist for Mathematics
Lincoln Public Schools
Lincoln, Nebraska

Thomasenia Lott Adams
Professor of Mathematics Education
University of Florida
Gainesville, Florida

Whole Number Operations

Developing understanding of multiplication and division and strategies for multiplication and division within 100

1 Addition and Subtraction Within 1,000 — 3

Domains Operations and Algebraic Thinking
Number and Operations in Base Ten

2 Represent and Interpret Data — 59

Domain Measurement and Data

DIGITAL PATH
Go online! Your math lessons are interactive. Use *i*Tools, Animated Math Models, the Multimedia eGlossary, and more.

Look for these:

Project Inventing Toys

REAL WORLD

H.O.T. Higher Order Thinking

Connect to Reading pp. 42, 90

Use every day for Standards Practice.

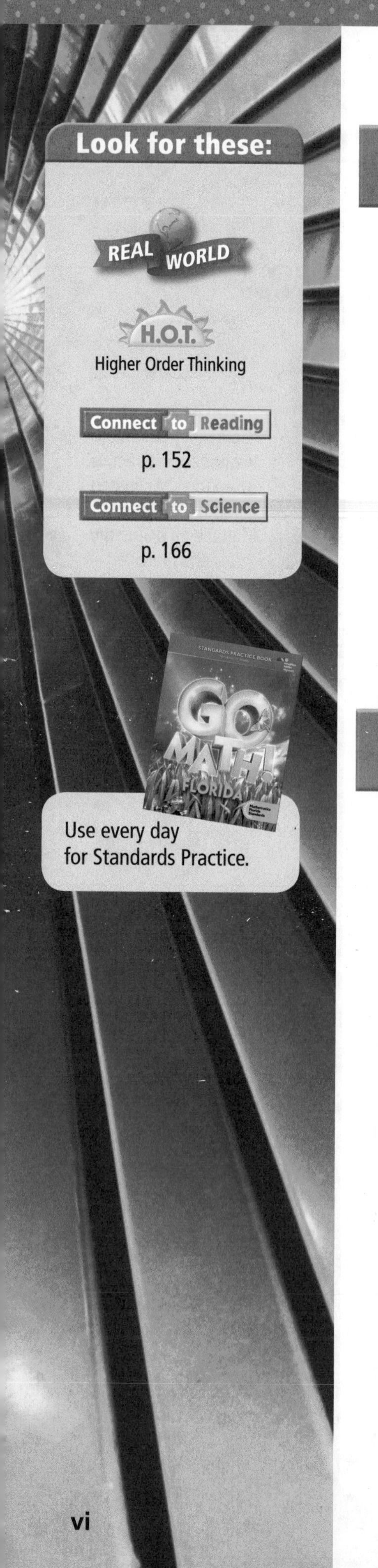

3 Understand Multiplication 95

Domain Operations and Algebraic Thinking

4 Multiplication Facts and Strategies 131

Domain Operations and Algebraic Thinking

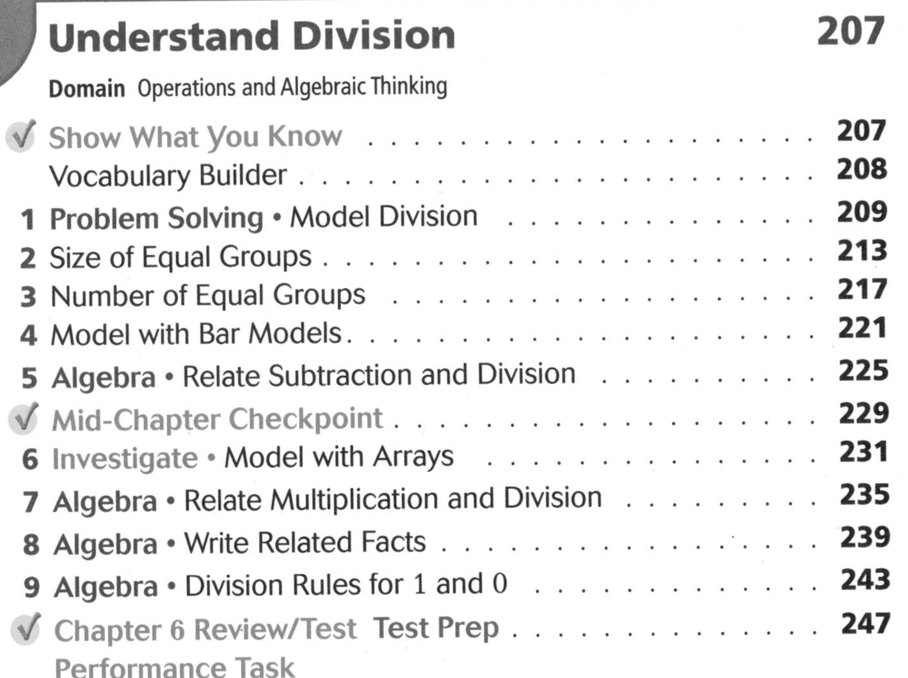

Look for these:

REAL WORLD

H.O.T. Higher Order Thinking

Connect to Reading

p. 246

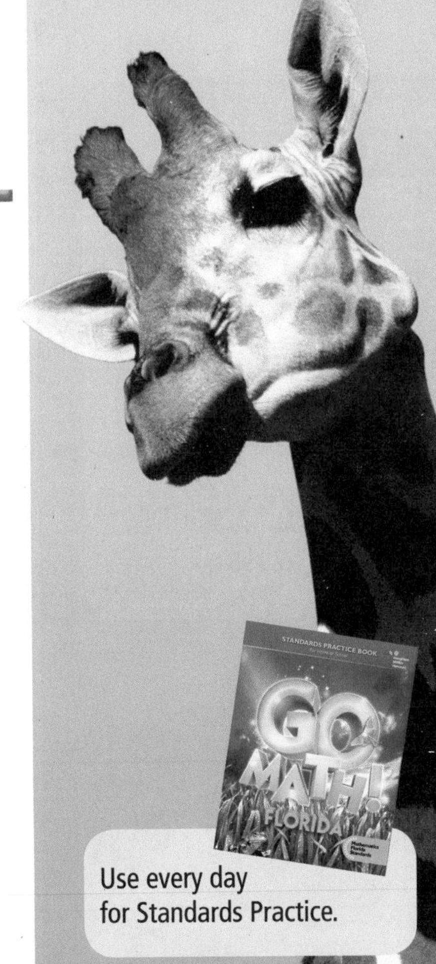

Use every day for Standards Practice.

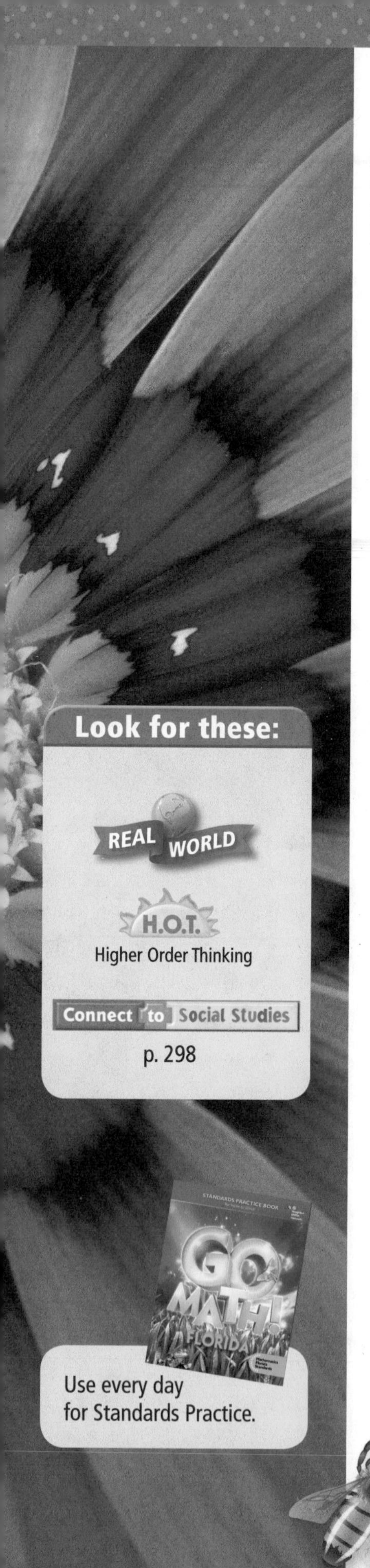

7 Division Facts and Strategies 251

Domain Operations and Algebraic Thinking

Fractions

Developing understanding of fractions, especially unit fractions (fractions with numerator 1)

8 Understand Fractions 305

Domain Number and Operations–Fractions

9 Compare Fractions 349

Domain Number and Operations–Fractions

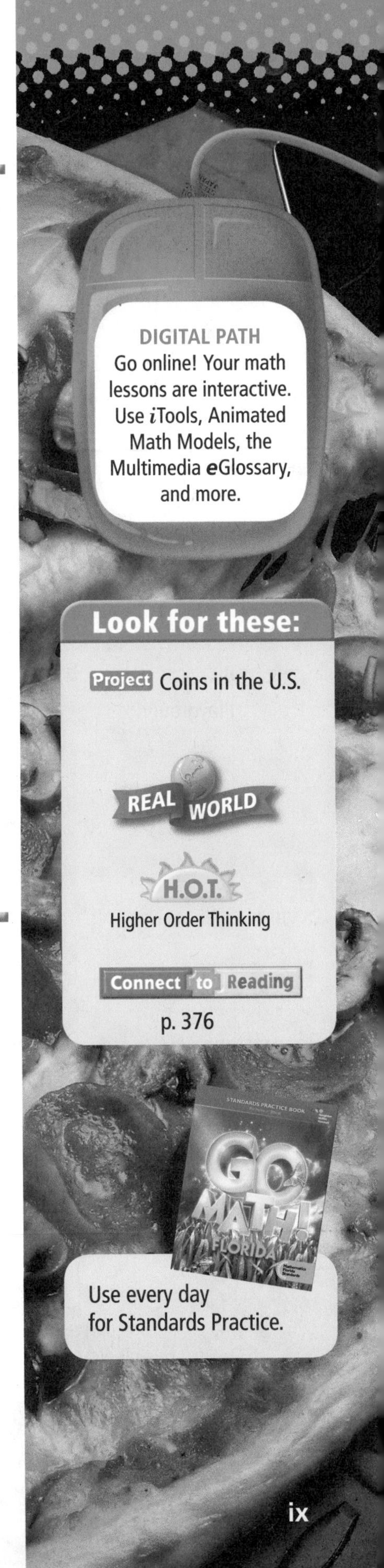

Measurement

Developing understanding of the structure of rectangular arrays and of area

10 Time, Length, Liquid Volume, and Mass 387

Domain Measurement and Data

11 Perimeter and Area 431

Domain Measurement and Data

DIGITAL PATH
Go online! Your math lessons are interactive. Use *i*Tools, Animated Math Models, the Multimedia **e**Glossary, and more.

Look for these:

Project Plan a Playground

REAL WORLD

Higher Order Thinking

Connect to Science

p. 404

Connect to Reading

p. 470

Use every day for Standards Practice.

Geometry

Describing and analyzing two-dimensional shapes

12 Two-Dimensional Shapes 481

Domain Geometry

DIGITAL PATH
Go online! Your math lessons are interactive. Use *i*Tools, Animated Math Models, the Multimedia **e**Glossary, and more.

Look for these:

Make a Mosaic

REAL WORLD

Higher Order Thinking

Connect to Reading
p. 504

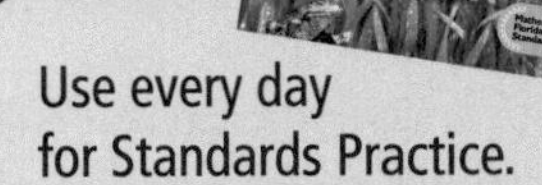

Use every day for Standards Practice.

Whole Number Operations

Developing understanding of multiplication and division and strategies for multiplication and division within 100

Some baby Abuelita dolls sing Spanish rhymes and lullabies.

Project

Inventing Toys

The dolls in the picture are called Abuelitos. Some of them are grandmother and grandfather dolls that were invented to sing lullabies. They and the grandchildren dolls have music boxes inside them. You squeeze their hands to start them singing!

Get Started

Suppose you and a partner work in a toy store. You want to order enough dolls to fill two shelves in the store. Each shelf is 72 inches long. How many cartons of dolls will fill the two shelves? Use the Important Facts to help you.

Important Facts

- Each Abuelito doll comes in a box that is 8 inches wide.
- There are 4 boxes in 1 carton.
- Abuelita Rosa sings 6 songs.
- Abuelito Pancho sings 4 songs.
- Baby Andrea and Baby Tita each sing 5 songs.
- Baby Mimi plays music but does not sing.

8 in.

Completed by ______________________

Chapter 1

Addition and Subtraction Within 1,000

Show What You Know

Check your understanding of important skills.

Name Oluwa Damilola

▶ Think Addition to Subtract Write the missing numbers.

1. $9 - 4 =$ 5

   Think: $4 + \square = 9$

   $4 +$ 5 $= 9$

   So, $9 - 4 =$ 5.

2. $13 - 7 =$ 6

   Think: $7 +$ 6 $= 13$

   $7 +$ 6 $= 13$

   So, $13 - 7 =$ 6.

3. $17 - 9 =$ 8

   Think: $9 +$ 8 $= 17$

   $9 +$ 8 $= 17$

   So, $17 - 9 =$ 8.

▶ Addition Facts Find the sum.

4. $4 + 3 =$ 7
5. $2 + 7 =$ 9
6. $8 + 6 =$ 14
7. $9 + 4 =$ 13
8. $7 + 9 =$ 16

▶ Subtraction Facts Find the difference.

9. $8 - 5 =$ 3
10. $11 - 2 =$ 9
11. $10 - 6 =$ 4
12. $18 - 9 =$ 9
13. $15 - 7 =$ 8

Manuel's puppy chewed part of this homework paper. Two of the digits in his math problem are missing. Be a Math Detective to help him figure out the missing digits. What digits are missing?

Vocabulary Builder

▶ Visualize It

Sort the review words with a ✓ into the Venn diagram.

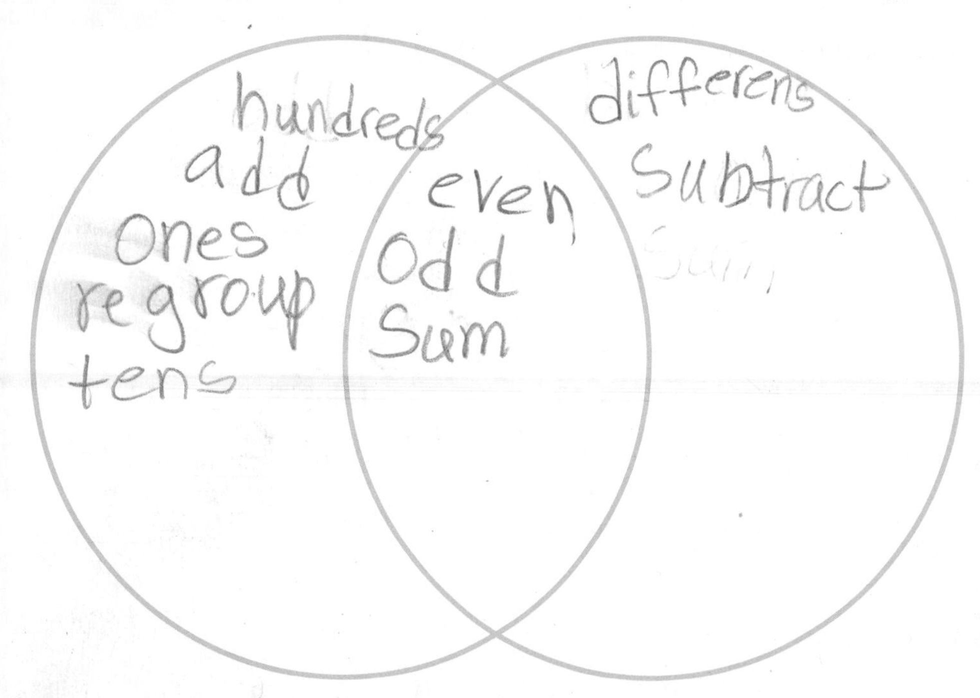

Addition Words **Subtraction Words**

Review Words

- ✓ add
- ✓ difference
- even
- ✓ hundreds
- odd
- ✓ ones
- ✓ regroup
- ✓ subtract
- ✓ sum
- ✓ tens

Preview Words

- Associative Property of Addition 4
- Commutative Property of Addition
- compatible numbers 3
- estimate 1
- Identity Property of Addition
- pattern
- round 2

▶ Understand Vocabulary

Complete the sentences by using preview words.

1. A number close to an exact number is called an estimate.

2. You can round a number to the nearest ten or hundred to find a number that tells *about* how much or *about* how many.

3. Compatiblee are numbers that are easy to compute mentally.

4. The Associative states that you can add two or more numbers in any order and get the same sum.

Name ______________________

Number Patterns

Essential Question How can you use properties to explain patterns on the addition table?

UNLOCK the Problem

A **pattern** is an ordered set of numbers or objects. The order helps you predict what will come next.

You can use the addition table to explore patterns.

+	0	1	2	3	4	5	6	7	8	9	10
0	0	1	2	3	4	5	6	7	8	9	10
1	1	2	3	4	5	6	7	8	9	10	11
2	2	3	4	5	6	7	8	9	10	11	12
3	3	4	5	6	7	8	9	10	11	12	13
4	4	5	6	7	8	9	10	11	12	13	14
5	5	6	7	8	9	10	11	12	13	14	15
6	6	7	8	9	10	11	12	13	14	15	16
7	7	8	9	10	11	12	13	14	15	16	17
8	8	9	10	11	12	13	14	15	16	17	18
9	9	10	11	12	13	14	15	16	17	18	19
10	10	11	12	13	14	15	16	17	18	19	20

Activity 1

Materials ■ orange and green crayons

- Look across each row and down each column. What pattern do you see?

  It go's by 2s.

- Shade the row and column orange for the addend 0. Compare the shaded squares to the yellow row and the blue column. What pattern do you see?

  What happens when you add 0 to a number?

  you get the same number

- Shade the row and column green for the addend 1. What pattern do you see?

  What happens when you add 1 to a number?

  it is like 2+1 you add one 3 it gives you

The **Identity Property of Addition** states that the sum of any number and zero is that number.

$7 + 0 = 7$

MATHEMATICAL PRACTICES

Math Talk What other patterns can you find in the addition table?

Activity 2

Materials ■ orange crayon

- Shade all the sums of 5 orange. What pattern do you see?

____

____

- Write two addition sentences for each sum of 5. The first two are started for you.

5 + 0 = ______ and 0 + 5 = ______

______ + ______ = ______ and ______ + ______ = ______

______ + ______ = ______ and ______ + ______ = ______

- What pattern do you see?

____

____

+	0	1	2	3	4	5	6	7	8	9	10
0	0	1	2	3	4	5	6	7	8	9	10
1	1	2	3	4	5	6	7	8	9	10	11
2	2	3	4	5	6	7	8	9	10	11	12
3	3	4	5	6	7	8	9	10	11	12	13
4	4	5	6	7	8	9	10	11	12	13	14
5	5	6	7	8	9	10	11	12	13	14	15
6	6	7	8	9	10	11	12	13	14	15	16
7	7	8	9	10	11	12	13	14	15	16	17
8	8	9	10	11	12	13	14	15	16	17	18
9	9	10	11	12	13	14	15	16	17	18	19
10	10	11	12	13	14	15	16	17	18	19	20

The **Commutative Property of Addition** states that you can add two or more numbers in any order and get the same sum.

$$3 + 4 = 4 + 3$$
$$7 = 7$$

Activity 3

Materials ■ orange and green crayons

- Shade a diagonal from left to right orange. Start with a square for 1. What pattern do you see?

____

- Shade a diagonal from left to right green. Start with a square for 2. What pattern do you see?

____

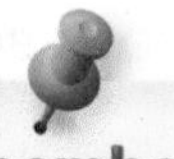

Remember

Even numbers end in 0, 2, 4, 6, or 8. Odd numbers end in 1, 3, 5, 7, or 9.

- Write addition sentences for the shaded boxes. Write *even* or *odd* under each addend.

______ + ______ = 6	______ + ______ = 7	______ + ______ = 8
↑ ↑ ↑	↑ ↑ ↑	↑ ↑ ↑
______ + ______ = even	______ + ______ = odd	______ + ______ = even

MATHEMATICAL PRACTICES

Math Talk **Explain** how you know when the sum of two numbers will be odd.

Name ____________________

Share and Show 

Use the addition table on page 6 for 1–15.

1. Complete the addition sentences to show the Commutative Property of Addition.

   $3 + \_\_\_ = \_\_\_$ $4 + \_\_\_ = \_\_\_$

MATHEMATICAL PRACTICES

Math Talk **Explain** why you can use the Commutative Property of Addition to write a related addition sentence.

Find the sum. Then use the Commutative Property of Addition to write the related addition sentence.

2. $8 + 5 = \_\_\_$
   $\_\_\_ + \_\_\_ = \_\_\_$
3. $7 + 9 = \_\_\_$
   $\_\_\_ + \_\_\_ = \_\_\_$
4. $10 + 4 = \_\_\_$
   $\_\_\_ + \_\_\_ = \_\_\_$

Is the sum even or odd? Write *even* or *odd*.

5. $8 + 1$ ________
6. $3 + 9$ ________
7. $4 + 8$ ________

On Your Own

Find the sum. Then use the Commutative Property of Addition to write the related addition sentence.

8. $0 + 8 = \_\_\_$
   $\_\_\_ + \_\_\_ = \_\_\_$
9. $7 + 3 = \_\_\_$
   $\_\_\_ + \_\_\_ = \_\_\_$
10. $8 + 7 = \_\_\_$
    $\_\_\_ + \_\_\_ = \_\_\_$

Is the sum even or odd? Write *even* or *odd*.

11. $6 + 4$ ________
12. $5 + 9$ ________
13. $2 + 3$ ________

Problem Solving

14. H.O.T. **Write Math** Look back at the shaded diagonals in Activity 2. Why does the orange diagonal show only odd numbers? **Explain.**

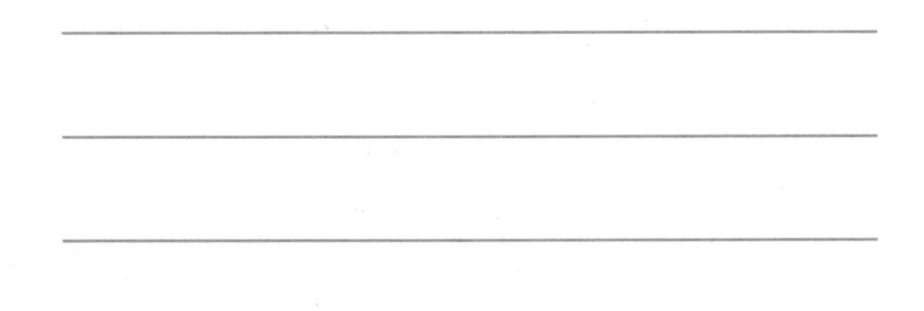

15. **Test Prep** Which describes the number sentence?

$$5 + 0 = 5$$

Ⓐ Commutative Property of Addition
Ⓑ Identity Property of Addition
Ⓒ even + even = even
Ⓓ odd + odd = odd

Sense or Nonsense?

16. Whose statement makes sense? Whose statement is nonsense? **Explain** your reasoning.

The sum of an odd number and an odd number is odd.

The sum of an even number and an even number is even.

Joey's Work

odd + odd = odd

5 + 7

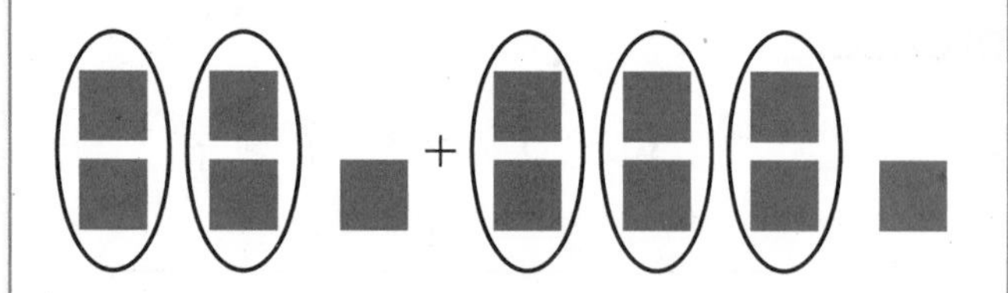

I can circle pairs of tiles in each addend and there is 1 left over in each addend. So, the sum will be odd.

____

____

____

Kayley's Work

even + even = even

4 + 6

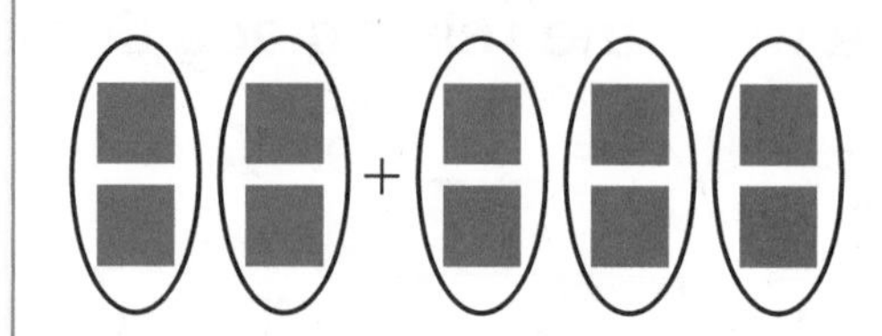

I can circle pairs of tiles with no tiles left over. So, the sum is even.

____

____

____

- For the statement that is nonsense, correct the statement.

____

FOR MORE PRACTICE:
Standards Practice Book, pp. P3–P4

Name _______________

Round to the Nearest Ten or Hundred

Essential Question How can you round numbers?

UNLOCK the Problem REAL WORLD

When you **round** a number, you find a number that tells you *about* how much or *about* how many.

Mia's baseball bat is 32 inches long. What is its length rounded to the nearest ten inches?

One Way Use a number line to round.

A **Round 32 to the nearest ten.**

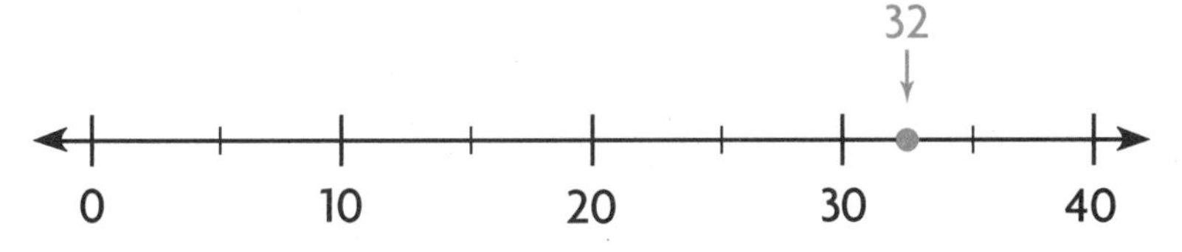

Find which tens the number is between.

32 is between ______ and ______.

32 is closer to ______ than it is to ______.

32 rounded to the nearest ten is ______.

So, the length of Mia's bat rounded to the nearest ten inches is ______ inches.

B **Round 174 to the nearest hundred.**

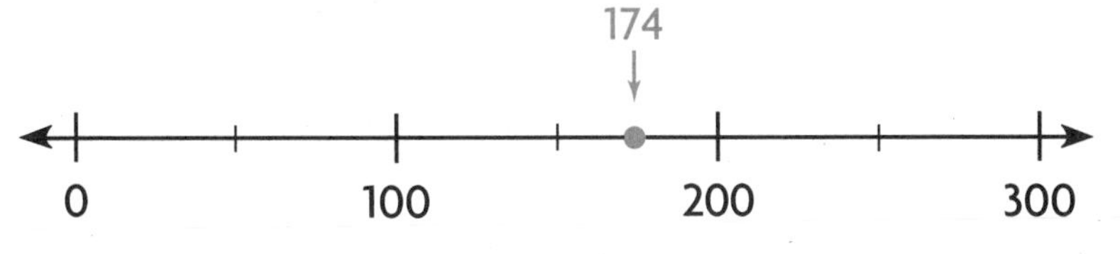

Find which hundreds the number is between.

174 is between ______ and ______.

174 is closer to ______ than it is to ______.

So, 174 rounded to the nearest hundred is ______.

Math Talk MATHEMATICAL PRACTICES Name three other numbers that round to 30 when rounded to the nearest ten. **Explain.**

Try This! **Round 718 to the nearest ten and hundred. Locate and label 718 on the number lines.**

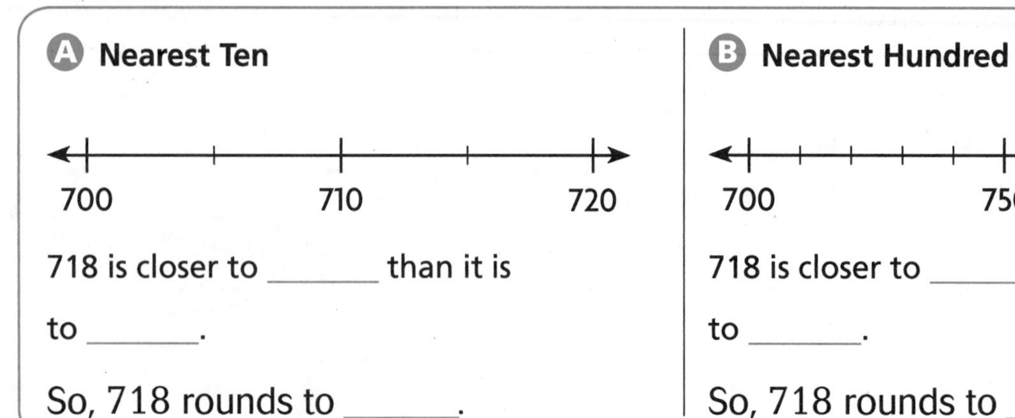

A Nearest Ten

718 is closer to ______ than it is to ______.

So, 718 rounds to ______.

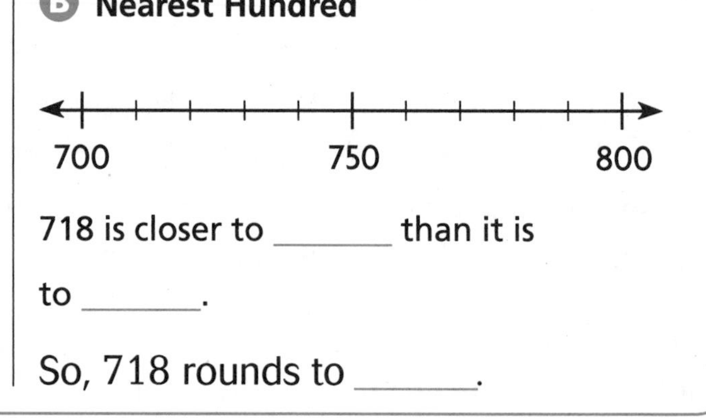

B Nearest Hundred

718 is closer to ______ than it is to ______.

So, 718 rounds to ______.

Another Way Use place value.

A Round 63 to the nearest ten.

Think: The digit in the ones place tells if the number is closer to 60 or 70.

63 ↑

3 ◯ 5

So, the tens digit stays the same. Write 6 as the tens digit.

Write zero as the ones digit.

So, 63 rounded to the nearest ten is ______.

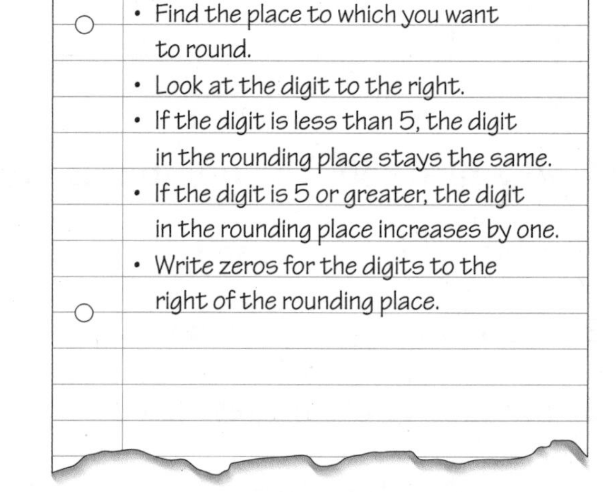

- Find the place to which you want to round.
- Look at the digit to the right.
- If the digit is less than 5, the digit in the rounding place stays the same.
- If the digit is 5 or greater, the digit in the rounding place increases by one.
- Write zeros for the digits to the right of the rounding place.

B Round 457 to the nearest hundred.

Think: The digit in the tens place tells if the number is closer to 400 or 500.

457 ↑

5 ◯ 5

So, the hundreds digit increases by one. Write 5 as the hundreds digit.

Write zeros as the tens and ones digits.

So, 457 rounded to the nearest hundred is ______.

MATHEMATICAL PRACTICES

Math Talk **Explain** how using place value is similar to using a number line.

Name ________________________

Share and Show 

Locate and label 46 on the number line. Round to the nearest ten.

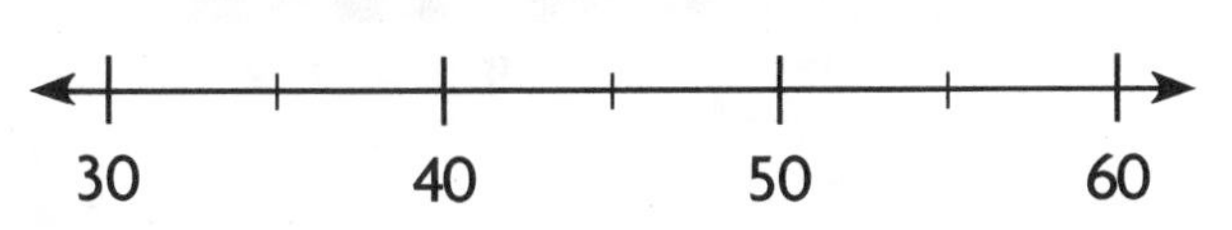

MATHEMATICAL PRACTICES

Math Talk What is the greatest number that rounds to 50 when rounded to the nearest ten? What is the least number? **Explain.**

1. 46 is between ______ and ______.
2. 46 is closer to ______ than it is to ______.
3. 46 rounded to the nearest ten is ______.

Round to the nearest ten.

4. 19 ______
5. 66 ______
6. 51 ______

Round to the nearest hundred.

7. 463 ______
8. 202 ______
9. 658 ______

On Your Own

Locate and label 548 on the number line. Round to the nearest hundred.

10. 548 is between ______ and ______.
11. 548 is closer to ______ than it is to ______.
12. 548 rounded to the nearest hundred is ______.

Round to the nearest ten and hundred.

13. 576 ______ ______
14. 298 ______ ______
15. 844 ______ ______

Problem Solving REAL WORLD

Use the table for 16–18.

Visitors to the Giraffe Exhibit

Day	Number of Visitors
Sunday	894
Monday	793
Tuesday	438
Wednesday	362
Thursday	839
Friday	725
Saturday	598

16. On which day did about 900 visitors come to the giraffe exhibit?

____

17. To the nearest ten, how many visitors came to the giraffe exhibit on Sunday?

____

18. On which two days did about 800 visitors come to the giraffe exhibit each day?

____

19. H.O.T. Write five numbers that round to 360 when rounded to the nearest ten.

____

20. Write Math **What's the Error?** Cole said that 555 rounded to the nearest ten is 600. What is Cole's error? **Explain.**

____

____

21. **Test Prep** What is 438 rounded to the nearest ten?

Ⓐ 450

Ⓑ 440

Ⓒ 430

Ⓓ 400

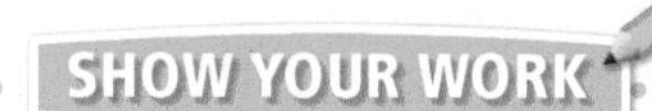

FOR MORE PRACTICE:
Standards Practice Book, pp. P5–P6

Name ____________________________________

Estimate Sums

Essential Question How can you use compatible numbers and rounding to estimate sums?

UNLOCK the Problem REAL WORLD

The table shows how many dogs went to Pine Lake Dog Park during the summer months. About how many dogs went to the park during June and August?

Pine Lake Dog Park

Month	Number of Dogs
June	432
July	317
August	489

You can estimate to find *about* how many or *about* how much. An **estimate** is a number close to an exact amount.

One Way Use compatible numbers.

Compatible numbers are numbers that are easy to compute mentally and are close to the real numbers.

$$\begin{array}{r} 432 \\ +\ 489 \\ \hline \end{array} \quad \rightarrow \quad \begin{array}{r} 425 \\ +\ 475 \\ \hline \end{array}$$

So, about __________ dogs went to Pine Lake Dog Park during June and August.

MATHEMATICAL PRACTICES

Math Talk Will the sum of the compatible numbers 425 and 475 be greater than or less than the exact sum? **Explain.**

1. What other compatible numbers could you have used?

____

2. About how many dogs went to the park during July and August? What compatible numbers could you use to estimate?

____

Another Way Use place value to round.

432 + 489 = ■

Remember

When you round a number, you find a number that tells *about* how many or *about* how much.

First, find the place to which you want to round. Round both numbers to the same place. The greatest place value of 432 and 489 is hundreds. So, round to the nearest hundred.

STEP 1 Round 432 to the nearest hundred.

- Look at the digit to the right of the hundreds place.
- Since 3 < 5, the digit 4 stays the same.
- Write zeros for the tens and ones digits.

4 3 2
↑

  4 3 2 → ____
+ 4 8 9 → + ____

STEP 2 Round 489 to the nearest hundred.

- Look at the digit to the right of the hundreds place.
- Since 8 > 5, the digit 4 increases by one.
- Write zeros for the tens and ones digits.

4 8 9
↑

  4 3 2 → 4 0 0
+ 4 8 9 → + ____

STEP 3 Find the sum of the rounded numbers.

  4 3 2 → 4 0 0
+ 4 8 9 → + 5 0 0

So, 432 + 489 is about ______.

MATHEMATICAL PRACTICES

Math Talk How would you round 432 and 489 to the nearest ten? What would be the estimated sum? **Explain.**

Try This! Estimate the sum.

A Use compatible numbers.

  47 → ____
+ 23 → + 25

B Use rounding.

  304 → 300
+ 494 → + ____

Name ______________________________

Share and Show 

1. Use compatible numbers to complete the problem. Then estimate the sum.

428	→	
+286	→	+

MATHEMATICAL PRACTICES

Math Talk What other compatible numbers could you use for 428 and 286?

Use rounding or compatible numbers to estimate the sum.

2. 65 + 23 → ___ + ___ = ___

3. 421 + 218 → ___ + ___ = ___

4. 369 + 480 → ___ + ___ = ___

On Your Own

Use rounding or compatible numbers to estimate the sum.

5. 19 + 54 → ___ + ___ = ___

6. 39 + 42 → ___ + ___ = ___

7. 327 + 581 → ___ + ___ = ___

8. 27 + 78 → ___ + ___ = ___

9. 267 + 517 → ___ + ___ = ___

10. 465 + 478 → ___ + ___ = ___

11. 186 + 460 → ___ + ___ = ___

12. 817 + 118 → ___ + ___ = ___

13. 632 + 244 → ___ + ___ = ___

14. 278 + 369

___ + ___ = ___

15. 523 + 195

___ + ___ = ___

Problem Solving REAL WORLD

Use the table for 16–18.

Month	Pet Bowls	Bags of Pet Food
June	91	419
July	57	370
August	76	228

Dan's Pet Supplies Sold

16. About how many pet bowls were sold in June and July altogether?

____

17. **What's the Question?** The answer is about 800.

____

____

18. H.O.T. Write Math Dan said the total number of bags of pet food sold in June, July, and August was about 1,000. How did Dan estimate? **Explain.**

____

____

____

____

____

19. **Test Prep** Tracy ordered 325 pet toys and 165 bags of dog food for her new pet store. Which is the best estimate of the total number of items Tracy ordered?

Ⓐ 400

Ⓑ 500

Ⓒ 600

Ⓓ 700

SHOW YOUR WORK

FOR MORE PRACTICE: Standards Practice Book, pp. P7–P8

Name ________________________

Mental Math Strategies for Addition

Essential Question What mental math strategies can you use to find sums?

UNLOCK the Problem REAL WORLD

The table shows how many musicians are in each section of a symphony orchestra. How many musicians play either string or woodwind instruments?

Orchestra Musicians

Section	Number
Brass	12
Percussion	13
String	57
Woodwind	15

One Way Count by tens and ones to find 57 + 15.

A Count on to the nearest ten. Then count by tens and ones.

Think: 3 + ■ = 15

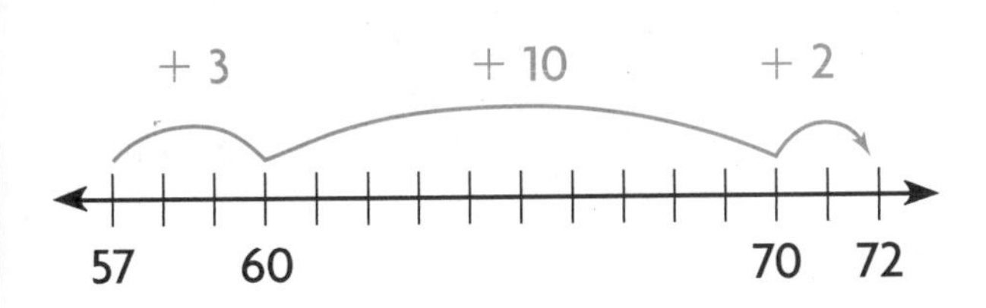

B Count by tens. Then count by ones.

Think: 10 + 5 = 15

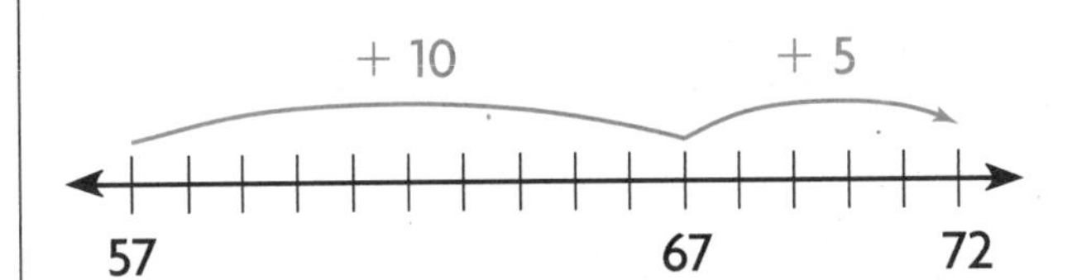

57 + 15 = ______

So, ______ musicians play either string or woodwind instruments.

Math Idea
Count on from the greater addend.

Try This! Find 43 + 28. Draw jumps and label the number line to show your thinking.

So, 43 + 28 = ______.

MATHEMATICAL PRACTICES

Math Talk Explain another way you can draw the jumps.

Other Ways

A Use compatible numbers to find 178 + 227.

STEP 1 Break apart the addends to make them compatible.

Think: 178 = 175 + 3
227 = 225 + 2

175 and 225 are compatible numbers.

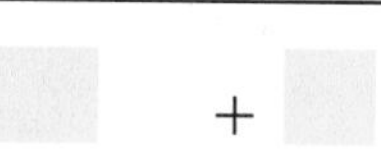

Remember
Compatible numbers are easy to compute mentally and are close to the real numbers.

STEP 2 Find the sums.

$$\begin{array}{r} 178 \\ +227 \\ \hline \end{array} \quad \begin{array}{c} \rightarrow \\ \rightarrow \end{array} \quad \begin{array}{rcr} 175 & + & 3 \\ 225 & + & 2 \\ \hline \square & + & \square \end{array}$$

STEP 3 Add the sums. ______ + ______ = ______

So, 178 + 227 = ______.

B Use friendly numbers and adjust to find 38 + 56.

STEP 1 Make a friendly number. 38 + 2 = ______

Think: Add to 38 to make a number with 0 ones.

STEP 2 Since you added 2 to 38, you have to subtract 2 from 56. 56 − 2 = ______

STEP 3 Find the sum. ______ + ______ = ______

So, 38 + 56 = ______.

MATHEMATICAL PRACTICES

Math Talk Describe another way to use friendly numbers to find the sum.

Share and Show

1. Count by tens and ones to find 63 + 27. Draw jumps and label the number line to show your thinking.

Think: Count by tens and ones from 63.

63

63 + 27 = ______

Name ____________________

2. Use compatible numbers to find 26 + 53.

Think: 26 = 25 + 1
53 = 50 + 3

26 + 53 = ____

MATHEMATICAL PRACTICES

Math Talk **Explain** how you could use friendly numbers to find 26 + 53.

Count by tens and ones to find the sum.
Use the number line to show your thinking.

3. 34 + 18 = ____

4. 22 + 49 = ____

On Your Own

Count by tens and ones to find the sum.
Use the number line to show your thinking.

5. 56 + 27 = ____

6. 28 + 35 = ____

Use mental math to find the sum.
Draw or describe the strategy you use.

7. 116 + 203 = ____

8. 18 + 57 = ____

9. **Write Math** On Friday, 376 people attended the school concert. On Saturday, 427 people attended. **Explain** how can you use mental math to find how many people attended the concert.

____

____

UNLOCK the Problem REAL WORLD

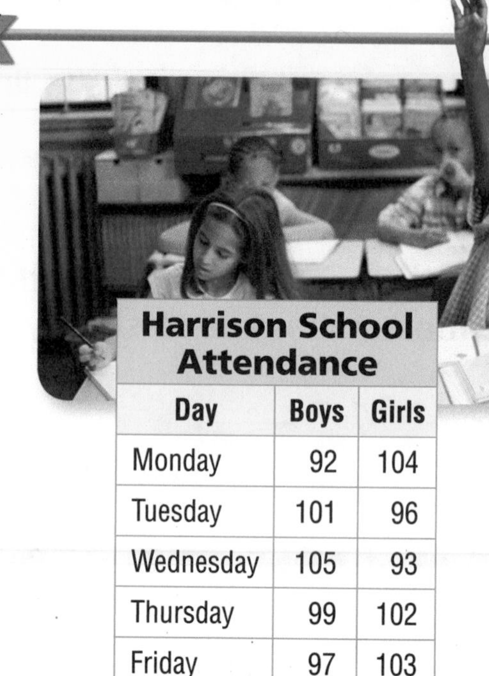

Harrison School Attendance

Day	Boys	Girls
Monday	92	104
Tuesday	101	96
Wednesday	105	93
Thursday	99	102
Friday	97	103

10. The table shows the attendance at Harrison School for one week. On which day did the most students attend school?

a. What do you need to find?

____

____

b. What information do you know?

____

____

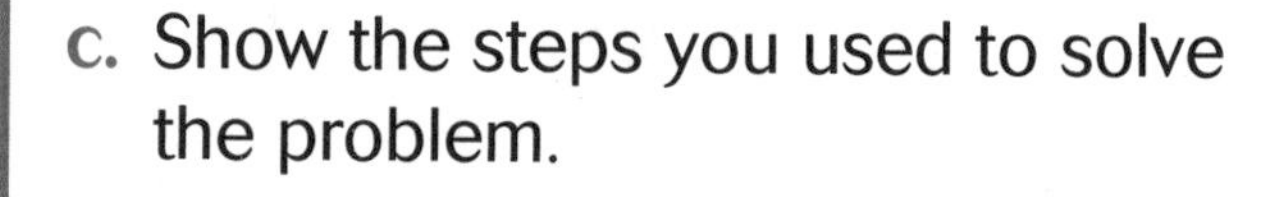

c. Show the steps you used to solve the problem.

d. Complete the sentences.

_____ students attended on Monday.

_____ students attended on Tuesday.

_____ students attended on Wednesday.

_____ students attended on Thursday.

_____ students attended on Friday.

So, the most students attended school on _________.

11. H.O.T. There are 14 more girls than boys in the school orchestra. There are 19 boys. How many students are in the school orchestra?

____

12. **Test Prep** On Monday, 46 boys and 38 girls bought lunch at school. How many students bought lunch?

Ⓐ 84

Ⓑ 76

Ⓒ 74

Ⓓ 73

FOR MORE PRACTICE:
Standards Practice Book, pp. P9–P10

Name ____________________

Use Properties to Add

Essential Question How can you add more than two addends?

CONNECT You have learned the Commutative Property of Addition. You can add two or more numbers in any order and get the same sum.

$$16 + 9 = 9 + 16$$

The **Associative Property of Addition** states that you can group addends in different ways and still get the same sum. It is also called the Grouping Property.

$$(16 + 7) + 23 = 16 + (7 + 23)$$

Math Idea

You can change the order or the grouping of the addends to make combinations that are easy to add.

UNLOCK the Problem REAL WORLD

Mrs. Gomez sold 23 cucumbers, 38 tomatoes, and 42 peppers at the Farmers' Market. How many vegetables did she sell in all?

- Will the sum be closer to 90 or 100?

____

Find 23 + 38 + 42.

Look for an easy way to add.

STEP 1 Line up the numbers by place value.

$$\begin{array}{r} 23 \\ 38 \\ +\ 42 \\ \hline \end{array}$$

STEP 2 Group the ones to make them easy to add.

Think: Make a ten.

$$\begin{array}{r} {}^{1}\ \ \\ 23 \\ 38 \\ +\ 42 \\ \hline 3 \end{array}$$

8 + 2 → 10

STEP 3 Group the tens to make them easy to add.

Think: Make doubles.

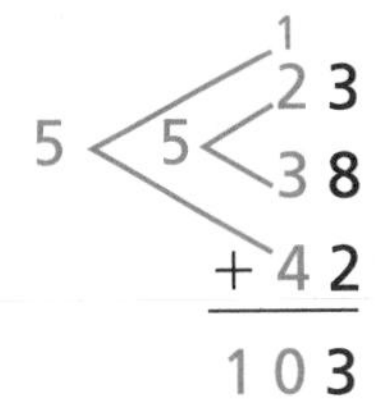

$$\begin{array}{r} {}^{1}\ \ \\ 23 \\ 38 \\ +\ 42 \\ \hline 103 \end{array}$$

23 + 38 + 42 = ____________

So, Mrs. Gomez sold ____________ vegetables in all.

MATHEMATICAL PRACTICES

Math Talk Explain how to group the digits to make them easy to add.

Example Use properties to find 36 + 37 + 51.

STEP 1 Line up the numbers by place value.

```
  36
  37
+ 51
----
```

STEP 2 Change the grouping.

Think: Adding 37 + 51 first would be easy because there is no regrouping needed.

```
  36
  37 \
+ 51 /  88
----
```

STEP 3 Add.

```
  36
+ 88
----
```

So, 36 + 37 + 51 = ________.

Try This! Use properties to add.

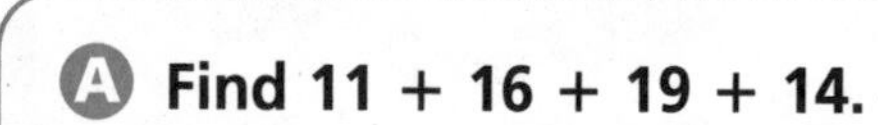

A Find 11 + 16 + 19 + 14.

Think: Use the Commutative Property of Addition to change the order.

```
  11          11 \
  16    →     19 /  10
  19          16 \
+ 14        + 14 /  10
----        ----
```

B Find 17 + (33 + 45).

Think: Use the Associative Property of Addition to change the grouping.

```
  17 \
  33 /  50           50
+ 45          →    + 45
----               ----
```

MATHEMATICAL PRACTICES

Math Talk Explain how the Commutative and Associative Properties of Addition are alike and how they are different.

Share and Show

1. Find the sum. Write the addition property you used.

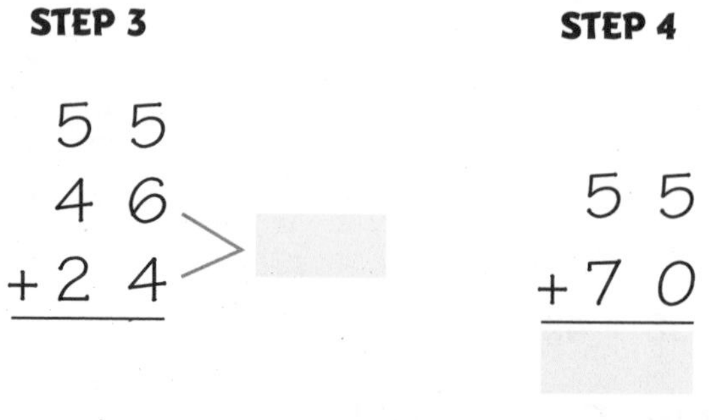

STEP 1

```
  46
  55
+ 24
----
```

STEP 2

```
  55
  
+ 24
----
```

________ Property of Addition

STEP 3

```
  55
  46 \
+ 24 /
----
```

STEP 4

```
  55
+ 70
----
```

________ Property of Addition

Name ______________________

Use addition properties and strategies to find the sum.

2. $13 + 26 + 54 =$ ______

3. $57 + 62 + 56 + 43 =$ ______

On Your Own

Use addition properties and strategies to find the sum.

4. $18 + 39 + 32 =$ ______

5. $13 + 49 + 87 =$ ______

6. $15 + 76 + 125 =$ ______

7. $33 + 71 + 56 + 29 =$ ______

8. Change the order and the grouping of the addends so that you can use mental math to find the sum. Then find the sum.

$43 + 39 + 43 + 11 =$ ______

______ + ______ + ______ + ______ = ______

Problem Solving REAL WORLD

9. Mr. Arnez bought 32 potatoes, 29 onions, 31 tomatoes and 28 peppers to make salads for his deli. How many vegetables did he buy?

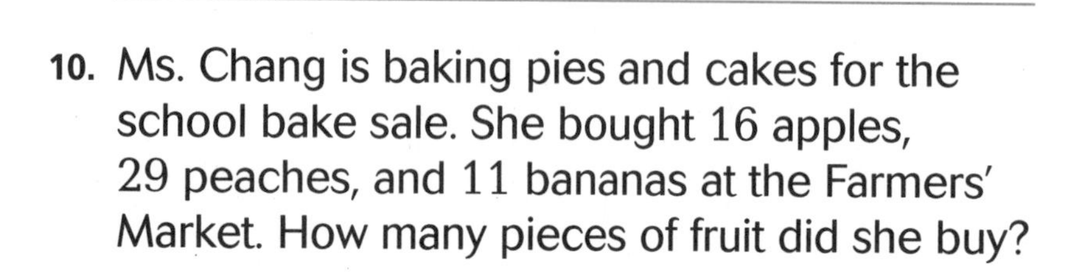

____

10. Ms. Chang is baking pies and cakes for the school bake sale. She bought 16 apples, 29 peaches, and 11 bananas at the Farmers' Market. How many pieces of fruit did she buy?

____

11. H.O.T. What is the unknown number? Which property did you use?

$$(\blacksquare + 8) + 32 = 49$$

____

12. Write Math Change the order or grouping to find the sum. **Explain** how you used properties to find the sum.

$$63 + 86 + 77$$

____

____

____

____

13. **Test Prep** Which shows the Associative Property of Addition?

Ⓐ $(86 + 7) + 93 = 86 + (7 + 93)$

Ⓑ $86 + 7 = 7 + 86$

Ⓒ $86 + 0 = 86$

Ⓓ $86 = 80 + 6$

FOR MORE PRACTICE:
Standards Practice Book, pp. P11–P12

Name ______________________

Use the Break Apart Strategy to Add

Essential Question How can you use the break apart strategy to add 3-digit numbers?

There are more zoos in Germany than in any other country. At one time, there were 355 zoos in the United States and 414 zoos in Germany. How many zoos were there in the United States and Germany altogether?

You can use the break apart strategy to find sums.

Math Talk MATHEMATICAL PRACTICES Do you think the sum will be greater than or less than 800? **Explain.**

Example 1 Add. 355 + 414

STEP 1 Estimate. 400 + 400 = ______

STEP 2 Break apart the addends.
Start with the hundreds.
Then add each place value.

355	=	300	+	____	+	5
+ 414	=	____	+	10	+	4
		700	+	60	+	9

STEP 3 Add the sums.

700 + 60 + 9 = ______

So, there were ______ zoos in the United States and Germany altogether.

Example 2 Add. 467 + 208

STEP 1 Estimate. 500 + 200 = ______

STEP 2 Break apart the addends.
Start with the hundreds.
Then add each place value.

467	=	400	+	____	+	____
+ 208	=	____	+	0	+	8
		600	+	60	+	15

STEP 3 Add the sums.

600 + 60 + 15 = ______

So, 467 + 208 = ______.

Try This! **Use the break apart strategy to find 343 + 259.**

Estimate. 300 + 300 = ____

$$
\begin{array}{rcl}
343 & = & 300 + \_\_\_ + \_\_\_ \\
+\,259 & = & \_\_\_ + \_\_\_ + \_\_\_ \\
\hline
 & & \_\_\_ + \_\_\_ + \_\_\_ = \_\_\_
\end{array}
$$

1. **Explain** why there is a zero in the tens place in the sum.

____

____

____

____

2. How do you know your answer is reasonable?

____

____

Share and Show

1. Complete.
   Estimate: 400 + 400 = ____

$$
\begin{array}{rcl}
425 & = & 400 + \_\_\_ + 5 \\
+\,362 & = & \_\_\_ + 60 + \_\_\_ \\
\hline
 & & 700 + \_\_\_ + 7 = \_\_\_
\end{array}
$$

So, 425 + 362 = ____.

2. Write the numbers the break apart strategy shows.

$$
\begin{array}{rcl}
\_\_\_ & = & 100 + 30 + 4 \\
+\,\_\_\_ & = & 200 + 40 + 9 \\
\hline
\_\_\_ & = & 300 + 70 + 13
\end{array}
$$

MATHEMATICAL PRACTICES

Math Talk **Explain** how the break apart strategy uses expanded forms of numbers.

Name ________________

Estimate. Then use the break apart strategy to find the sum.

3. Estimate: ______

142 =
+ 436 =

4. Estimate: ______

459 =
+ 213 =

5. Estimate: ______

291 =
+ 420 =

6. Estimate: ______

654 =
+ 243 =

On Your Own

Estimate. Then use the break apart strategy to find the sum.

7. Estimate: ______

435 =
+ 312 =

8. Estimate: ______

163 =
+ 205 =

9. Estimate: ______

634 =
+ 251 =

10. Estimate: ______

526 =
+ 357 =

Practice: Copy and Solve **Estimate. Then solve.**

11. 163 + 205

12. 543 + 215

13. 213 + 328

14. 372 + 431

15. 152 + 304

16. 268 + 351

17. 413 + 257

18. 495 + 312

Problem Solving REAL WORLD

Use the table for 19–20.

Number of Students	
School	**Number**
Harrison	304
Montgomery	290
Bryant	421

19. Which two schools together have fewer than 600 students? **Explain.**

____

____

____

20. The number of students in Collins School is more than double the number of students in Montgomery School. What is the least number of students that could attend Collins School?

____

SHOW YOUR WORK

21. H.O.T. **What's the Error?** Lexi used the break apart strategy to find 145 + 203. Describe her error. What is the correct sum?

$$\begin{array}{r} 100 + 40 + 5 \\ +\ 200 + 30 + 0 \\ \hline 300 + 70 + 5 \end{array} = 375$$

____

____

22. Write Math Is the sum of 425 and 390 less than or greater than 800? How do you know?

____

____

____

23. **Test Prep** What is the sum of 421 and 332?

Ⓐ 653 Ⓑ 751 Ⓒ 753 Ⓓ 763

FOR MORE PRACTICE:
Standards Practice Book, pp. P13–P14

Name ______________________________

Use Place Value to Add

Essential Question How can you use place value to add 3-digit numbers?

UNLOCK the Problem REAL WORLD

Dante is planning a trip to Illinois. His airplane leaves from Dallas, Texas and stops in Tulsa, Oklahoma. Then it flies from Tulsa to Chicago, Illinois. How many miles does Dante fly?

Use place value to add two addends.

Add. 236 + 585

Estimate. 200 + 600 = ______

STEP 1

Add the ones. Regroup the ones as tens and ones.

$$\begin{array}{r} {\scriptstyle 1} \\ 2\,3\,6 \\ +\,5\,8\,5 \\ \hline \square \end{array}$$

STEP 2

Add the tens. Regroup the tens as hundreds and tens.

$$\begin{array}{r} {\scriptstyle 1\,1} \\ 2\,3\,6 \\ +\,5\,8\,5 \\ \hline \square\,1 \end{array}$$

STEP 3

Add the hundreds.

$$\begin{array}{r} {\scriptstyle 1\,1} \\ 2\,3\,6 \\ +\,5\,8\,5 \\ \hline \square\,2\,1 \end{array}$$

236 + 585 = ______

So, Dante flies ______ miles.

Since ______ is close to the estimate of ______, the answer is reasonable.

ERROR Alert

Remember to add the regrouped ten and hundred.

- You can also use the Commutative Property of Addition to check your work. Change the order of the addends and find the sum.

$$\begin{array}{r} 5\,8\,5 \\ +\,2\,3\,6 \\ \hline \square \end{array}$$

Try This! **Find 563 + 48 in two ways.**

Estimate. 550 + 50 = ____

A **Use the break apart strategy.**

$$
\begin{array}{rcl}
563 & = & 500 + \square + \square \\
+\ 48 & = & 40 + \square \\
\hline
 & & \square + \square + \square = \square
\end{array}
$$

B **Use place value.**

$$
\begin{array}{r}
563 \\
+\ 48 \\
\hline
\square
\end{array}
$$

Use place value to add three addends.

A **Add.** 140 + 457 + 301

Estimate. 150 + 450 + 300 = ____

STEP 1 Add the ones.

$$
\begin{array}{r}
140 \\
457 \\
+\ 301 \\
\hline
\square
\end{array}
$$

STEP 2 Add the tens.

$$
\begin{array}{r}
140 \\
457 \\
+\ 301 \\
\hline
\square 8
\end{array}
$$

STEP 3 Add the hundreds.

$$
\begin{array}{r}
140 \\
457 \\
+\ 301 \\
\hline
\square 98
\end{array}
$$

So, 140 + 457 + 301 = ____.

B **Add.** 173 + 102 + 328

Estimate. 200 + 100 + 300 = ____

STEP 1 Add the ones. Regroup the ones as tens and ones.

$$
\begin{array}{r}
1 \\
173 \\
102 \\
+\ 328 \\
\hline
\square
\end{array}
$$

STEP 2 Add the tens. Regroup the tens as hundreds and tens.

$$
\begin{array}{r}
1\,1 \\
173 \\
102 \\
+\ 328 \\
\hline
\square 3
\end{array}
$$

STEP 3 Add the hundreds.

$$
\begin{array}{r}
1\,1 \\
173 \\
102 \\
+\ 328 \\
\hline
\square 03
\end{array}
$$

So, 173 + 102 + 328 = ____.

Name ______________________________

Share and Show

1. Circle the problem in which you need to regroup. Use the strategy that is easier to find the sum.

   a. 496 + 284

   b. 482 + 506

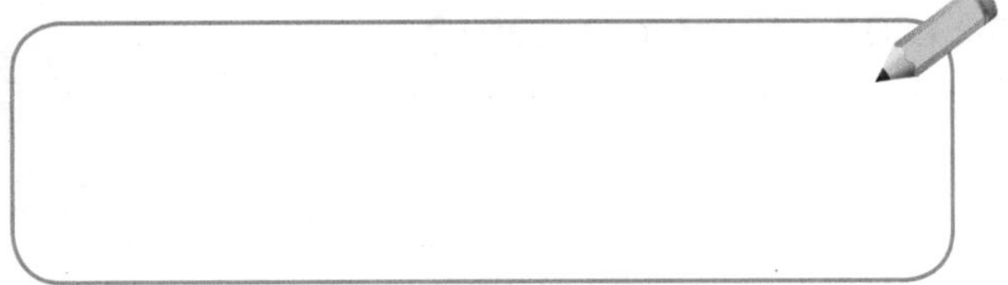

Estimate. Then find the sum.

2. Estimate: ______

   251
   + 345

3. Estimate: ______

   479
   + 395

4. Estimate: ______

   686
   + 314

5. Estimate: ______

   231
   410
   + 158

On Your Own

Estimate. Then find the sum.

6. Estimate: ______

   572
   + 124

7. Estimate: ______

   163
   + 205

8. Estimate: ______

   334
   + 218

9. Estimate: ______

   357
   101
   + 467

Practice: Copy and Solve Estimate. Then solve.

10. 253 + 376
11. 654 + 263
12. 321 + 439
13. 482 + 323
14. 182 + 321
15. 701 + 108
16. 543 + 372
17. 516 + 326

H.O.T. Algebra Find the unknown digits.

18.
   1 ☐ 4
   + ☐ 3 ☐
   257

19.
   ☐ 7 ☐
   + 6 ☐ 4
   986

20.
   2 ☐ ☐
   + ☐ 2 9
   682

21.
   3 ☐ ☐
   + ☐ 1 7
   903

UNLOCK the Problem REAL WORLD

22. A plane flew 187 miles from New York City, New York to Boston, Massachusetts. It then flew 273 miles from Boston to Philadelphia, Pennsylvania. The plane flew the same distance on the return trip. How many miles did the plane fly?

Ⓐ 460 miles　Ⓒ 900 miles
Ⓑ 820 miles　Ⓓ 920 miles

a. What do you need to find? ______

b. What is an estimate of the total distance?

______

c. Show the steps you used to solve the problem.

d. Complete the sentences.

The plane flew ______ miles from New York City to Boston.

Then it flew ______ miles from Boston to Philadelphia.

It flew ______ miles from New York City to Boston to Philadelphia.

The total distance is ______ miles round trip.

e. Fill in the bubble for the correct answer choice above.

23. The Colorado River is 119 miles longer than the Kansas River. The Kansas River is 743 miles long. How long is the Colorado River?

Ⓐ 852 miles　Ⓒ 862 miles
Ⓑ 861 miles　Ⓓ 962 miles

24. Mr. Miller drove 568 miles last week. He drove 308 miles this week. What was the number of miles he drove in these two weeks?

Ⓐ 865 miles　Ⓒ 987 miles
Ⓑ 876 miles　Ⓓ 988 miles

FOR MORE PRACTICE:
Standards Practice Book, pp. P15–P16

Name ____________________

Mid-Chapter Checkpoint

Vocabulary

Choose the best term from the box.

Vocabulary
Commutative Property of Addition
compatible numbers
Identity Property of Addition
pattern

1. A ________ is an ordered set of numbers or objects in which the order helps you predict what comes next. (p. 5)

2. The ____________________ states that when you add zero to any number, the sum is that number. (p. 5)

Concepts and Skills

Is the sum even or odd? Write *even* or *odd*.

3. $8 + 5$ ________

4. $9 + 7$ ________

5. $4 + 6$ ________

Use rounding or compatible numbers to estimate the sum.

6. $\begin{array}{r} 56 \\ +32 \\ \hline \end{array}$ $\begin{array}{r} \square \\ +\ \square \\ \hline \square \end{array}$

7. $\begin{array}{r} 271 \\ +425 \\ \hline \end{array}$ $\begin{array}{r} \square \\ +\ \square \\ \hline \square \end{array}$

8. $\begin{array}{r} 328 \\ +127 \\ \hline \end{array}$ $\begin{array}{r} \square \\ +\ \square \\ \hline \square \end{array}$

Use mental math to find the sum.

9. $46 + 14 =$ ______

10. $39 + 243 =$ ______

11. $326 + 402 =$ ______

Estimate. Then find the sum.

12. Estimate: ______

$\begin{array}{r} 356 \\ +442 \\ \hline \end{array}$

13. Estimate: ______

$\begin{array}{r} 164 \\ +230 \\ \hline \end{array}$

14. Estimate: ______

$\begin{array}{r} 545 \\ +139 \\ \hline \end{array}$

15. Estimate: ______

$\begin{array}{r} 437 \\ +184 \\ \hline \end{array}$

Fill in the bubble for the correct answer choice.

16. Nancy planted 48 roses and 39 tulips. Which is the best estimate for the number of flowers she planted?

    Ⓐ about 50

    Ⓑ about 60

    Ⓒ about 65

    Ⓓ about 90

17. Tomas collected 139 cans for recycling on Monday, and twice that number on Tuesday. How many cans did he collect on Tuesday?

    Ⓐ 278

    Ⓑ 279

    Ⓒ 280

    Ⓓ 288

18. There are 294 boys and 332 girls in the Hill School. How many students are in the school?

    Ⓐ 172

    Ⓑ 526

    Ⓒ 626

    Ⓓ 637

19. On Monday, 76 students played soccer. On Tuesday, 62 students played soccer. On Wednesday, 68 students played soccer. How many students played soccer on those three days?

    Ⓐ 216

    Ⓑ 207

    Ⓒ 206

    Ⓓ 196

Name ______________________________

Estimate Differences

Essential Question How can you use compatible numbers and rounding to estimate differences?

UNLOCK the Problem REAL WORLD

The largest yellowfin tuna caught by fishers weighed 387 pounds. The largest grouper caught weighed 436 pounds. About how much more did the grouper weigh than the yellowfin tuna?

You can estimate to find *about* how much more.

- Does the question ask for an exact answer? How do you know?

____

____

- Circle the numbers you need to use.

Yellowfin tuna

Grouper

One Way Use compatible numbers.

Think: Compatible numbers are numbers that are easy to compute mentally and are close to the real numbers.

$$\begin{array}{r} 436 \\ -\ 387 \\ \hline \end{array} \rightarrow \begin{array}{r} 425 \\ -\ 375 \\ \hline \end{array}$$

So, the grouper weighed about _____ pounds more than the yellowfin tuna.

- What other compatible numbers could you have used?

____

Try This! Estimate. Use compatible numbers.

A

$$\begin{array}{r} 73 \\ -\ 22 \\ \hline \end{array} \rightarrow \begin{array}{r} 75 \\ -\ \square \\ \hline \end{array}$$

B

$$\begin{array}{r} 376 \\ -\ 148 \\ \hline \end{array} \rightarrow \begin{array}{r} \square \\ -\ 150 \\ \hline \end{array}$$

Another Way Use place value to round.

436 − 387 = ■

STEP 1 Round 436 to the nearest ten.

Think: Find the place to which you want to round. Look at the digit to the right.

- Look at the digit in the ones place.
- Since 6 > 5, the digit 3 increases by one.
- Write a zero for the ones place.

436
↑

436 →
− 387

−

STEP 2 Round 387 to the nearest ten.

- Look at the digit in the ones place.
- Since 7 > 5, the digit 8 increases by one.
- Write a zero for the ones place.

387
↑

436 → 440
− 387 → −

STEP 3 Find the difference of the rounded numbers.

436 → 440
− 387 → − 390

So, 436 − 387 is about _____.

Try This! Estimate. Use place value to round.

A
761 → 800
− 528 → −

Think: Round both numbers to the same place value.

B
642 →
− 287 → − 300

MATHEMATICAL PRACTICES

Math Talk Explain a different way you can round each number in Example B to find another estimate.

Name ___

Share and Show 

1. Use compatible numbers to complete the problem. Then estimate the difference.

546	→	550
−209	→	− ___
		___

MATHEMATICAL PRACTICES

Math Talk Explain another way you can estimate 546 − 209.

Use rounding or compatible numbers to estimate the difference.

2. 57 − 21 → ___ − ___ = ___

3. 642 − 137 → ___ − ___ = ___

4. 374 − 252 → ___ − ___ = ___

On Your Own

Use rounding or compatible numbers to estimate the difference.

5. 67 − 24 → ___ − ___ = ___

6. 81 − 39 → ___ − ___ = ___

7. 936 − 421 → ___ − ___ = ___

8. 804 − 259 → ___ − ___ = ___

9. 584 − 208 → ___ − ___ = ___

10. 442 − 36 → ___ − ___ = ___

11. 96 − 63 → ___ − ___ = ___

12. 528 − 274 → ___ − ___ = ___

13. 761 − 489 → ___ − ___ = ___

14. 429 − 51

___ − ___ = ___

15. 491 − 270

___ − ___ = ___

Problem Solving REAL WORLD

Use the table for 16–19.

Type of Fish	Weight in Pounds
Pacific Halibut	459
Conger	133
Yellowfin Tuna	387

Largest Saltwater Fish Caught

16. About how many more pounds does the Pacific halibut weigh than the yellowfin tuna? **Explain.**

17. **What's the Question?** The answer is about 500 pounds.

18. Sean said the three fish weigh about 1,000 pounds altogether. **Explain** why you agree or disagree.

19. H.O.T. Write Math About how much more is the total weight of the Pacific halibut and conger than the weight of the yellowfin tuna? **Explain.**

20. **Test Prep** A total of 926 people went to a fishing tournament. Of these people, 607 arrived before noon. Which is the best estimate of the number of people who arrived in the afternoon?

Ⓐ 150 Ⓑ 250 Ⓒ 300 Ⓓ 500

SHOW YOUR WORK

FOR MORE PRACTICE:
Standards Practice Book, pp. P17–P18

Name ______________________________

Mental Math Strategies for Subtraction

Essential Question What mental math strategies can you use to find differences?

UNLOCK the Problem REAL WORLD

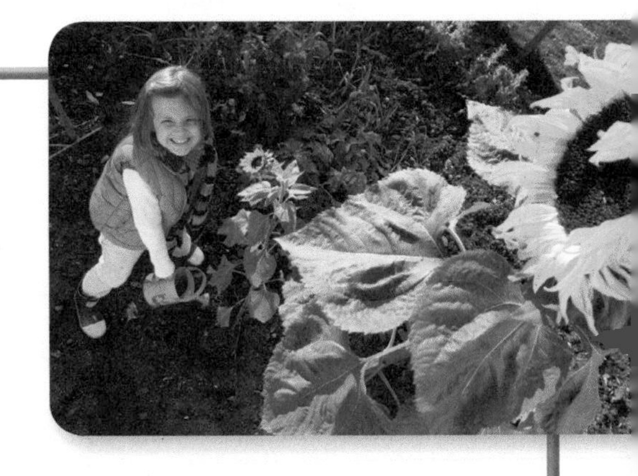

A sunflower can grow to be very tall. Dylan is 39 inches tall. She watered a sunflower that grew to be 62 inches tall. How many inches shorter was Dylan than the sunflower?

One Way Use a number line to find 62 − 39.

A Count up by tens and then ones.

Think: Start at 39. Count up to 62.

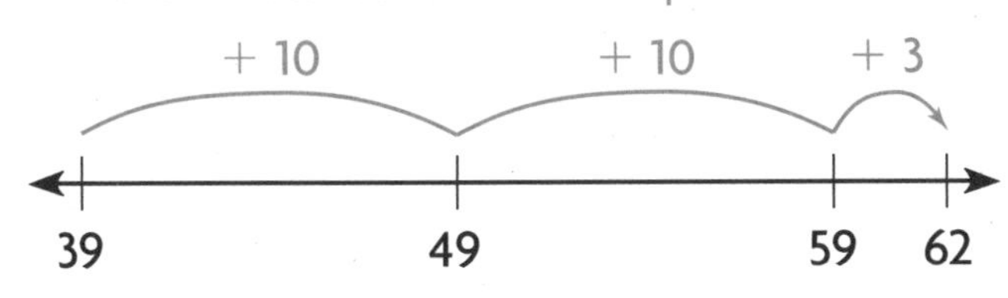

Add the lengths of the jumps to find the difference.

$10 + 10 + 3 =$ ______

$62 - 39 =$ ______

So, Dylan was ______ inches shorter than the sunflower.

B Take away tens and ones.

Think: Start at 62. Count back 39.

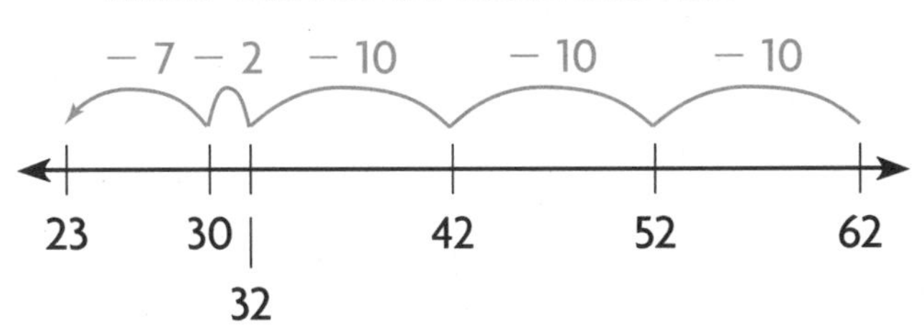

Take away lengths of jumps to end on the difference.

MATHEMATICAL PRACTICES

Math Talk Compare the number lines. **Explain** where the answer is on each one.

Other Ways

A Use friendly numbers and adjust to find 74 − 28.

STEP 1 Make the number you subtract a friendly number. $28 + 2 =$ ______

Think: Add to 28 to make a number with 0 ones.

STEP 2 Since you added 2 to 28, you have to add 2 to 74. $74 + 2 =$ ______

STEP 3 Find the difference. ______ − ______ = ______

So, $74 - 28 =$ ______.

Try This! **Use friendly numbers to subtract 9 and 99.**

- **Find 36 − 9.**

  Think: 9 is 1 less than 10.

  Subtract 10. 36 − 10 = ______

  Then add 1. ______ + 1 = ______

  So, 36 − 9 = ______.

- **Find 423 − 99.**

  Think: 99 is 1 less than 100.

  Subtract 100. 423 − 100 = ______

  Then add 1. ______ + 1 = ______

  So, 423 − 99 = ______.

B Use the break apart strategy to find 458 − 136.

STEP 1 Subtract the hundreds. 400 − 100 = ______

STEP 2 Subtract the tens. 50 − 30 = ______

STEP 3 Subtract the ones. 8 − 6 = ______

STEP 4 Add the differences. ______ + ______ + ______ = ______

So, 458 − 136 = ______.

Share and Show

1. Find 61 − 24. Draw jumps and label the number line to show your thinking.

   Think: Take away tens and ones.

61

61 − 24 = ______

2. Use friendly numbers to find the difference.

   86 − 42 = ______

   Think: 42 − 2 = 40
   86 − 2 = 84

MATHEMATICAL PRACTICES

Math Talk Explain how you can use the break apart strategy to find 86 − 42.

Name ______________________

Use mental math to find the difference.
Draw or describe the strategy you use.

3. $56 - 38 =$ ________

4. $435 - 121 =$ ________

On Your Own

Use mental math to find the difference.
Draw or describe the strategy you use.

5. $82 - 49 =$ ________

6. $152 - 99 =$ ________

7. $367 - 225 =$ ________

8. $96 - 47 =$ ________

Problem Solving REAL WORLD

9. H.O.T. **What's the Error?** Erica used friendly numbers to find $43 - 19$. She added 1 to 19 and subtracted 1 from 43. What is Erica's error? **Explain.**

____

____

10. **Test Prep** There were 87 sunflowers at the flower shop in the morning. There were 56 sunflowers left at the end of the day. How many sunflowers were sold?

Ⓐ 13 Ⓑ 31 Ⓒ 41 Ⓓ 143

Connect to Reading

Compare and Contrast

Emus and ostriches are the world's largest birds. They are alike in many ways and different in others.

When you compare things, you decide how they are alike. When you contrast things, you decide how they are different.

The table shows some facts about emus and ostriches. Use the information on this page to compare and contrast the birds.

Facts About Emus and Ostriches

	Emus	Ostriches
Can they fly?	No	No
Where do they live?	Australia	Africa
How much do they weigh?	About 120 pounds	About 300 pounds
How tall are they?	About 72 inches	About 108 inches
How fast can they run?	About 40 miles per hour	About 40 miles per hour

Ostrich

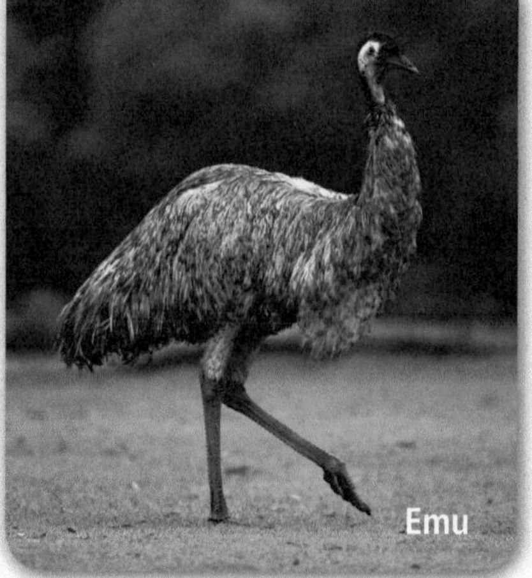
Emu

11. How are emus and ostriches alike? How are they different?

Alike: 1. ____________

2. ____________

Different: 1. ____________

2. ____________

3. ____________

12. How much taller is an ostrich than an emu?

____

13. **What if** an emu weighs 117 pounds and an ostrich weighs 274 pounds? How much more does the ostrich weigh?

____

FOR MORE PRACTICE:
Standards Practice Book, pp. P19–P20

Name ______________________________

Use Place Value to Subtract

Essential Question How can you use place value to subtract 3-digit numbers?

UNLOCK the Problem REAL WORLD

Ava sold 473 tickets for the school play. Kim sold 294 tickets. How many more tickets did Ava sell than Kim?

- Do you need to combine or compare the number of tickets sold?

____

- Circle the numbers you will need to use.

Use place value to subtract.

Subtract. 473 − 294

Estimate. 475 − 300 = __________

STEP 1

Subtract the ones.
3 < 4, so regroup.

7 tens 3 ones =

6 tens ______ ones

$$\begin{array}{r} {\scriptstyle 6\,13} \\ 4\not{7}\not{3} \\ -\,294 \\ \hline \end{array}$$

STEP 2

Subtract the tens.
6 < 9, so regroup.

4 hundreds 6 tens =

3 hundreds ______ tens

$$\begin{array}{r} {\scriptstyle 16} \\ {\scriptstyle 3\,\not{6}\,13} \\ \not{4}\not{7}\not{3} \\ -\,2\,9\,4 \\ \hline 9 \end{array}$$

STEP 3

Subtract the hundreds.
Add to check your answer.

$$\begin{array}{r} {\scriptstyle 16} \\ {\scriptstyle 3\,\not{6}\,13} \\ \not{4}\not{7}\not{3} \\ -\,2\,9\,4 \\ \hline 7\,9 \end{array} \qquad \begin{array}{r} {\scriptstyle 1\,1} \\ 179 \\ +\,294 \\ \hline 473 \end{array}$$

So, Ava sold __________ more tickets than Kim.

Since __________ is close to the estimate of __________, the answer is reasonable.

Math Idea

Addition and subtraction undo each other. So you can use addition to check subtraction.

Try This! **Use place value to subtract. Use addition to check your work.**

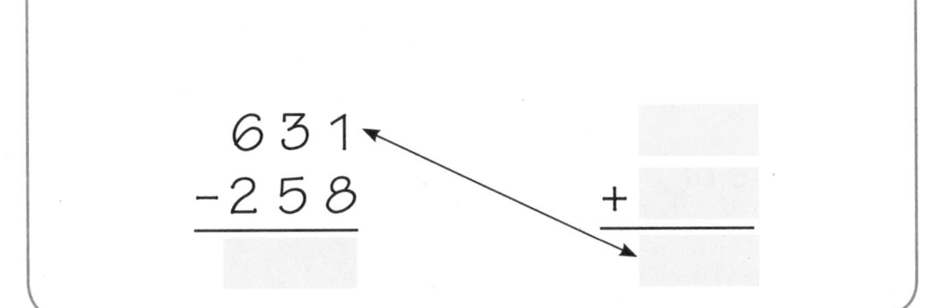

$$\begin{array}{r} 631 \\ -\,258 \\ \hline \end{array} \qquad \begin{array}{r} \square \\ +\,\square \\ \hline \square \end{array}$$

Example Use place value to find 890 − 765.

Estimate. 900 − 750 = ________

STEP 1

Subtract the ones.
Regroup the tens as tens and ones.

```
  8 10
  8 9 0
− 7 6 5
```

STEP 2

Subtract the tens.

```
  8 10
  8 9 0
− 7 6 5
      5
```

STEP 3

Subtract the hundreds.
Add to check your answer.

```
  8 10
  8 9 0        1 2 5
− 7 6 5      + 7 6 5
    2 5
```

So, 890 − 765 = ________.

MATHEMATICAL PRACTICES

Math Talk Explain how you know your answer is reasonable.

Try This! Circle the problem in which you need to regroup. Find the difference.

A	B	C
894 − 583	521 − 301	918 − 427

Share and Show

1. Estimate. Then use place value to find 627 − 384. Add to check your answer.

Estimate. ________ − ________ = ________

```
  6 2 7
− 3 8 4      + 3 8 4
```

Since ________ is close to the estimate of ________, the answer is reasonable.

MATHEMATICAL PRACTICES

Math Talk Did you need to regroup to find the difference? Explain.

Name ______________________

Estimate. Then find the difference.

2. Estimate: ______

$$\begin{array}{r} 386 \\ -123 \\ \hline \end{array}$$

3. Estimate: ______

$$\begin{array}{r} 519 \\ -205 \\ \hline \end{array}$$

4. Estimate: ______

$$\begin{array}{r} 456 \\ -217 \\ \hline \end{array}$$

5. Estimate: ______

$$\begin{array}{r} 642 \\ -159 \\ \hline \end{array}$$

6. Estimate: ______

$$\begin{array}{r} 242 \\ -220 \\ \hline \end{array}$$

7. Estimate: ______

$$\begin{array}{r} 870 \\ -492 \\ \hline \end{array}$$

8. Estimate: ______

$$\begin{array}{r} 654 \\ -263 \\ \hline \end{array}$$

9. Estimate: ______

$$\begin{array}{r} 937 \\ -618 \\ \hline \end{array}$$

MATHEMATICAL PRACTICES

Math Talk Which exercises can you compute mentally? **Explain** why.

On Your Own

Estimate. Then find the difference.

10. Estimate: ______

$$\begin{array}{r} 435 \\ -312 \\ \hline \end{array}$$

11. Estimate: ______

$$\begin{array}{r} 617 \\ -501 \\ \hline \end{array}$$

12. Estimate: ______

$$\begin{array}{r} 893 \\ -268 \\ \hline \end{array}$$

13. Estimate: ______

$$\begin{array}{r} 750 \\ -276 \\ \hline \end{array}$$

Practice: Copy and Solve **Estimate. Then solve.**

14. $568 - 276$
15. $761 - 435$
16. $829 - 765$
17. $974 - 285$

H.O.T. **Algebra** **Find the unknown number.**

18. $\begin{array}{r} 86 \\ -\square \\ \hline 62 \end{array}$

19. $\begin{array}{r} 372 \\ -\square \\ \hline 240 \end{array}$

20. $\begin{array}{r} 537 \\ -\square \\ \hline 172 \end{array}$

21. $\begin{array}{r} 629 \\ -\square \\ \hline 335 \end{array}$

Problem Solving REAL WORLD

Use the table for 22–24.

School Play Tickets Sold	
Student	**Number of Tickets**
Jenna	282
Matt	178
Sonja	331

22. About how many tickets did Jenna, Matt, and Sonja sell altogether?

____

23. How many more tickets did Sonja sell than Matt?

____

24. Alicia sold 59 fewer tickets than Jenna and Matt sold together. How many tickets did Alicia sell?

____

25. H.O.T. **Sense or Nonsense?** Nina says to check subtraction, add the difference to the number you started with. Does this statement make sense? **Explain.**

____

26. Write Math Do you have to regroup to find 523 − 141? **Explain.** Then solve.

____

27. **Test Prep** There are 842 seats in the school auditorium. 138 seats need repairs. How many seats do not need repairs?

Ⓐ 980 Ⓑ 804 Ⓒ 716 Ⓓ 704

SHOW YOUR WORK

FOR MORE PRACTICE:
Standards Practice Book, pp. P21–P22

Name ____________________

Combine Place Values to Subtract

Essential Question How can you use the combine place values strategy to subtract 3-digit numbers?

UNLOCK the Problem REAL WORLD

Elena collected 431 bottles for recycling. Pete collected 227 fewer bottles than Elena. How many bottles did Pete collect?

- What do you need to find?

____

- Circle the numbers you need to use.

Combine place values to find the difference.

A **Subtract.** 431 − 227

Estimate. 400 − 200 = ________

STEP 1 Look at the ones place. Since 7 > 1, combine place values. Combine the tens and ones places. There are 31 ones and 27 ones. Subtract the ones. Write 0 for the tens.

$$\begin{array}{r} 431 \\ -\ 227 \\ \hline \end{array}$$

Think: 31 − 27

STEP 2 Subtract the hundreds.

$$\begin{array}{r} 431 \\ -\ 227 \\ \hline 04 \end{array}$$

So, Pete collected ________ bottles.

Since ________ is close to the estimate of ________, the answer is reasonable.

MATHEMATICAL PRACTICES

Math Talk Explain why there is a zero in the tens place.

B **Subtract.** 513 − 482

Estimate. 510 − 480 = ________

STEP 1 Subtract the ones.

$$\begin{array}{r} 513 \\ -\ 482 \\ \hline \end{array}$$

STEP 2 Look at the tens place. Since 8 > 1, combine place values. Combine the hundreds and tens places. There are 51 tens and 48 tens. Subtract the tens.

$$\begin{array}{r} 513 \\ -\ 482 \\ \hline 1 \end{array}$$

Think: 51 − 48

So, 513 − 482 = ________.

Example Combine place values to find 500 − 173.

Estimate. 500 − 175 = ______

STEP 1 Look at the ones and tens places. Since 3 > 0 and 7 > 0, combine the hundreds and tens.

There are 50 tens. Regroup 50 tens as 49 tens 10 ones.

```
 4 9 10
 5 0 0
-1 7 3
------
```

STEP 2 Subtract the ones.

Think: 10 − 3

```
 4 9 10
 5 0 0
-1 7 3
------
```

STEP 3 Subtract the tens.

Think: 49 − 17

```
 4 9 10
 5 0 0
-1 7 3
------
     7
```

So, 500 − 173 = ______.

MATHEMATICAL PRACTICES

Math Talk Explain why you combined the hundreds and tens.

Try This! **Find 851 − 448 in two ways.**

Estimate. 850 − 450 = ______

A **Use place value.**

```
 851
-448
----
```

B **Combine place values.**

```
 851
-448
----
```

Think: Combine tens and ones.

1. When does the combine place values strategy make it easier to find the difference? **Explain.**

______

______

______

2. Which strategy would you use to find 431 − 249? **Explain.**

______

______

Name ______________________

Share and Show MATH BOARD

1. Combine place values to find $406 - 274$.

$$\begin{array}{r} 406 \\ -274 \\ \hline \end{array}$$

Think: Subtract the ones. Then combine the hundreds and tens places.

MATHEMATICAL PRACTICES **Math Talk** Explain how to combine place values.

Estimate. Then find the difference.

2. Estimate: ______ $\begin{array}{r} 595 \\ -286 \\ \hline \end{array}$

3. Estimate: ______ $\begin{array}{r} 728 \\ -515 \\ \hline \end{array}$

4. Estimate: ______ $\begin{array}{r} 543 \\ -307 \\ \hline \end{array}$

5. Estimate: ______ $\begin{array}{r} 600 \\ -453 \\ \hline \end{array}$

On Your Own

Estimate. Then find the difference.

6. Estimate: ______ $\begin{array}{r} 438 \\ -257 \\ \hline \end{array}$

7. Estimate: ______ $\begin{array}{r} 706 \\ -681 \\ \hline \end{array}$

8. Estimate: ______ $\begin{array}{r} 839 \\ -754 \\ \hline \end{array}$

9. Estimate: ______ $\begin{array}{r} 916 \\ -558 \\ \hline \end{array}$

10. Estimate: ______ $\begin{array}{r} 537 \\ -428 \\ \hline \end{array}$

11. Estimate: ______ $\begin{array}{r} 528 \\ -297 \\ \hline \end{array}$

12. Estimate: ______ $\begin{array}{r} 734 \\ -327 \\ \hline \end{array}$

13. Estimate: ______ $\begin{array}{r} 800 \\ -789 \\ \hline \end{array}$

Practice: Copy and Solve **Estimate. Then solve.**

14. $457 - 364$
15. $652 - 341$
16. $700 - 648$
17. $963 - 256$

Problem Solving REAL WORLD

Use the table for 18–22.

Roller Coaster Heights

Roller Coaster	State	Height in Feet
Titan	Texas	245
Kingda Ka	New Jersey	456
Intimidator 305	Virginia	305
Top Thrill Dragster	Ohio	420

18. The table shows the heights of some roller coasters in the United States. How much taller is Kingda Ka than Titan?

It is 456

19. How many feet shorter is Intimidator 305 than Top Thrill Dragster?

20. How much shorter is Top Thrill Dragster than the heights of Titan and Kingda Ka together?

21. **What if** a roller coaster was 500 feet tall? Which roller coaster would be 195 feet shorter?

22. H.O.T. Write Math ▸ **Pose a Problem** Look back at Problem 21. Write and solve a similar problem by using different numbers.

23. **Test Prep** The Goliath roller coaster is 235 feet tall. The Steel Dragon roller coaster is 318 feet tall. How many feet taller is Steel Dragon than Goliath?

Ⓐ 73 feet　Ⓒ 123 feet

Ⓑ 83 feet　Ⓓ 183 feet

SHOW YOUR WORK

FOR MORE PRACTICE:
Standards Practice Book, pp. P23–P24

Name ____________________________________

Problem Solving • Model Addition and Subtraction

Essential Question How can you use the strategy *draw a diagram* to solve one- and two-step addition and subtraction problems?

UNLOCK the Problem REAL WORLD

Sami scored 84 points in the first round of a new computer game. He scored 21 more points in the second round than in the first round. What was Sami's total score?

You can use a bar model to solve the problem.

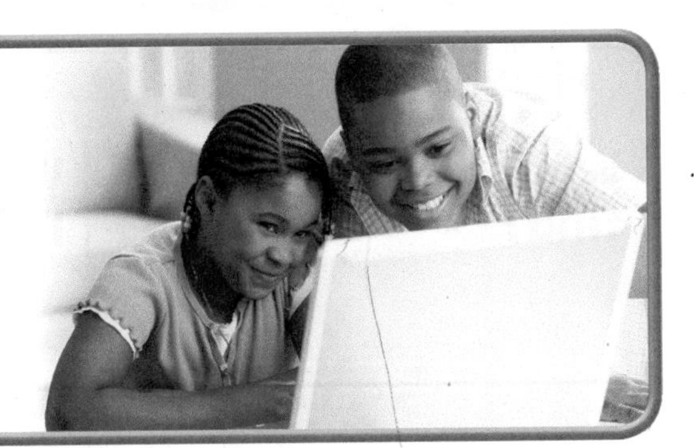

Read the Problem

What do I need to find?	What information do I need to use?	How will I use the information?
I need to find ______________.	Sami scored _____ points in the first round. He scored _____ more points than that in the second round.	I will draw a bar model to show the number of points Sami scored in each round. Then I will use the bar model to decide which operation to use.

Solve the Problem

- Complete the bar model to show the number of points Sami scored in the second round.

_____ points	_____ points

■ points

_______ + _______ = ■

_______ = ■

- Complete another bar model to show Sami's total score.

_____ points	_____ points

▲ points

_______ + _______ = ▲

_______ = ▲

1. How many points did Sami score in the second round? _______

2. What was Sami's total score? _______

Try Another Problem

Anna scored 265 points in a computer game. Greg scored 142 points. How many more points did Anna score than Greg?

You can use a bar model to solve the problem.

Read the Problem

What do I need to find?	What information do I need to use?	How will I use the information?

Solve the Problem

Record the steps you used to solve the problem.

Anna ______ points

Greg ______ points

■ points

3. How many more points did Anna score than Greg?

______

4. How do you know your answer is reasonable?

______

______

______

5. How did your drawing help you solve the problem?

______

______

MATHEMATICAL PRACTICES

Math Talk **Explain** how the length of each bar in the model would change if Greg scored more points than Anna but the totals remained the same.

Name ____________________

Share and Show 

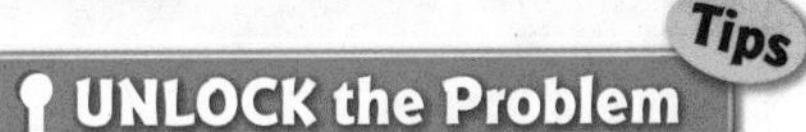

√ Use the problem solving MathBoard.

√ Choose a strategy you know.

1. Sara received 73 votes in the school election. Ben received 25 fewer votes than Sara. How many students voted?

   First, find how many students voted for Ben.

   Think: 73 − 25 = ■

   Write the numbers in the bars.

   So, Ben received _____ votes.

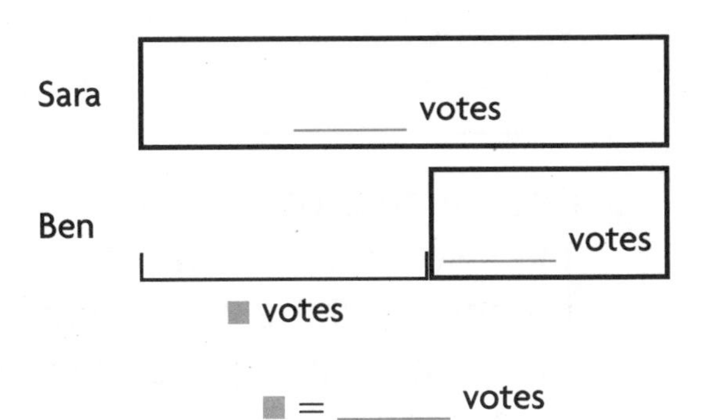

■ = _____ votes

   Next, find the total number of votes.

   Think: 73 + 48 = ▲

   Write the numbers in the bars.

   So, __________ students voted.

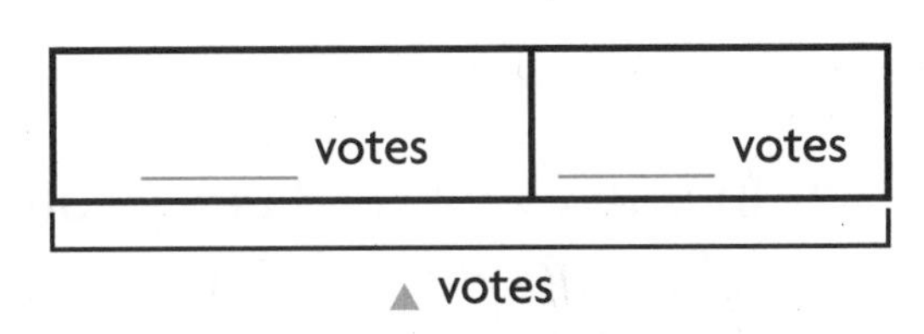

▲ = _____ votes

2. H.O.T. **What if** there were 3 students in another election and the total number of votes was the same? What would the bar model for the total number of votes look like? How many votes might each student get?

   ____

   ____

3. **Pose a Problem** Use the bar model at the right. Write a problem to match it.

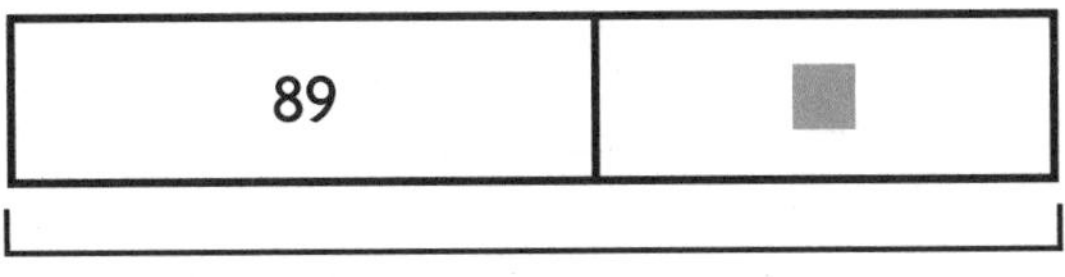

   ____

   ____

   ____

4. Solve your problem. Will you add or subtract?

   ____

On Your Own

Choose a STRATEGY

- Act It Out
- Draw a Diagram
- Find a Pattern
- Make a Table

5. Tony's Tech Store is having a sale. The store had 142 computers in stock. During the sale, 91 computers were sold. How many computers were not sold?

51

6. The number of computer games sold during the sale was 257. This is 162 more than the number sold the week before the sale. How many computer games were sold the week before the sale?

SHOW YOUR WORK

7. In one week, 128 cell phones were sold. The following week, 37 more cell phones were sold than the week before. How many cell phones were sold in those two weeks?

8. H.O.T. **Write Math** On Monday, the number of customers in the store, rounded to the nearest hundred, was 400. What is the greatest number of customers that could have been in the store? **Explain.**

9. **Test Prep** The number of laptop computers sold in one day was 42. That is 18 fewer than the number of desktop computers sold. How many desktop computers were sold?

   Ⓐ 24 Ⓑ 50 Ⓒ 60 Ⓓ 61

FOR MORE PRACTICE:
Standards Practice Book, pp. P25–P26

Name ____________________

Chapter Review/Test

Vocabulary

Choose the best term from the box to complete the sentence.

Vocabulary
Commutative Property of Addition
compatible numbers
Identity Property of Addition

1. The ____________________ states that you can add two or more numbers in any order and get the same sum. (p. 6)

2. ____________________ are numbers that are easy to compute mentally. (p. 13)

Concepts and Skills

Is the sum even or odd? Write *even* or *odd*.

3. $6 + 4$ __________

4. $3 + 8$ __________

5. $9 + 7$ __________

Use rounding or compatible numbers to estimate the sum or difference.

6. $267 + 621$ → ____ + ____

7. $672 - 431$ → ____ − ____

8. $800 - 632$ → ____ − ____

Estimate. Then find the sum or difference.

9. Estimate: ______ $218 + 342$

10. Estimate: ______ $766 + 125$

11. Estimate: ______ $367 + 351$

12. Estimate: ______ $532 + 402$

13. Estimate: ______ $261 - 150$

14. Estimate: ______ $948 - 532$

15. Estimate: ______ $919 - 838$

16. Estimate: ______ $706 - 574$

GO Online Assessment Options Chapter Test

Fill in the bubble for the correct answer choice.

Use the table for 17–20.

Annie's Apple Farm	
Day	**Number of Visitors**
Saturday	243
Sunday	345

17. The table shows the number of visitors to Annie's Apple Farm. About how many people visited the farm on Saturday?

Ⓐ about 400

Ⓑ about 350

Ⓒ about 300

Ⓓ about 200

18. Rounded to the nearest ten, how many people visited the farm on Sunday?

Ⓐ 350

Ⓑ 340

Ⓒ 300

Ⓓ 240

19. Which is the best estimate of the number of visitors to the farm on Saturday and Sunday?

Ⓐ about 400

Ⓑ about 450

Ⓒ about 600

Ⓓ about 700

20. There were 186 fewer visitors to the farm on Friday than on Sunday. How many people visited the farm on Friday?

Ⓐ 159

Ⓑ 169

Ⓒ 241

Ⓓ 269

Name ________________________________

Fill in the bubble for the correct answer choice.

Use the table for 21–23.

Susie's Sweater Shop	
Month	**Number of Sweaters Sold**
January	402
February	298
March	171

21. Susie's Sweater Shop sells sweaters online. The table shows the number of sweaters sold in three months. How many sweaters were sold in January and February?

Ⓐ 800

Ⓑ 700

Ⓒ 690

Ⓓ 600

22. How many more sweaters were sold in January than in March?

Ⓐ 231

Ⓑ 331

Ⓒ 371

Ⓓ 573

23. How many more sweaters were sold in February and March than in January?

Ⓐ 871

Ⓑ 167

Ⓒ 67

Ⓓ 66

24. Susie sold only 28 sweaters in June, 19 in July, and 11 in August. How many sweaters did she sell in June, July, and August?

Ⓐ 48

Ⓑ 56

Ⓒ 57

Ⓓ 58

▶ Constructed Response

25. Diana sold 336 muffins at the bake sale. Bob sold 287 muffins. Bob estimates that he sold 50 fewer muffins than Diana. How did he estimate? **Explain.**

____

____

____

26. Sunnyday Elementary School is having its annual Read-a-thon. The third graders have read 573 books so far. Their goal is to read more than 900 books. What is the least number of books they need to read to reach their goal? **Explain.**

____

____

____

▶ Performance Task

27. There are 318 fiction books in the class library. The number of nonfiction books is 47 less than double the number of fiction books.

A About how many nonfiction books are there in the class library? **Explain.**

____

____

____

____

____

B How many fiction and nonfiction books are there in the class library altogether? Show your work.

____

Chapter 2

Represent and Interpret Data

Show What You Know

Check your understanding of important skills.

Name ______________________

▶ Numbers to 20 Circle the number word. Write the number.

1. 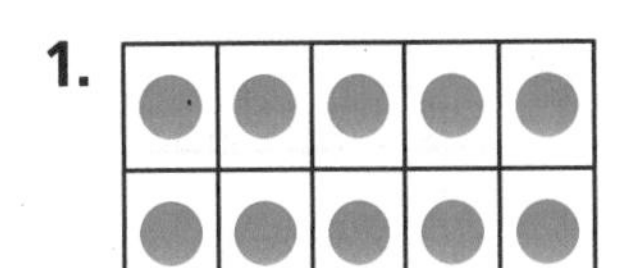 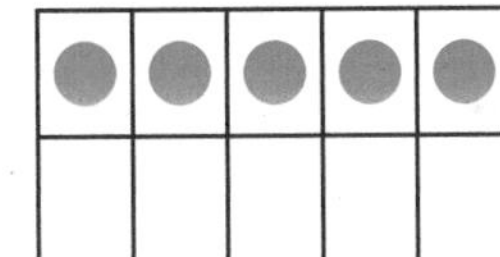

fourteen

fifteen ____

2. 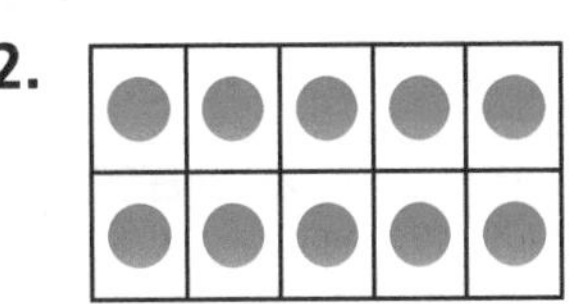 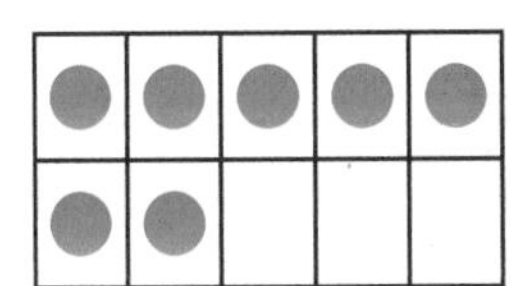

seventeen

eighteen ____

▶ Skip Count Skip count to find the missing numbers.

3. Count by twos. 2, 4, ____, ____, 10, ____, ____, 16

4. Count by fives. 5, 10, ____, ____, ____, 30, ____

▶ Addition and Subtraction Facts Find the sum or difference.

5. $12 - 4 =$ ____

6. $9 + 8 =$ ____

7. $11 - 7 =$ ____

Paige helps to sell supplies in the school store. Each month she totals all the sales and makes a bar graph. The graph shows sales through December. Be a Math Detective to find the month during which the hundredth sale was made.

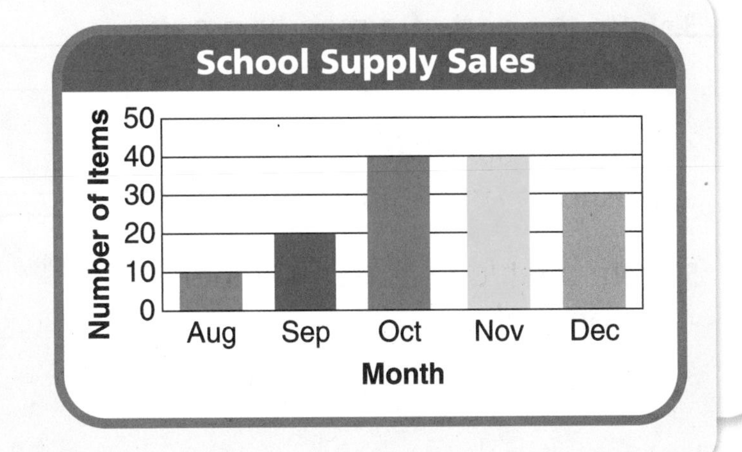

GO Online Assessment Options: Soar to Success Math

Vocabulary Builder

▶ Visualize It

Complete the bubble map by using the words with a ✓.

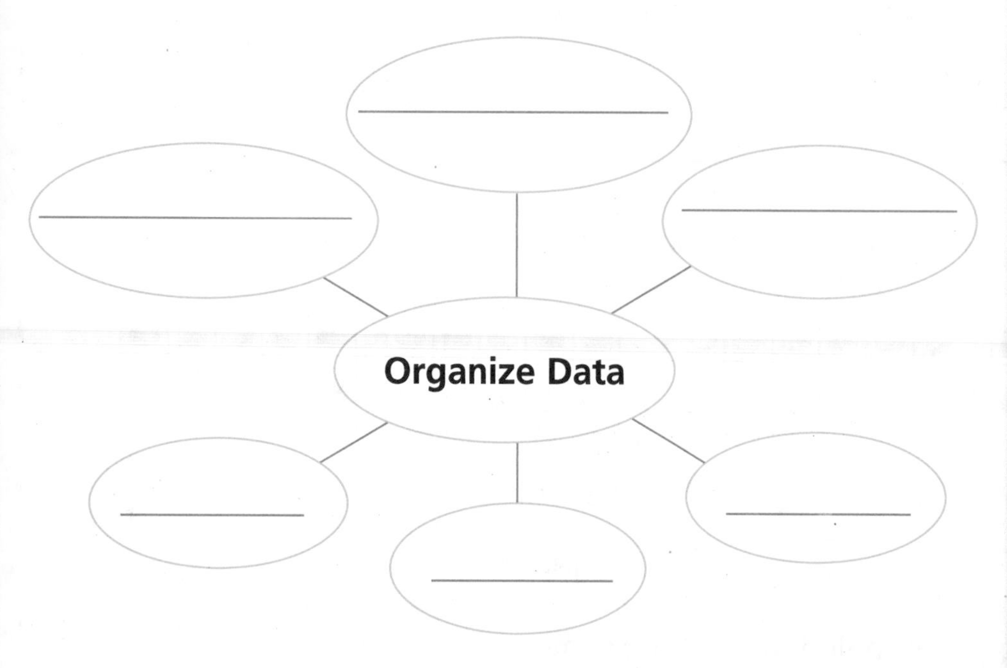

Review Words

- compare
- data
- fewer
- more
- survey
- ✓tally table

Preview Words

- ✓frequency table
- ✓horizontal bar graph
- key
- ✓line plot
- ✓picture graph
- scale
- ✓vertical bar graph

▶ Understand Vocabulary

Write the review word or preview word that answers the riddle.

1. I am a graph that records each piece of data above a number line. ____________

2. I am the numbers that are placed at fixed distances on a graph to help label the graph. ____________

3. I am the part of a map or graph that explains the symbols. ____________

4. I am a graph that uses pictures to show and compare information. ____________

5. I am a table that uses numbers to record data. ____________

GO Online • eStudent Edition • Multimedia eGlossary

PROBLEM SOLVING
Lesson 2.1

Name ______________________

Problem Solving • Organize Data

Essential Question How can you use the strategy *make a table* to organize data and solve problems?

UNLOCK the Problem REAL WORLD

The students in Alicia's class voted for their favorite ice cream flavor. They organized the data in this tally table. How many more students chose chocolate than strawberry?

Another way to show the data is in a frequency table. A **frequency table** uses numbers to record data.

Favorite Ice Cream Flavor	
Flavor	**Tally**
Vanilla	𝍸 ||
Chocolate	𝍸 |||
Strawberry	||||

Read the Problem

What do I need to find?

How many more students chose __________ than __________ ice cream as their favorite?

What information do I need to use?

the data about favorite __________ in the tally table

How will I use the information?

I will count the __________. Then I will put the numbers in a frequency table and compare the number of students who chose __________ to the number of students who chose __________.

Solve the Problem

Favorite Ice Cream Flavor	
Flavor	**Number**
Vanilla	|||
Strabery	||
Chocolate	|

Count the tally marks. Record ____ for vanilla. Write the other flavors and record the number of tally marks.

To compare the number of students who chose strawberry and the number of students who chose chocolate, subtract.

____ − ____ = ____

So, ____ more students chose chocolate as their favorite flavor.

MATHEMATICAL PRACTICES

Math Talk **Explain** why you would record data in a frequency table.

Try Another Problem

Two classes in Carter's school grew bean plants for a science project. The heights of the plants after six weeks are shown in the tally table. The plants were measured to the nearest inch. How many fewer bean plants were 9 inches tall than 7 inches and 8 inches combined?

Bean Plant Heights	
Height in Inches	Tally
7	卌 IIII
8	卌 III
9	卌 卌 II
10	卌 IIII

Read the Problem	Solve the Problem
What do I need to find?	**Record the steps you used to solve the problem.**
What information do I need to use?	
How will I use the information?	

- Suppose the number of 3-inch plants was half the number of 8-inch plants. How many 3-inch bean plants were there?

____

Math Talk MATHEMATICAL PRACTICES Explain another strategy you could use to solve the problem.

Name ____________________________

Share and Show 

Use the Shoe Lengths table for 1–4.

Shoe Lengths		
Length in Centimeters	Tally	
	Boys	Girls
18	𝍸 |	||||
19	𝍸	||||
20	𝍸 |||	𝍸 ||||
21	𝍸 ||	𝍸
22	𝍸 ||||	𝍸 ||

1. The students in three third-grade classes recorded the lengths of their shoes to the nearest centimeter. The data are in the tally table. How many more shoes were 18 or 22 centimeters long combined than 20 centimeters long?

   You can put the data in a table and compare the lengths of the shoes to solve the problem.

   First, count the tally marks and put the data in a frequency table.

   To find the number of shoes that were 18 or 22 centimeters long, add

   6 + _____ + _____ + _____ = _____.

   To find the number of shoes that were 20 centimeters long, add _____ + _____ = _____.

   To find the difference between the shoes that were 18 or 22 centimeters long and the shoes that were 20 centimeters long, subtract the sums.

   _____ − _____ = _____.

   So, _____ more shoes were 18 or 22 centimeters long than 20 centimeters long.

Shoe Lengths		
Length in Centimeters	Number	
	Boys	Girls
18		
19		
20		
21		
22		

2. How many fewer girls' shoes than boys' shoes were measured? ______________

3. The length of the least number of shoes was _____ centimeters long.

4. H.O.T. **What if** the length of 5 more boys' shoes measured 21 centimeters? **Describe** how the table would look.

____

____

On Your Own

Choose a STRATEGY

- Act It Out
- Draw a Diagram
- Find a Pattern
- Make a Table

5. Isabel is thinking of an even number between 234 and 250. The sum of the digits is double the digit in the ones place. What is Isabel's number?

____

6. Ben needs 400 points to win a prize at the arcade. He has 237 points. How many more points does he need?

____

7. There were 428 visitors at the pet show on Saturday and 395 visitors on Sunday. How many visitors in all went to the pet show?

____

8. Write Math ▸ Heather has 6 dimes and 10 pennies. Jason has 3 quarters. Who has more money? **Explain** your answer.

____

____

9. H.O.T. Andrew has 10 more goldfish than Todd. Together, they have 50 goldfish. How many goldfish does each boy have?

____

10. **Test Prep** Jade made a tally table to record how many people have dogs. Her table shows

Dog 𝍸 𝍸 ||||

How many people have dogs?

Ⓐ 4 Ⓒ 14

Ⓑ 9 Ⓓ 15

SHOW YOUR WORK

FOR MORE PRACTICE:
Standards Practice Book, pp. P31–P32

Name ______________________

Use Picture Graphs

Essential Question How can you read and interpret data in a picture graph?

UNLOCK the Problem REAL WORLD

A **picture graph** uses small pictures or symbols to show and compare information.

Nick has a picture graph that shows how some students get to school. How many students ride the bus?

- Underline the words that tell you where to find the information to answer the question.
- How many ☺ are shown for bus?

____

How We Get to School

Walk	☺ ☺ ☺
Bike	☺ ☺ ☺ ☺
Bus	☺ ☺ ☺ ☺ ☺ ☺ ☺ ☺
Car	☺ ☺ ☺ ☺ ☺ ☺

Key: Each ☺ = 10 students.

Each row has a label that names one way students get to school.

The title says that the picture graph is about how some students get to school.

The **key** tells that each picture or symbol stands for the way 10 students get to school.

To find the number of students who ride the bus, count each ☺ as 10 students.

10, 20, _____, _____, _____, _____, _____, _____

So, _____ students ride the bus to school.

1. How many fewer students walk than ride the bus? _____
2. How many students were surveyed? _____
3. **What if** the symbol stands for 5 students? How many symbols will you need to show the number of students who walk to school? _____

Use a Half Symbol

How many students chose a cupcake as their favorite dessert?

Our Favorite Dessert	
Ice Cream	☺ ☺ ☺ ☺ ☺
Cake	☺ ☺ ☺
Pie	☺ ☺
Cupcake	☺ ☺ ☺ ☺ ◖
Key: Each ☺ = 2 students.	

Math Idea

Half of the picture stands for half the value of the whole picture.

☺ = 2 students

◖ = 1 student

Count the ☺ in the cupcake row by twos. Then add 1 for the half symbol.

2, 4, _____, _____   _____ + _____ = _____

So, _____ students chose a cupcake as their favorite dessert.

Share and Show

Use the Number of Books Students Read picture graph for 1–3.

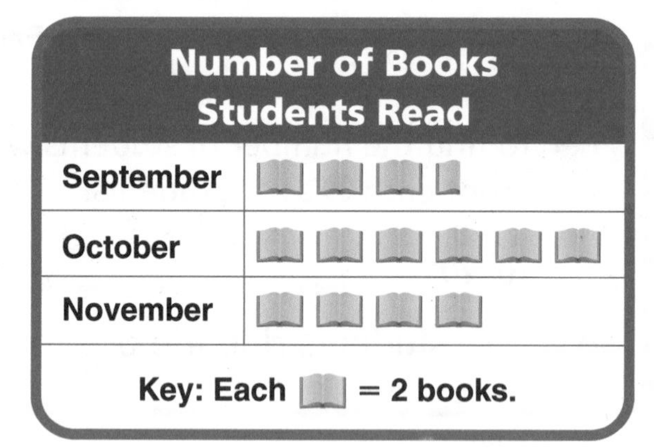

Number of Books Students Read	
September	📖 📖 📖 ▯
October	📖 📖 📖 📖 📖 📖
November	📖 📖 📖 📖
Key: Each 📖 = 2 books.	

1. What does ▯ stand for?

   Think: Half of 2 is 1.

   ____

2. How many books did the students read in September?

   ____

3. How many more books did the students read in October than in November?

   ____

Math Talk MATHEMATICAL PRACTICES

Explain how to find the number of books the students read.

Name ______________________________

On Your Own

Use the Favorite Game picture graph for 4–10.

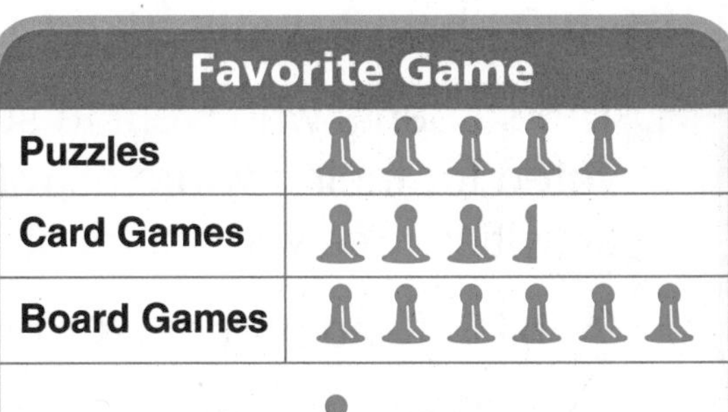

Favorite Game	
Puzzles	
Card Games	
Board Games	
Key: Each = 4 students.	

4. How many students chose puzzles?

____

5. How many fewer students chose card games than board games?

____

6. Which two types of games did a total of 34 students choose?

____

7. How many students were surveyed?

____

8. How many students did not choose card games?

____

9. **Write Math** **What's the Error?** Jacob said one more student chose board games than puzzles. **Explain** his error.

____

____

____

10. **H.O.T.** **What if** computer games were added as a choice and more students chose it than puzzles, but fewer students chose it than board games? How many students would choose computer games?

____

UNLOCK the Problem REAL WORLD

Use the picture graph for 11–13.

Favorite Camp Activity	
Activity	**Number**
Biking	☼ ☼ ☼ ◐
Hiking	☼ ☼ ☼ ☼
Boating	☼ ☼ ☼
Fishing	☼ ◐
Key: Each ☼ = 6 students.	

11. The students who went to summer camp voted for their favorite activity. Which two activities received a total of 39 votes?

Ⓐ biking and hiking Ⓒ biking and boating

Ⓑ hiking and boating Ⓓ fishing and hiking

a. What do you need to find?

____

b. What steps will you use to solve the problem?

____

c. Show the steps you used to solve the problem.

d. Complete the sentences.

Each ☼ = _____ students.

Each ◐ = _____ students.

votes for biking + hiking = _____

votes for hiking + boating = _____

votes for biking + boating = _____

votes for fishing + hiking = _____

e. Fill in the bubble for the correct answer.

12. How many students voted for their favorite camp activity?

Ⓐ 91

Ⓑ 81

Ⓒ 72

Ⓓ 61

13. **Test Prep** How many fewer students voted for fishing than for hiking?

Ⓐ 33

Ⓑ 24

Ⓒ 15

Ⓓ 9

FOR MORE PRACTICE:
Standards Practice Book, pp. P33–P34

Name ______________________

Make Picture Graphs

Essential Question How can you draw a picture graph to show data in a table?

UNLOCK the Problem REAL WORLD

Delia made the table at the right. She used it to record the places the third grade classes would like to go during a field trip. How can you show the data in a picture graph?

Field Trip Choices

Place	Number
Museum	6
Science Center	15
Aquarium	12
Zoo	9

Make a picture graph.

STEP 1

Write the title at the top of the picture graph. Write the name of a place in each row.

STEP 2

Look at the numbers in the table. Choose a picture for the key, and tell how many students each picture represents. Write the key at the bottom of the graph.

STEP 3

Draw the correct number of pictures for each field trip choice.

Museum	

Key: Each ____ = ____ students.

- How did you decide how many pictures to draw for the Science Center?

____

____

Try This! Make a picture graph from data you collect. Take a survey or observe a subject that interests you. Collect and record the data in a frequency table. Then make a picture graph. Decide on a symbol and a key. Include a title and labels.

Key:	

Share and Show

Jeremy pulled marbles from a bag one at a time, recorded their color, and then put them back. Make a picture graph of the data. Use this key:

Each ◯ = 2 marbles.

Jeremy's Marble Experiment	
Color	**Number**
Blue	4
Green	11
Red	8

Key:	

Use your picture graph above for 1–2.

1. How many more times did Jeremy pull out a red marble than a blue marble?

   ______

2. How many fewer times did Jeremy pull out green marbles than blue and red marbles combined?

   ______

MATHEMATICAL PRACTICES

Math Talk **Explain** how you knew how many pictures to draw for green.

Name ________________________________

On Your Own

3. Two classes from Delia's school visited the Science Center. They recorded their favorite exhibit in the tally table. Use the data in the table to make a picture graph. Use this key:

   Each ☼ = 4 votes.

Favorite Exhibit

Exhibit	Tally
Nature	𝍸 |
Solar System	𝍸 |||
Light and Sound	𝍸 𝍸 ||||
Human Body	𝍸 |||

Use your picture graph above for 4–6.

4. Which exhibits received the same number of votes?

   ____

5. **What if** a weather exhibit received 22 votes? **Explain** how many pictures you would draw.

   ____

6. H.O.T. **What if** the Solar System exhibit received 15 votes? Would it make sense to use the key Each ☼ = 4 votes to represent 15 votes? **Explain**.

   ____

   ____

Problem Solving REAL WORLD

Teeth in Mammals

Animal	Number
Hamster	16
Cat	30
Dog	42
Cow	32

7. While at the Science Center, Delia's classmates learned how many teeth some mammals have. Use the data in the table to make a picture graph. Use this key:

Each △ = 4 teeth.

Key:

Use your picture graph above for 8–10.

8. **Pose a Problem** Write a question that can be answered by using the data in your picture graph. Then answer the question.

____

____

9. Write Math ▸ **What if** the only data in the table was for the hamster and cow? **Explain** why you may use a different key.

____

____

10. **Test Prep** How many pictures would you draw for Cat if each △ = 5 teeth?

Ⓐ 4 Ⓑ 5 Ⓒ 6 Ⓓ 8

FOR MORE PRACTICE:
Standards Practice Book, pp. P35–P36

Name ____________________

Mid-Chapter Checkpoint

▶ Vocabulary

Choose the best term from the box.

Vocabulary
frequency table
key
picture graph

1. A ________________ uses numbers to record data. (p. 61)

2. A ________________ uses small pictures or symbols to show and compare information. (p. 65)

▶ Concepts and Skills

Use the Favorite Season table for 3–6.

Favorite Season	
Season	**Number**
Spring	19
Summer	28
Fall	14
Winter	22

3. Which season got the most votes?

____

4. Which season got 3 fewer votes than winter?

____

5. How many more students chose summer than fall?

____

6. How many students were surveyed?

____

Use the Our Pets picture graph for 7–9.

Our Pets	
Bird	🐾🐾🐾🐾
Cat	🐾🐾🐾🐾🐾
Dog	🐾🐾🐾🐾🐾🐾 and a half 🐾
Fish	🐾🐾🐾
Key: Each 🐾 = 2 students.	

7. How many students have cats as pets?

____

8. Five more students have dogs than which other pet? ________________

9. How many pets in all do students have?

____

Fill in the bubble for the correct answer choice.

Use the Favorite Summer Activity picture graph for 10–14.

10. Some students in Brooke's school chose their favorite summer activity. The results are in the picture graph at the right. How many students chose camping?

    Ⓐ 5
    Ⓑ 10
    Ⓒ 40
    Ⓓ 50

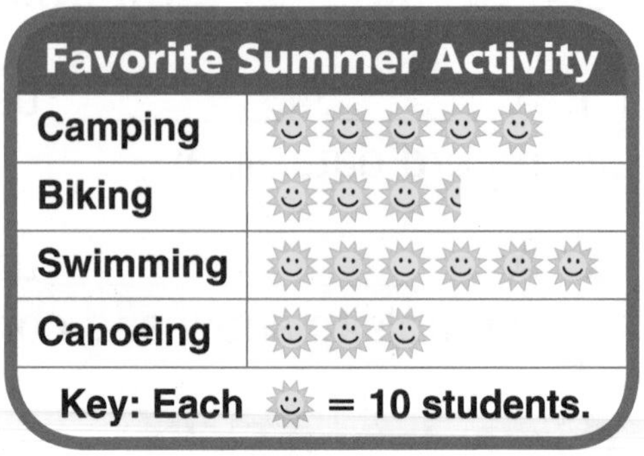

Favorite Summer Activity	
Camping	
Biking	
Swimming	
Canoeing	
Key: Each = 10 students.	

11. How many more students chose swimming than canoeing?

    Ⓐ 3
    Ⓑ 20
    Ⓒ 30
    Ⓓ 60

12. Which activity did 15 fewer students choose than camping?

    Ⓐ biking
    Ⓑ swimming
    Ⓒ canoeing
    Ⓓ hiking

13. How many pictures would you draw for biking if each = 5 students?

    Ⓐ 3 Ⓒ 5
    Ⓑ 4 Ⓓ 7

14. How many more students chose biking or canoeing combined than swimming?

    Ⓐ 3 Ⓒ 5
    Ⓑ 4 Ⓓ 7

Name ______________________

Use Bar Graphs

Essential Question How can you read and interpret data in a bar graph?

UNLOCK the Problem REAL WORLD

A **bar graph** uses bars to show data. A **scale** of equally spaced numbers helps you read the number each bar shows.

The students in the reading group made a bar graph to record the number of books they read in October. How many books did Seth read?

- Underline the words that tell you where to find the information to answer the question.

The title tells what the bar graph is about.

Each bar is labeled with a student's name.

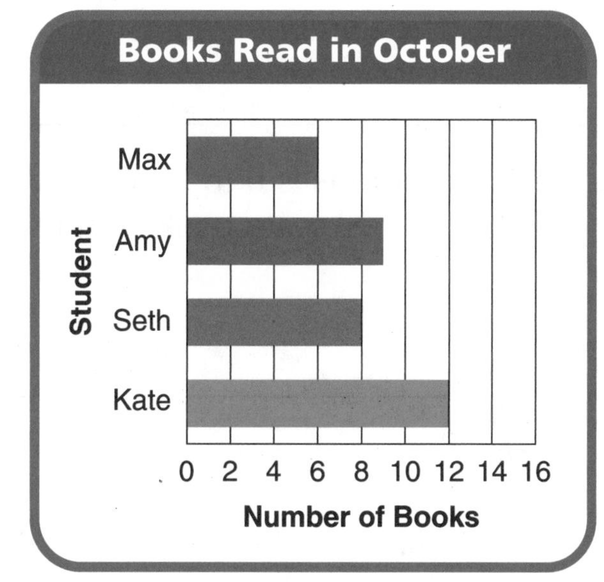

The length of a bar tells how many books each student read.

The scale is 0–16 by twos.

MATHEMATICAL PRACTICES **Math Talk** Explain how to read the bar that tells how many books Amy read.

Find the bar for Seth. It ends at ______.

So, Seth read ______ books in October.

1. How many books did Max read? ________
2. Who read 4 fewer books than Kate? ________
3. **What if** Amy read 5 more books? How many books did Amy read? ________ Shade the graph to show how many she read.

More Examples These bar graphs show the same data.

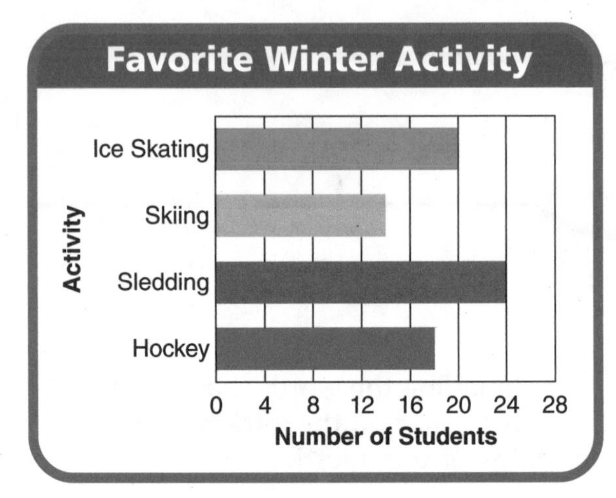

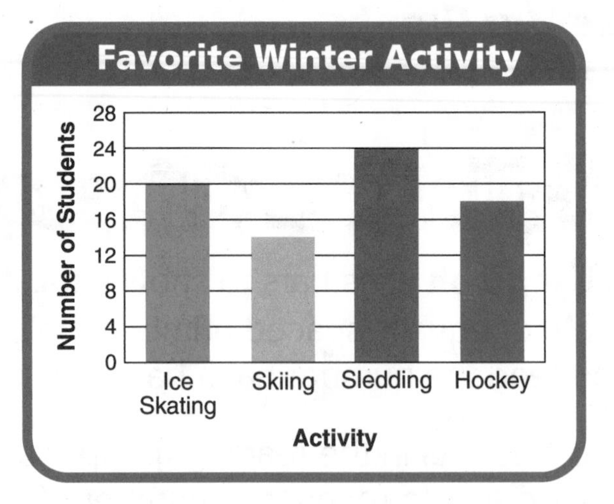

In a **horizontal bar graph**, the bars go across from left to right. The length of the bar shows the number.

In a **vertical bar graph**, the bars go up from the bottom. The height of the bar shows the number.

4. What does each space between the numbers represent?

____

5. Why do you think the scale in the graphs is 0 to 28 by fours instead of 0 to 28 by ones? What other scale could you use?

____

____

Share and Show 

Use the Favorite Way to Exercise bar graph for 1–3.

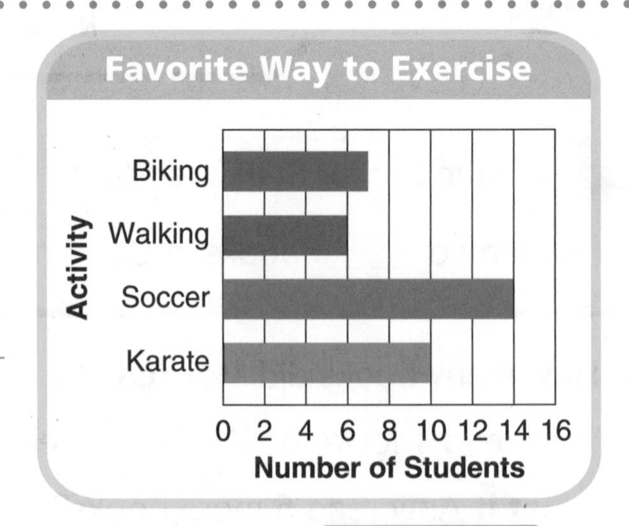

1. Which activity did the most students choose?

   Think: Which bar is the longest?

   ____

2. How many students answered the survey? __________

3. Which activity received 7 fewer votes than soccer? __________

MATHEMATICAL PRACTICES

Math Talk What can you tell just by comparing the lengths of the bars in the graph? **Explain.**

Name ______________________

On Your Own

Use the Favorite Kind of Book bar graph for 4–5.

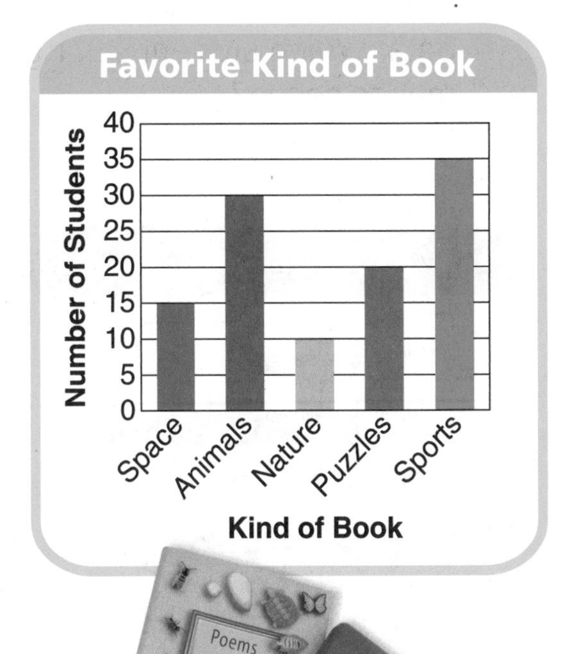

4. Which kind of book was chosen by half the number of students as books about animals?

   Space

5. Did more students choose books about sports or books about animals and nature together? **Explain.**

   ____

   ____

Problem Solving

Use the Favorite Kind of Book bar graph for 6–9.

6. Which two kinds of books together did students choose as often as books about sports?

   ____

7. **Write Math** **Pose a Problem** Write and solve a problem that matches the data in the graph.

   ____

   ____

8. **H.O.T.** **What if** 10 more students were asked and they chose books about animals? Describe what the bar graph would look like.

   ____

   ____

9. **Test Prep** Which kind of book did 15 fewer students choose than books about sports?

   Ⓐ space   Ⓑ animals   Ⓒ nature   Ⓓ puzzles

H.O.T. Sense or Nonsense?

10. The table shows data about some students' favorite amusement park rides. Four students graphed the data. Which student's bar graph makes sense?

____

Favorite Amusement Ride

Ride	Number of Students
Super Slide	11
Ferris Wheel	14
Bumper Cars	18
Roller Coaster	23

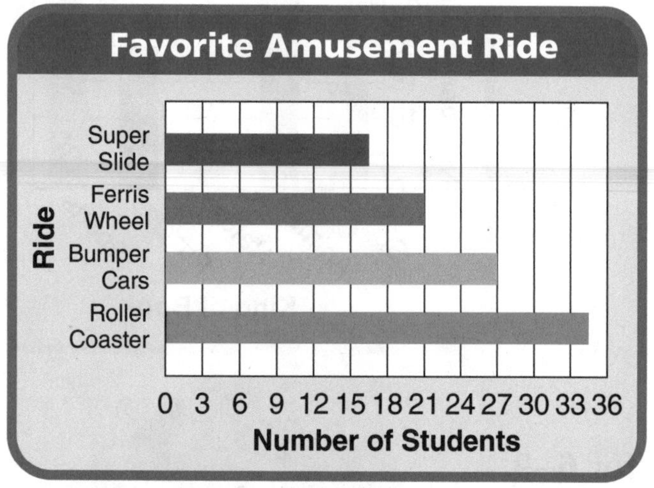

Alicia

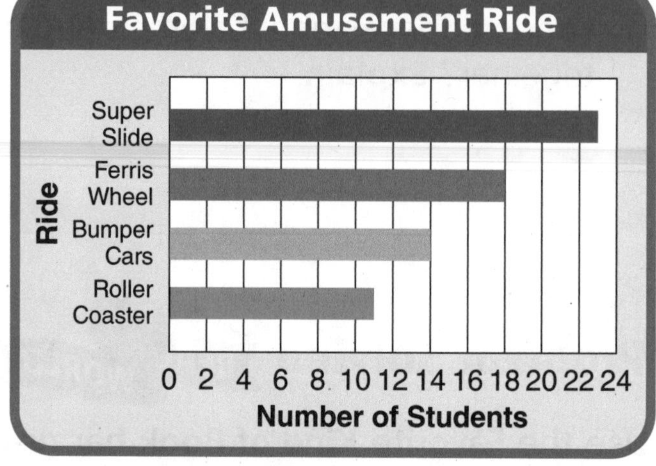

Spencer

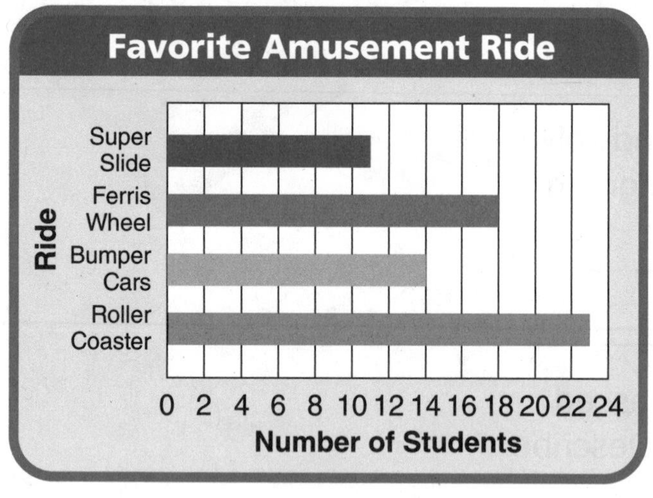

Tyler

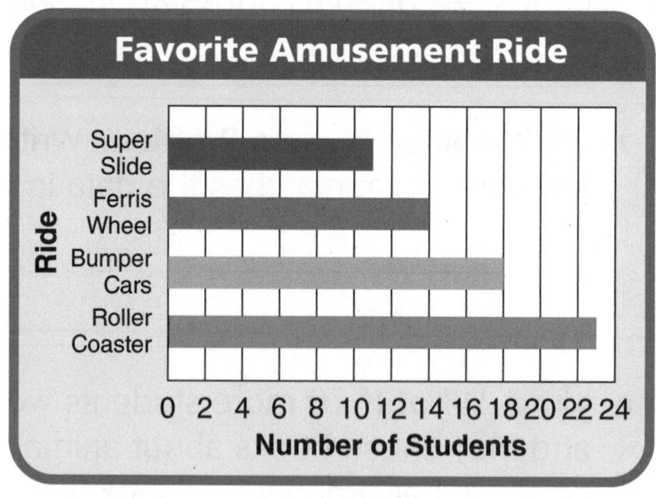

Kate

- **Explain** why the other bar graphs do not make sense.

____

____

____

FOR MORE PRACTICE:
Standards Practice Book, pp. P37–P38

Name ____________________

Make Bar Graphs

Essential Question How can you draw a bar graph to show data in a table or picture graph?

UNLOCK the Problem REAL WORLD

Jordan took a survey of his classmates' favorite team sports. He recorded the results in the table at the right. How can he show the results in a bar graph?

Favorite Team Sport

Sport	Tally
Soccer	~~IIII~~ ~~IIII~~ II
Basketball	IIII
Baseball	~~IIII~~ ~~IIII~~ IIII
Football	~~IIII~~ IIII

Make a bar graph.

STEP 1

Write a title at the top to tell what the graph is about. Label the side of the graph to tell about the bars. Label the bottom of the graph to explain what the numbers tell.

STEP 2

Choose numbers for the bottom of the graph so that most of the bars will end on a line. Since the least number is 4 and the greatest number is 14, make the scale 0–16. Mark the scale by twos.

STEP 3

Draw and shade a bar to show the number for each sport.

MATHEMATICAL PRACTICES

Math Talk How did you know how long to draw the bar for football?

Share and Show 

Matt's school is having a walk-a-thon to raise money for the school library. Matt made a picture graph to show the number of miles some students walked. Make a bar graph of Matt's data. Use a scale of

0–______, and mark the scale by ______.

School Walk-a-Thon	
Sam	👕 👕 👕 👕 👕
Matt	👕 👕 (half)
Ben	👕
Erica	👕 👕 👕 👕
Key: Each 👕 = 2 miles.	

Use your bar graph for 1–4.

1. Which student walked the most miles? ______

   Think: Which student's bar is the tallest?

2. How many more miles would Matt have had to walk to equal the number of miles Erica walked? ______

3. How many miles did the students walk? ______

4. Write the number of miles the students walked in order from greatest to least. ______

MATHEMATICAL PRACTICES

Math Talk **Explain** how the graph would have to change if another student, Daniel, walked double the number of miles Erica walked.

Name ______________________

On Your Own

5. Lydia and Joey did an experiment with a spinner. Lydia recorded the result of each spin in the table at the right. Use the data in the table to make a bar graph. Choose numbers and a scale and decide how to mark your graph.

Spinner Results

Color	Tally
Red	卌 卌 卌 I
Yellow	卌 III
Blue	卌 卌 II
Green	卌 卌

ERROR Alert

Be sure to draw the bars correctly when you transfer data from a table.

Use your bar graph for 6–8.

6. The pointer stopped on __________ half the number of times that it stopped on __________.

7. The pointer stopped on green __________ fewer times than it stopped on blue.

8. **Explain** why you chose the scale you did.

____

____

Problem Solving REAL WORLD

9. Susie recorded the number of points some basketball players scored. Use the data in the table to make a bar graph. Choose numbers so that most of the bars will end on a line.

Points Scored	
Player	**Number of Points**
Billy	10
Dwight	30
James	15
Raul	25
Sean	10

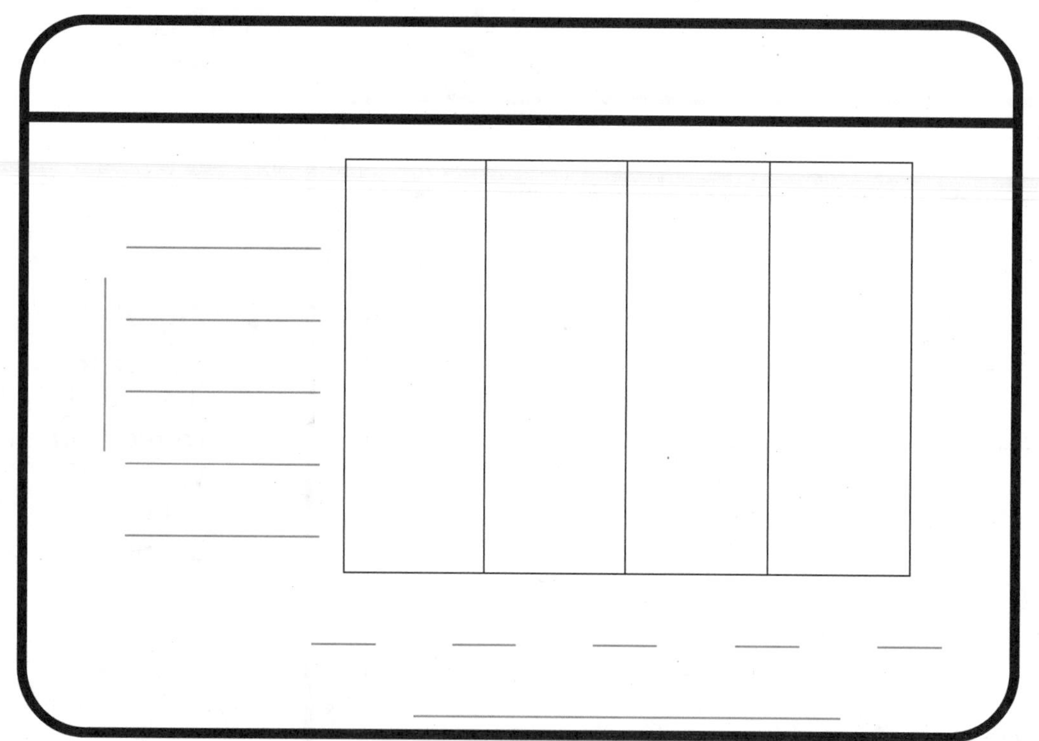

Use your bar graph for 10–12.

10. Which player scored more points than James but fewer points than Dwight? ____________

11. H.O.T. Write Math Write and solve a new question that matches the data in your bar graph.

____

____

12. **Test Prep** Which player scored 10 more points than James?

Ⓐ Billy Ⓑ Dwight Ⓒ Raul Ⓓ Sean

FOR MORE PRACTICE:
Standards Practice Book, pp. P39–P40

Name ______________________________

Solve Problems Using Data

Essential Question How can you solve problems using data represented in bar graphs?

UNLOCK the Problem REAL WORLD

CONNECT Answering questions about data helps you better understand the information.

Derek's class voted on a topic for the school bulletin board. The bar graph shows the results. How many more votes did computers receive than space?

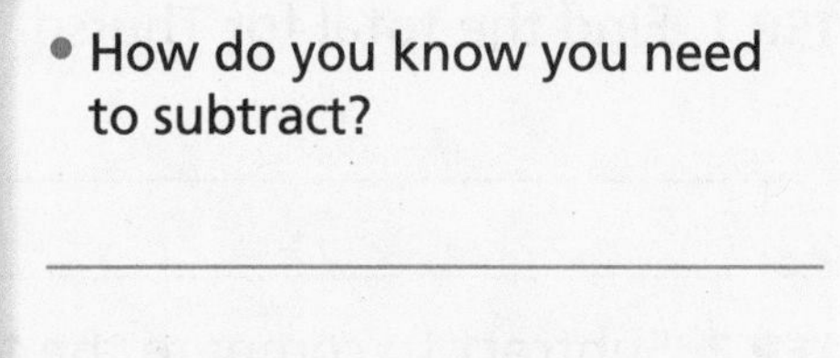

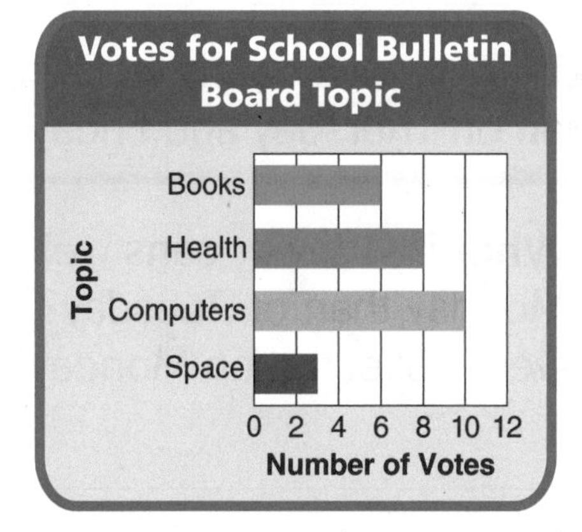

One Way Use a model.

Count back along the scale to find the difference between the bars.

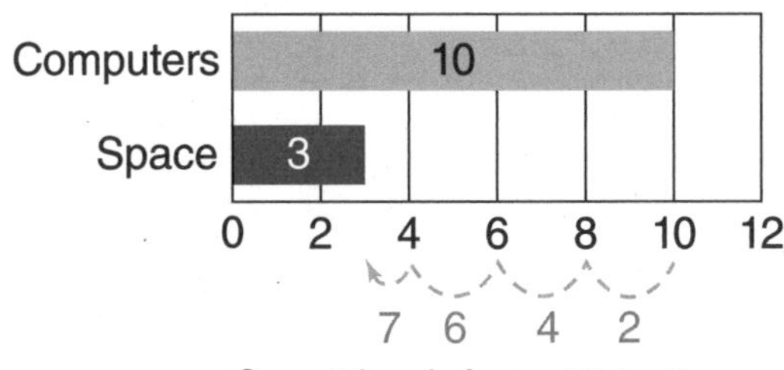

Count back from 10 to 3.

The difference is _____ votes.

MATHEMATICAL PRACTICES

Math Talk Explain another way you can skip count to find the difference.

Another Way Write a number sentence.

Think: There are 10 votes for computers. There are 3 votes for space. Subtract to compare the number of votes.

____

So, computers received _____ more votes than space.

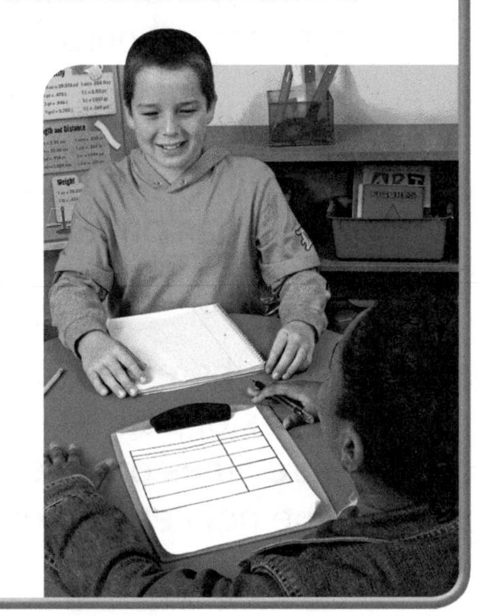

Example

Brooke's school collected cans of food. The bar graph at the right shows the number of cans. How many fewer cans were collected on Tuesday than on Thursday and Friday combined?

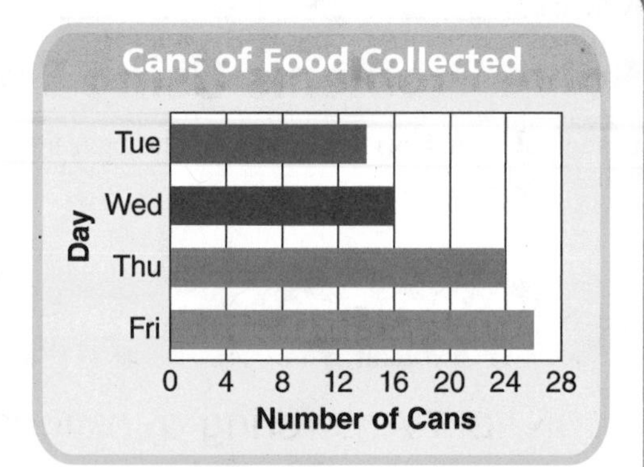

STEP 1 Find the total for Thursday and Friday.

____

____

STEP 2 Subtract to compare the total for Thursday and Friday to Tuesday and to find the difference.

____

So, _____ fewer cans were collected on Tuesday than on Thursday and Friday combined.

- **What if** 4 fewer cans were collected on Monday than on Tuesday? How many cans were collected on Monday? **Explain.**

____

Share and Show 

Use the Spinner Results bar graph for 1–3.

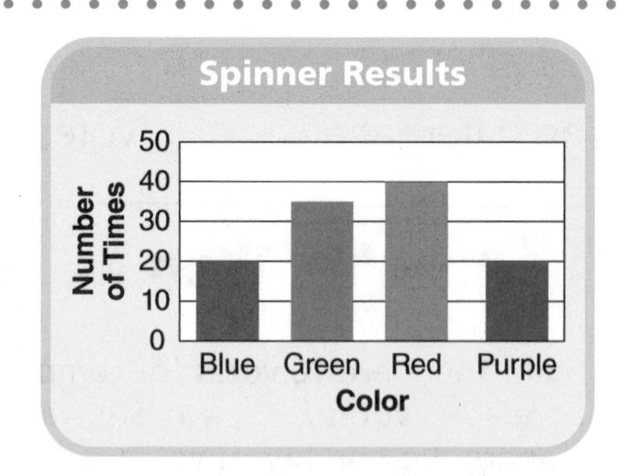

1. How many more times did the pointer stop on green than on purple?

   _____ more times

2. How many fewer times did the pointer stop on blue than on red and green combined?

   _____ fewer times

3. **What if** there were 15 more spins and the pointer stopped 10 more times on green and 5 more times on blue? How many more times did the pointer stop on green than blue?

   ____

Math Talk MATHEMATICAL PRACTICES What can you tell just by comparing the lengths of the bars in the graphs? **Explain.**

Name ________________________________

On Your Own

Use the Diego's DVDs bar graph for 4–6.

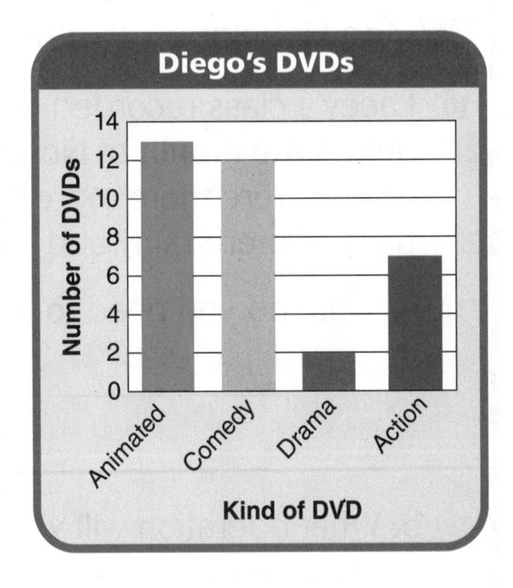

4. Diego has 5 fewer of this kind of DVD than comedy. Which kind of DVD is this?

____

5. How many more animated than action DVDs does Diego have?

____

6. Is the number of comedy and action DVDs greater than or less than the number of animated and drama DVDs? **Explain.**

____

____

Problem Solving REAL WORLD

Use the Science Fair Projects bar graph for 7–9.

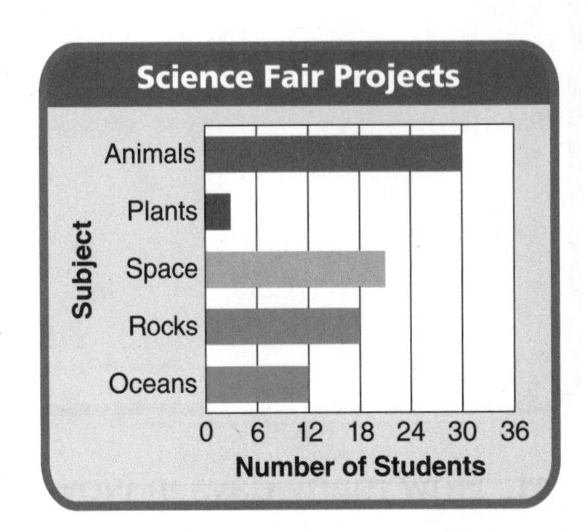

7. How many more students would have to do a project on plants to equal the number of projects on space?

____

8. Write Math **What's the Question?** The answer is animals, space, rocks, oceans, and plants.

____

____

9. H.O.T. **What if** 3 fewer students did a project on weather than did a project on rocks? **Describe** what the bar graph would look like.

____

____

UNLOCK the Problem REAL WORLD

Use the November Weather bar graph for 10–12.

10. Lacey's class recorded the kinds of weather during the month of November in a bar graph. Were there more cloudy and sunny days or more rainy and snowy days?

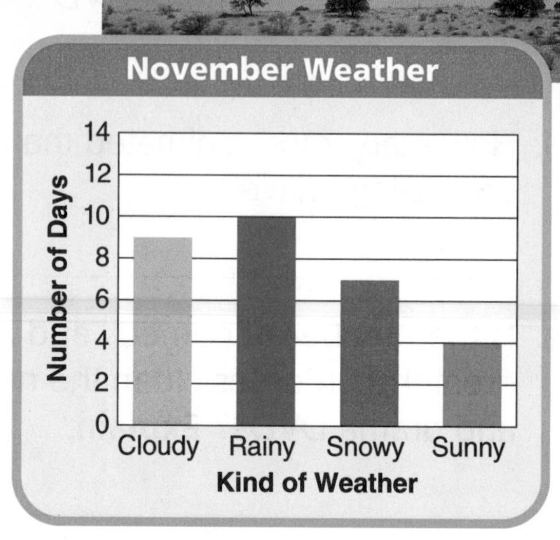

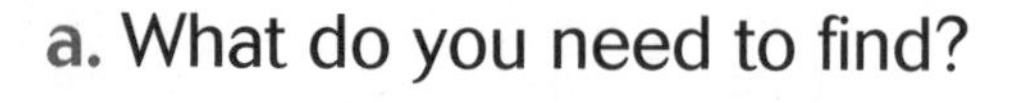

a. What do you need to find?

____

____

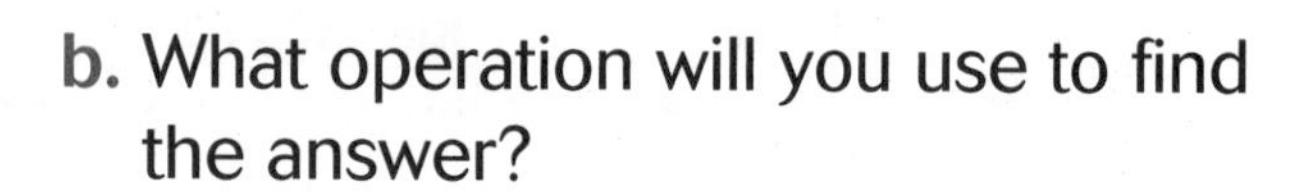

b. What operation will you use to find the answer?

____

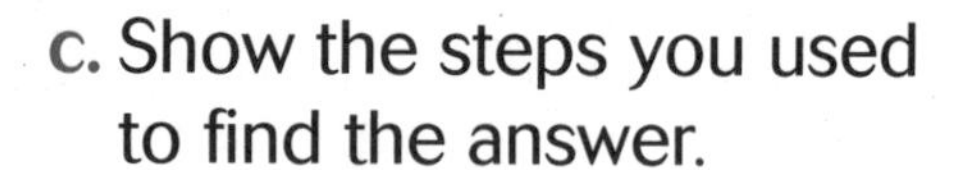

c. Show the steps you used to find the answer.

d. Complete the sentences.

_____ cloudy days +

_____ sunny days = _____ days

_____ rainy days +

_____ snowy days = _____ days

_____ > _____

So, there were more ______________ days.

11. How many days in November were NOT cloudy?

Think: There are 30 days in November.

____

12. **Test Prep** How many fewer snowy and sunny days were there than cloudy and rainy days?

Ⓐ 3 days

Ⓑ 4 days

Ⓒ 8 days

Ⓓ 30 days

FOR MORE PRACTICE:
Standards Practice Book, pp. P41–P42

Name ______

Use and Make Line Plots

Essential Question How can you read and interpret data in a line plot and use data to make a line plot?

UNLOCK the Problem REAL WORLD

A **line plot** uses marks to record each piece of data above a number line. It helps you see groups in the data.

Some students took a survey of the number of letters in their first names. Then they recorded the data in a line plot.

How many students have 6 letters in their first names?

Each ✗ stands for 1 student. →

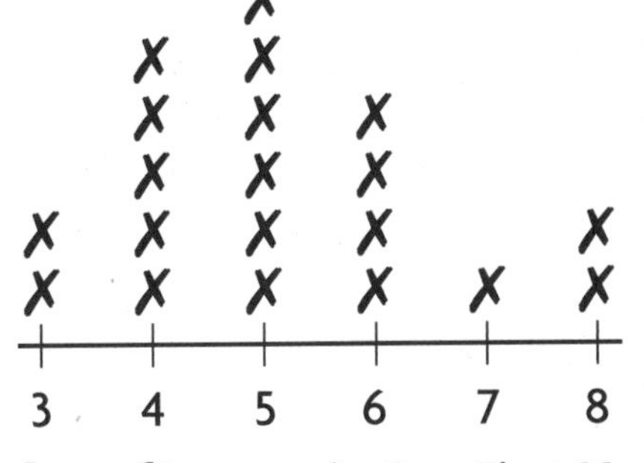

Number of Letters in Our First Names

← The numbers show the number of letters in a name.

Find 6 on the number line. The 6 stands for 6 ______.

There are ______ ✗s above the 6.

So, ______ students have 6 letters in their first names.

1. Which number of letters was found most often? ______
2. Write a sentence to describe the data. ______
3. How many letters are in your first name? ______
4. Put an ✗ above the number of letters in your first name.

MATHEMATICAL PRACTICES

Math Talk What does the shape of the data show you?

Activity Make a line plot.

Materials ■ ruler ■ measuring tape

Measure the height of four classmates to the nearest inch. Combine your data with other groups. Make a line plot to show the data you collected.

STEP 1 Record the heights in the table.

STEP 2 Write a title below the number line to describe your line plot.

STEP 3 Write the number of inches in order from left to right above the title.

STEP 4 Draw *X*s above the number line to show each student's height.

Heights in Inches	
Number of Inches	Tally

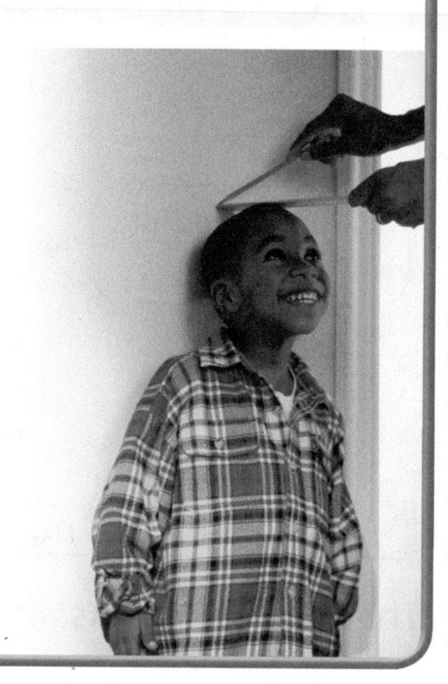

5. Which height appears most often? ______________

Think: Which height has the most *X*s?

6. Which height appears least often? ______________

7. Complete the sentence. Most of the students in the class are ______ inches tall or taller.

8. H.O.T. Is there any height for which there are no data? Explain. ______________

MATHEMATICAL PRACTICES

Math Talk Explain what the shape of the data tells you.

Name ______________________

Share and Show 

Use the Number of Pets line plot for 1–3.

1. How many students have 1 pet?

   Think: How many Xs are above 1?

   ____

2. What number of pets do no students have?

   ____

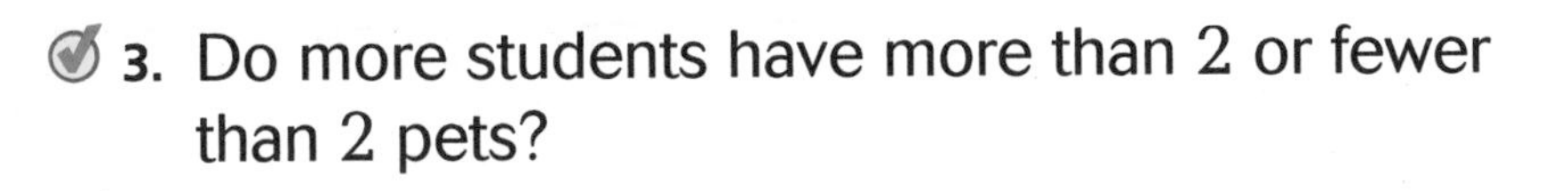

3. Do more students have more than 2 or fewer than 2 pets?

   ____

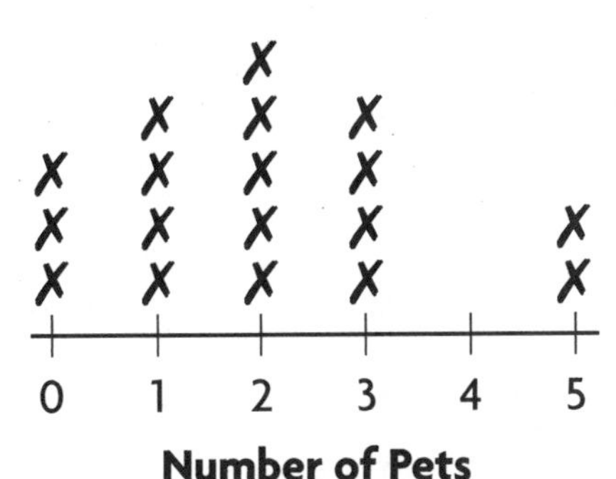

MATHEMATICAL PRACTICES

Math Talk Explain why you think the line plot stops at 5.

On Your Own

4. Use the data in the table to make a line plot.

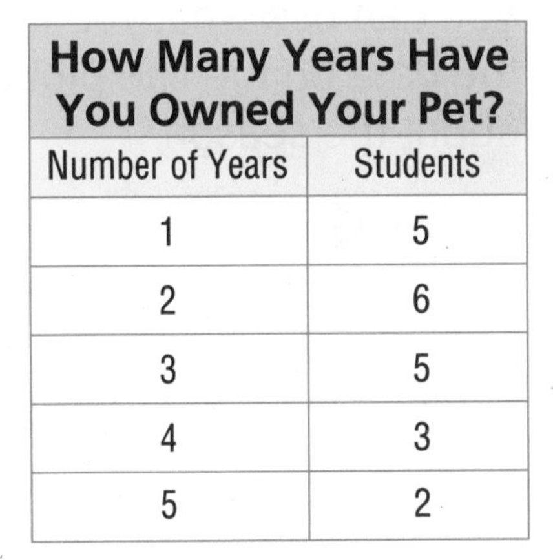

How Many Years Have You Owned Your Pet?	
Number of Years	Students
1	5
2	6
3	5
4	3
5	2

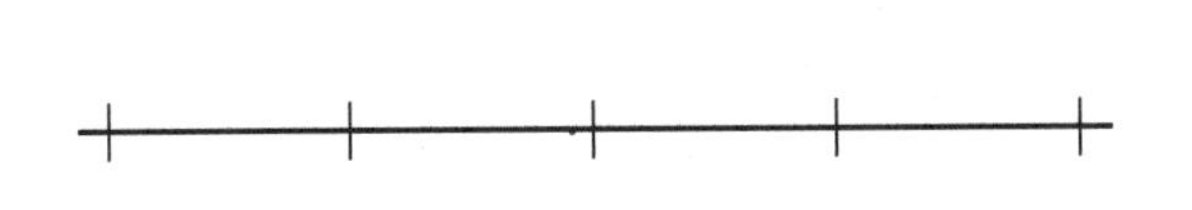

How Many Years Have You Owned Your Pet?

5. How many students have owned a pet 1 year? ______________________

6. Which number of years is found most often? ______________________

7. **Test Prep** How many students have owned a pet 2 years or more?

   Ⓐ 21 Ⓑ 16 Ⓒ 10 Ⓓ 5

Connect to Reading

Make an Inference

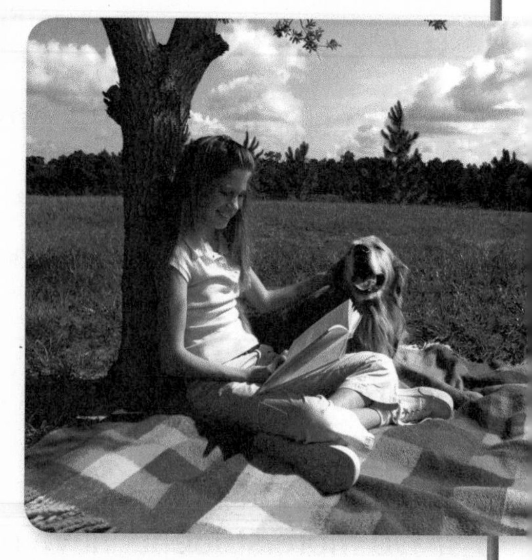

Addison made the line plot below to show the high temperature every day for one month. What *inference* can you make about what season this is?

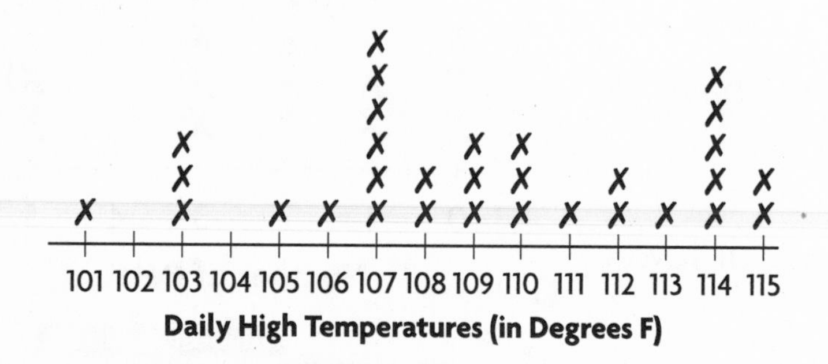

When you combine what you see with what you already know to come up with an idea, you are making an inference.

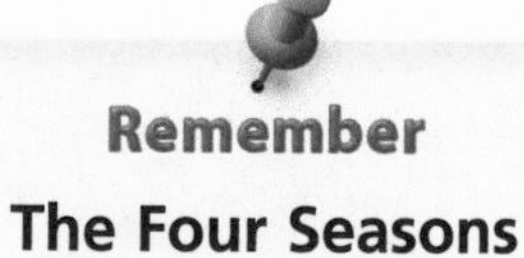

Remember

The Four Seasons

spring

summer

fall

winter

You can use what you know about weather and the data on the line plot to make an inference about the season.

You know that the numbers on the line plot are the high temperatures recorded during the month.

The highest temperature recorded was ______________.

The lowest temperature recorded was ______________.

The temperature recorded most often was ______________.

Since all the high temperatures are greater than 100, you know the days were hot. This will help you make an inference about the season.

So, you can infer that the season is ______________.

FOR MORE PRACTICE:
Standards Practice Book, pp. P43–P44

Name ___________________________________

Vocabulary

Choose the best term from the box.

Vocabulary
bar graph
frequency table
line plot
picture graph

1. A ________________ uses marks to record each piece of data above a number line. (p. 87)
2. A ________________ uses pictures to show and compare information. (p. 65)
3. A ________________ uses bars to show data. (p. 75)

Concepts and Skills

Use the Dolphins Max Saw table for 4–6.

Dolphins Max Saw

Day	Number
Friday	12
Saturday	15
Sunday	19

4. Max recorded in a table the number of dolphins he saw. How many dolphins did he see?

   ____

5. How many more dolphins did Max see on Sunday than on Friday? ________________

6. If you made a bar graph of the data in the table, what labels would you use? ________________

   ____

Use the Number of Goals Scored line plot for 7–9.

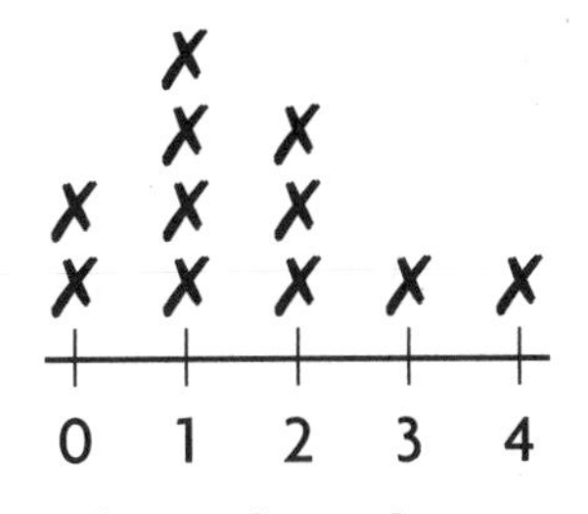

7. Katie recorded the number of goals the players on her team scored during soccer practice. How many players scored 2 goals?

   ____

8. What does each ✗ stand for? ________________

9. What was the least number of goals scored?

   ____

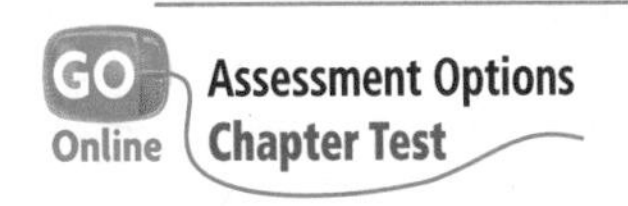

Fill in the bubble for the correct answer choice.

Use the Musical Instruments bar graph for 10–14.

10. Three more students play piano than which other instrument?

Ⓐ flute
Ⓑ drums
Ⓒ violin
Ⓓ guitar

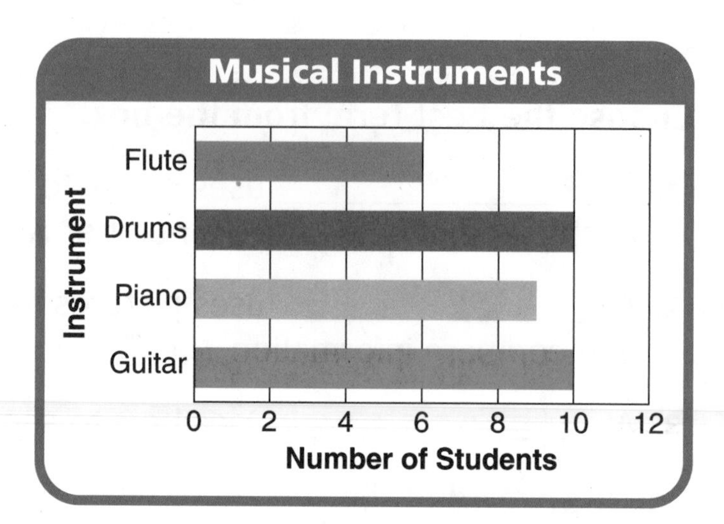

11. The same number of students play which two instruments?

Ⓐ guitar and drums
Ⓑ piano and drums
Ⓒ drums and flute
Ⓓ flute and piano

12. How many fewer students play flute than play drums?

Ⓐ 2
Ⓑ 3
Ⓒ 4
Ⓓ 5

13. How many more students play piano and guitar combined than play drums?

Ⓐ 29
Ⓑ 19
Ⓒ 10
Ⓓ 9

14. Which of the following statements is true?

Ⓐ All of the students play flute.
Ⓑ None of the students play guitar.
Ⓒ Some of the students play piano.
Ⓓ Most of the students play flute.

Name ________________________________

Fill in the bubble for the correct answer choice.

Use the Jim's Books picture graph for 15–18.

15. Jim made a picture graph to show the types of books he has. Which part of the picture graph tells you what each symbol stands for?

Ⓐ label

Ⓑ key

Ⓒ title

Ⓓ scale

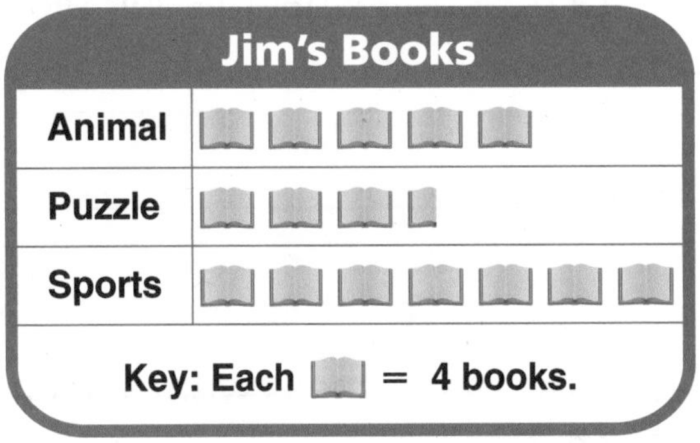

Jim's Books	
Animal	📖 📖 📖 📖 📖
Puzzle	📖 📖 📖 ½📖
Sports	📖 📖 📖 📖 📖 📖 📖
Key: Each 📖 = 4 books.	

16. How many puzzle books does Jim have?

Ⓐ 3

Ⓑ 4

Ⓒ 14

Ⓓ 16

17. How many fewer animal books than sports books does Jim have?

Ⓐ 2

Ⓑ 4

Ⓒ 8

Ⓓ 14

18. Which statement is true about the books Jim has?

Ⓐ All of the books are puzzle books.

Ⓑ None of the books is an animal book.

Ⓒ Some of the books are computer books.

Ⓓ Most of the books are sports or animal books.

▶ Constructed Response

Use the line plot for 19–20.

19. Belle collected shells during her vacation. She recorded the length of the shells to the nearest inch in a line plot. Write a sentence to describe what the line plot shows.

5	6	7	8	9
X X X	X X X X X	X X	X	X X X

Length of Shells in Inches

____

____

____

20. How many shells were 7 inches long or shorter?

____

▶ Performance Task

Use the bar graph for 21.

21. Suppose you are a reporter for the school newspaper.

A Write a story that summarizes the data about the After-School Clubs. Say something about each club in your story.

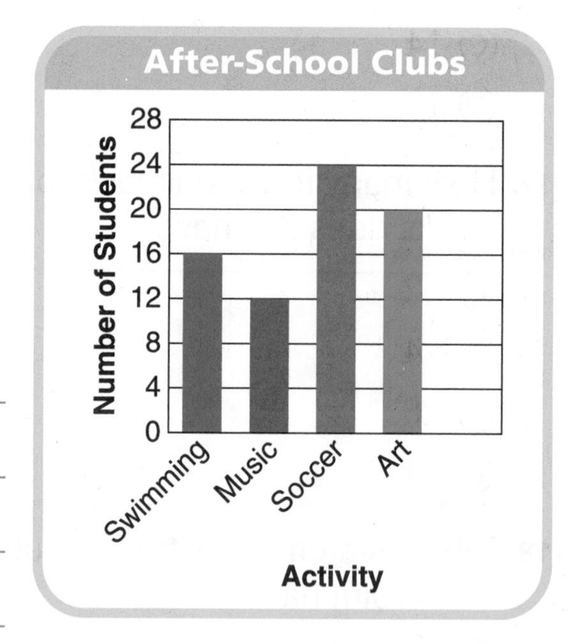

____

____

____

____

B There are more students in the gymnastics club than in the swimming club, but fewer than in the soccer club. Use the information to add a bar for the new gymnastics club. **Explain** how you decided how tall to make the bar.

____

____

Chapter 3

Understand Multiplication

Show What You Know

Check your understanding of important skills.

Name ______________________________

▶ Count On to Add Use the number line. Write the sum.

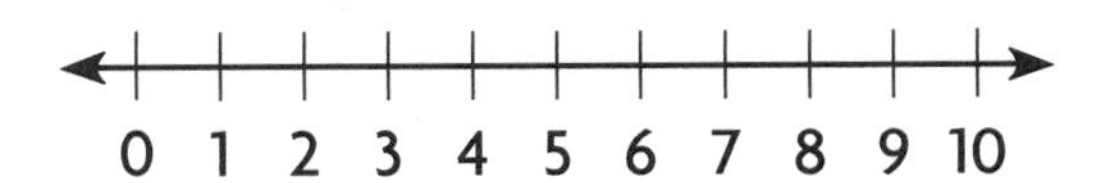

1. $6 + 2 =$ _____

2. $3 + 7 =$ _____

▶ Skip Count by Twos and Fives Skip count. Write the missing numbers.

3. 2, 4, 6, _____, _____, _____

4. 5, 10, 15, _____, _____, _____

▶ Model with Arrays Use the array. Complete.

5. 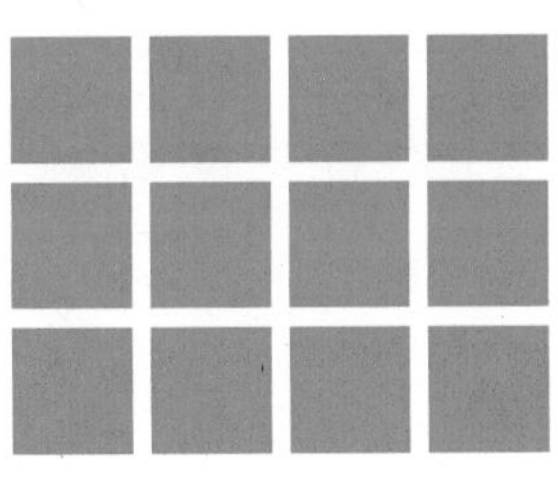

_____ + _____ + _____ = _____

6. 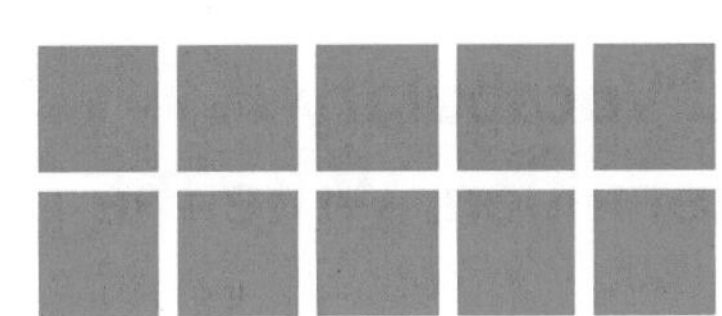

_____ + _____ = _____

Ryan's class went on a field trip to a farm. They saw 5 cows and 6 chickens. Be a Math Detective to find how many legs were on all the animals they saw.

Vocabulary Builder

▶ Visualize It

Complete the tree map by using the review words.

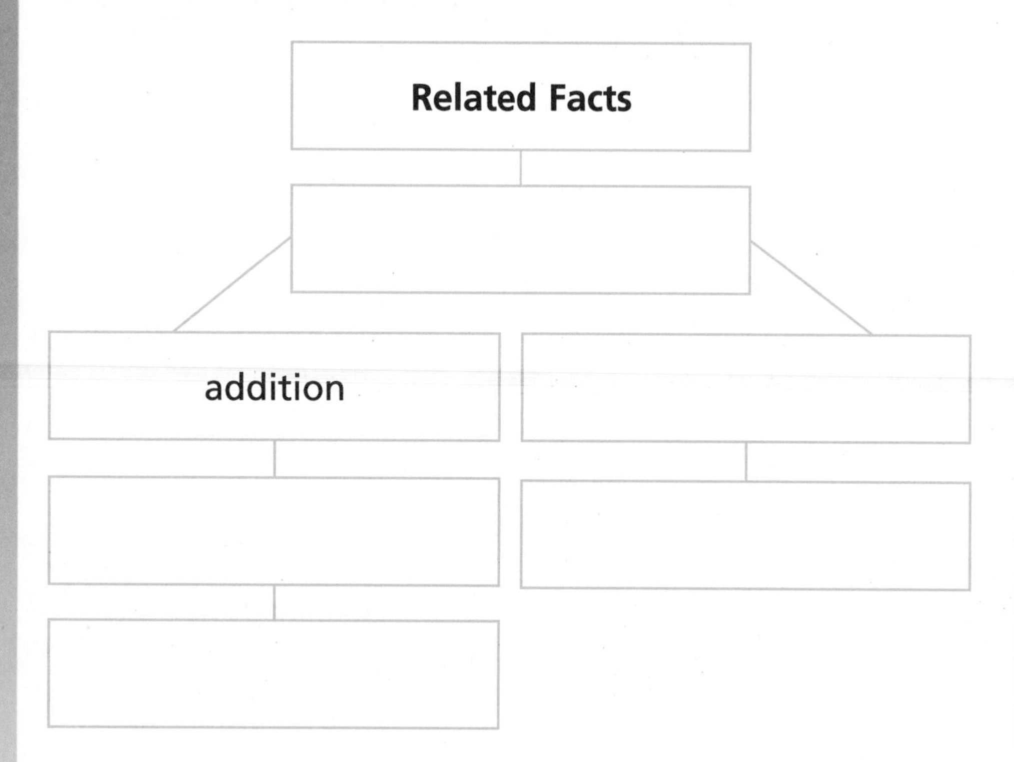

Review Words
addend
addition
difference
number sentences
related facts
subtraction
sum

Preview Words
array
equal groups
factor
multiply
product

▶ Understand Vocabulary

Read the definition. Write the preview word that matches it.

1. A set of objects arranged in rows and columns ______________

2. The answer in a multiplication problem ______________

3. When you combine equal groups to find how many in all ______________

4. A number that is multiplied by another number to find a product ______________

GO Online • eStudent Edition • Multimedia eGlossary

Name ______________________________

Count Equal Groups

Essential Question How can you use equal groups to find how many in all?

UNLOCK the Problem REAL WORLD

Equal groups have the same number of objects in each group.

Matchbox® cars were invented in 1952. Tim has 6 Matchbox cars. Each car has 4 wheels. How many wheels are there in all?

- How many wheels are on each car?

  ____

- How many equal groups of wheels are there?

  ____

- How can you find how many wheels in all?

  ____

Activity Use counters to model the equal groups.

Materials ■ counters

STEP 1 Draw 4 counters in each group.

STEP 2 Skip count to find how many wheels in all. Skip count by 4s until you say 6 numbers.

number of equal groups → 1 2 3 4 5 6

4, ____, 12, ____, ____, ____

There are ____ groups with ____ wheels in each group.

So, there are ____ wheels in all.

MATHEMATICAL PRACTICES

Math Talk What if Tim had 8 cars? How could you find the total number of wheels?

Example Count equal groups to find the total.

Sam, Kyla, and Tia each have 5 pennies.
How many pennies do they have in all?

How many pennies does each person have? ______

How many equal groups of pennies are there? ______

Draw 5 counters in each group.

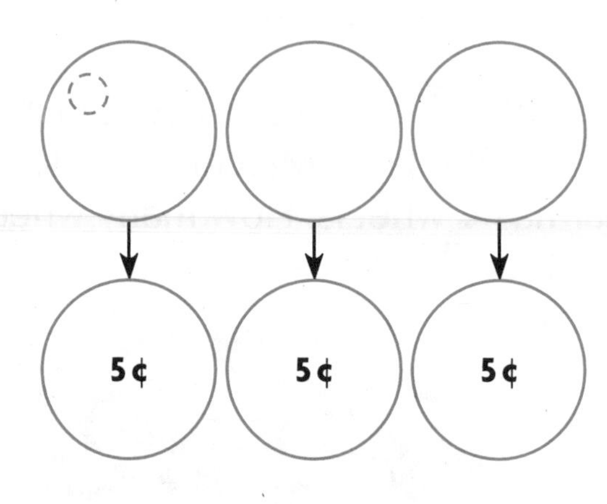

Think: There are ______ groups of 5 pennies.

Think: There are ______ fives.

Skip count to find how many pennies in all. ______, ______, ______

So, they have ______ pennies in all.

-  **Explain** why you can skip count by 5s to find how many in all.

______

Share and Show

1. Complete. Use the picture. Skip count to find how many wheels in all.

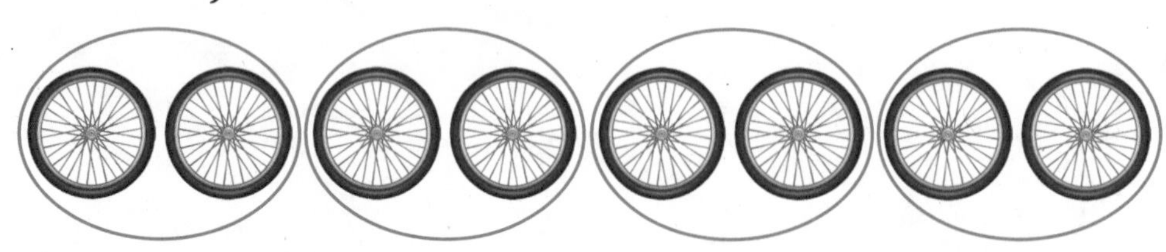

______ groups of 2

______ twos

Skip count by 2s. 2, 4, ______, ______

So, there are ______ wheels in all.

MATHEMATICAL PRACTICES

Math Talk What if there were 2 groups of 4 wheels? Would your answers change? If so, how?

Name ______________________________

Draw equal groups. Skip count to find how many.

2. 2 groups of 6 ____

3. 3 groups of 2 ____

Count equal groups to find how many.

4.  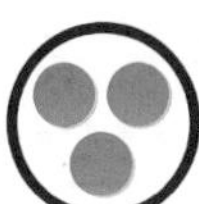 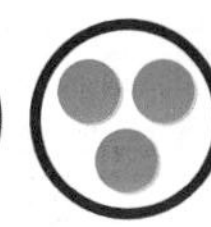 

____ groups of ____

____ in all

5. 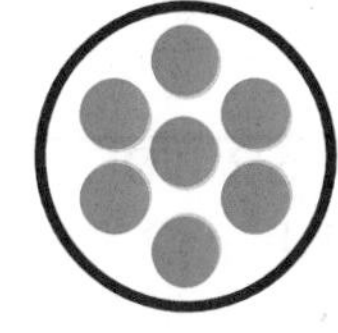 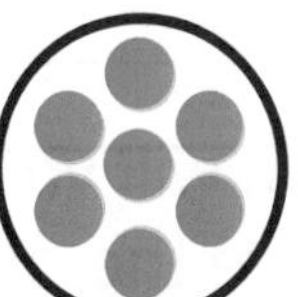 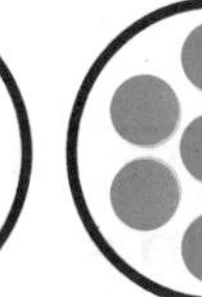

____ groups of ____

____ in all

On Your Own

Draw equal groups. Skip count to find how many.

6. 3 groups of 3 ____

7. 2 groups of 9 ____

Count equal groups to find how many.

8. 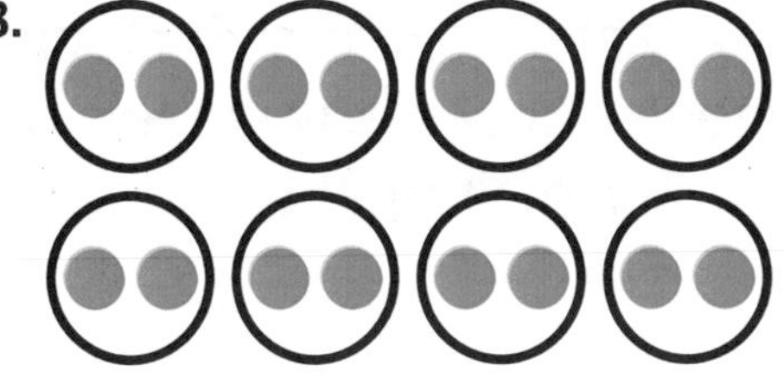

____ groups of ____

____ in all

9. 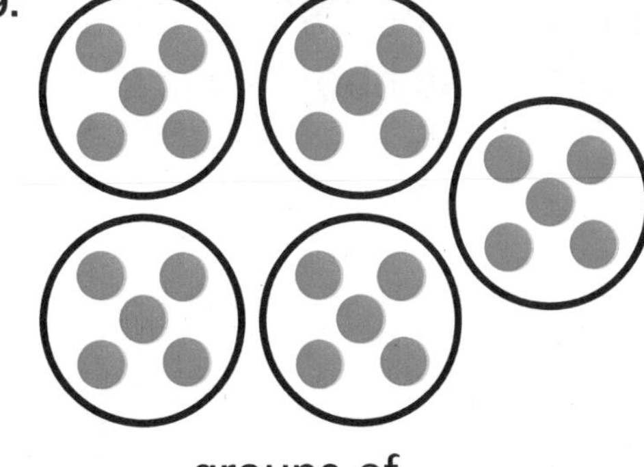

____ groups of ____

____ in all

UNLOCK the Problem REAL WORLD

10. Tina, Charlie, and Amber have Matchbox cars. Each car has 4 wheels. How many wheels do their cars have altogether?

Ⓐ 10 Ⓒ 40

Ⓑ 22 Ⓓ 44

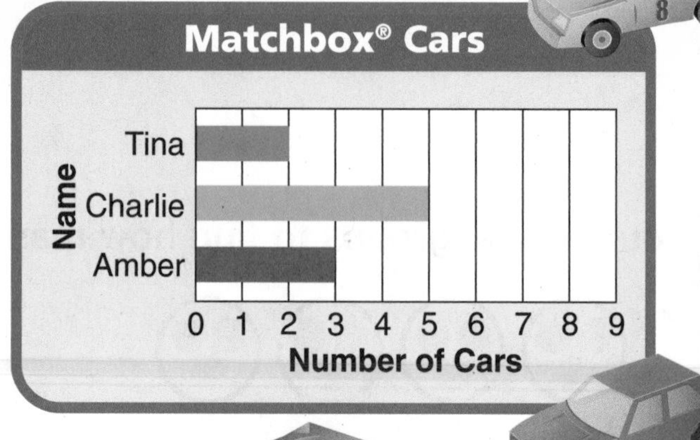

a. What do you need to find?

____

b. What information will you use from the graph to solve the problem?

____

c. Show the steps you used to solve the problem.

d. Complete the sentences.

Tina has _____ cars.

Charlie has _____ cars.

Amber has _____ cars.

There are _____ groups of 4 wheels.

The cars have _____ wheels altogether.

e. Fill in the bubble for the correct answer choice.

11. A Matchbox car costs $3. What is the price of 4 Matchbox® cars?

Ⓐ $7

Ⓑ $12

Ⓒ $14

Ⓓ $20

12. Nita has 7 Matchbox cars. Each car has 4 wheels. How many wheels do Nita's cars have altogether?

Ⓐ 4

Ⓑ 7

Ⓒ 11

Ⓓ 28

FOR MORE PRACTICE:
Standards Practice Book, pp. P49–P50

Name ______________________________

Relate Addition and Multiplication

Essential Question How is multiplication like addition? How is it different?

UNLOCK the Problem REAL WORLD

Tomeka needs 3 apples to make one apple cake. Each cake has the same number of apples. How many apples does Tomeka need to make 4 cakes?

- How many cakes is Tomeka making?

____

- How many apples are in each cake?

____

- How can you solve the problem?

____

One Way Add equal groups.

Use the 4 circles to show the 4 cakes.

Draw 3 counters in each circle to show the apples Tomeka needs for each cake.

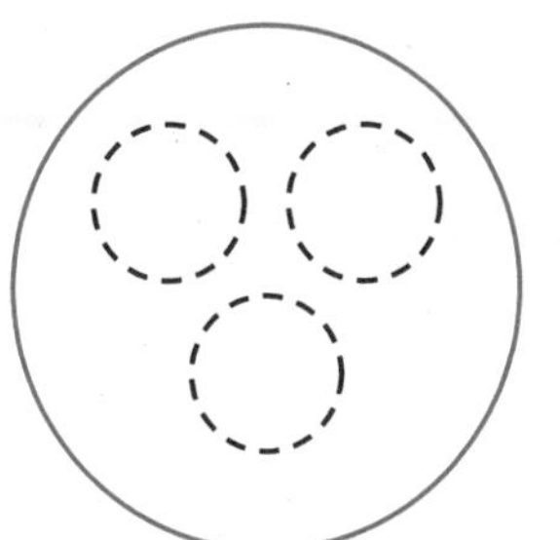

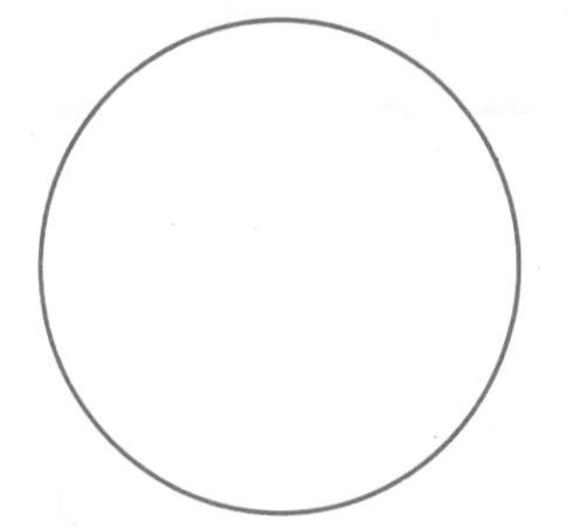

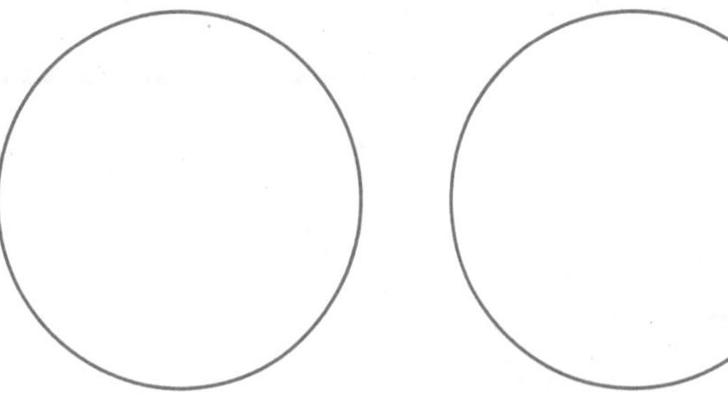

Find the number of counters in all.
Complete the addition sentence.

3 + ______ + ______ + ______ = ______

So, Tomeka needs ______ apples to

make ______ cakes.

MATHEMATICAL PRACTICES

Math Talk How is the picture you drew like the addition sentence you wrote?

Another Way **Multiply.**

When you combine equal groups, you can **multiply** to find how many in all.

Think: 4 groups of 3

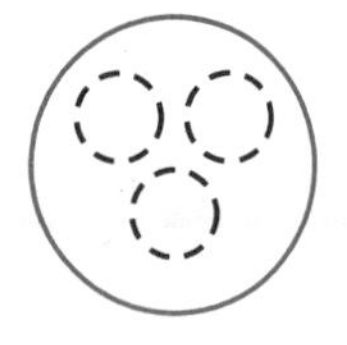  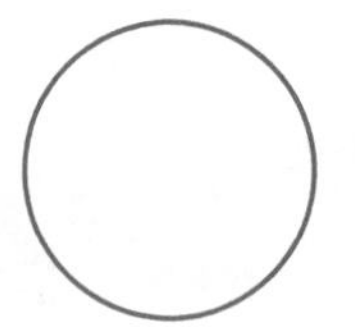 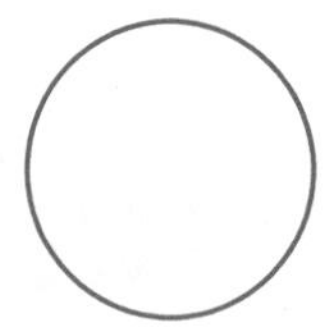

Draw 3 counters in each circle.

Since there are the same number of counters in each circle, you can multiply to find how many in all.

Multiplication is another way to find how many there are altogether in equal groups.

Write: $4 \times 3 = 12$ — 4 ↑ factor, 3 ↑ factor, 12 ↑ product

or

$$\begin{array}{r} 3 \\ \times\ 4 \\ \hline 12 \end{array}$$

3 ← factor
× 4 ← factor
12 ← product

Read: Four times three equals twelve.

The **factors** are the numbers multiplied.
The **product** is the answer to a multiplication problem.

Share and Show

1. Write related addition and multiplication sentences for the model.

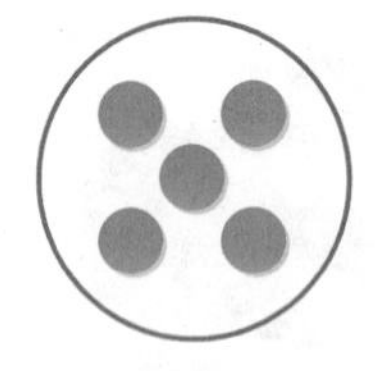 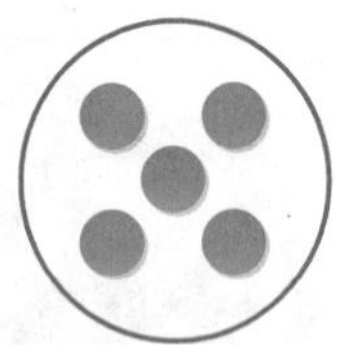 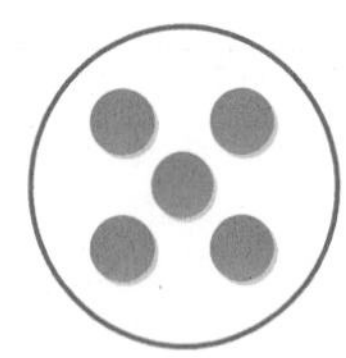 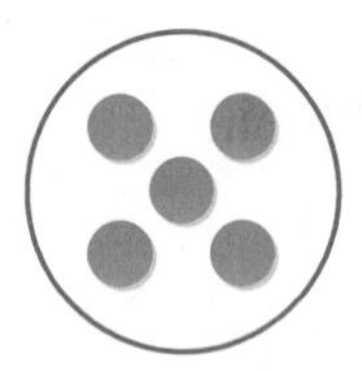

_____ + _____ + _____ + _____ = _____

_____ × _____ = _____

MATHEMATICAL PRACTICES

Math Talk How would you change this model so you could write a multiplication sentence to match it?

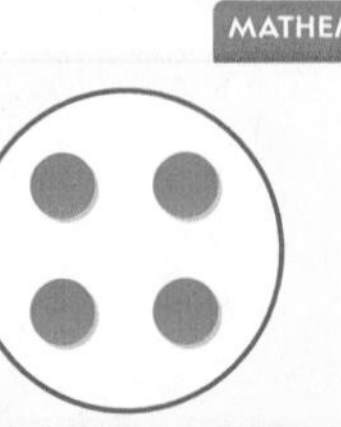

Name ______________________________

Draw a quick picture to show the equal groups. Then write related addition and multiplication sentences.

2. 3 groups of 6

___ + ___ + ___ = ___

___ × ___ = ___

3. 2 groups of 3

___ + ___ = ___

___ × ___ = ___

On Your Own

Draw a quick picture to show the equal groups. Then write related addition and multiplication sentences.

4. 4 groups of 2

___ + ___ + ___ + ___ = ___

___ × ___ = ___

5. 5 groups of 4

___ + ___ + ___ + ___ + ___ = ___

___ × ___ = ___

Complete. Write a multiplication sentence.

6.  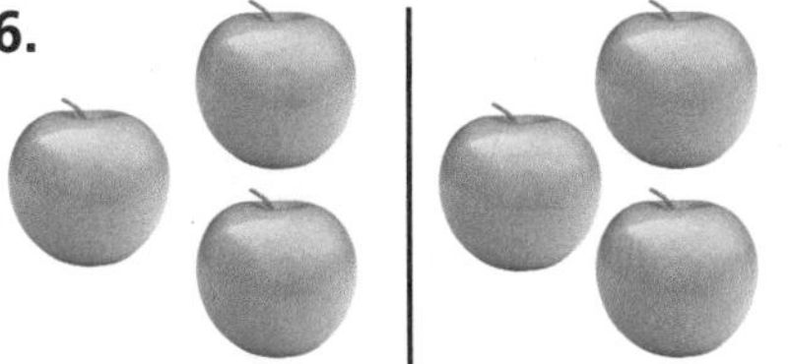

___ × ___ = ___

7.

___ × ___ = ___

8. 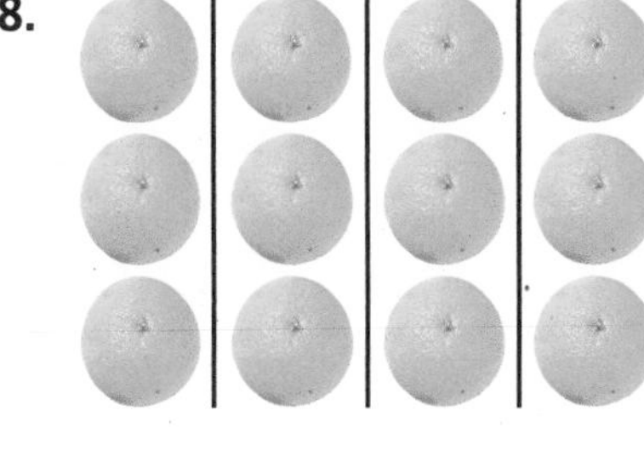

___ × ___ = ___

9. $2 + 2 + 2 + 2 =$ ___

___ × ___ = ___

10. $4 + 4 + 4 + 4 =$ ___

___ × ___ = ___

11. $9 + 9 + 9 =$ ___

___ × ___ = ___

Problem Solving REAL WORLD

Use the table for 12–13.

Average Weight of Fruits	
Fruit	**Weight in Ounces**
Apple	6
Orange	5
Peach	3
Banana	4

12. Morris bought 4 peaches. How much do the peaches weigh in all? Write a multiplication sentence to find the weight of the peaches.

    ______ × ______ = ______ ounces

13. H.O.T. Thomas bought 2 apples. Sydney bought 4 bananas. Which weighed more—the 2 apples or the 4 bananas? How much more? **Explain** how you know.

    ____

    ____

    ____

14. **Sense or Nonsense?** Shane said that he could write related multiplication and addition sentences for $6 + 4 + 3$. Does Shane's statement make sense? **Explain.**

    ____

    ____

    ____

15. Write Math ▸ Write a word problem that can be solved using 3×4. Solve the problem.

    ____

    ____

    ____

16. **Test Prep** Which is another way to show $2 + 2 + 2 + 2$?

    Ⓐ 2×2 Ⓒ 4×2

    Ⓑ 5×2 Ⓓ 2×8

SHOW YOUR WORK

 FOR MORE PRACTICE:
Standards Practice Book, pp. P51–P52

Name ______________________________

Skip Count on a Number Line

Essential Question How can you use a number line to skip count and find how many in all?

UNLOCK the Problem REAL WORLD

Caleb wants to make 3 balls of yarn for his cat to play with. He uses 6 feet of yarn to make each ball. How many feet of yarn does Caleb need in all?

- How many equal groups of yarn will Caleb make?

  ____

- How many feet of yarn will be in each group?

  ____

- What do you need to find?

  ____

Use a number line to count equal groups.

How many feet of yarn does Caleb need for each ball? __________

How many equal lengths of yarn does he need? __________

Begin at 0. Skip count by 6s by drawing jumps on the number line.

How many jumps did you make? __________

How long is each jump? __________

Multiply. $3 \times 6 =$ ______

So, Caleb needs ______ feet of yarn in all.

MATHEMATICAL PRACTICES

Math Talk What if Caleb made 4 balls of yarn with 5 feet of yarn in each ball? What would you do differently to find the total number of feet of yarn?

- Why did you jump by 6s on the number line?

  ____

Share and Show 

1. Skip count by drawing jumps on the number line. Find how many in 5 jumps of 4. Then write the product.

Think: 1 jump of 4 shows 1 group of 4.

$5 \times 4 =$ _____

Draw jumps on the number line to show equal groups. Find the product.

2. 3 groups of 8

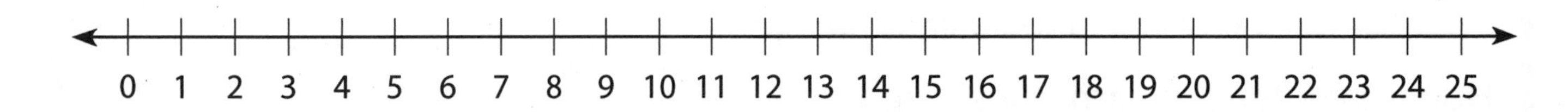

$3 \times 8 =$ _____

3. 8 groups of 3

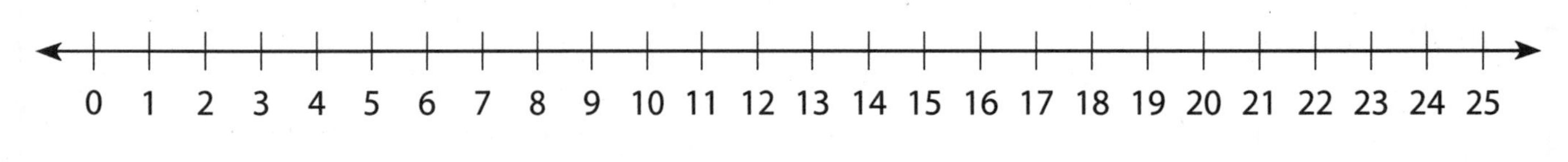

$8 \times 3 =$ _____

Write the multiplication sentence the number line shows.

4.

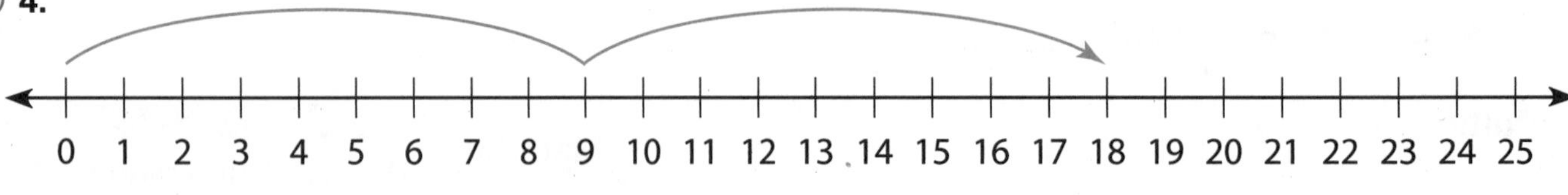

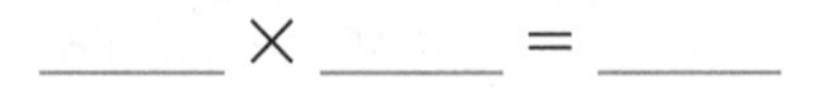

_____ × _____ = _____

MATHEMATICAL PRACTICES

Math Talk How do equal jumps on the number line show equal groups?

Name ___________________________

On Your Own

Draw jumps on the number line to show equal groups. Find the product.

5. 6 groups of 4

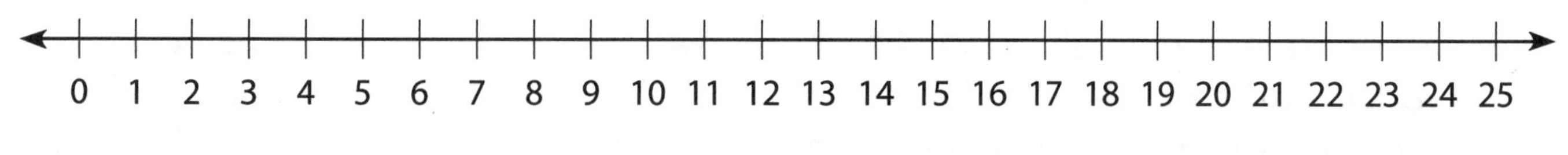

$6 \times 4 =$ _____

6. 7 groups of 3

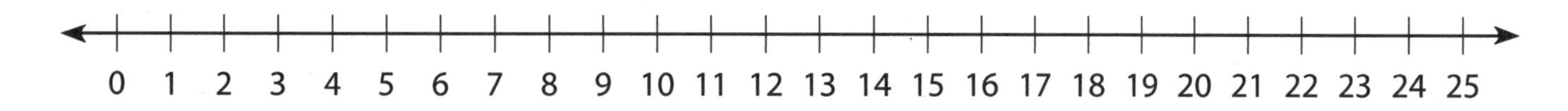

$7 \times 3 =$ _____

7. 2 groups of 10

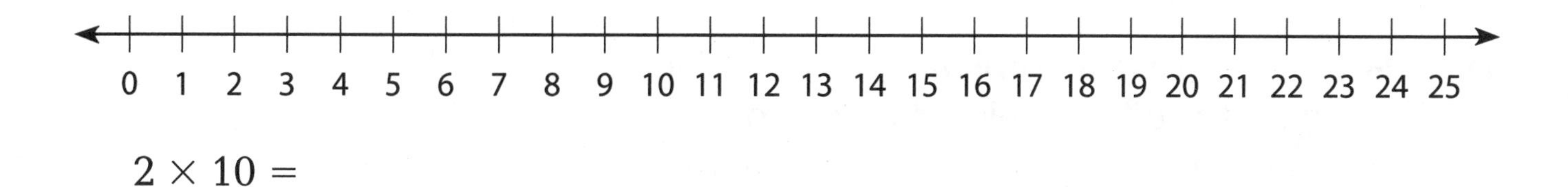

$2 \times 10 =$ _____

Write the multiplication sentence the number line shows.

8.

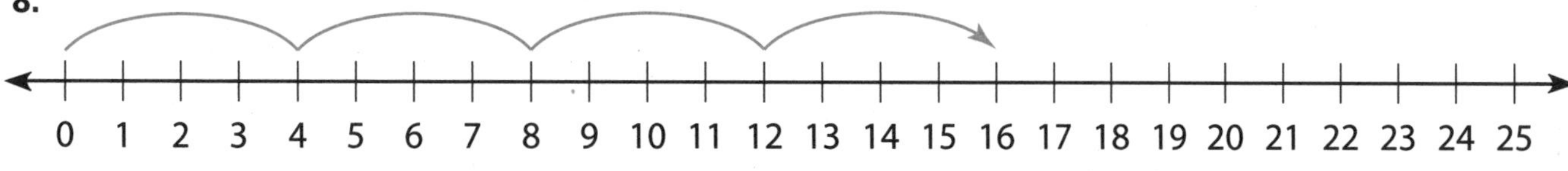

_____ × _____ = _____

9.

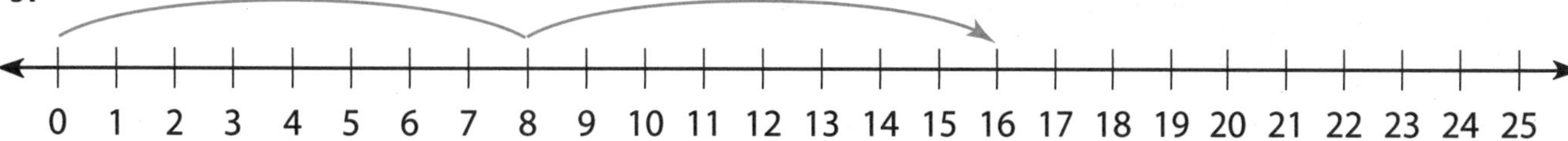

_____ × _____ = _____

Problem Solving REAL WORLD

10. Erin displays her toy cat collection on 3 shelves. She puts 8 cats on each shelf. If she collects 3 more cats, how many cats will she have?

____

SHOW YOUR WORK

11. H.O.T. Write two multiplication sentences that have a product of 12. Draw jumps on the number line to show the multiplication.

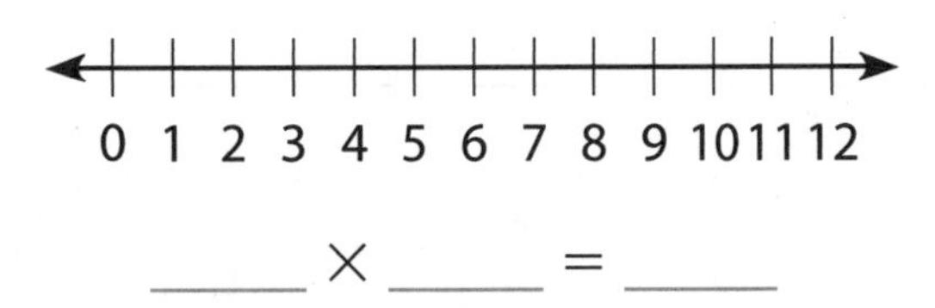

_____ × _____ = _____

0 1 2 3 4 5 6 7 8 9 10 11 12

_____ × _____ = _____

12. Write Math ▸ **Pose a Problem** Write a problem that can be solved by finding 8 groups of 5. Write a multiplication sentence to solve the problem. Then solve.

____

____

____

____

13. **Test Prep** Brooke needs 4 lengths of string, each 2 inches long. How much string does she need altogether?

Ⓐ 2 Ⓒ 6

Ⓑ 4 Ⓓ 8

Name ______________________________

Mid-Chapter Checkpoint

▶ Vocabulary

Choose the best term from the box.

Vocabulary
equal groups
factors
multiply
product

1. When you combine equal groups, you can ____________ to find how many in all. (p. 102)
2. The answer in a multiplication problem is called the ____________. (p. 102)
3. The numbers you multiply are called the ____________. (p. 102)

▶ Concepts and Skills

Count equal groups to find how many.

4. 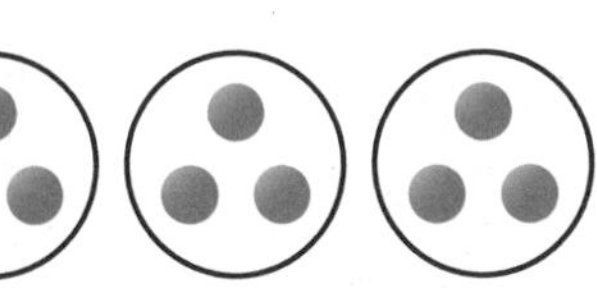

___ groups of ___

___ in all

5. 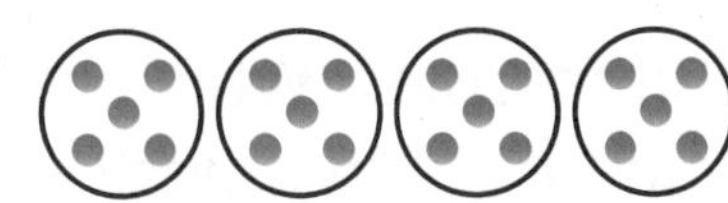

___ groups of ___

___ in all

6. 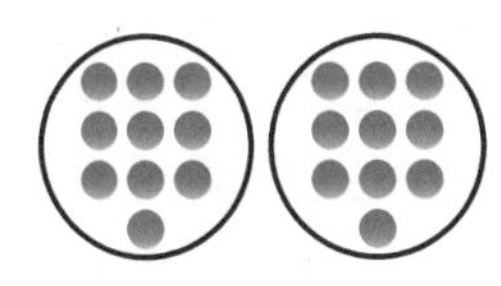

___ groups of ___

___ in all

Write related addition and multiplication sentences.

7. 3 groups of 9

___ + ___ + ___ = ___

___ × ___ = ___

8. 5 groups of 7

___ + ___ + ___ + ___ + ___ = ___

___ × ___ = ___

Draw jumps on the number line to show equal groups. Find the product.

9. 6 groups of 3

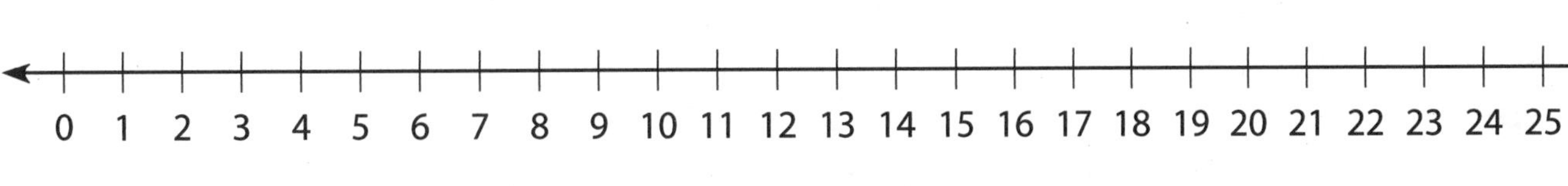

____ × ____ = ____

Fill in the bubble for the correct answer choice.

10. Beth's mother made some cookies. She put 4 cookies on each of 8 plates. Which multiplication sentence can be used to find the number of cookies Beth's mother made?

Ⓐ $8 + 4 = \blacksquare$ Ⓒ $8 \times 4 = \blacksquare$

Ⓑ $4 \times 4 = \blacksquare$ Ⓓ $4 + 4 + 4 + 4 + 4 + 4 + 4 + 4 = \blacksquare$

11. Avery gave 5 animal stickers to each of 10 friends. Which number sentence can be used to find the number of stickers she gave away?

Ⓐ $10 + 5 = \blacksquare$

Ⓑ $10 - 5 = \blacksquare$

Ⓒ $10 \times 5 = \blacksquare$

Ⓓ $10 \times 10 = \blacksquare$

12. Matt made 2 equal groups of marbles. Which multiplication sentence shows the total number of marbles?

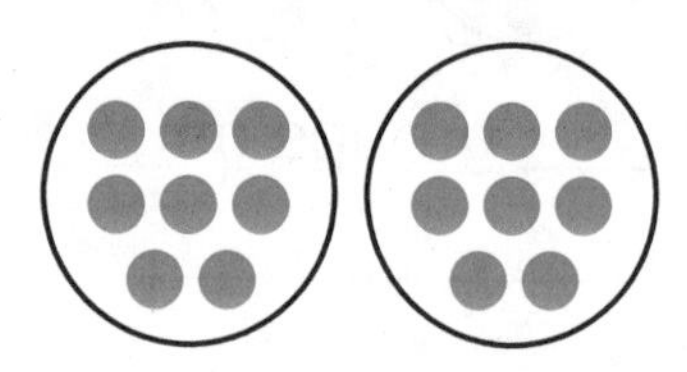

Ⓐ $2 \times 2 = 4$

Ⓑ $2 \times 8 = 16$

Ⓒ $3 \times 8 = 24$

Ⓓ $8 \times 8 = 64$

13. Lindsey needs 3 lengths of ribbon each 5 inches long. How much ribbon does she need altogether?

Ⓐ 3 inches

Ⓑ 5 inches

Ⓒ 15 inches

Ⓓ 20 inches

Name ________________________________

Problem Solving • Model Multiplication

Essential Question How can you use the strategy *draw a diagram* to solve one- and two-step problems?

UNLOCK the Problem REAL WORLD

Three groups of students are taking drum lessons. There are 8 students in each group. How many students are taking lessons in all?

Read the Problem

What do I need to find?

I need to find how many ____________ are taking drum lessons.

What information do I need to use?

There are ____________ groups of students taking drum lessons. There are ____________ students in each group.

How will I use the information?

I will draw a bar model to help me see

____

____.

Solve the Problem

Complete the bar model to show the drummers.

Write 8 in each box to show the 8 students in each of the 3 groups.

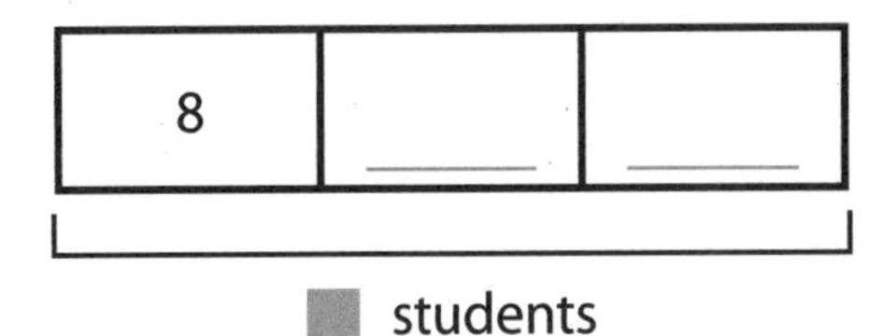

Since there are equal groups, I can multiply to find the number of students taking lessons.

_____ × _____ = ■

_____ = ■

So, there are _____ students in all.

MATHEMATICAL PRACTICES

Math Talk How would the bar model change if there were 6 groups of 4 students? Solve.

Try Another Problem

Twelve students in Mrs. Taylor's class want to start a band. Seven students each made a drum. The rest made 2 maracas each. How many maracas in all were made?

Read the Problem	Solve the Problem
What do I need to find?	**Record the steps you used to solve the problem.** | 7 | ___ | 12 students
What information do I need to use?	
How will I use the information?	

1. How many maracas in all did the students make? ______

2. How do you know your answer is reasonable? ______

MATHEMATICAL PRACTICES

Math Talk Why wouldn't you draw 2 boxes and write 5 in each box?

Name ________________________________

Share and Show 

1. There are 6 groups of 4 students who play the trumpet in the marching band. How many students play the trumpet in the band?

   First, draw a bar model to show each group of students.

   Draw ______ boxes and write ______ in each box.

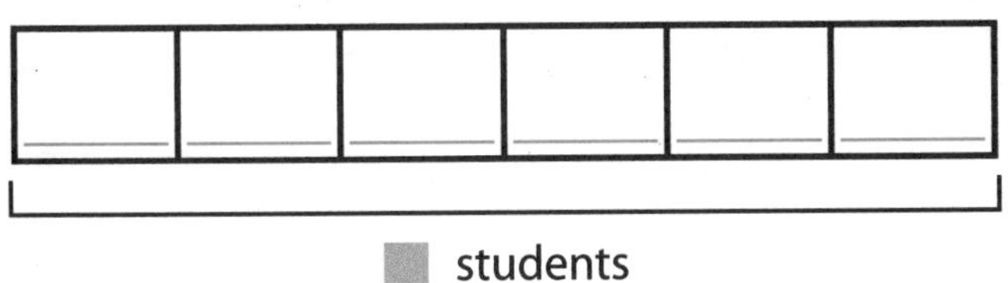

   Then, multiply to find the total number of trumpet players.

   ______ × ______ = ■

   ______ = ■

   So, ______ students play the trumpet in the marching band.

2. H.O.T. **What if** there are 4 groups of 7 students who play the saxophone? How many students play the saxophone or trumpet?

   ____

3. Suppose there are 5 groups of 4 trumpet players. In front of the trumpet players are 18 saxophone players. How many students play the trumpet or saxophone?

   ____

4. There are 3 rows of flute players in the marching band. There are 7 students in each row. How many flute players are in the marching band?

   ____

Choose a STRATEGY

- Act It Out
- Draw a Diagram
- Find a Pattern
- Make a Table

SHOW YOUR WORK

On Your Own

Use the picture graph for 5–7.

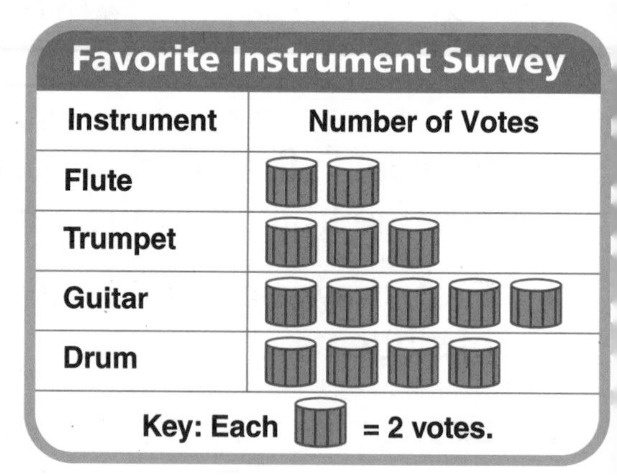

5. The picture graph shows how students in Jillian's class voted for their favorite instrument. How many students voted for the guitar?

____

6. **Write Math** On the day of the survey, two students were absent. The picture graph shows the votes of all the other students in the class, including Jillian. How many students are in the class? **Explain** your answer.

____

____

____

7. H.O.T. Jillian added the number of votes for two instruments and got a total of 12 votes. For which two instruments did she add the votes?

_______________ and _______________

8. The flute was invented 26 years after the harmonica. The electric guitar was invented 84 years after the flute. How many years was the electric guitar invented after the harmonica?

____

9. **Test Prep** After music class, the students filled 4 cabinets with their instruments. Each cabinet has space for 8 instruments. How many instruments did the students put away?

Ⓐ 32

Ⓑ 24

Ⓒ 16

Ⓓ 12

SHOW YOUR WORK

FOR MORE PRACTICE:
Standards Practice Book, pp. P55–P56

Name ______________________________

Model with Arrays

Essential Question How can you use arrays to model multiplication and find factors?

UNLOCK the Problem REAL WORLD

Many people grow tomatoes in their gardens. Lee plants 3 rows of tomato plants with 6 plants in each row. How many tomato plants are there?

▲ Tomatoes are a great source of vitamins.

Activity 1

Materials ■ square tiles ■ MathBoard

- You make an **array** by placing the same number of tiles in each row. Make an array with 3 rows of 6 tiles to show the tomato plants.
- Now draw the array you made.

- Find the total number of tiles.

Multiply. $3 \times 6 =$ ______

3 ↑ number of rows

6 ↑ number in each row

So, there are ______ tomato plants.

MATHEMATICAL PRACTICES

Math Talk Does the number of tiles change if you turn the array to show 6 rows of 3? **Explain.**

Activity 2 **Materials** ■ square tiles ■ MathBoard

Use 8 tiles. Make as many different arrays as you can, using all 8 tiles. Draw the arrays. The first one is done for you.

A

1 row of 8

$1 \times 8 = 8$

B

8 rows of ______

$8 \times$ ______ $= 8$

C

______ rows of ______

______ $\times$ ______ $= 8$

D

______ rows of ______

______ $\times$ ______ $= 8$

You can make ______ different arrays using 8 tiles.

Share and Show

1. Complete. Use the array.

______ rows of ______ = ______

______ $\times$ ______ = ______

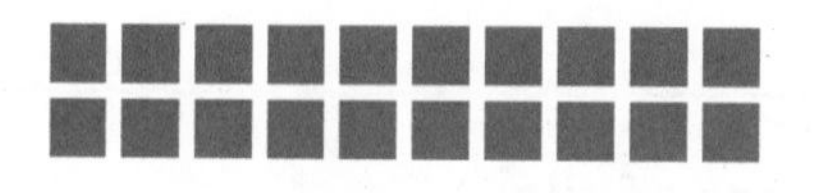

Write a multiplication sentence for the array.

2. 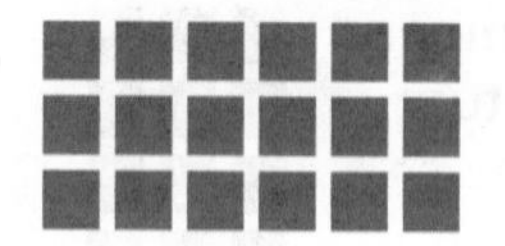

____

3. 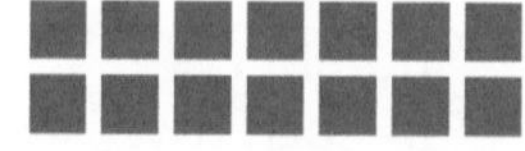

____

MATHEMATICAL PRACTICES

Math Talk Suppose you make an array with 12 tiles and you want to put 3 tiles in each row. How do you decide how many rows you need to make?

Name ____________________

On Your Own

Write a multiplication sentence for the array.

4. 

____

5. 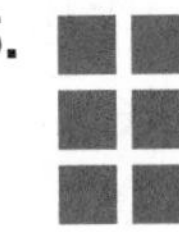

____

Draw an array to find the product.

6. $3 \times 6 =$ ____

7. $4 \times 7 =$ ____

8. $3 \times 5 =$ ____

9. $4 \times 4 =$ ____

10. Use 6 tiles. Make as many different arrays as you can using all the tiles. Draw the arrays. Then write a multiplication sentence for each array.

____

____

Problem Solving REAL WORLD

Use the table to solve 11–12.

Mr. Bloom's Garden

Vegetable	Planted In
Beans	4 rows of 6
Carrots	2 rows of 8
Corn	5 rows of 9
Beets	4 rows of 7

11. Mr. Bloom grows vegetables in his garden. Draw an array and write the multiplication sentence to show how many corn plants Mr. Bloom has in his garden.

____

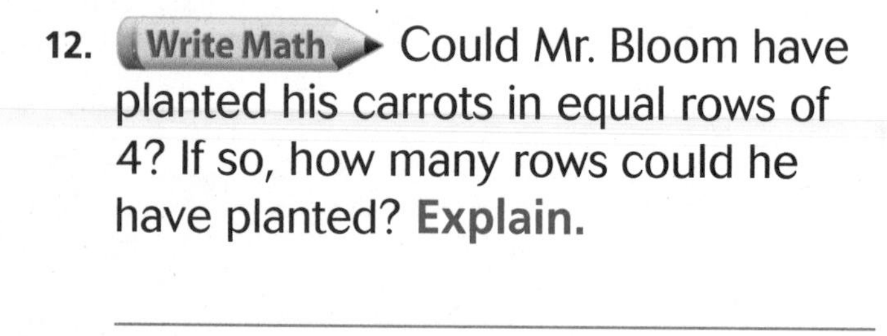

12. Write Math ▸ Could Mr. Bloom have planted his carrots in equal rows of 4? If so, how many rows could he have planted? **Explain.**

____

____

____

____

SHOW YOUR WORK

13. H.O.T. Mr. Bloom has 12 strawberry plants. Describe all of the different arrays that Mr. Bloom could make using all of his strawberry plants. The first one is done for you.

2 rows of 6; ______________________________

____

____

14. **Test Prep** What multiplication sentence does this array show?

Ⓐ $2 \times 3 = 6$ Ⓒ $2 \times 5 = 10$

Ⓑ $4 \times 1 = 4$ Ⓓ $2 \times 4 = 8$

Name ______________________________

ALGEBRA
Lesson 3.6

Commutative Property of Multiplication

Essential Question How can you use the Commutative Property of Multiplication to find products?

UNLOCK the Problem REAL WORLD

Dave works at the Bird Store. He arranges 15 boxes of birdseed in rows on the shelf. What are two ways he can arrange the boxes in equal rows?

- Circle the number that is the product.

Activity Make an array.

Materials ■ square tiles ■ MathBoard

Arrange 15 tiles in 5 equal rows.
Draw a quick picture of your array.

How many tiles are in each row? __________

What multiplication sentence does your array show? ______________

Suppose Dave arranges the boxes in 3 equal rows.
Draw a quick picture of your array.

How many tiles are in each row? __________

What multiplication sentence does your array show? ______________

So, two ways Dave can arrange the 15 boxes are in _____ rows of 3 or in 3 rows of _____.

MATHEMATICAL PRACTICES

Math Talk Why do 5 rows of 3 and 3 rows of 5 both equal the same number?

Multiplication Property The **Commutative Property of Multiplication** states that when you change the order of the factors, the product stays the same. You can think of it as the Order Property of Multiplication.

$2 \times$ ______ $=$ ______

$3 \times$ ______ $=$ ______

Math Idea

Facts that show the Commutative Property of Multiplication have the same factors in a different order.

$2 \times 3 = 6$ and $3 \times 2 = 6$

So, $2 \times$ ______ $= 3 \times$ ______.

- **Explain** how the models are alike and how they are different.

____

____

Try This! Draw a quick picture on the right that shows the Commutative Property of Multiplication. Then complete the multiplication sentences.

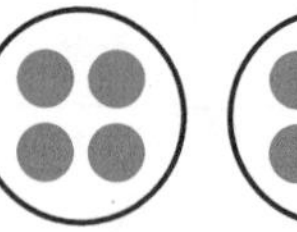 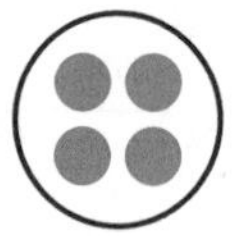 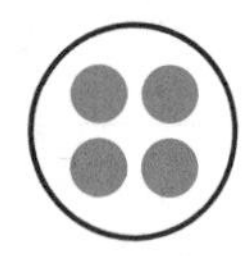

______ $\times 4 =$ ______ ______ $\times 3 =$ ______

B 

$2 \times$ ______ $=$ ______ $5 \times$ ______ $=$ ______

Name ______________________

Share and Show 

1. Write a multiplication sentence for the array.

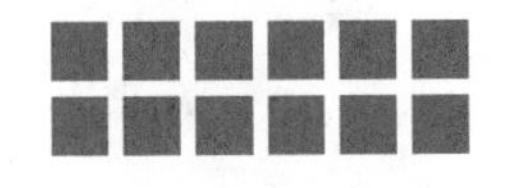

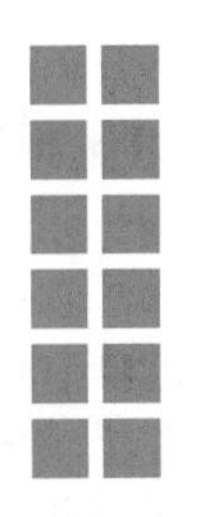

______________ ______________

MATHEMATICAL PRACTICES

Math Talk Explain what the factor 2 means in each multiplication sentence.

Write a multiplication sentence for the model. Then use the Commutative Property of Multiplication to write a related multiplication sentence.

2. 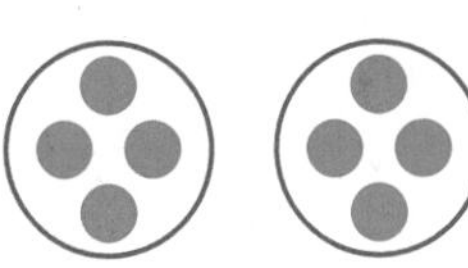

____ × ____ = ____

____ × ____ = ____

3. 

____ × ____ = ____

____ × ____ = ____

4. 

____ × ____ = ____

____ × ____ = ____

On Your Own

Write a multiplication sentence for the model. Then use the Commutative Property of Multiplication to write a related multiplication sentence.

5. 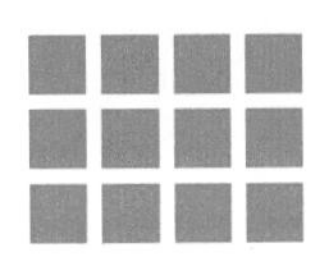

____ × ____ = ____

____ × ____ = ____

6. 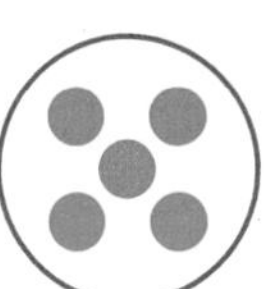 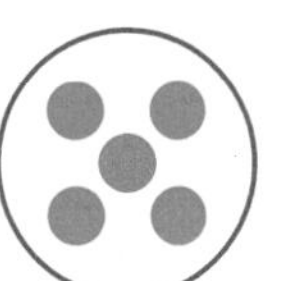

____ × ____ = ____

____ × ____ = ____

7. 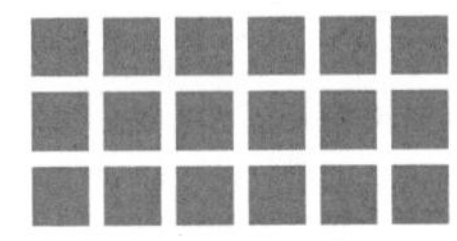

____ × ____ = ____

____ × ____ = ____

Algebra Write the unknown factor.

8. $3 \times 7 =$ ______ $\times 3$

9. $4 \times 5 = 10 \times$ ______

10. $3 \times 6 =$ ______ $\times 9$

11. $6 \times$ ______ $= 4 \times 9$

12. ______ $\times 8 = 4 \times 6$

13. $5 \times 8 = 8 \times$ ______

Problem Solving REAL WORLD

14. Jenna used pinecones to make 18 peanut butter bird feeders. She hung the same number of feeders in each of 6 trees. Draw an array to show how many feeders she put in each tree.

    She put ______ bird feeders in each tree.

SHOW YOUR WORK

15. **What if** Jenna hung the same number of feeders in each of 9 trees? How many feeders would she put in each tree?

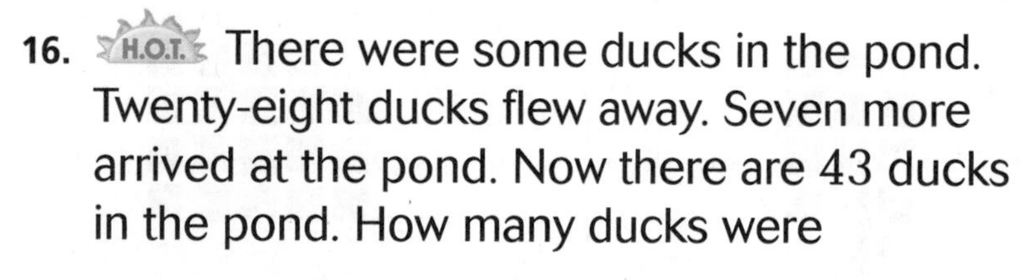

    ____

16. H.O.T. There were some ducks in the pond. Twenty-eight ducks flew away. Seven more arrived at the pond. Now there are 43 ducks in the pond. How many ducks were in the pond to start? ______________

17. Write Math Write two different word problems about 12 birds to show 2×6 and 6×2. Solve each problem.

    ____

    ____

    ____

    ____

    ____

18. **Test Prep** Which is an example of the Commutative Property of Multiplication?

    Ⓐ $2 + 3 = 3 \times 2$

    Ⓑ $4 \times 1 = 4 \times 1$

    Ⓒ $2 \times 4 = 4 \times 2$

    Ⓓ $2 \times 6 = 3 \times 4$

FOR MORE PRACTICE:
Standards Practice Book, pp. P59–P60

Name ______________________________

Multiply with 1 and 0

Essential Question What happens when you multiply a number by 0 or 1?

UNLOCK the Problem REAL WORLD

Luke sees 4 birdbaths. Each birdbath has 2 birds in it. What multiplication sentence tells how many birds there are?

- How many birdbaths are there? ______
- How many birds are in each birdbath to start? ______

Draw a quick picture to show the birds in the birdbaths.

____ × ____ = ____

One bird flies away from each birdbath. Cross out 1 bird in each birdbath above. What multiplication sentence shows the total number of birds left?

____	×	____	=	____
↑ birdbaths		↑ bird left in each birdbath		↑ total number of birds

Now cross out another bird in each birdbath. What multiplication sentence shows the total number of birds left in the birdbaths?

____	×	____	=	____
↑ birdbaths		↑ birds left in each birdbath		↑ total number of birds

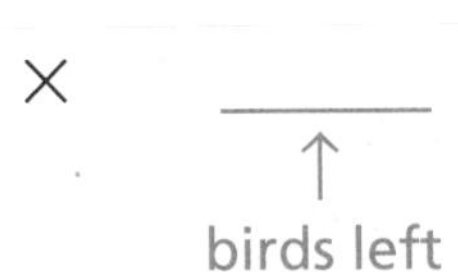

- How do the birdbaths look now? ______________________

MATHEMATICAL PRACTICES

Math Talk What if there were 5 birdbaths with 0 birds in each of them? What would be the product? **Explain.**

Example

Jenny has 2 pages of bird stickers. There are 4 stickers on each page. How many stickers does she have in all?

$2 \times 4 =$ ______ **Think:** 2 groups of 4

So, Jenny has ______ stickers in all.

Suppose Jenny uses 1 page of the stickers. What fact shows how many stickers are left?

______ $\times$ ______ $=$ ______ **Think:** 1 group of 4

So, Jenny has ______ stickers left.

Now, Jenny uses the rest of the stickers. What fact shows how many stickers Jenny has left?

ERROR Alert

A 0 in a multiplication sentence means 0 groups or 0 things in a group, so the product is always 0.

______ $\times$ ______ $=$ ______ **Think:** 0 groups of 4

So, Jenny has ______ stickers left.

- What does each number in $0 \times 4 = 0$ tell you?

____

____

1. What pattern do you see when you multiply numbers with 1 as a factor?

Think: $1 \times 2 = 2$ $1 \times 3 = 3$ $1 \times 4 = 4$

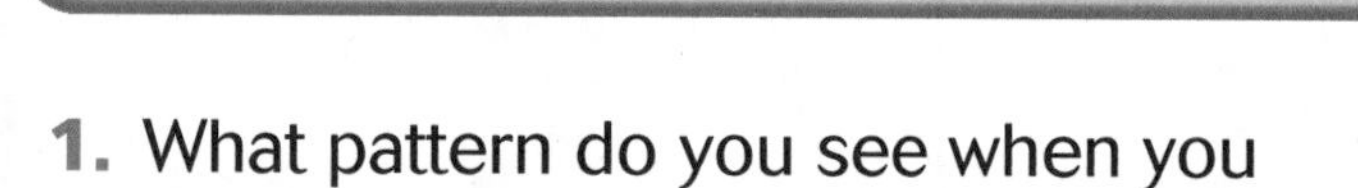

____

____

The **Identity Property of Multiplication** states that the product of any number and 1 is that number.

$7 \times 1 = 7$ $6 \times 1 = 6$

$1 \times 7 = 7$ $1 \times 6 = 6$

2. What pattern do you see when you multiply numbers with 0 as a factor?

Think: $0 \times 1 = 0$ $0 \times 2 = 0$ $0 \times 5 = 0$

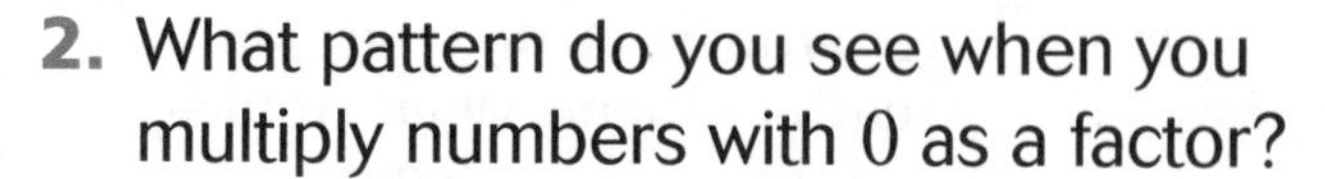

____

____

The **Zero Property of Multiplication** states that the product of zero and any number is zero.

$0 \times 5 = 0$ $0 \times 8 = 0$

$5 \times 0 = 0$ $8 \times 0 = 0$

Name ______________________

Share and Show 

1. What multiplication sentence matches this picture? Find the product.

____

Find the product.

2. $5 \times 1 =$ ____
3. $0 \times 2 =$ ____
4. $4 \times 0 =$ ____
5. $1 \times 6 =$ ____
6. $3 \times 0 =$ ____
7. $1 \times 2 =$ ____
8. $0 \times 6 =$ ____
9. $8 \times 1 =$ ____

MATHEMATICAL PRACTICES

Math Talk Explain how 3×1 and $3 + 1$ are different.

On Your Own

Find the product.

10. $3 \times 1 =$ ____
11. $8 \times 0 =$ ____
12. $1 \times 9 =$ ____
13. $0 \times 7 =$ ____
14. $0 \times 4 =$ ____
15. $10 \times 1 =$ ____
16. $1 \times 3 =$ ____
17. $6 \times 1 =$ ____
18. $1 \times 0 =$ ____
19. $1 \times 7 =$ ____
20. $6 \times 0 =$ ____
21. $1 \times 4 =$ ____
22. $0 \times 8 =$ ____
23. $9 \times 1 =$ ____
24. $0 \times 9 =$ ____
25. $1 \times 1 =$ ____
26. $4 \times 1 =$ ____
27. $1 \times 5 =$ ____
28. $0 \times 0 =$ ____
29. $5 \times 0 =$ ____

H.O.T. **Algebra** **Complete the multiplication sentence.**

30. ____ $\times 1 = 15$
31. $1 \times 28 =$ ____
32. $0 \times 46 =$ ____
33. $36 \times 0 =$ ____
34. ____ $\times 5 = 5$
35. $19 \times$ ____ $= 0$
36. ____ $\times 0 = 0$
37. $7 \times$ ____ $= 7$

Problem Solving REAL WORLD

Use the table for 38–40.

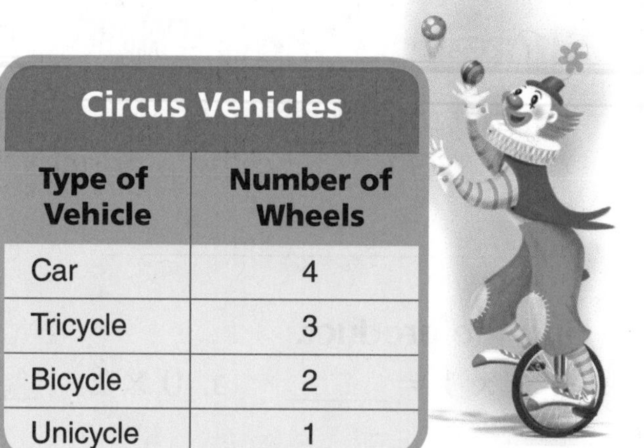

Circus Vehicles	
Type of Vehicle	**Number of Wheels**
Car	4
Tricycle	3
Bicycle	2
Unicycle	1

38. At the circus Jon saw 5 unicycles. How many wheels are on the unicycles in all? Write a multiplication sentence.

_____ $\times$ _____ $=$ __________

39. H.O.T. Brian saw some circus vehicles. He saw 17 wheels in all. If 2 of the vehicles are cars, how many vehicles are bicycles and tricycles?

____

40. **What's the Question?** Julia used multiplication with 1 and the information in the table. The answer is 3.

____

____

41. **Write Math** Write a word problem that uses multiplying with 1 or 0. Show how to solve your problem.

____

____

____

____

42. **Test Prep** Eric has 1 pencil box at school. He has 6 pencils in the box. Which number sentence shows how many pencils Eric has in the pencil box?

Ⓐ $6 + 1 = 7$ Ⓒ $6 - 1 = 5$

Ⓑ $0 \times 6 = 0$ Ⓓ $1 \times 6 = 6$

SHOW YOUR WORK

FOR MORE PRACTICE:
Standards Practice Book, pp. P61–P62

Name ______________________________

Chapter Review/Test

▶ Vocabulary

Choose the best term from the box.

Vocabulary
array
Commutative Property of Multiplication
Identity Property of Multiplication
product
Zero Property of Multiplication

1. The ______________________ states that the product of any number and 1 is that number. (p. 124)
2. When you place the same number of tiles in each row, you make an ______________. (p. 115)
3. The ______________________ states that the product of zero and any number is zero. (p. 124)
4. The ______________________ states that when you change the order of the factors, the product is the same. (p. 120)

▶ Concepts and Skills

Write a multiplication sentence for the model. Then use the Commutative Property of Multiplication to write a related multiplication sentence.

5.

6.

7.

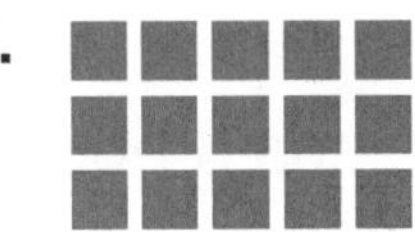

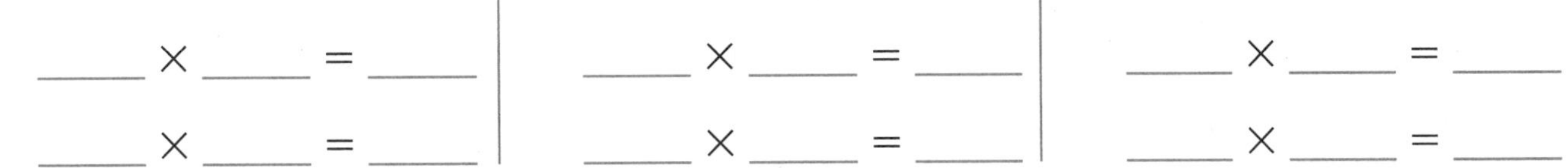

5. ____ × ____ = ____
   ____ × ____ = ____

6. ____ × ____ = ____
   ____ × ____ = ____

7. ____ × ____ = ____
   ____ × ____ = ____

Find the product.

8. $5 \times 2 =$ ____
9. $4 \times 7 =$ ____
10. $8 \times 1 =$ ____
11. $2 \times 0 =$ ____
12. $2 \times 1 =$ ____
13. $5 \times 0 =$ ____
14. $9 \times 6 =$ ____
15. $7 \times 3 =$ ____

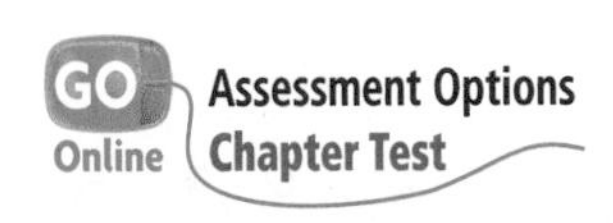

Fill in the bubble for the correct answer choice.

16. John made 3 groups of 8 marbles. Which number sentence could be used to find the number of marbles?

Ⓐ $3 + 5 = ■$

Ⓑ $3 + 8 = ■$

Ⓒ $3 \times 5 = ■$

Ⓓ $3 \times 8 = ■$

17. Lucy and her mother made tacos. They put 2 tacos on each of 7 plates. Which number sentence shows the number of tacos they made?

Ⓐ $2 + 2 + 2 + 2 + 2 + 2 + 2 = 14$

Ⓑ $2 + 7 = 9$

Ⓒ $7 - 2 = 5$

Ⓓ $7 + 7 + 7 + 7 = 28$

18. Morgan has 6 toy cars. Each car has 4 wheels. How many wheels do the cars have in all?

Ⓐ 2

Ⓑ 10

Ⓒ 12

Ⓓ 24

19. Sonya needs 3 equal lengths of wire to make 3 bracelets. The jump on the number line shows the length of one wire in inches. How much wire will she need altogether?

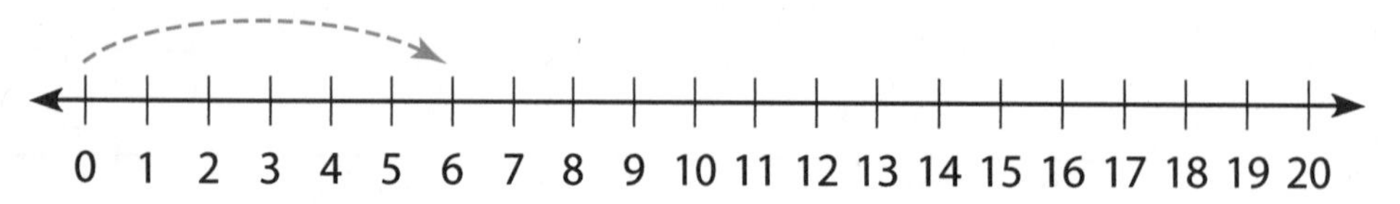

Ⓐ 3 inches

Ⓑ 6 inches

Ⓒ 18 inches

Ⓓ 19 inches

Name ____________________

Fill in the bubble for the correct answer choice.

20. A unicycle has only 1 wheel. How many wheels are there on 9 unicycles?

Ⓐ 1

Ⓑ 9

Ⓒ 10

Ⓓ 11

21. Dan and his dad baked some cookies. They put 5 cookies on each of 4 plates. Which number sentence shows how many cookies they put on plates?

Ⓐ $4 \times 5 = 20$

Ⓑ $4 \times 4 = 16$

Ⓒ $3 \times 4 = 12$

Ⓓ $2 \times 5 = 10$

22. Josh has 4 dogs. Each dog gets 2 dog biscuits every day. How many biscuits will Josh need for all of his dogs for Saturday and Sunday?

Ⓐ 4

Ⓑ 8

Ⓒ 12

Ⓓ 16

23. Lacy planted 4 rows of 9 flowers. How many flowers did she plant?

Ⓐ 13

Ⓑ 35

Ⓒ 36

Ⓓ 49

▶ Constructed Response

24. Write a problem that you can use 7×0 to solve.

____

____

25. James made this array. Draw another array to show the Commutative Property of Multiplication. Then write a multiplication sentence for each array.

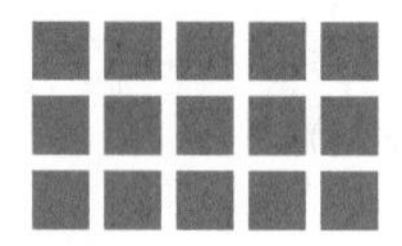

____

▶ Performance Task

26. Sharon is putting 16 cookies on plates. She will put an equal number of cookies on each plate.

A Draw two quick pictures to show how she could arrange the 16 cookies two different ways. Then write a multiplication sentence to match each drawing.

Picture 1	Picture 2
_____ × _____ = _____	_____ × _____ = _____

B **Explain** how you knew what multiplication sentence to write.

____

____

____

Chapter 4

Multiplication Facts and Strategies

Show What You Know 

Check your understanding of important skills.

Name ______________________________

▶ Doubles and Doubles Plus One

Write the doubles and doubles plus one facts.

1. ____ + ____ = ____ ____ + ____ = ____

2. ____ + ____ = ____ ____ + ____ = ____

▶ Equal Groups **Complete.**

3. 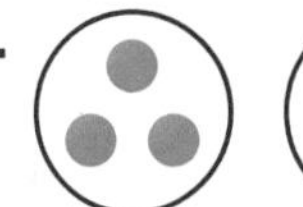 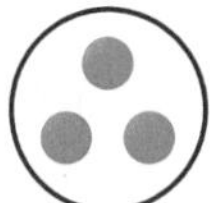 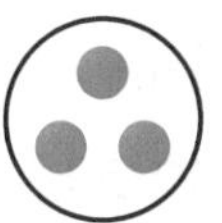 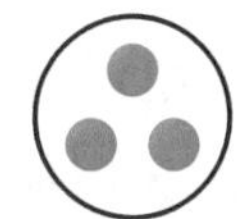 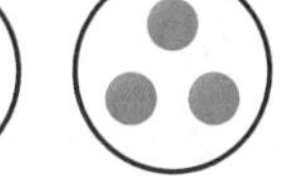

____ groups of ____

____ in all

4.  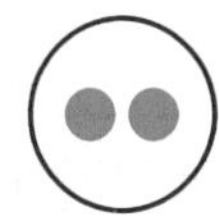   

____ groups of ____

____ in all

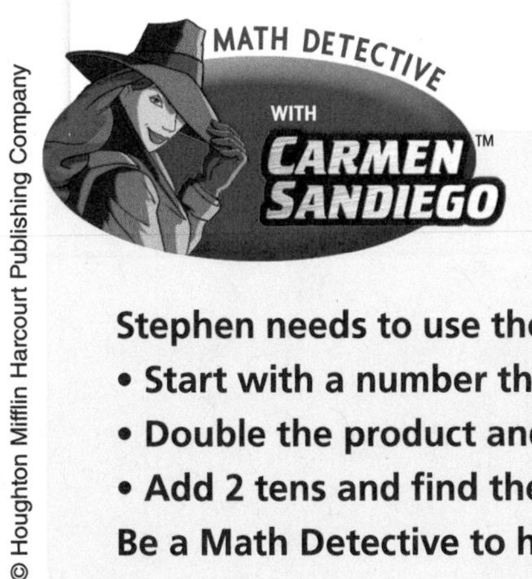

Stephen needs to use these clues to find a buried time capsule.

- **Start with a number that is the product of 3 and 4.**
- **Double the product and go to that number.**
- **Add 2 tens and find the number that is 1 less than the sum.**

Be a Math Detective to help him find the time capsule.

GO Online Assessment Options: **Soar to Success Math**

Vocabulary Builder

▶ Visualize It

Complete the tree map by using the words with a ✓.

Multiplication Properties

______ Property of Multiplication	______ Property of Multiplication	______ Property of Multiplication
$1 \times 4 = 4$	$(4 \times 2) \times 3 = 4 \times (2 \times 3)$	$3 \times 2 = 2 \times 3$
______ ______		______

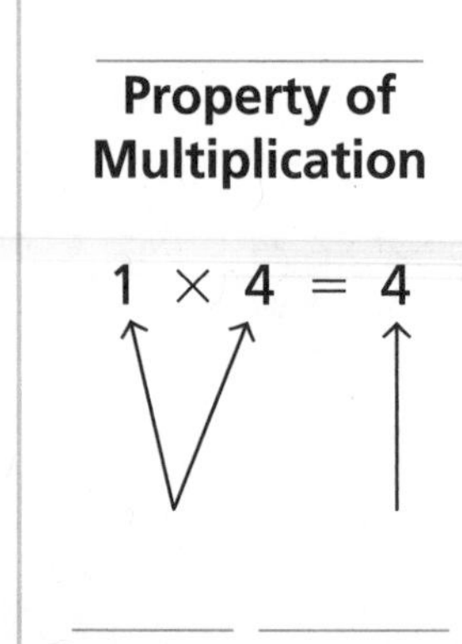

Review Words

✓arrays
✓Commutative Property of Multiplication
even
✓factors
✓Identity Property of Multiplication
odd
✓product

Preview Words

✓Associative Property of Multiplication
Distributive Property
multiple

▶ Understand Vocabulary

Complete the sentences by using the preview words.

1. The ______ Property of Multiplication states that when the grouping of factors is changed, the product is the same.

2. A ______ of 5 is any product that has 5 as one of its factors.

3. The ______ Property states that multiplying a sum by a number is the same as multiplying each addend by the number and then adding the products.

Example: $2 \times 8 = 2 \times (4 + 4)$

$2 \times 8 = (2 \times 4) + (2 \times 4)$

$2 \times 8 = 8 + 8$

$2 \times 8 = 16$

GO Online • eStudent Edition • Multimedia eGlossary

Name ____________________

Multiply with 2 and 4

Essential Question How can you multiply with 2 and 4?

UNLOCK the Problem REAL WORLD

Two students are in a play. Each of the students has 3 costumes. How many costumes do they have in all?

Multiplying when there are two equal groups is like adding doubles.

- What does the word "each" tell you?

____

- How can you find the number of costumes the 2 students have?

____

Find 2×3.

MODEL	THINK	RECORD
Draw counters to show the costumes.	2 groups of 3 $3 + 3$ 6	$2 \times 3 = 6$ 2: how many groups 3: how many in each group 6: how many in all

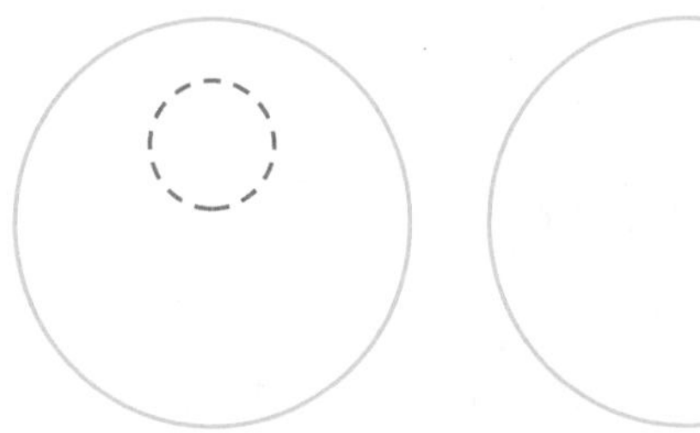

So, the 2 students have _____ costumes in all.

Try This!

$2 \times 1 = 1 + 1 = 2$

$2 \times 2 = 2 + 2 = 4$

$2 \times$ _____ $= 3 +$ _____ $= 6$

$2 \times$ _____ $= 4 +$ _____ $= 8$

$2 \times$ _____ $= 5 +$ _____ $=$ _____

$2 \times$ _____ $= 6 +$ _____ $=$ _____

$2 \times$ _____ $= 7 +$ _____ $=$ _____

$2 \times$ _____ $= 8 +$ _____ $=$ _____

$2 \times$ _____ $= 9 +$ _____ $=$ _____

MATHEMATICAL PRACTICES

Math Talk What do you notice about the product when you multiply by 2?

Count by 2s.

When there are 2 in each group, you can count by 2s to find how many there are in all.

There are 4 students with 2 costumes each. How many costumes do they have in all?

Skip count by drawing the jumps on the number line.

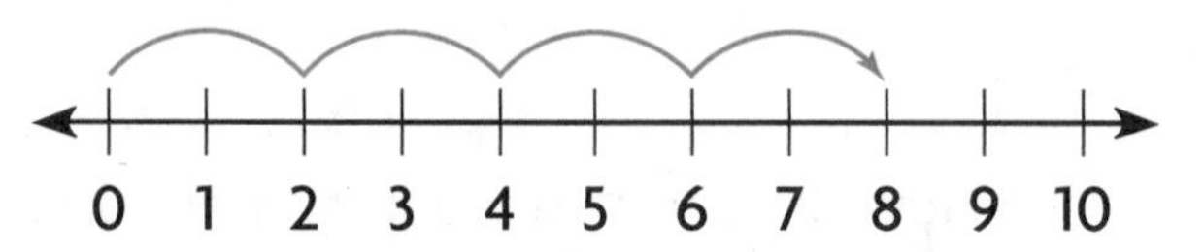

So, the 4 students have ___________ in all.

- How can you decide whether to count by 2s or double?

____

Example Use doubles to find 4×5.

When you multiply with 4, you can multiply with 2 and then double the product.

	MULTIPLY WITH 2	DOUBLE THE PRODUCT
4×5	$2 \times 5 = 10$	$10 + 10 = 20$

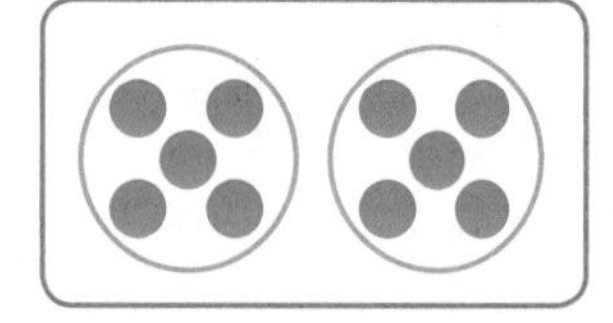

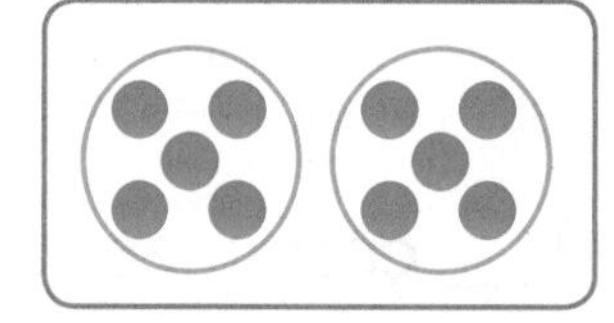

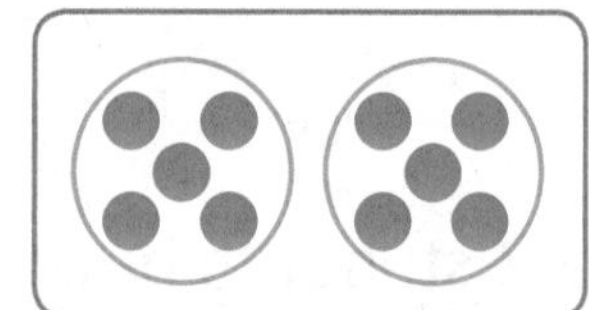

So, $4 \times 5 =$ _____.

Share and Show

1. Double 2×7 to find 4×7.

Multiply with 2. $2 \times 7 =$ _____

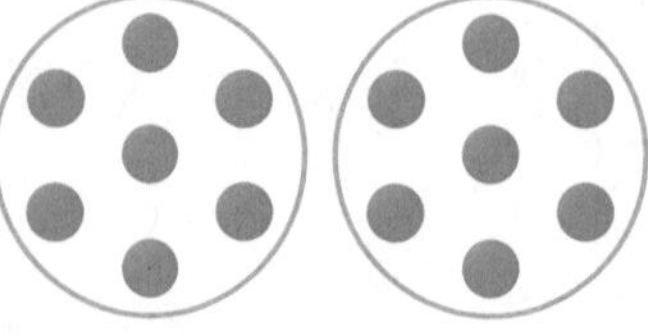

Double the product. $14 + 14 =$ _____

So, $4 \times 7 =$ _____.

MATHEMATICAL PRACTICES

Math Talk Explain how knowing the product for 2×8 helps you find the product for 4×8.

Name ______________________________

Write a multiplication sentence for the model.

2. 

___ × ___ = ___

3. 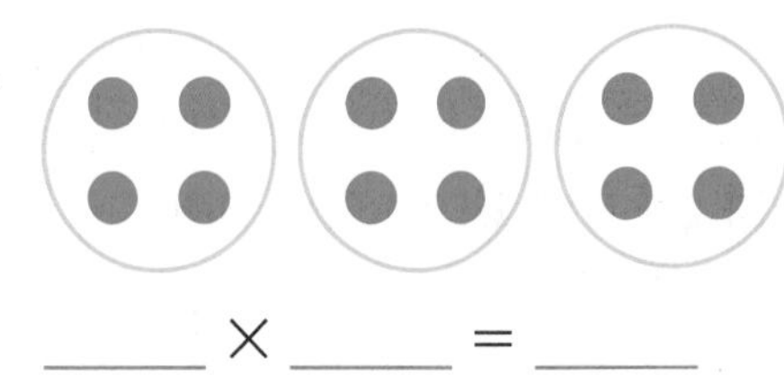

___ × ___ = ___

Find the product.

4. 6×2

5. 9×4

6. 2×7

7. 8×4

8. 5×2

On Your Own

Write a multiplication sentence for the model.

9. 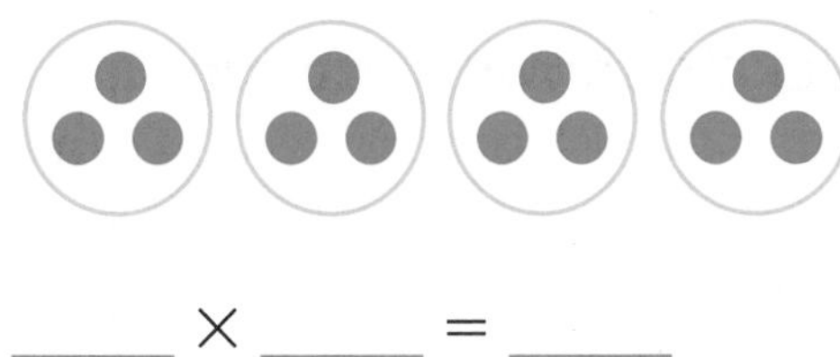

___ × ___ = ___

10. 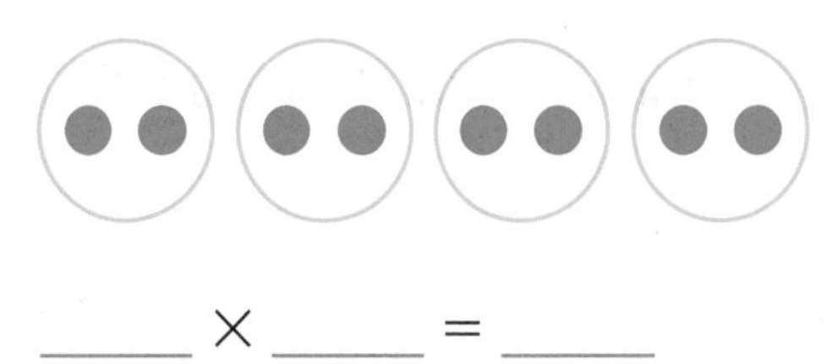

___ × ___ = ___

Find the product. Use your MathBoard.

11. 10×4

12. 2×9

13. 4×6

14. 7×2

15. 2×0

16. 4×3

17. 2×8

18. 4×4

19. 10×2

20. 4×5

Algebra Complete the table for the factors 2 and 4.

	×	1	2	3	4	5	6	7	8	9	10
21.	2										
22.	4										

Algebra Write the unknown number.

23. $4 \times 8 = 16 +$ ___

24. $20 = 2 \times$ ___

25. $8 \times 2 = 10 +$ ___

UNLOCK the Problem REAL WORLD

26. Ms. Peterson's class sold tickets for the class play. How many tickets in all did Brandon and Haylie sell?

Play Tickets

Name	Tickets Sold
Brandon	 🎟🎟🎟🎟
Haylie	 🎟🎟🎟🎟🎟🎟🎟
Elizabeth	 🎟🎟🎟🎟🎟🎟🎟🎟

Key: Each 🎟 = 2 tickets sold.

a. What do you need to find? ______________________________

____

b. Why should you multiply to find the number of tickets shown? **Explain.**

____

____

c. Show the steps you used to solve the problem.

d. Complete the sentences.

Brandon sold _____ tickets.

Haylie sold _____ tickets.

Together, Brandon and Haylie sold _____ tickets.

27. H.O.T. Suppose Sam sold 20 tickets to the school play. How many tickets should be on the picture graph above to show his sales? **Explain.**

____

____

____

28. **Test Prep** Lindsey, Louis, Sally, and Matt each brought 5 guests to the school play. Which number sentence shows their total number of guests?

Ⓐ $4 + 5 = 9$

Ⓑ $5 + 4 = 9$

Ⓒ $4 \times 4 = 16$

Ⓓ $4 \times 5 = 20$

FOR MORE PRACTICE:
Standards Practice Book, pp. P67–P68

Name ___________________________

Multiply with 5 and 10

Essential Question How can you multiply with 5 and 10?

UNLOCK the Problem REAL WORLD

Marcel is making 6 doll-sized banjos. He needs 5 strings for each banjo. How many strings does he need in all?

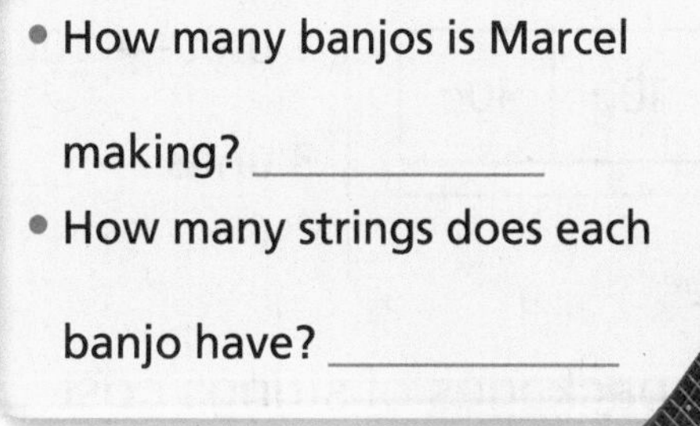

- How many banjos is Marcel making? ________
- How many strings does each banjo have? ________

Use skip counting.

Skip count by 5s until you say 6 numbers.

5, _____, _____, _____, _____, _____

$6 \times 5 =$ _____

So, Marcel needs _____ strings in all.

Example 1 Use a number line.

Each string is 10 inches long. How many inches of string will Marcel use for each banjo?

Think: 1 jump = 10 inches

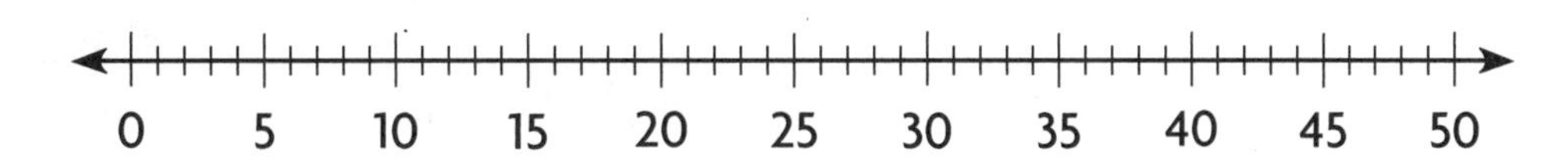

- Draw 5 jumps for the 5 strings. Jump 10 spaces at a time for the length of each string.
- You land on 10, _____, _____, _____, and _____. $5 \times 10 =$ _____

The numbers 10, 20, 30, 40, and 50 are multiples of 10.

So, Marcel will use _____ inches of string for each banjo.

A **multiple** of 10 is any product that has 10 as one of its factors.

MATHEMATICAL PRACTICES

Math Talk What do you notice about the multiples of 10?

Example 2 Use a bar model.

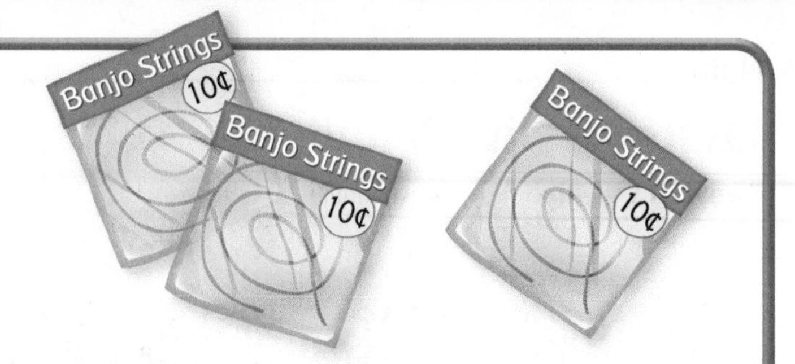

Marcel bought 3 packages of strings. Each package cost 10¢. How much did the packages cost in all?

MODEL	THINK	RECORD
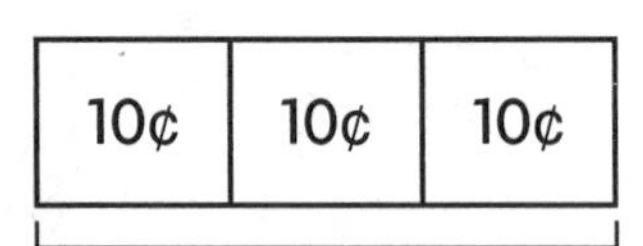 10¢ | 10¢ | 10¢ 	1 unit → 10¢ 3 units → ____ × ____	____ × ____ = ____

So, the packages of strings cost ____ in all.

Share and Show

MATH BOARD

1. How can you use this number line to find 8×5?

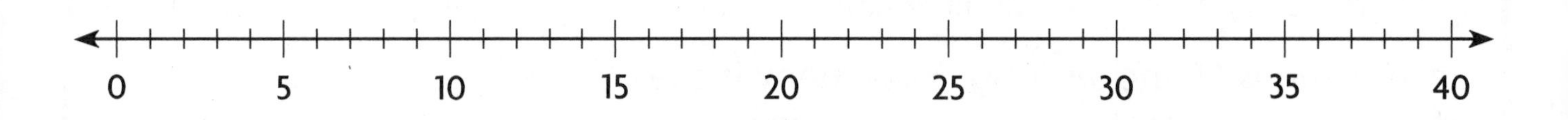

____

____

MATHEMATICAL PRACTICES

Math Talk Explain how knowing 4×5 can help you find 4×10.

Find the product.

2. $2 \times 5 =$ ____

3. ____ $= 6 \times 10$

4. ____ $= 5 \times 5$

5. $10 \times 7 =$ ____

6. 10×4

7. 5×6

8. 10×0

9. 5×3

10. 7×5

11. 5×10

12. 4×5

13. 9×10

Name ___________________

On Your Own

Find the product.

14. $5 \times 1 =$ ____

15. ____ $= 10 \times 2$

16. ____ $= 4 \times 5$

17. $10 \times 10 =$ ____

18. $10 \times 0 =$ ____

19. $10 \times 5 =$ ____

20. ____ $= 1 \times 5$

21. ____ $= 5 \times 9$

22. $\begin{array}{r} 3 \\ \times 4 \\ \hline \end{array}$

23. $\begin{array}{r} 5 \\ \times 0 \\ \hline \end{array}$

24. $\begin{array}{r} 4 \\ \times 8 \\ \hline \end{array}$

25. $\begin{array}{r} 10 \\ \times 5 \\ \hline \end{array}$

26. $\begin{array}{r} 10 \\ \times 9 \\ \hline \end{array}$

27. $\begin{array}{r} 10 \\ \times 1 \\ \hline \end{array}$

28. $\begin{array}{r} 10 \\ \times 8 \\ \hline \end{array}$

29. $\begin{array}{r} 9 \\ \times 2 \\ \hline \end{array}$

30. $\begin{array}{r} 4 \\ \times 10 \\ \hline \end{array}$

31. $\begin{array}{r} 5 \\ \times 9 \\ \hline \end{array}$

32. $\begin{array}{r} 5 \\ \times 0 \\ \hline \end{array}$

33. $\begin{array}{r} 5 \\ \times 7 \\ \hline \end{array}$

H.O.T. **Algebra** **Use the pictures to find the unknown numbers.**

34. 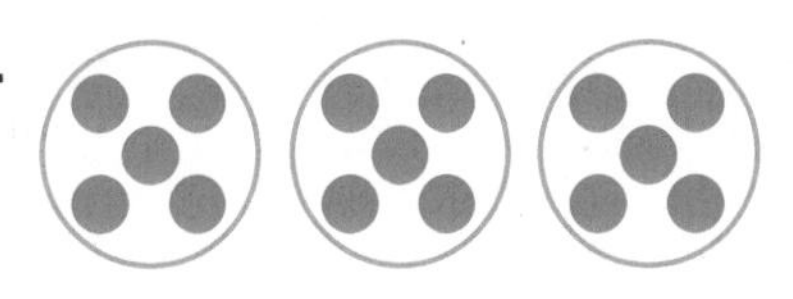

$3 \times$ ____ $=$ ____

35. 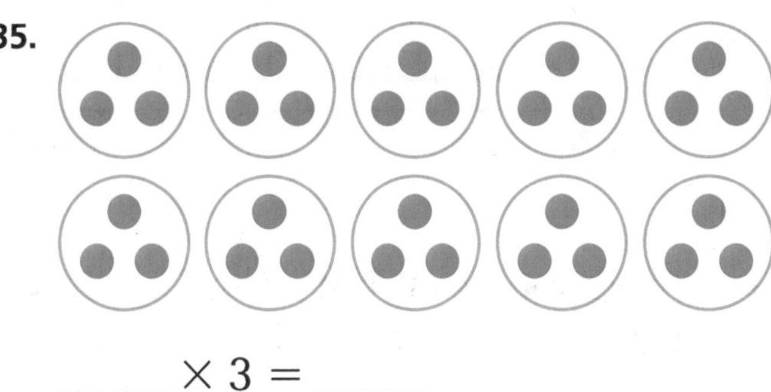

____ $\times 3 =$ ____

Complete the bar model to solve.

36. Marcel played 5 songs on the banjo. If each song lasted 8 minutes, how long did he play?

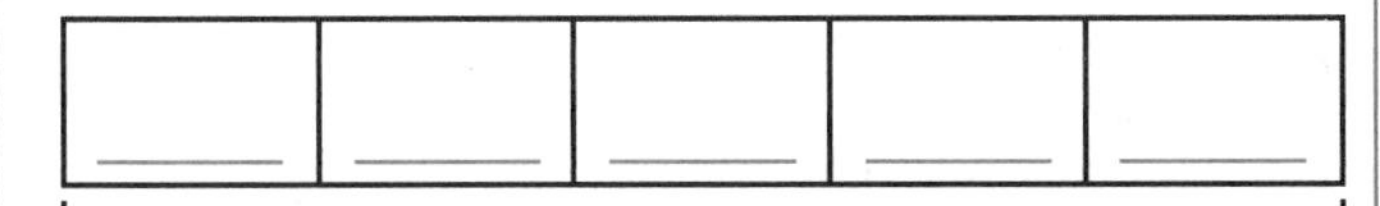

____ minutes

____

37. There are 6 banjo players. If each player needs 10 sheets of music, how many sheets of music are needed?

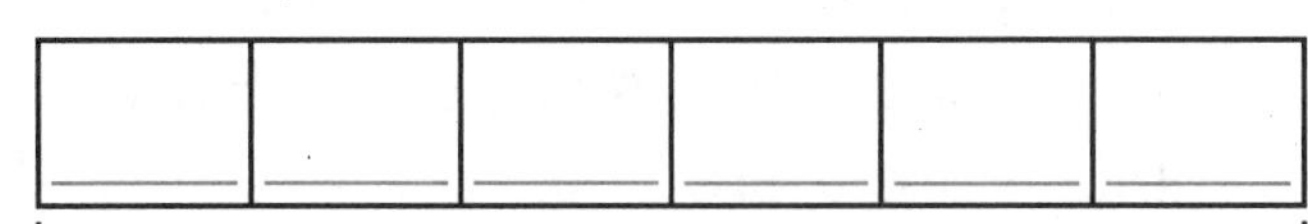

____ sheets

____

Problem Solving REAL WORLD

Use the table for 38–39.

Stringed Instruments

Instrument	Strings
Guitar	6
Banjo	5
Mandolin	8
Violin	4

38. John and his dad own 7 banjos. They want to replace the strings on all of them. How many strings should they buy? Write a multiplication sentence to solve.

____

39. H.O.T. Mr. Lemke has 5 guitars, 4 banjos, and 2 mandolins. What is the total number of strings on Mr. Lemke's instruments?

____

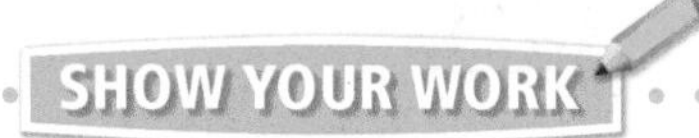

40. Mrs. Thompson has 4 shelves with 5 violins on each shelf. How many violins does Mrs. Thompson have? Complete the bar model to solve.

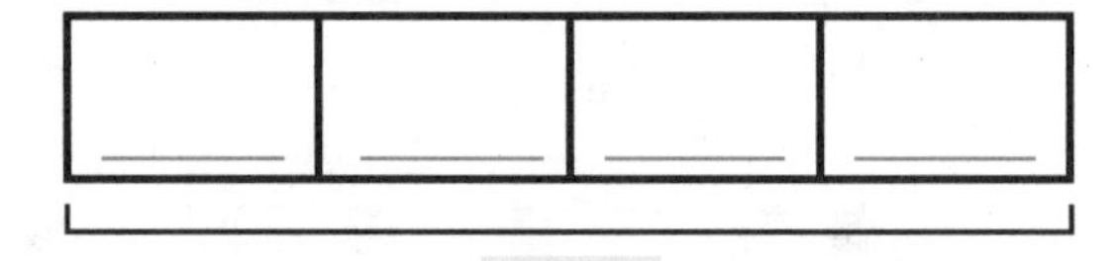

violins

41. Write Math **What's the Error?** Mr. James has 3 banjos. Mr. Lewis has 5 times the number of banjos Mr. James has. Riley says Mr. Lewis has 12 banjos. Describe her error.

____

____

42. **Test Prep** There are 5 guitars on each of 5 shelves at a music store. How many guitars are there?

Ⓐ 10 Ⓒ 20

Ⓑ 15 Ⓓ 25

Name ____________________

Multiply with 3 and 6

Essential Question What are some ways to multiply with 3 and 6?

UNLOCK the Problem REAL WORLD

Sabrina is making triangles with toothpicks. She uses 3 toothpicks for each triangle. She makes 4 triangles. How many toothpicks does Sabrina use?

- Why does Sabrina need 3 toothpicks for each triangle?

____

____

 **Draw a picture.**

STEP 1

Complete the 4 triangles.

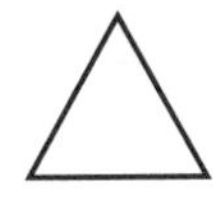   

STEP 2

Skip count by the number of sides. _____, _____, _____, _____

How many triangles are there in all? _____

How many toothpicks are in each triangle? _____

How many toothpicks are there in all?

$4 \times$ _____ $=$ _____

4 triangles have _____ toothpicks.

So, Sabrina uses _____ toothpicks.

MATHEMATICAL PRACTICES

Math Talk How can you use what you know about the number of toothpicks needed for 4 triangles to find the number of toothpicks needed for 8 triangles? **Explain.**

Try This! Find the number of toothpicks needed for 6 triangles.

Draw a quick picture to help you. How did you find the answer?

____

____

____

Jessica is using craft sticks to make 6 octagons. How many craft sticks will she use in all?

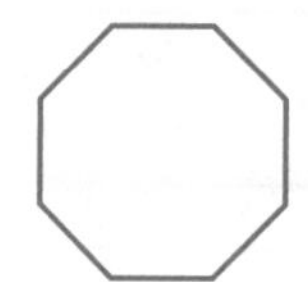

▲ An octagon has 8 sides.

One Way Use 5s facts and addition.

To multiply a factor by 6, multiply the factor by 5, and then add the factor.

$6 \times 7 = 5 \times 7 + 7 = 42$

$6 \times 6 = 5 \times 6 +$ ______ $=$ ______

$6 \times 8 = 5 \times$ ______ $+$ ______ $=$ ______

$6 \times 9 =$ ______ $\times$ ______ $+$ ______ $=$ ______

So, Jessica will use ______ craft sticks.

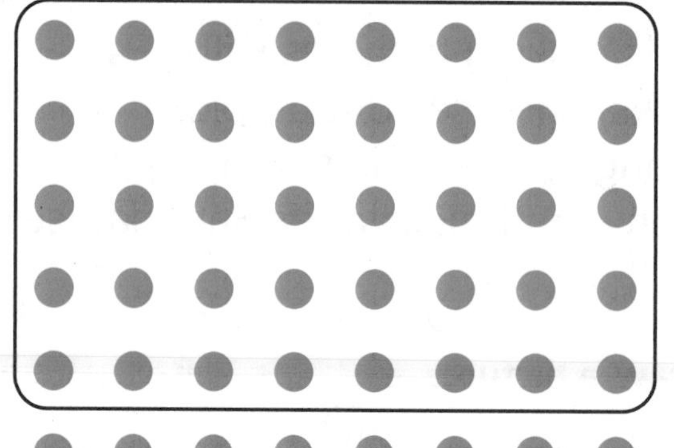

5×8

$+ 8$

Other Ways

A Use doubles.

When at least one factor is an even number, you can use doubles.

First multiply with half of an even number.

After you multiply, double the product.

$6 \times 8 = \square$

$3 \times 8 =$ ______

______ $+ 24 =$ ______

$6 \times 8 =$ ______

B Use a multiplication table.

Find the product 6×8 where row 6 and column 8 meet.

$6 \times 8 =$ ______

- Shade the row for 3 in the table. Then, compare the rows for 3 and 6. What do you notice about their products?

____

____

×	0	1	2	3	4	5	6	7	8	9	10
0	0	0	0	0	0	0	0	0	0	0	0
1	0	1	2	3	4	5	6	7	8	9	10
2	0	2	4	6	8	10	12	14	16	18	20
3	0	3	6	9	12	15	18	21	24	27	30
4	0	4	8	12	16	20	24	28	32	36	40
5	0	5	10	15	20	25	30	35	40	45	50
6	0	6	12	18	24	30	36	42	48	54	60
7	0	7	14	21	28	35	42	49	56	63	70
8	0	8	16	24	32	40	48	56	64	72	80
9	0	9	18	27	36	45	54	63	72	81	90
10	0	10	20	30	40	50	60	70	80	90	100

Name ______________________

Share and Show

1. Use 5s facts and addition to find $6 \times 4 = \blacksquare$.

$6 \times 4 =$ ____ $\times$ ____ $+$ ____ $=$ ____

$6 \times 4 =$ ____

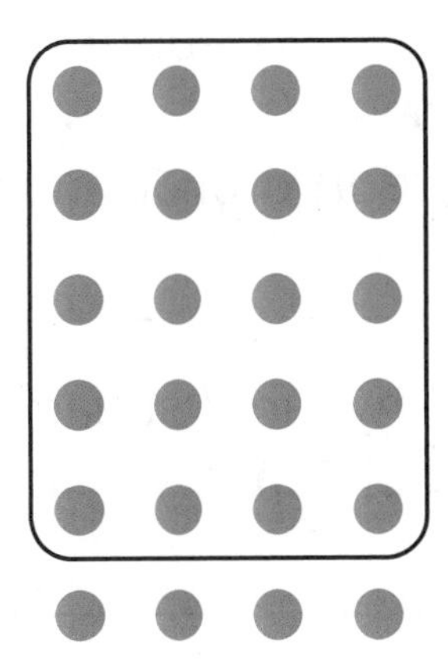

MATHEMATICAL PRACTICES

Math Talk Explain how you would use 5s facts and addition to find 6×3.

Find the product.

2. $6 \times 1 =$ ____
3. ____ $= 3 \times 7$
4. ____ $= 6 \times 5$
5. $3 \times 9 =$ ____

On Your Own

Find the product.

6. $2 \times 3 =$ ____
7. ____ $= 3 \times 6$
8. ____ $= 3 \times 0$
9. $1 \times 6 =$ ____

10. 3×6
11. 8×3
12. 6×7
13. 3×3
14. 10×6

H.O.T. **Algebra** Complete the table.

	Multiply by 3.	
	Factor	Product
15.	4	
16.		18

	Multiply by 6.	
	Factor	Product
17.	5	
18.	7	

19.	Multiply by ▢.	
	Factor	Product
	3	15
20.	2	

Problem Solving REAL WORLD

Use the table for 21–22.

Quilt Pieces	
Shape	**Number in One Quilt Piece**
Square	6
Triangle	4
Circle	4

21. The table tells about quilt pieces Jenna has made. How many squares are there in 6 of Jenna's quilt pieces?

____

22. How many more squares than triangles are in 3 of Jenna's quilt pieces?

____

SHOW YOUR WORK

23. Write Math **What if** you used craft sticks to make shapes? If you used one craft stick for each side, would you use more craft sticks for 5 squares or 6 triangles? **Explain.**

____

____

____

____

24. H.O.T. Draw a picture and use words to explain the Commutative Property of Multiplication with the factors 3 and 4.

____

____

____

25. **Test Prep** There are 8 crayons in each of 6 packages. How many crayons are there?

Ⓐ 2 Ⓑ 14 Ⓒ 24 Ⓓ 48

FOR MORE PRACTICE:
Standards Practice Book, pp. P71–P72

Name ______________________________

Distributive Property

Essential Question How can you use the Distributive Property to find products?

UNLOCK the Problem REAL WORLD

Mark bought 6 new fish for his aquarium. He paid $7 for each fish. How much did he spend in all?

Find $6 \times \$7$.

You can use the Distributive Property to solve the problem.

The **Distributive Property** states that multiplying a sum by a number is the same as multiplying each addend by the number and then adding the products.

- Describe the groups in this problem.

____

- Circle the numbers you will use to solve the problem.

Remember

sum—the answer to an addition problem

addends—the numbers being added

Activity **Materials** ■ square tiles

Make an array with tiles to show 6 rows of 7.

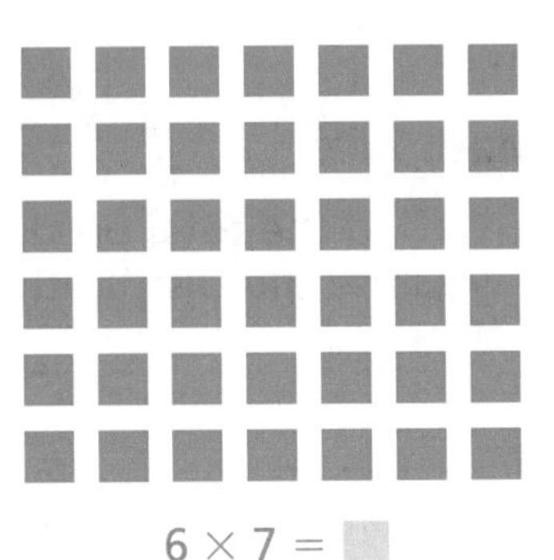

$6 \times 7 =$ ■

Break apart the array to make two smaller arrays for facts you know.

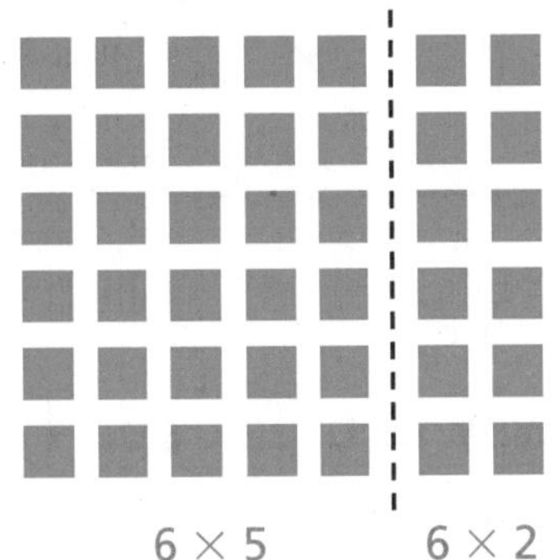

6×5 6×2

$6 \times 7 =$ ■

$6 \times 7 = 6 \times (5 + 2)$ **Think:** $7 = 5 + 2$

$6 \times 7 = (6 \times 5) + (6 \times 2)$ Multiply each addend by 6.

$6 \times 7 =$ _____ + _____ Add the products.

$6 \times 7 =$ _____

So, Mark spent $_____ for his new fish.

Math Talk MATHEMATICAL PRACTICES What other ways could you break apart the 6×7 array?

Try This!

Suppose Mark bought 9 fish for $6 each.

You can break apart a 9×6 array into two smaller arrays for facts you know. One way is to think of 9 as $5 + 4$. Draw a line to show this way. Then find the product.

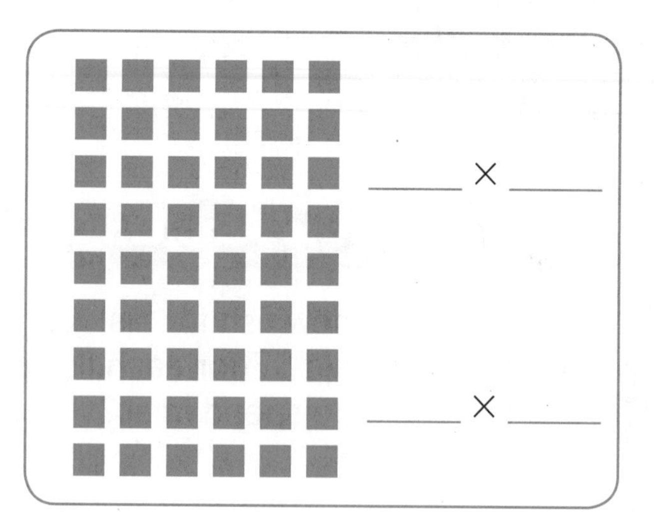

$9 \times 6 = (\_\_\_\_ \times \_\_\_\_) + (\_\_\_\_ \times \_\_\_\_)$

$9 \times 6 = \_\_\_\_ + \_\_\_\_$

So, Mark spent $______ for 9 fish.

Share and Show

1. Draw a line to show how you could break apart this 6×8 array into two smaller arrays for facts you know.

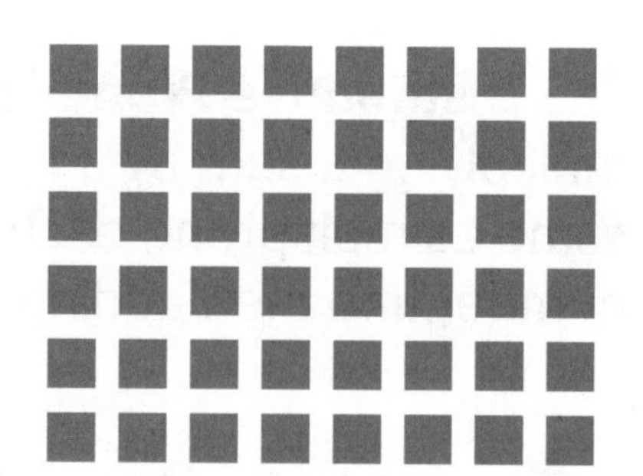

- What numbers do you multiply? ______ and ______

  ______ and ______

- What numbers do you add? ______ + ______

$6 \times 8 = 6 \times (\_\_\_\_ + \_\_\_\_)$

$6 \times 8 = (\_\_\_\_ \times \_\_\_\_) + (\_\_\_\_ \times \_\_\_\_)$

$6 \times 8 = \_\_\_\_ + \_\_\_\_$

$6 \times 8 = \_\_\_\_$

MATHEMATICAL PRACTICES

Math Talk Why do you have to add to find the total product when you use the Distributive Property?

Write one way to break apart the array. Then find the product.

2.

____

____

3.

____

____

Name ___

On Your Own

Write one way to break apart the array. Then find the product.

4. 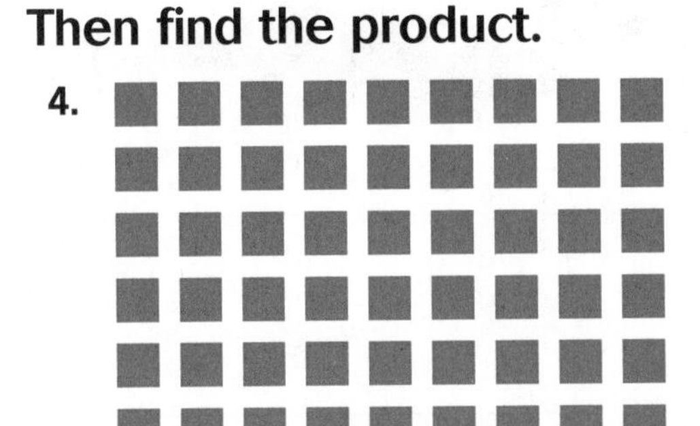

___

___

5. 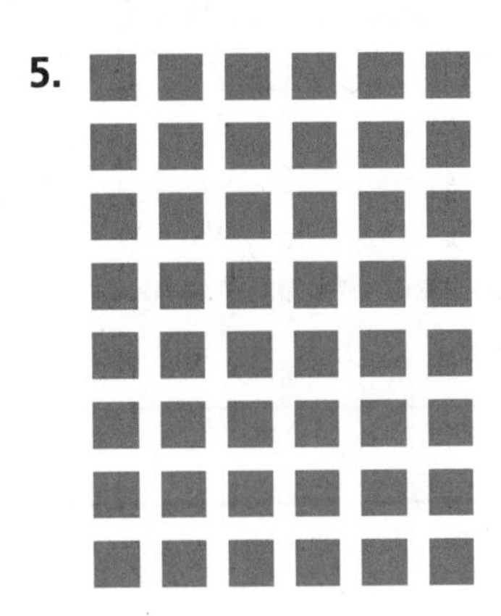

___

___

6. Shade tiles to make an array that shows a fact with 7, 8, or 9 as a factor. Write the fact. **Explain** how you found the product.

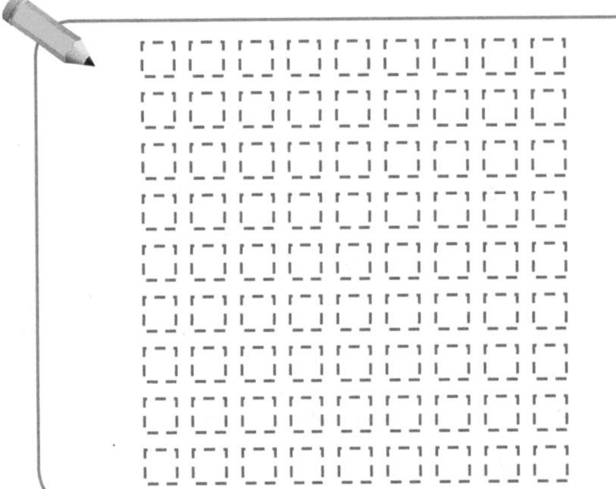

___

___

___

___

7. H.O.T. **Sense or Nonsense?** Robin says, "I can find 8×7 by multiplying 3×7 and doubling it." Does her statement make sense? **Explain** your answer.

___

___

___

___

8. **Test Prep** Which number sentence below is an example of the Distributive Property?

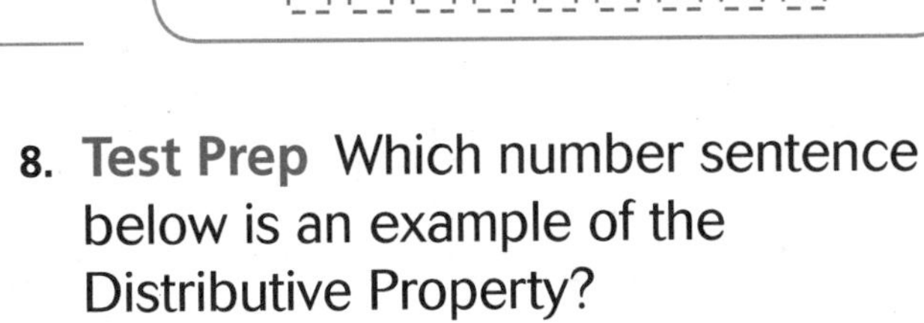

Ⓐ $6 \times 8 = 40 + 8$

Ⓑ $(8 \times 2) \times 3 = 8 \times (2 \times 3)$

Ⓒ $6 \times 8 = 8 \times 6$

Ⓓ $6 \times 8 = (6 \times 2) + (6 \times 6)$

H.O.T. **What's the Error?**

9. Brandon needs 8 boxes of spinners for his fishing club. The cost of each box is \$9. How much will Brandon pay.

$8 \times \$9 = ■$

Look at how Brandon solved the problem. Find and describe his error.

$8 \times 9 = (4 \times 9) + (5 \times 9)$

$8 \times 9 = 36 + 45$

$8 \times 9 = 81$

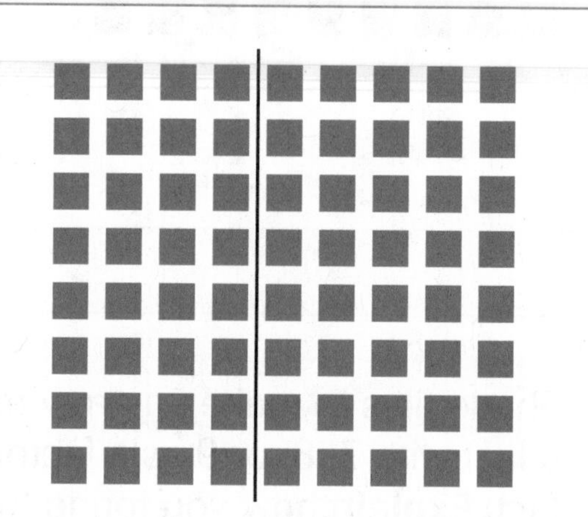

____

____

Use the array to help solve the problem and correct his error.

$8 \times 9 = (4 + 4) \times 9$

$8 \times 9 = (\_\_\_\_ \times \_\_\_\_) + (\_\_\_\_ \times \_\_\_\_)$

$8 \times 9 = \_\_\_\_ + \_\_\_\_$

$8 \times 9 = \_\_\_\_$

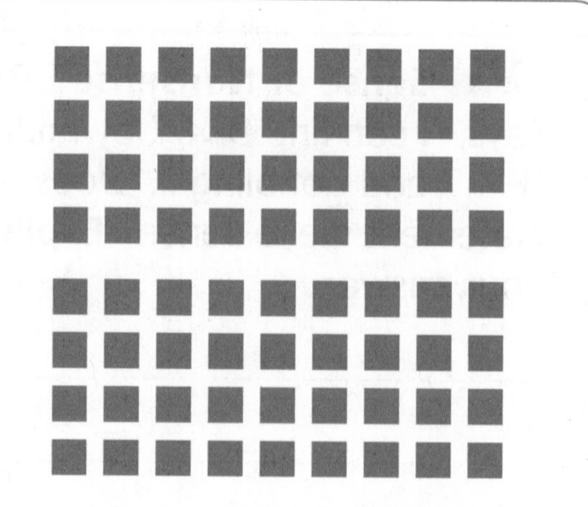

So, Brandon will pay \$ ______ for the spinners.

 FOR MORE PRACTICE:
Standards Practice Book, pp. P73–P74

Name ______________________________

Multiply with 7

Essential Question What strategies can you use to multiply with 7?

UNLOCK the Problem REAL WORLD

Jason's family has a new puppy. Jason takes a turn walking the puppy once a day. How many times will Jason walk the puppy in 4 weeks?

Find 4×7.

- How often does Jason walk the puppy?

____

- How many days are in 1 week?

____

One Way Use the Commutative Property of Multiplication.

If you know 7×4, you can use that fact to find 4×7. You can change the order of the factors and the product is the same.

$7 \times 4 =$ ______, so $4 \times 7 =$ ______.

So, Jason will walk the puppy ______ times in 4 weeks.

Other Ways

A Use the Distributive Property.

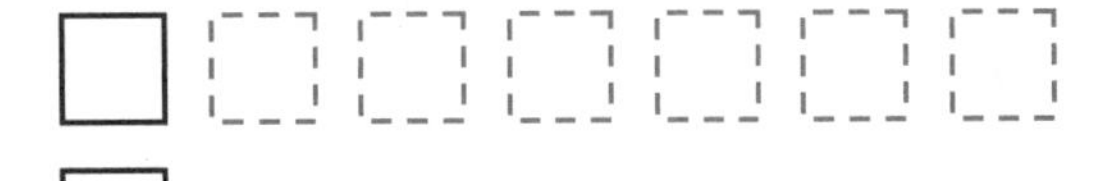

STEP 1 Complete the array to show 4 rows of 7.

STEP 2 Draw a line to break the array into two smaller arrays for facts you know.

STEP 3 Multiply the facts for the smaller arrays. Add the products.

$4 \times$ ______ $=$ ______ $4 \times$ ______ $=$ ______

______ $+$ ______ $=$ ______

So, $4 \times 7 =$ ______.

MATHEMATICAL PRACTICES

Math Talk Explain two other ways you can break apart the 4×7 array.

B **Use a fact you know.**

Multiply. $4 \times 7 = \blacksquare$

- Start with a fact you know. $2 \times 7 =$ _____
- Add a group of 7 for 3×7. $2 \times 7 + 7 =$ _____
- Then add 7 more for 4×7. $3 \times 7 + 7 =$ _____

So, $4 \times 7 =$ _____.

Share and Show

1. **Explain** how you could break apart an array to find 6×7. Draw an array to show your work.

____

____

MATHEMATICAL PRACTICES

Math Talk How can you use doubles to find 8×7?

Find the product.

2. $9 \times 7 =$ _____
3. _____ $= 5 \times 7$
4. _____ $= 7 \times 3$
5. $1 \times 7 =$ _____

On Your Own

Find the product.

6. _____ $= 7 \times 7$
7. $6 \times 7 =$ _____
8. _____ $= 7 \times 10$
9. _____ $= 7 \times 2$

10. 7×3
11. 6×7
12. 9×7
13. 8×7
14. 1×7
15. 4×7

16. 10×4
17. 0×7
18. 2×7
19. 5×7
20. 6×9
21. 7×8

Problem Solving REAL WORLD

Use the table for 22–24.

Rusty's Care	
Food	3 cups a day
Water	4 cups a day
Bath	2 times a month

22. Lori has a dog named Rusty. How many baths will Rusty have in 7 months? ____________

23. How many more cups of water than food will Rusty get in 1 week?

____

24. H.O.T. Write Math Tim's dog, Midnight, eats 28 cups of food in a week. Midnight eats the same amount each day. In one day, how many more cups of food will Midnight eat than Rusty? **Explain.**

____

____

____

____

25. José walks his dog 10 miles every week. How many miles do they walk in 7 weeks? ____________

26. Dave takes Zoey, his dog, for a 3-mile walk twice a day. How many miles do they walk in one week?

____

27. **Test Prep** Sam walks his dog 5 miles a day. How many miles does he walk in one week?

Ⓐ 5 miles
Ⓑ 25 miles
Ⓒ 35 miles
Ⓓ 40 miles

SHOW YOUR WORK

Connect to Reading

Summarize

To help you stay healthy, you should eat a balanced diet and exercise every day.

The table shows the recommended daily servings for third graders. You should eat the right amounts of the food groups.

Suppose you want to share with your friends what you learned about healthy eating. How could you summarize what you learned?

When you ***summarize***, you restate the most important information in a shorter way to help you understand what you have read.

Recommended Daily Servings

Food Group	Servings
Whole Grains (bread, cereal)	6 ounces
Vegetables (beans, corn)	2 cups
Fruits (apples, oranges)	1 cup
Dairy Products (milk, cheese)	3 cups
Meat, Beans, Fish, Eggs, Nuts	5 ounces
8 ounces = 1 cup	

- To stay healthy, you should eat a balanced ________________ and ________________ every day.
- A third grader should eat 3 cups of ________________, such as milk and cheese, each day.
- A third grader should eat ________________ of vegetables and fruits each day.

  How many cups of vegetables and fruits should a third grader eat in 1 week? ________________

  Remember: 1 week = 7 days
- A third grader should eat ________________ of whole grains, such as bread and cereal, each day.

  How many ounces of whole grains should a third grader eat in 1 week? ________________

FOR MORE PRACTICE:
Standards Practice Book, pp. P75–P76

Name ________________________________

Mid-Chapter Checkpoint

▶ Vocabulary

Choose the best term from the box to complete the sentence.

Vocabulary
Commutative Property of Multiplication
Distributive Property
multiple

1. A ______________________ of 4 is any product that has 4 as one of its factors. (p. 137)

2. This is an example of the ______________________ Property.

$$3 \times 8 = (3 \times 6) + (3 \times 2)$$

This property states that multiplying a sum by a number is the same as multiplying each addend by the number and then adding the products. (p. 145)

▶ Concepts and Skills

Write one way to break apart the array. Then find the product.

3. 

____

____

4. 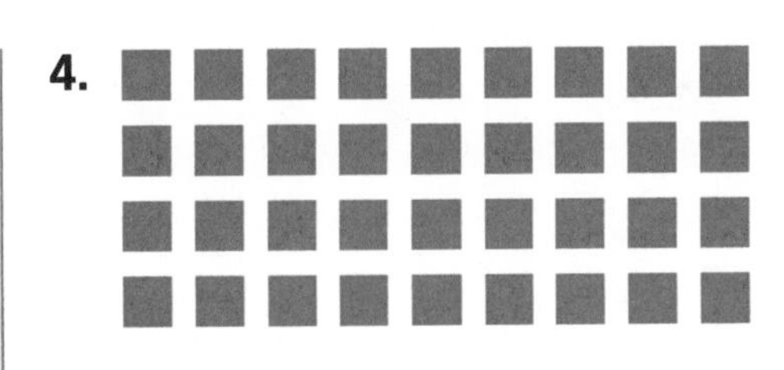

____

____

Find the product.

5. $3 \times 1 =$ ______

6. $5 \times 6 =$ ______

7. $7 \times 7 =$ ______

8. $2 \times 10 =$ ______

9. $\begin{array}{r} 2 \\ \times 1 \\ \hline \end{array}$

10. $\begin{array}{r} 6 \\ \times 6 \\ \hline \end{array}$

11. $\begin{array}{r} 8 \\ \times 7 \\ \hline \end{array}$

12. $\begin{array}{r} 6 \\ \times 0 \\ \hline \end{array}$

13. $\begin{array}{r} 3 \\ \times 8 \\ \hline \end{array}$

Fill in the bubble for the correct answer choice.

14. Lori saw 6 lightning bugs. They each had 6 legs. How many legs did the lightning bugs have in all?

Ⓐ 36

Ⓑ 24

Ⓒ 12

Ⓓ 6

15. Zach gave his dog 2 dog biscuits a day, for 5 days. How many biscuits did Zach give to his dog?

Ⓐ 3

Ⓑ 7

Ⓒ 10

Ⓓ 30

16. Annette buys 4 boxes of pencils. If there are 8 pencils in each box, how many pencils does she have?

Ⓐ 32

Ⓑ 24

Ⓒ 16

Ⓓ 12

17. Shelly can paint 4 pictures in a day. How many pictures can she paint in 7 days?

Ⓐ 3

Ⓑ 11

Ⓒ 28

Ⓓ 32

Name ______________________________

Associative Property of Multiplication

Essential Question How can you use the Associative Property of Multiplication to find products?

CONNECT You have learned the Associative Property of Addition. When the grouping of the addends is changed, the sum stays the same.

$$(2 + 3) + 4 = 2 + (3 + 4)$$

Math Idea

Always multiply the numbers inside the parentheses first.

The **Associative Property of Multiplication** states that when the grouping of the factors is changed, the product is the same. It is also called the Grouping Property of Multiplication.

UNLOCK the Problem REAL WORLD

Each car on the roller coaster has 2 rows of seats. Each row has 2 seats. There are 3 cars in each train. How many seats are on each train?

- Underline what you need to find.
- Describe the grouping of the seats.

____

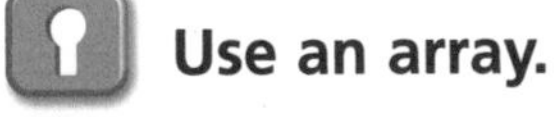

Use an array.

You can use an array to show $3 \times (2 \times 2)$.

$3 \times (2 \times 2) = \square$

$3 \times$ _____ $=$ _____

So, there are 3 cars with 4 seats in each car.

There are _____ seats on each roller coaster train.

You can change the grouping with parentheses and the product is the same.

$(3 \times 2) \times 2 = \square$

_____ $\times 2 =$ _____

MATHEMATICAL PRACTICES

Math Talk Explain why the products $3 \times (2 \times 2)$ and $(3 \times 2) \times 2$ are the same.

Example Use the Commutative and Associative Properties.

You can also change the order of the factors.
The product is the same.

$(4 \times 3) \times 2 = \square$

$4 \times (3 \times 2) = \square$ Associative Property

$4 \times$ ______ = ______

$4 \times (3 \times 2) = \square$

$4 \times (2 \times 3) = \square$ Commutative Property

$(4 \times 2) \times 3 = \square$ Associative Property

______ $\times 3 =$ ______

Share and Show

1. Find the product of 5, 2, and 3. Write another way to group the factors. Is the product the same? Why?

______

______

Write another way to group the factors. Then find the product.

2. $(2 \times 1) \times 7$

______

______

3. $3 \times (3 \times 4)$

______

______

4. $5 \times (2 \times 5)$

______

______

5. $3 \times (2 \times 6)$

______

______

6. $2 \times (2 \times 5)$

______

______

7. $(1 \times 3) \times 6$

______

______

MATHEMATICAL PRACTICES

Math Talk Choose one answer from Exercises 2–7. **Explain** why you multiplied those factors first.

Name ______________________________

On Your Own

Write another way to group the factors. Then find the product.

8. $(2 \times 3) \times 3$

9. $(8 \times 3) \times 2$

10. $2 \times (5 \times 5)$

11. $(3 \times 2) \times 4$

12. $(6 \times 1) \times 4$

13. $2 \times (2 \times 6)$

14. $2 \times (4 \times 2)$

15. $5 \times (2 \times 4)$

16. $9 \times (1 \times 2)$

Practice: Copy and Solve Use parentheses and multiplication properties. Then, find the product.

17. $6 \times 5 \times 2$

18. $2 \times 3 \times 5$

19. $3 \times 1 \times 6$

20. $2 \times 5 \times 6$

21. $2 \times 0 \times 8$

22. $1 \times 9 \times 4$

23. $2 \times 2 \times 2$

24. $4 \times 2 \times 2$

25. $2 \times 4 \times 5$

26. $2 \times 6 \times 1$

27. $2 \times 9 \times 3$

28. $2 \times 7 \times 2$

Algebra Find the unknown factor.

29. $7 \times (2 \times$ ______ $) = 56$

30. $30 = 6 \times (5 \times$ ______ $)$

31. ______ $\times (2 \times 2) = 32$

32. $42 = 7 \times (2 \times$ ______ $)$

33. $8 \times (5 \times$ ______ $) = 40$

34. $0 =$ ______ $\times (25 \times 1)$

35. $(2 \times 9) \times$ ______ $= 18$

36. $60 = (2 \times$ ______ $) \times 6$

37. $4 \times (3 \times$ ______ $) = 24$

Problem Solving REAL WORLD

Use the graph for 38–39.

38. Each car on the Steel Force train has 3 rows with 2 seats in each row. How many seats are on the train? Draw a quick picture.

____

39. H.O.T. A Kingda Ka train has 4 seats per car, but the last car has only 2 seats. How many seats are on one Kingda Ka train?

____

40. Write Math ▸ **Sense or Nonsense?** Each week, Kelly works 2 days for 4 hours each day and earns $5 an hour. Len works 5 days for 2 hours each day and earns $4 an hour. Kelly says they both earn the same amount. Does this statement make sense? **Explain.**

____

____

41. **Test Prep** Mark has 2 rows of 3 car models on each of his 3 shelves. How many models does he have?

Ⓐ 8 Ⓒ 11

Ⓑ 9 Ⓓ 18

SHOW YOUR WORK

FOR MORE PRACTICE:
Standards Practice Book, pp. P77–P78

ALGEBRA
Lesson 4.7

Name ______________________

Patterns on the Multiplication Table

Essential Question How can you use properties to explain patterns on the multiplication table?

UNLOCK the Problem REAL WORLD

You can use a multiplication table to explore number patterns.

×	0	1	2	3	4	5	6	7	8	9	10
0											
1											
2											
3											
4											
5											
6											
7											
8											
9											
10											

Activity 1

Materials ■ MathBoard

- Write the products for the green squares. What do you notice about the products?

____

____

____

Write the multiplication sentences for the products on your MathBoard. What do you notice about the factors?

____

____

- Will this be true in the yellow squares? **Explain** using a property you know.

____

Write the products for the yellow squares.

- Complete the columns for 1, 5, and 6. Look across each row and compare the products. What do you notice?

____

What property does this show?

____

MATHEMATICAL PRACTICES

Math Talk **Explain** how you can use these patterns to find other products.

Activity 2

Materials ■ yellow and blue crayons

- Shade the rows for 0, 2, 4, 6, 8, and 10 yellow.
- What pattern do you notice about each shaded row? ______________________

____

- Compare the rows for 2 and 4. What do you notice about the products? ______________

____

×	0	1	2	3	4	5	6	7	8	9	10
0	0	0	0	0	0	0	0	0	0	0	0
1	0	1	2	3	4	5	6	7	8	9	10
2	0	2	4	6	8	10	12	14	16	18	20
3	0	3	6	9	12	15	18	21	24	27	30
4	0	4	8	12	16	20	24	28	32	36	40
5	0	5	10	15	20	25	30	35	40	45	50
6	0	6	12	18	24	30	36	42	48	54	60
7	0	7	14	21	28	35	42	49	56	63	70
8	0	8	16	24	32	40	48	56	64	72	80
9	0	9	18	27	36	45	54	63	72	81	90
10	0	10	20	30	40	50	60	70	80	90	100

- Shade the columns for 1, 3, 5, 7, and 9 blue.
- What do you notice about the products for each shaded column?

____

- Compare the products for the green squares. What do you notice? What do you notice about the factors?

____

- What other patterns do you see?

____

____

Share and Show

1. Use the table to write the products for the row for 2.

_____, _____, _____, _____, _____,

_____, _____, _____, _____, _____, _____

Describe a pattern you see.

____

MATHEMATICAL PRACTICES

Math Talk What do you notice about the product of any number and 2?

Is the product even or odd? Write *even* or *odd*.

2. 5×8 _____
3. 6×3 _____
4. 3×5 _____
5. 4×4 _____

Name ______________________

Use the multiplication table. Describe a pattern you see.

6. in the column for 10

____

____

7. in the column for 8

____

____

On Your Own

Is the product even or odd? Write *even* or *odd*.

8. 4×8 ______ 9. 5×5 ______ 10. 7×4 ______ 11. 2×9 ______

Use the multiplication table. Describe a pattern you see.

12. in the row for 3

____

____

13. in the columns for 4 and 8

____

____

14. H.O.T. Correct the pattern. Rewrite your pattern.

6, 12, 18, 22, 30, 36 ______________________

Problem Solving

Complete the table. Then describe a pattern you see in the products.

15.

×	2	4	6	8	10
5					

____

____

16.

×	1	3	5	7	9
5					

____

____

17. Write Math **Explain** how patterns of the ones digits in the products relate to the factors in Exercises 15 and 16.

____

____

18. **Test Prep** Which of the following describes this pattern?

12, 16, 20, 24, 28

Ⓐ Multiply by 4. Ⓑ Add 4. Ⓒ Multiply by 5. Ⓓ Subtract 4.

Sense or Nonsense?

19. Whose statement makes sense? Whose statement is nonsense? **Explain** your reasoning.

The product of an odd number and an even number is even.

The product of two even numbers is even.

Gunter's Work

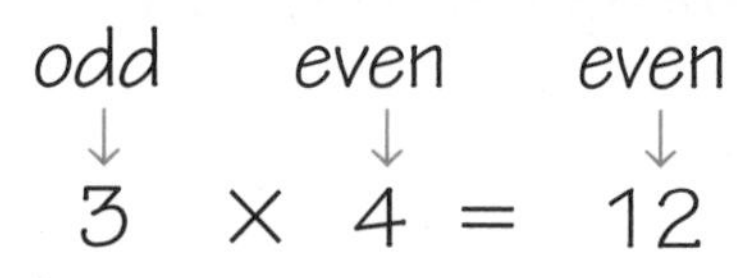

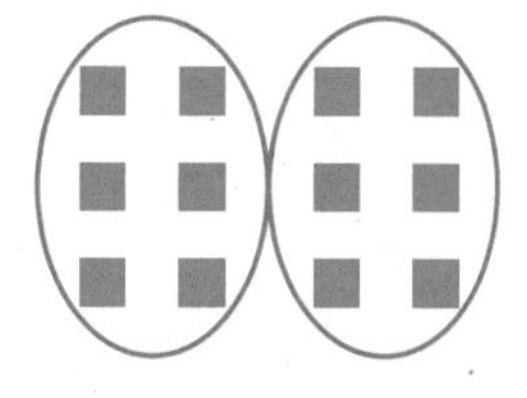

I can circle 2 equal groups of 6 with no tiles left over. So, the product is even.

Giselle's Work

even ↓ 2 × even ↓ 6 = even ↓ 12

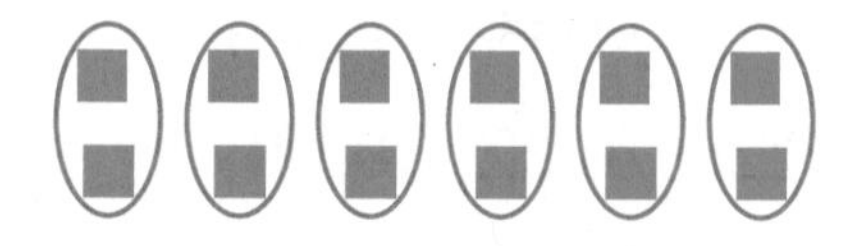

I can circle 6 pairs with no tiles left over. So, the product is even.

____

____

- **Write** a statement about the product of two odd numbers. Give an example to show why your statement is true.

FOR MORE PRACTICE:
Standards Practice Book, pp. P79–P80

Name ______________________________

Multiply with 8

Essential Question What strategies can you use to multiply with 8?

UNLOCK the Problem REAL WORLD

A scorpion has 8 legs. How many legs do 5 scorpions have?

- How many legs does one scorpion have?

____

- What are you asked to find?

____

____

Find 5×8.

One Way Use doubles.

$5 \times 8 = \square$

$4 + 4$

Think: The factor 8 is an even number. $4 + 4 = 8$

$5 \times 4 =$ _____

20 doubled is _____.

$5 \times 8 =$ _____

So, 5 scorpions have _____ legs.

Another Way Use a number line.

Use the number line to show 5 jumps of 8.

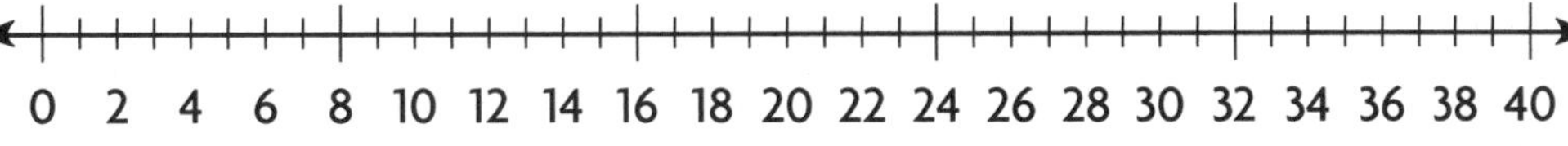

So, 5 jumps of 8 is _____.   _____ $\times$ _____ $=$ _____

ERROR Alert

Be sure to count the spaces between the tick marks, not the tick marks.

- **Describe** two different ways you can use doubles to find 6×8.

____

____

Example Use the Associative Property of Multiplication.

Scorpions have two eyes on the top of the head, and usually two to five pairs along the front corners of the head. If each scorpion has 6 eyes, how many eyes would 8 scorpions have?

$8 \times 6 = \blacksquare$

$8 \times 6 = (2 \times 4) \times 6$ Think: $8 = 2 \times 4$

$8 \times 6 = 2 \times (4 \times 6)$ Use the Associative Property.

$8 \times 6 = 2 \times$ _____ Multiply. 4×6

$8 \times 6 =$ _____ $+$ _____ Double the product.

$8 \times 6 =$ _____

MATHEMATICAL PRACTICES

Math Talk When you multiply with 8, will the product always be even? **Explain.**

Share and Show

1. **Explain** one way you can find 4×8.

____

____

Find the product.

2. $3 \times 8 =$ _____
3. _____ $= 8 \times 2$
4. _____ $= 7 \times 8$
5. $9 \times 8 =$ _____

MATHEMATICAL PRACTICES

Math Talk **Explain** why using doubles is a good strategy for finding 8×9.

On Your Own

Find the product.

6. _____ $= 6 \times 8$
7. $10 \times 8 =$ _____
8. _____ $= 8 \times 3$
9. $1 \times 8 =$ _____

10. $4 \times 8 =$ _____
11. $5 \times 8 =$ _____
12. $0 \times 8 =$ _____
13. $8 \times 8 =$ _____

14. 6×8
15. 8×2
16. 5×8
17. 3×8
18. 10×8
19. 7×8

Name ______________________

Algebra Complete the table.

	Multiply by 4.	
20.	8	
21.	5	
22.	10	

	Multiply by 8.	
23.	4	
24.	2	
25.	7	

26.	Multiply by ___ .	
	3	9
	6	18
27.	10	

Problem Solving REAL WORLD

Use the table for 28–31.

Average Yearly Rainfall in North American Deserts

Desert	Inches
Chihuahuan	8
Great Basin	9
Mojave	4
Sonoran	9

28. About how much rain falls in the Chihuahuan Desert in 6 years? **Explain** how you can use doubles to find the answer.

____

____

29. **Predict** In 2 years, about how many more inches of rain will fall in the Sonoran Desert than in the Chihuahuan Desert? **Explain.**

____

____

30. Write Math ▸ **Pose a Problem** Look back at Exercise 29. Write and show how to solve a similar problem by comparing two different deserts.

____

____

____

31. H.O.T. How can you find about how many inches of rain will fall in the Mohave Desert in 20 years?

____

32. **Test Prep** A black widow spider has 8 legs. How many legs do 7 black widow spiders have?

Ⓐ 1 Ⓑ 15 Ⓒ 48 Ⓓ 56

Connect to Science

There are 90 species of scorpions that live in the United States. Only 3 species of scorpions live in Arizona. They are the Arizona bark scorpion, the Desert hairy scorpion, and the Stripe-tailed scorpion.

Facts About Scorpions

Scorpions:

- are between 1 and 4 inches long
- mostly eat insects
- glow under ultraviolet light

They have:

- 8 legs for walking
- 2 long, claw-like pincers used to hold their food
- a curled tail held over their body with a stinger on the tip

▲ Scorpions glow under ultraviolet light.

33. How many species of scorpions do *not* live in Arizona?

____

34. Students saw 8 scorpions. What multiplication sentences can help you find how many pincers and legs the 8 scorpions had?

____

35. Three scorpions were in a display with ultraviolet light. Eight groups of 4 students saw the display. How many students saw the glowing scorpions?

____

FOR MORE PRACTICE:
Standards Practice Book, pp. P81–P82

Name ______________________________

Multiply with 9

Essential Question What strategies can you use to multiply with 9?

UNLOCK the Problem

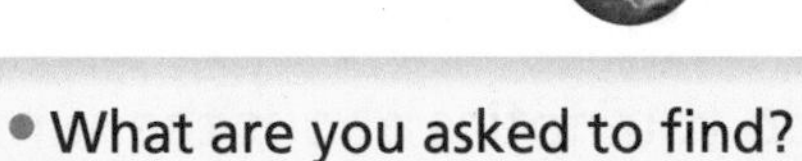

Lindsey's class is studying the solar system. Seven students are making models of the solar system. Each model has 9 spheres (eight for the planets and one for Pluto, a dwarf planet). How many spheres do the 7 students need for all the models?

Find 7×9.

- What are you asked to find?

____

____

- How many students are making models? ________

One Way Use the Distributive Property.

A With multiplication and addition

$7 \times 9 = \square$

Think: $9 = 3 + 6$ — $7 \times 9 = 7 \times (3 + 6)$

Multiply each addend by 7. — $7 \times 9 = (7 \times 3) + (7 \times 6)$

Add the products. — $7 \times 9 =$ ______ + ______

$7 \times 9 =$ ______

B With multiplication and subtraction

$7 \times 9 = \square$

Think: $9 = 10 - 1$ — $7 \times 9 = 7 \times (10 - 1)$

Multiply each number by 7. — $7 \times 9 = (7 \times 10) - (7 \times 1)$

Subtract the products. — $7 \times 9 =$ ______ − ______

$7 \times 9 =$ ______

So, 7 students need ______ spheres for all the models.

Another Way Use patterns of 9.

The table shows the 9s facts.

- What do you notice about the tens digit in the product?

  The tens digit is _____ less than the factor that is multiplied by 9.

- What do you notice about the sum of the digits in the product?

  The sum of the digits in the product is always _____.

So, to multiply 7×9, think the tens digit is _____ and the ones digit is _____. The product is _____.

Multiply by 9.	
Factors	**Product**
1×9	9
2×9	18
3×9	27
4×9	36
5×9	45
6×9	54
7×9	
8×9	
9×9	

Try This! Complete the table above.

Use the patterns to find 8×9 and 9×9.

Share and Show

1. What is the tens digit in the product 3×9? _____

   Think: What number is 1 less than 3?

MATHEMATICAL PRACTICES

Math Talk Explain how you know the ones digit in the product 3×9.

Find the product.

2. $9 \times 8 =$ _____
3. _____ $= 2 \times 9$
4. _____ $= 6 \times 9$
5. $9 \times 1 =$ _____

On Your Own

Find the product.

6. $4 \times 9 =$ _____
7. $5 \times 9 =$ _____
8. $10 \times 9 =$ _____
9. $1 \times 9 =$ _____
10. $9 \times 2 =$ _____
11. $9 \times 9 =$ _____
12. $8 \times 9 =$ _____
13. $7 \times 9 =$ _____
14. 9×5
15. 9×3
16. 6×9
17. 7×9
18. 4×9

Name ____________________

Algebra **Compare. Write $<$, $>$, or $=$.**

19. $2 \times 9 \bigcirc 3 \times 6$

20. $5 \times 9 \bigcirc 6 \times 7$

21. $1 \times 9 \bigcirc 3 \times 3$

22. $9 \times 4 \bigcirc 7 \times 5$

23. $9 \times 0 \bigcirc 2 \times 3$

24. $5 \times 8 \bigcirc 3 \times 9$

Problem Solving REAL WORLD

Use the table for 25–28.

Moons	
Planet	Number of Moons
Earth	1
Mars	2
Jupiter	63
Saturn	47
Uranus	27
Neptune	13

25. The number of moons for one of the planets can be found by multiplying 7×9. Which planet is it?

____

26. H.O.T. Uranus has 27 moons. What multiplication fact with 9 can be used to find the number of moons Uranus has? **Describe** how you can find the fact.

____

____

____

____

27. H.O.T. This planet has 9 times the number of moons that Mars and Earth have together. Which planet is it? **Explain** your answer.

____

____

____

28. Write Math **What's the Question?** Nine students made models of Mars and its moons. The answer is 18. What's the question?

____

UNLOCK the Problem REAL WORLD

29. The school library has 97 books about space. John and 3 of his friends each check out 9 books. How many space books are still in the school library?

Ⓐ 27   Ⓑ 36   Ⓒ 61   Ⓓ 70

a. What do you need to find? ____________________

____

b. Describe one way you can find the answer. ____________________

____

c. Show the steps you used to solve the problem.

d. Complete the sentences.

The library has _____ space books.

Multiply _____ × _____ to find how many books John and his 3 friends check out in all.

After you find the number of books they check out,

____________________ to find the number of books still in the library.

So, there are _____ space books still in the library.

e. Fill in the bubble for the answer choice above.

30. Mark read 5 pages of a space book every day for a week. How many pages did he read?

Ⓐ 5   Ⓒ 35
Ⓑ 12   Ⓓ 40

31. Joel has 5 shelves of model airplanes. He has 9 airplanes on each shelf. How many model airplanes does he have?

Ⓐ 45   Ⓒ 9
Ⓑ 36   Ⓓ 5

FOR MORE PRACTICE:
Standards Practice Book, pp. P83–P84

Name ________________________________

Problem Solving • Multiplication

Essential Question How can you use the strategy *make a table* to solve multiplication problems?

UNLOCK the Problem REAL WORLD

Scott has a stamp album. Some pages have 1 stamp on them, and other pages have 2 stamps on them. If Scott has 18 stamps, show how many different ways he could put them in the album. Use the graphic organizer below to solve the problem.

Read the Problem

What do I need to find?

____

____

____

What information do I need to use?

Scott has _____ stamps. Some of the pages have _____ stamp on them, and the other pages have _____ stamps.

How will I use the information?

I will make a _______ showing all the different ways of arranging the stamps in the album.

Solve the Problem

Make a table to show the number of pages with 1 stamp and with 2 stamps. Each row must equal

_____, the total number of stamps.

Pages with 2 Stamps	Pages with 1 Stamp	Total Stamps
8	2	18
7	4	18
6	6	18
5		18
	10	18
3	12	
2		

So, there are _____ different ways.

1. What number patterns do you see in the table?

____

____

Try Another Problem

What if Scott bought 3 more stamps and now has 21 stamps? Some album pages have 1 stamp and some pages have 2 stamps. Show how many different ways he could put the odd number of stamps in the album.

Read the Problem	Solve the Problem
What do I need to find?	
What information do I need to use?	
How will I use the information?	So, there are _____ different ways.

2. What patterns do you see in this table? ______________________

3. How are these patterns different from the patterns in the table on page 171? ______________________

Name ________________________________

Share and Show 

1. Aaron's mother is making lemonade. For each pitcher, she uses 1 cup of lemon juice, 1 cup of sugar, and 6 cups of water. What is the total number of cups of ingredients she will use to make 5 pitchers of lemonade?

First, make a table to show the number of cups of lemon juice, sugar, and water that are in 1 pitcher of lemonade.

Next, multiply to find the number of cups of water needed for each pitcher of lemonade.

Last, use the table to solve the problem.

Think: For every pitcher, the number of cups of water increases by 6.

Number of Pitchers	1	2	3		5
Cups of Lemon Juice	1		3		
Cups of Sugar	1	2			
Cups of Water	6	12		24	
Total Number of Cups of Ingredients	8				

So, in 5 pitchers of lemonade, there are _____ cups of lemon juice, _____ cups of sugar, and _____ cups of water.

This makes a total of _____ cups of ingredients.

2. **What if** it takes 4 lemons to make 1 cup of lemon juice? How many lemons would it take to make 5 pitchers? **Explain** how you can use the table to help you find the answer.

____

____

3. What pattern do you see in the total number of cups of ingredients?

____

____

On Your Own

Choose a STRATEGY

Act It Out
Draw a Diagram
Find a Pattern
Make a Table

4. Julie saw 3 eagles each day she went bird-watching. How many eagles did Julie see in 6 days?

______

5. H.O.T. Greg has a dollar bill, quarters, and dimes. How many ways can he make $1.75?

Name the ways. ______

______

______

SHOW YOUR WORK

6. Write Math Cammi needs 36 postcards. She buys 4 packages of 10 postcards. How many postcards will Cammi have left over? **Explain.**

______

______

______

7. Phillip has 8 books on 3 bookshelves. How many books does Phillip have? Complete the bar model.

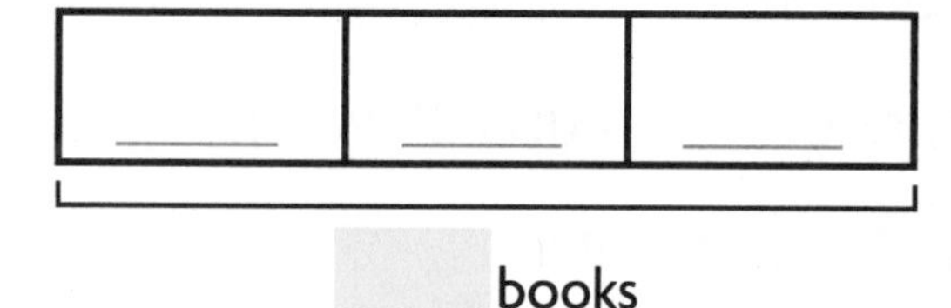

books

______

8. **Test Prep** Each year for 8 years, the Williams family visited 6 different states on vacation. How many more states do they need to visit before they have seen all 50 states?

Ⓐ 48 Ⓒ 24

Ⓑ 36 Ⓓ 2

FOR MORE PRACTICE:
Standards Practice Book, pp. P85–P86

Name ______________________________

Chapter Review/Test

▶ Vocabulary

Choose the best term from the box to complete the sentence.

Vocabulary
Associative Property of Multiplication
Distributive Property
multiple

1. The ______________ Property of Multiplication states that when the grouping of factors is changed, the product is the same. (p. 155)

2. A __________ of 9 is any product that has 9 as one of its factors. (p. 137)

▶ Concepts and Skills

Use arrays to show each property.

3. Distributive Property

   Break apart the array to show $8 \times 6 = (4 \times 6) + (4 \times 6)$.

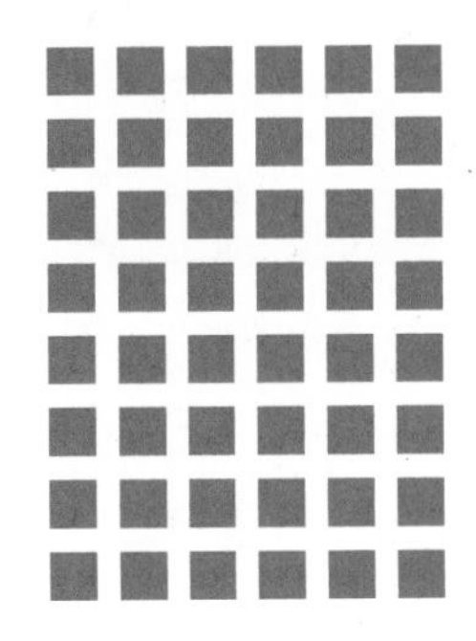

4. Associative Property of Multiplication

   Circle groups to show $3 \times (2 \times 3)$.

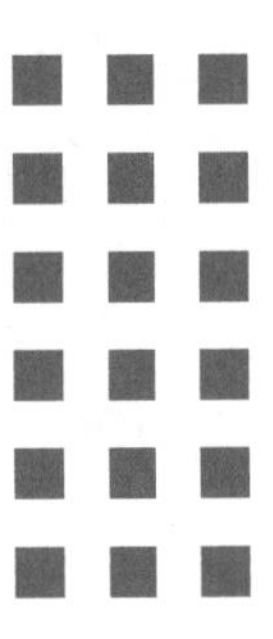

Find the product.

5. 6×8

6. 10×4

7. 5×9

8. 6×3

9. 5×8

10. 8×8

11. $9 \times 0 =$ _____

12. $7 \times 8 =$ _____

13. $4 \times 8 =$ _____

14. $2 \times 6 =$ _____

15. $9 \times 9 =$ _____

16. $3 \times 4 =$ _____

17. $5 \times 10 =$ _____

18. $7 \times 3 =$ _____

GO Online Assessment Options Chapter Test

Fill in the bubble for the correct answer choice.

19. Which number sentence is an example of the Distributive Property?

Ⓐ $5 \times 9 = (5 \times 5) + (5 \times 4)$

Ⓑ $(3 \times 2) \times 5 = 3 \times (2 \times 5)$

Ⓒ $5 \times 9 = 9 \times 5$

Ⓓ $5 \times 9 = 45$

20. Adel needs 5 pieces of ribbon, each 8 centimeters long. How much ribbon does she need altogether?

Ⓐ 13 centimeters Ⓒ 35 centimeters

Ⓑ 32 centimeters Ⓓ 40 centimeters

21. When Chloe finds the multiplication facts for 2, which digit will NOT be in the ones place of the products?

Ⓐ 8 Ⓒ 4

Ⓑ 6 Ⓓ 3

22. Vicky went to the store and bought 3 pairs of shorts. They each cost \$8. How much did she spend?

Ⓐ \$3 Ⓑ \$8 Ⓒ \$21 Ⓓ \$24

23. A honeybee is an insect. It has 6 legs. How many more legs do 7 honeybees have than 5 honeybees?

Ⓐ 5 Ⓒ 12

Ⓑ 6 Ⓓ 30

24. Jody has bags of shells. Each bag has 6 shells. She gives 3 bags to each of 2 friends. How many shells did Jody give away?

Ⓐ 36 Ⓑ 18 Ⓒ 12 Ⓓ 6

25. The camping club rents 4 rafts. Each raft can hold 6 people. How many people can 4 rafts hold?

Ⓐ 16 Ⓑ 20 Ⓒ 24 Ⓓ 30

Name ______________________________

26. James made an array with 6 rows of 8 blocks. Which number sentence shows one way to break apart his array to find the product?

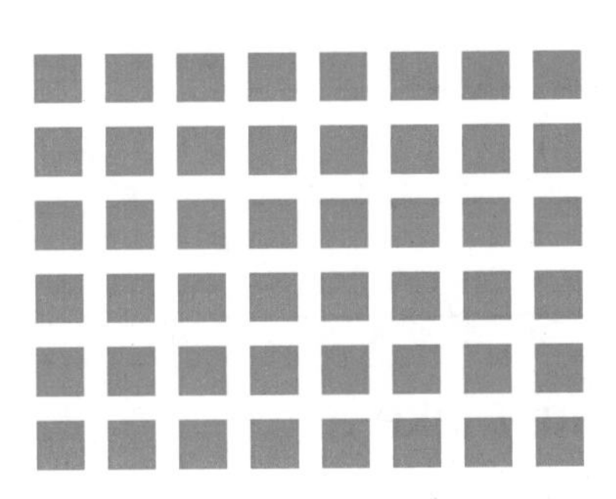

Ⓐ $6 \times 8 = (6 + 4) + (6 + 4)$

Ⓑ $6 \times 8 = (6 \times 4) + (6 \times 4)$

Ⓒ $6 \times 8 = (3 \times 4) + (3 \times 4)$

Ⓓ $6 \times 8 = (6 \times 8) + (6 \times 8)$

27. Zach and his dad baked some cupcakes for his class. They put 6 cupcakes on each of 8 plates. How many cupcakes did they put on the plates?

Ⓐ 12

Ⓑ 14

Ⓒ 24

Ⓓ 48

28. Sydnee's class is studying animals that hibernate, or go into a sleep-like state during the winter. A black bear's heartbeat slows to about 9 beats per minute during hibernation. About how many times will a black bear's heart beat in 5 minutes?

Ⓐ 45

Ⓑ 36

Ⓒ 18

Ⓓ 9

▶ Constructed Response

29. Terre was on summer vacation for 7 weeks. She spent 3 weeks at band camp and the rest of the time at home. How many days did she spend at home? **Explain.**

____

____

____

30. There are 5 pounds of apples in one bag. Make a table to show how many pounds of apples are in 6 bags. How do you know your answer is reasonable? **Explain.**

____

____

▶ Performance Task

31. Haylie and Justin went camping with their families for 7 days. They each took their own bottles of water.

A Haylie drank 6 bottles of water each day. She took home 6 bottles of water. How many bottles of water did Haylie take on the trip? **Explain** your answer.

____

____

B Justin had 5 packages with 8 bottles of water in each package. He drank some of the bottles before the trip. During the trip, he drank 5 bottles a day. He drank all the bottles before he went home. How many bottles of water did Justin drink before the trip? **Explain.**

____

____

____

Chapter 5

Use Multiplication Facts

Show What You Know

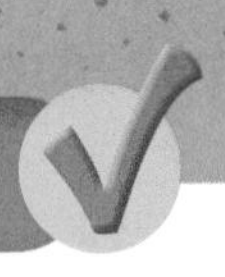

Check your understanding of important skills.

Name ______________________

▶Add Tens Write how many tens. Then add.

1. $30 + 30 = \blacksquare$

_____ tens + _____ tens =

_____ tens

$30 + 30 =$ _____

2. $40 + 50 = \blacksquare$

_____ tens + _____ tens =

_____ tens

$40 + 50 =$ _____

▶Regroup Tens as Hundreds Write the missing numbers.

3. 35 tens = _____ hundreds _____ tens
4. 52 tens = _____ hundreds _____ tens
5. 97 tens = _____ hundreds _____ tens

▶Multiplication Facts Through 9 Find the product.

6. $3 \times 9 =$ _____
7. $4 \times 5 =$ _____
8. $7 \times 6 =$ _____
9. $8 \times 2 =$ _____

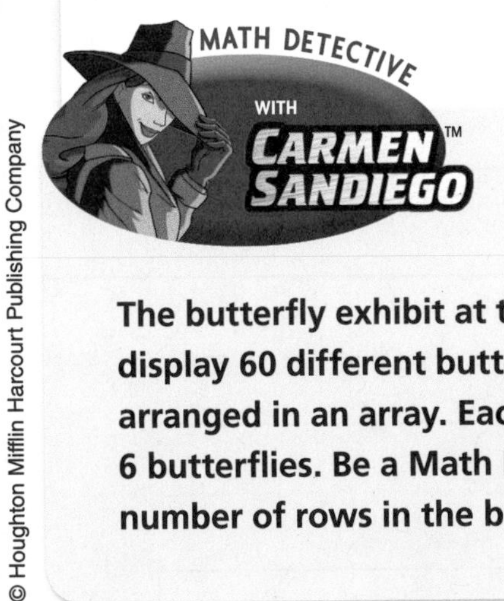

The butterfly exhibit at the museum will display 60 different butterfly species arranged in an array. Each row has 6 butterflies. Be a Math Detective to find the number of rows in the butterfly exhibit.

GO Online Assessment Options: Soar to Success Math

Vocabulary Builder

▶ Visualize It

Complete the tree map by using the words with a ✓.

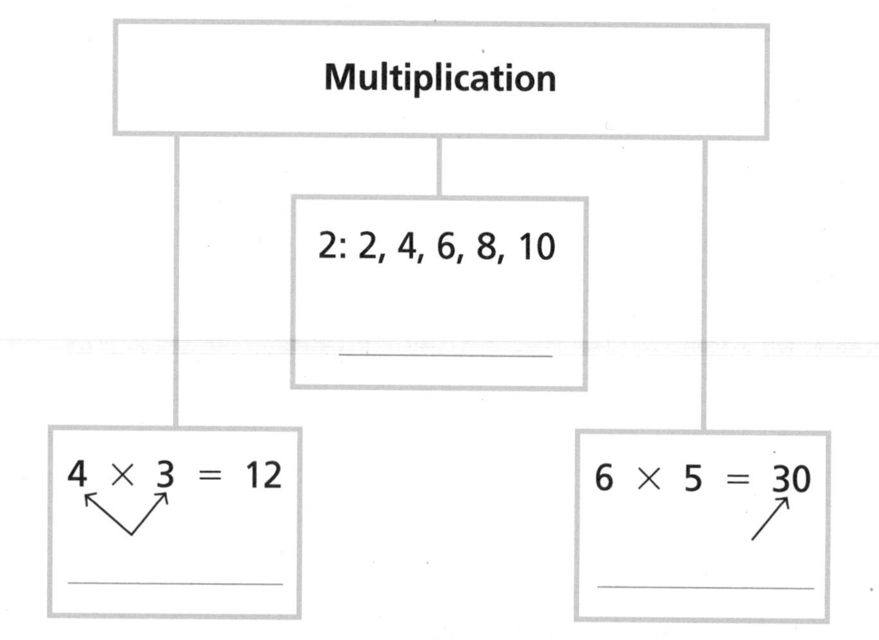

Review Words

array
Commutative Property of Multiplication
Distributive Property
✓ factors
hundreds
✓ multiples
ones
pattern
place value
✓ product
tens

Preview Word

equation

▶ Understand Vocabulary

Read the definition. Write the preview word or review word that matches it.

1. An ordered set of numbers or objects in which the order helps you predict what will come next. ______________
2. A set of objects arranged in rows and columns. ______________
3. A number sentence that uses the equal sign to show that two amounts are equal. ______________
4. The property that states that multiplying a sum by a number is the same as multiplying each addend by the number and then adding the products. ______________
5. The value of each digit in a number, based on the location of the digit. ______________

GO Online • eStudent Edition • Multimedia eGlossary

Name ________________________________

Describe Patterns

Essential Question What are some ways you can describe a pattern in a table?

UNLOCK the Problem REAL WORLD

The outdoor club is planning a camping trip. Each camper will need a flashlight. One flashlight uses 4 batteries. How many batteries are needed for 8 flashlights?

You can describe a pattern in a table.

Flashlights	1	2	3	4	5	6	7	8
Batteries	4	8	12	16	20	24	28	■

Think: Count by 1s.

Think: Count by 4s.

One Way Describe a pattern across the rows.

STEP 1 Look for a pattern to complete the table. As you look across the rows, you can see that the number of batteries increases by 4 for each flashlight.

So, for every flashlight add _______ batteries.

STEP 2 Use the pattern to find the number of batteries in 8 flashlights.

Add _______ to 28 batteries. $28 + 4 =$ _______

So, _______ batteries are needed for 8 flashlights.

ERROR Alert

Check that your pattern will work for all the numbers in the table.

Another Way Describe a pattern in the columns.

STEP 1 Look for a pattern by comparing the columns in the table. You can multiply the number of flashlights by 4 to find the number of batteries that are needed.

STEP 2 Use the pattern to find how many batteries are needed for 8 flashlights.

$8 \times 4 =$ _______

MATHEMATICAL PRACTICES

Math Talk Why is it important to know how many batteries are needed for 1 flashlight?

Try This! **Describe a pattern. Then complete the table.**

The campers need 5 packs of batteries. If there are 8 batteries in each pack, how many batteries will be in 5 packs?

Packs of Batteries	Number of Batteries
1	8
2	16
3	
4	32
5	

Use addition.

Describe a pattern.

Add ______ batteries for each pack.

Use multiplication.

Describe a pattern.

Multiply the number of packs of batteries by ______.

So, there will be ______ batteries in 5 packs.

Share and Show 

1. How can you describe a pattern to find the cost of 4 packs of batteries?

____

Packs of Batteries	1	2	3	4
Cost	$3	$6	$9	

Describe a pattern for the table. Then complete the table.

2.

Tents	Lanterns
2	4
3	6
4	8
5	10
6	
7	

____

3.

Adults	1	2	3	4	5
Campers	6	12	18		

____

MATHEMATICAL PRACTICES

Math Talk Explain how you use your description for a pattern to complete a table.

Name ______________________________

On Your Own

Describe a pattern for the table. Then complete the table.

4.

Hours	1	2	3	4	5
Miles Hiked	2	4	6		

____

5.

Cabins	3	4	5	6	7
Campers	27	36	45		

____

6.

Cabins	Beds
1	5
2	10
3	
4	20
5	
6	

____

7.

Adults	Students
2	12
3	18
4	
5	30
6	
7	

____

8.

Canoes	4	5	6	7	8
Campers	12	15	18		

____

9.

Canoes	2	3	4	5	6
Paddles	4	6	8		

____

10. H.O.T. Students made a craft project at camp. They used 2 small pine cone patterns and 1 large pine cone pattern. Complete the table to find how many patterns were used for the different numbers of projects.

Projects	1	2	3	4	5	6	7	8	9	10
Small Pattern	2									
Large Pattern	1									

Problem Solving REAL WORLD

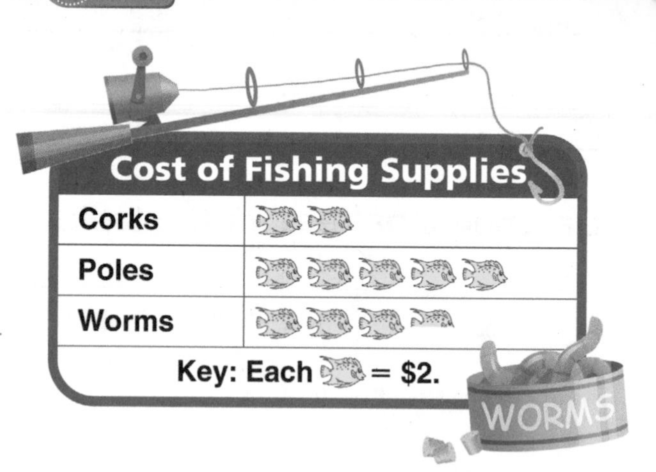

Use the picture graph for 11–13.

11. Jena bought 3 fishing poles. How much did she spend?

____

12. Noah bought 1 fishing pole, 2 corks, and 1 carton of worms. What was the total cost?

____

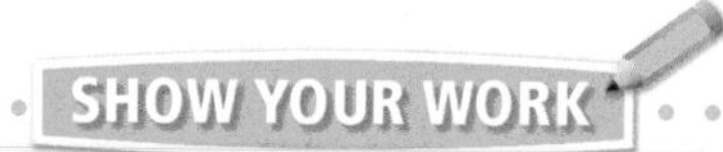

13. Write Math ▸ Ryan bought 8 corks. **Explain** how you can use the Commutative Property to find the cost.

____

____

14. H.O.T. The cost to rent a raft is $7 per person. A raft can hold up to 6 people. There is a $3 launch fee per raft. What is the total cost for a group of 6? **Explain.**

____

15. Taylor bought 4 boxes of granola bars. There are 6 bars in each box. How many granola bars did Taylor buy?

____

16. **Test Prep** Which of the following describes a pattern in the table?

Lifeguards	1	2	3	4	5
Swimmers	10	20	30	40	50

Ⓐ Add 9.
Ⓑ Multiply by 2.
Ⓒ Subtract 9.
Ⓓ Multiply by 10.

FOR MORE PRACTICE:
Standards Practice Book, pp. P91–P92

Name ___________________________

Find Unknown Factors

Essential Question How can you use an array or a multiplication table to find an unknown factor?

UNLOCK the Problem REAL WORLD

Tanisha plans to invite 24 people to a picnic. The invitations come in packs of 8. How many packs of invitations does Tanisha need to buy?

- How many people is Tanisha inviting? __________
- How many invitations are in 1 pack? __________

An **equation** is a number sentence that uses the equal sign to show that two amounts are equal.

A symbol or letter can stand for an unknown factor. You can write the equation, $n \times 8 = 24$, to find how many packs of invitations Tanisha needs. Find the number, n, that makes the equation true.

One Way Use an array.

- Show an array with 24 tiles with 8 tiles in each row by completing the drawing.

Math Talk MATHEMATICAL PRACTICES Explain how the array represents the problem. How do the factors relate to the array?

n	$\times$	8	$=$	24
↑		↑		↑
factor number of rows		factor number in each row		product total number

- Count how many rows of 8 tiles there are. **Think:** What number times 8 equals 24?

There are ______ rows of 8 tiles. The unknown factor is ______. $n =$ ______

______ $\times 8 = 24$ Check.

______ $= 24$ ✓ The equation is true.

So, Tanisha needs ______ packs of invitations.

Another Way Use a multiplication table.

$\blacksquare \times 8 = 24$

Think: The symbol, $\blacksquare$, stands for the unknown factor.

Start at the column for the factor 8, since 8 is the second factor.

Look down to find the product, 24.

Look left across the row from 24.

The unknown factor is ______.

$\blacksquare$ = ______

______ $\times 8 = 24$ Check.

______ $= 24$ ✓ The equation is true.

×	0	1	2	3	4	5	6	7	8	9	10
0	0	0	0	0	0	0	0	0	0	0	0
1	0	1	2	3	4	5	6	7	8	9	10
2	0	2	4	6	8	10	12	14	16	18	20
3	0	3	6	9	12	15	18	21	24	27	30
4	0	4	8	12	16	20	24	28	32	36	40
5	0	5	10	15	20	25	30	35	40	45	50
6	0	6	12	18	24	30	36	42	48	54	60
7	0	7	14	21	28	35	42	49	56	63	70
8	0	8	16	24	32	40	48	56	64	72	80
9	0	9	18	27	36	45	54	63	72	81	90
10	0	10	20	30	40	50	60	70	80	90	100

MATHEMATICAL PRACTICES

Math Talk How can you use the Commutative Property to check your answer?

Share and Show

1. What is the unknown factor shown by this array?

   $5 \times \blacksquare = 35$

   $\blacksquare$ = ______

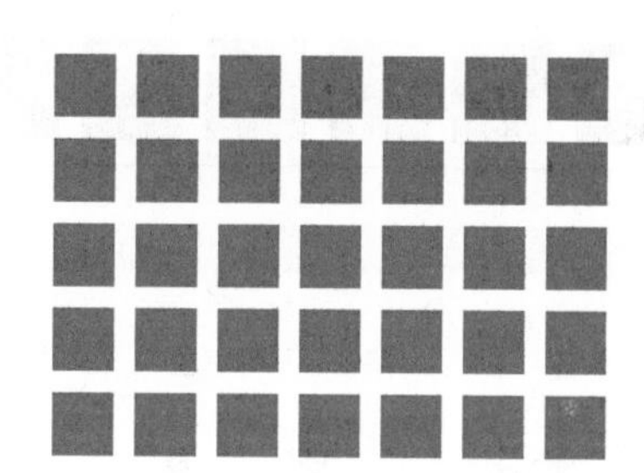

Find the unknown factor.

2. $d \times 3 = 27$

   d = ______

3. $6 \times \blacktriangle = 30$

   $\blacktriangle$ = ______

4. $20 = c \times 5$

   c = ______

5. $\blacksquare \times 2 = 14$

   $\blacksquare$ = ______

6. $36 = b \times 9$

   b = ______

7. $8 \times e = 64$

   e = ______

8. $7 \times \bigstar = 42$

   $\bigstar$ = ______

9. $z \times 9 = 72$

   z = ______

MATHEMATICAL PRACTICES

Math Talk **Explain** how you know if you are looking for the number of rows or the number in each row when you make an array to find an unknown factor.

Name ______________________

On Your Own

Find the unknown factor.

10. ■ $\times 2 = 18$

■ = ______

11. $28 = 4 \times m$

$m =$ ______

12. $y \times 3 = 9$

$y =$ ______

13. $7 \times g = 63$

$g =$ ______

14. $5 \times p = 40$

$p =$ ______

15. $8 \times w = 56$

$w =$ ______

16. $36 =$ ◆ $\times 6$

◆ = ______

17. $8 \times e = 72$

$e =$ ______

18. $9 \times$ ★ $= 27$

★ = ______

19. $6 \times a = 60$

$a =$ ______

20. $d \times 5 = 10$

$d =$ ______

21. $32 = 8 \times n$

$n =$ ______

22. $a \times 4 = 24$

$a =$ ______

23. $7 = 7 \times n$

$n =$ ______

24. $w \times 3 = 15$

$w =$ ______

25. ★ $\times 8 = 48$

★ = ______

26. $35 = h \times 5$

$h =$ ______

27. $54 =$ ▲ $\times 6$

▲ = ______

28. $7 \times p = 49$

$p =$ ______

29. $y \times 4 = 36$

$y =$ ______

H.O.T. **Algebra** **Find the unknown factor.**

30. $3 \times 6 = k \times 9$

$k =$ ______

31. $4 \times y = 2 \times 6$

$y =$ ______

32. $5 \times g = 36 - 6$

$g =$ ______

33. $6 \times 4 =$ ■ $\times 3$

■ = ______

34. $9 \times d = 70 + 2$

$d =$ ______

35. $8 \times h = 60 - 4$

$h =$ ______

Problem Solving REAL WORLD

Use the table for 36–40.

Item	Number in 1 Pack	Cost
Bowls	6	$10
Cups	8	$3
Tablecloth	1	$2
Napkins	36	$2
Forks	50	$3

Picnic Supplies

36. Tanisha needs 40 cups for the picnic. How many packs of cups should she buy?

37. H.O.T. Write Math **What if** Tanisha needs 40 bowls for the picnic? **Explain** how to write an equation with a letter for an unknown factor to find the number of packs she should buy. Then find the unknown factor.

38. Ms. Hill buys 3 tablecloths and 2 packs of napkins. How much does she spend?

39. H.O.T. **What if** Randy needs an equal number of bowls and cups for his picnic? How many packs of each will he need to buy?

40. H.O.T. Lori buys 200 forks. She gives the cashier $20. How much change should she get?

41. **Test Prep** What is the unknown factor?

$$m \times 6 = 48$$

Ⓐ 8 Ⓑ 7 Ⓒ 6 Ⓓ 5

SHOW YOUR WORK

FOR MORE PRACTICE: Standards Practice Book, pp. P93–P94

Name ___________________________________

Mid-Chapter Checkpoint

▶ Vocabulary

Choose the best term from the box.

Vocabulary
array
equation

1. An ______________ is a number sentence that uses the equal sign to show that two amounts are equal. (p. 185)

▶ Concepts and Skills

Describe a pattern for the table. Then complete the table.

2.

Weeks	1	2	3	4	5
Days	7	14	21		

____

3.

Tickets	2	3	4	5	6
Cost	\$8	\$12	\$16		

____

4.

Project Teams	Members
3	9
4	12
5	
6	18
7	

____

5.

Tables	Chairs
1	8
2	16
3	
4	32
5	

____

Find the unknown factor.

6. $m \times 5 = 30$

$m =$ ______

7. ■ $\times 6 = 48$

■ $=$ ______

8. $20 = 2 \times n$

$n =$ ______

9. $4 \times p = 32$

$p =$ ______

10. $25 = y \times 5$

$y =$ ______

11. ◆ $\times 10 = 10$

◆ $=$ ______

Fill in the bubble for the correct answer choice.

12. Which of the following describes a pattern in the table?

Packages	1	2	3	4	5
Stickers	6	12	18	24	30

Ⓐ Add 5.
Ⓑ Subtract 6.
Ⓒ Multiply by 2.
Ⓓ Multiply by 6.

13. What number makes the equation true?

$$a \times 8 = 72$$

Ⓐ 6
Ⓑ 7
Ⓒ 8
Ⓓ 9

14. Mia bought 2 copies of the same book. She spent \$18. What was the cost of one book?

Ⓐ \$2
Ⓑ \$9
Ⓒ \$10
Ⓓ \$18

15. Kyle saves \$10 every week for 6 weeks. How much money will Kyle have in Week 6?

Weeks	1	2	3	4	5	6
Amount	\$10	\$20	\$30	■	■	■

Ⓐ \$16
Ⓑ \$40
Ⓒ \$60
Ⓓ \$70

16. There are 24 students in the class. They arrange their desks in rows with 6 desks in each row. How many rows are there?

Ⓐ 8
Ⓑ 6
Ⓒ 4
Ⓓ 3

Name ______________________________

Problem Solving • Use the Distributive Property

Essential Question How can you use the strategy *draw a diagram* to multiply with multiples of 10?

UNLOCK the Problem REAL WORLD

The school assembly room has 5 rows of chairs with 20 chairs in each row. If the third-grade classes fill 3 rows of chairs, how many third graders are at the assembly?

Read the Problem

What do I need to find?

I need to find how many __________ are at the assembly.

What information do I need to use?

There are _____ chairs in each row.

The third graders fill _____ rows of chairs.

How will I use the information?

The Distributive Property tells me I can __________ the factor 20 to multiply.

$3 \times 20 = 3 \times (10 + \_\_\_\_)$

Solve the Problem

Draw a diagram. Finish the shading to show 3 rows of 20 chairs.

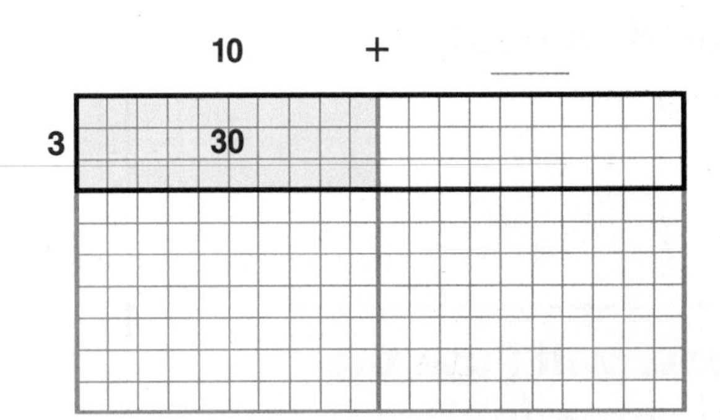

I can use the sum of the products of the smaller rectangles to find how many third graders are at the assembly.

$3 \times 10 = \_\_\_\_$ $\quad 3 \times 10 = \_\_\_\_$

$\_\_\_\_ + \_\_\_\_ = \_\_\_\_$

$3 \times 20 = \_\_\_\_$

So, _____ third graders are at the assembly.

1. **Explain** how breaking apart the factor 20 makes finding the product easier. ______________________________

Try Another Problem

Megan is watching a marching band practice. The band marches by with 4 rows of people playing instruments. She counts 30 people in each row. How many people march in the band?

Read the Problem	Solve the Problem
What do I need to find?	**Record the steps you used to solve the problem.**
What information do I need to use?	
How will I use the information?	

2. How can you check to see if your answer is reasonable?

____

____

3. **Explain** how you can use the Distributive Property to help you find a product.

____

____

Name ___________________________

Share and Show 

1. People filled all the seats in the front section of the theater. The front section has 6 rows with 40 seats in each row. How many people are in the front section of the theater?

Tips

UNLOCK the Problem

- Circle the numbers you will use.
- Use the Distributive Property and break apart a greater factor to use facts you know.
- Draw a diagram to help you solve the problem.

First, write the problem you need to solve. ______

Next, draw and label a diagram to break apart the problem into easier parts to solve.

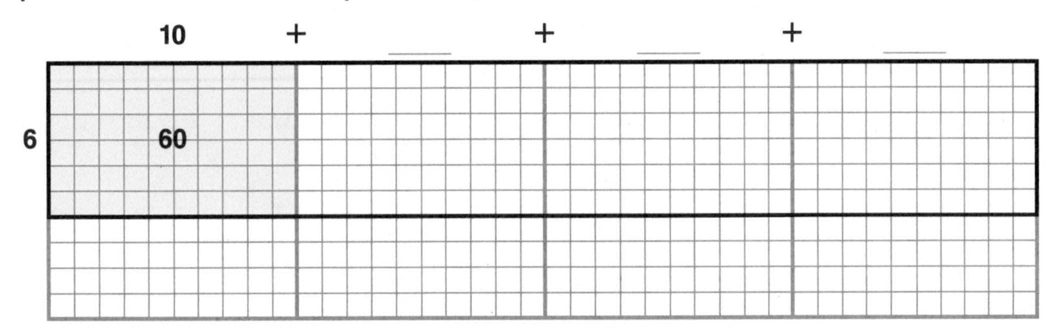

Then, find the products of the smaller rectangles.

$6 \times 10 =$ ______   ______ $\times$ ______ $=$ ______

______ $\times$ ______ $=$ ______   ______ $\times$ ______ $=$ ______

SHOW YOUR WORK

Last, find the sum of the products.

______ + ______ + ______ + ______ = __________

So, there are __________ people in the front section of the theater.

2. **H.O.T.** **What if** seats are added to the front section of the theater so that there are 6 rows with 50 seats in each row? How many seats are in the front section?

____

3. Tova sewed 60 pieces of ribbon together to make a costume. Each piece of ribbon was 2 meters long. How many meters of ribbon did Tova use to make the costume?

____

On Your Own

Choose a STRATEGY

- Act It Out
- Draw a Diagram
- Find a Pattern
- Make a Table

SHOW YOUR WORK

4. H.O.T. **What's the Error?** Carina draws this diagram to show that $8 \times 30 = 210$. **Explain** her error.

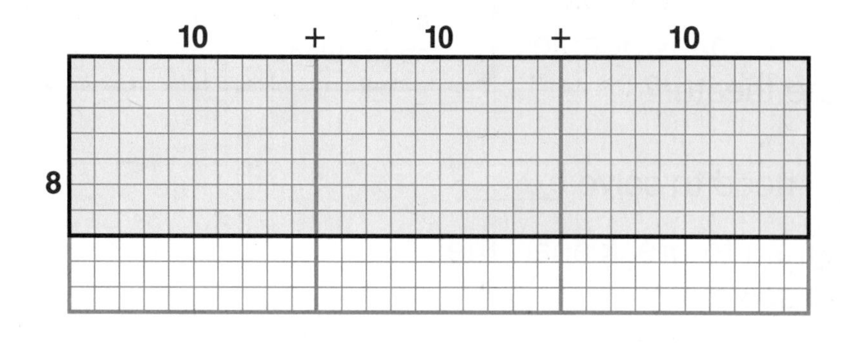

5. H.O.T. **Algebra** Mr. Chang orders 160 pencils. The pencils come in packs of 40. How many packs does Mr. Chang order?

6. **Write Math** Tamika wants to display 10 trophies on a table in a rectangular array. How many different ways can Tamika arrange the trophies? **Explain** your answer.

7. The drama club has 350 tickets to sell. They sell 124 tickets on Monday and 98 tickets on Tuesday. How many tickets does the drama club have left to sell?

8. **Test Prep** Steffen orders shirts for each of the 5 members in his family. If each shirt costs \$20, what is the total cost for the shirts?

Ⓐ \$25 Ⓑ \$30 Ⓒ \$70 Ⓓ \$100

 FOR MORE PRACTICE: Standards Practice Book, pp. P95–P96

Name ______________________________

Multiplication Strategies with Multiples of 10

Essential Question What strategies can you use to multiply with multiples of 10?

UNLOCK the Problem REAL WORLD

You can use models and place value to multiply with multiples of 10.

- What is a product of 10 and the counting numbers 1, 2, 3, and so on?

____

Activity Model multiples of 10.

Materials ■ base-ten blocks

Model the first nine multiples of 10.

1×10
1×1 ten
1 ten
10

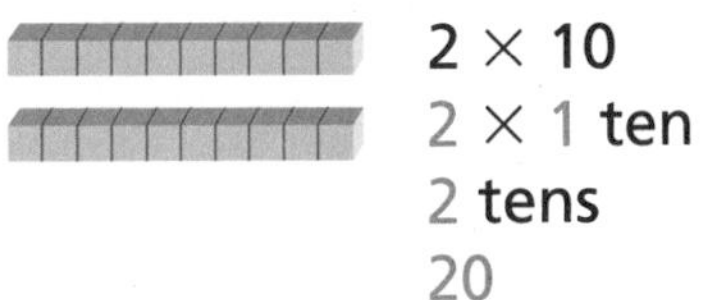

2×10
2×1 ten
2 tens
20

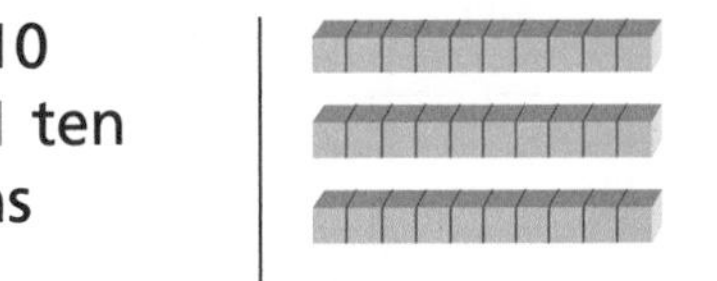

3×10
3×1 ten
3 tens
30

What are the first nine multiples of 10?

10, 20, 30, _____, _____, _____, _____, _____, _____

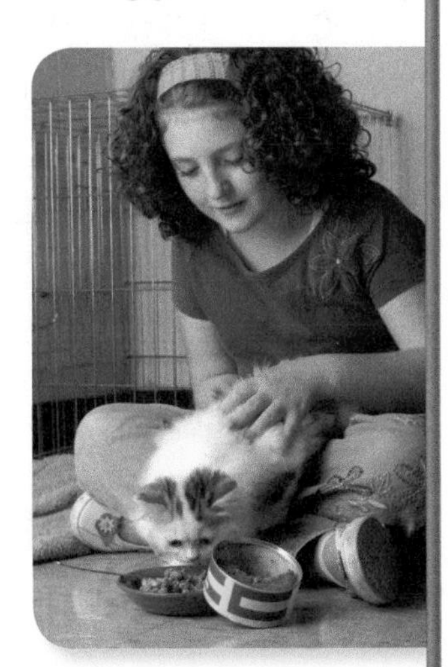

Best Care Veterinary Clinic offered free pet care classes for 5 days. Erin attended the pet care class for 30 minutes each day. How many minutes did Erin attend the class?

One Way Use a number line.

$5 \times 30 = \blacksquare$ Think: 30 = 3 tens

STEP 1 Complete the number line. Write the labels for the multiples of 10.

STEP 2 Draw jumps on the number line to show 5 groups of 3 tens.

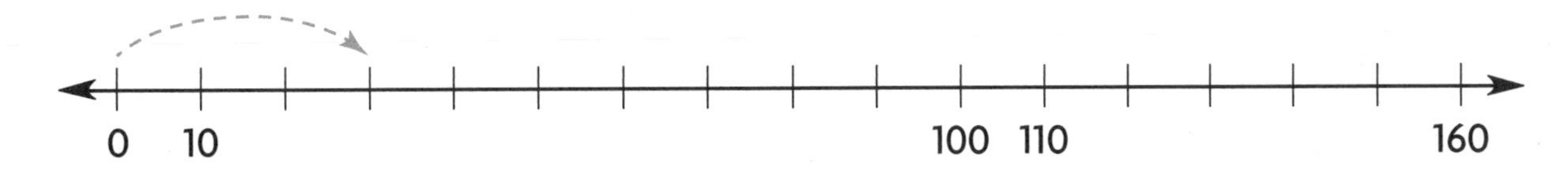

$5 \times 30 =$ _____

So, Erin attended the pet care class for _____ minutes.

Another Way Use place value.

MODEL

THINK

$5 \times 30 = 5 \times$ ______ tens

$=$ ______ tens $=$ ______

So, $5 \times 30 =$ ______.

Try This!

$4 \times 50 =$ ______ $\times$ ______ tens

$=$ ______ tens $=$ ______

MATHEMATICAL PRACTICES

Math Talk Explain why 5×30 has one zero in the product and 4×50 has two zeros in the product.

Share and Show

1. Use a number line to find the product.

$3 \times 40 =$ ______ **Think:** There are 3 jumps of 40.

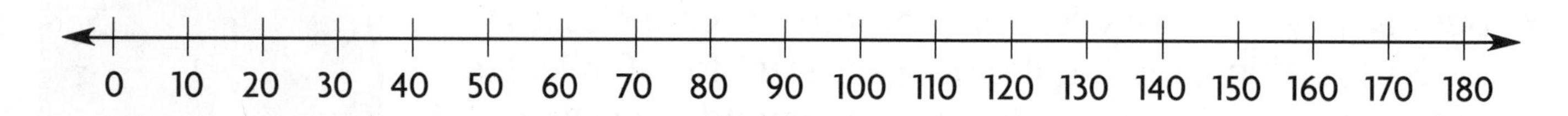

Use a number line to find the product.

2. $8 \times 20 =$ ______

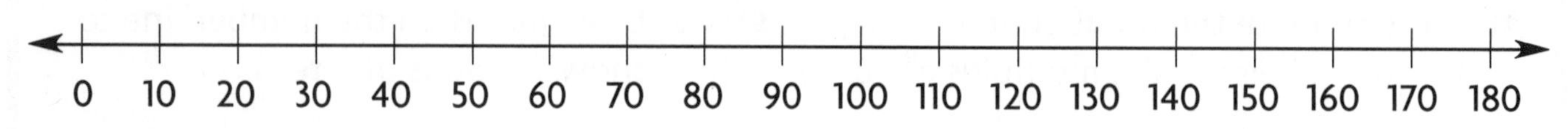

Use place value to find the product.

3. $3 \times 70 = 3 \times$ ______ tens

$=$ ______ tens $=$ ______

4. $50 \times 2 =$ ______ tens $\times 2$

$=$ ______ tens $=$ ______

MATHEMATICAL PRACTICES

Math Talk Will the product of 50×2 be the same as the product of 2×50? **Explain.**

Name ______________________________

On Your Own

Use a number line to find the product.

5. $7 \times 20 =$ _____

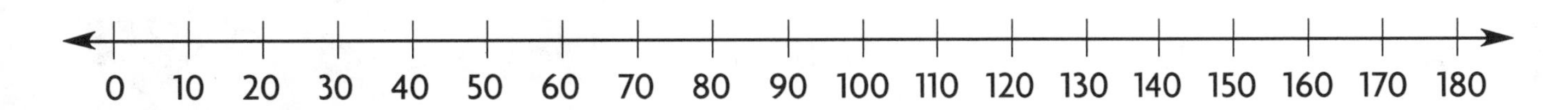

6. $3 \times 50 =$ _____

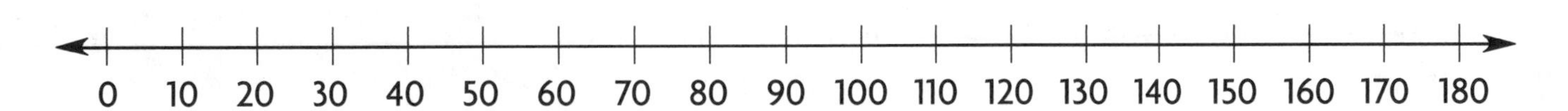

Use place value to find the product.

7. $6 \times 60 = 6 \times$ _____ tens

$=$ _____ tens $=$ _____

8. $50 \times 7 =$ _____ tens $\times 7$

$=$ _____ tens $=$ _____

Problem Solving REAL WORLD

Use the table for 9–11.

9. The cost of a bottle of shampoo is \$9. If the clinic sells their entire supply of shampoo, how much money will they receive?

____

10. **What's the Question?** Each bag of treats has 30 treats. The answer is 240.

____

11. H.O.T. There are 4 bottles of vitamins in each box of vitamins. Each bottle of vitamins has 20 vitamins. If the clinic wants to have a supply of 400 vitamins, how many more boxes should they order?

____

Best Care Clinic Pet Supplies

Item	Amount
Cat toys	10 packs
Treats	8 bags
Shampoo	20 bottles
Vitamins	3 boxes

UNLOCK the Problem REAL WORLD

12. Hiromi needs to set up chairs for 155 people to attend the school career day program. So far she has set up 6 rows with 20 chairs in each row. How many more chairs does Hiromi need to set up?

Ⓐ 25 Ⓑ 35 Ⓒ 75 Ⓓ 95

a. What do you need to find?

____

b. What operations will you use to find how many more chairs Hiromi needs to set up?

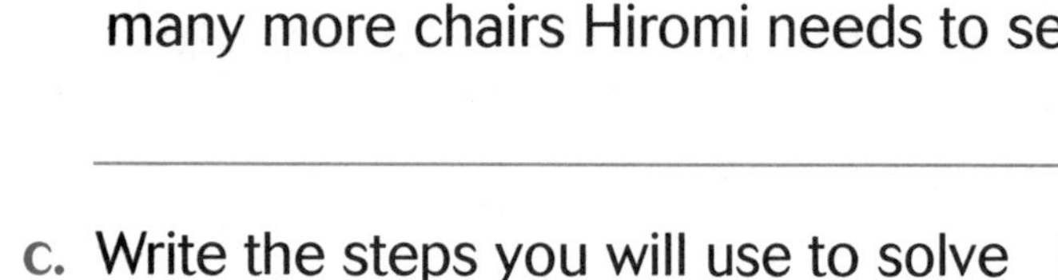

____

c. Write the steps you will use to solve the problem.

d. Complete the sentences.

Hiromi needs to set up _____ chairs for people to attend the program.

She has set up _____ rows with _____ chairs in each row.

So, Hiromi needs to set up _____ more chairs.

e. Fill in the bubble for the correct answer above.

13. Last week, Dr. Newman examined the paws of 30 dogs at her clinic. She examined the paws on 20 cats. What is the total number of paws Dr. Newman examined last week?

Ⓐ 50

Ⓑ 80

Ⓒ 120

Ⓓ 200

14. For the career day program, 8 people give presentations. If each person answers 10 questions during the presentations, how many questions are answered?

Ⓐ 18

Ⓑ 80

Ⓒ 88

Ⓓ 90

FOR MORE PRACTICE:
Standards Practice Book, pp. P97–P98

Name ______________________________

Multiply Multiples of 10 by 1-Digit Numbers

Essential Question How can you model and record multiplying multiples of 10 by 1-digit whole numbers?

UNLOCK the Problem REAL WORLD

The community center offers 4 dance classes. If 30 students sign up for each class, how many students sign up for dance class altogether?

- How many equal groups are there? ______
- How many are in each group? ______

Activity Use base-ten blocks to model 4 × 30.

Materials ■ base-ten blocks

STEP 1 Model 4 groups of 30.

STEP 2 Combine the tens. Regroup 12 tens as 1 hundred 2 tens.

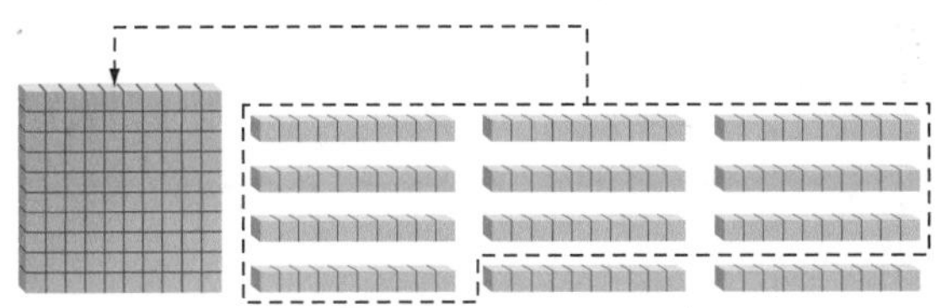

$4 \times 30 =$ ______

So, ______ students sign up for dance class altogether.

Math Idea
If one factor is a multiple of 10, then the product will also be a multiple of 10.

Try This! Find 7 × 40.

Use a quick picture to record your model. Draw a stick for each ten. Draw a square for each hundred.

STEP 1 Model ______ groups of ______.

STEP 2 Combine the tens. Regroup 28 tens as ______ hundreds ______ tens.

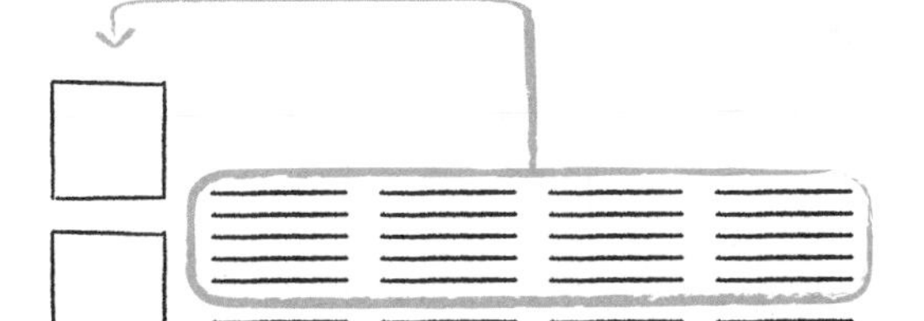

So, $7 \times 40 =$ ______.

MATHEMATICAL PRACTICES
Math Talk Will the product of 7 × 40 be the same as 4 × 70? **Explain.**

 **Example** **Use place value and regrouping.**

Find 9×50.

	MODEL	THINK	RECORD
STEP 1		Multiply the ones. 9×0 ones = _____ ones	$\begin{array}{r} 50 \\ \times\ 9 \\ \hline 0 \end{array}$
STEP 2	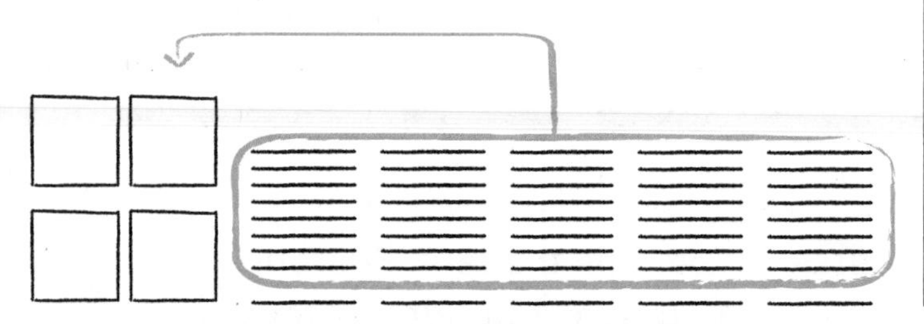	Multiply the tens. 9×5 tens = 45 tens Regroup the _____ tens as _____ hundreds _____ tens.	$\begin{array}{r} 50 \\ \times\ 9 \\ \hline 450 \end{array}$

So, $9 \times 50 =$ _____.

Share and Show 

1. Use the quick picture to find 5×40.

$5 \times 40 =$ _____

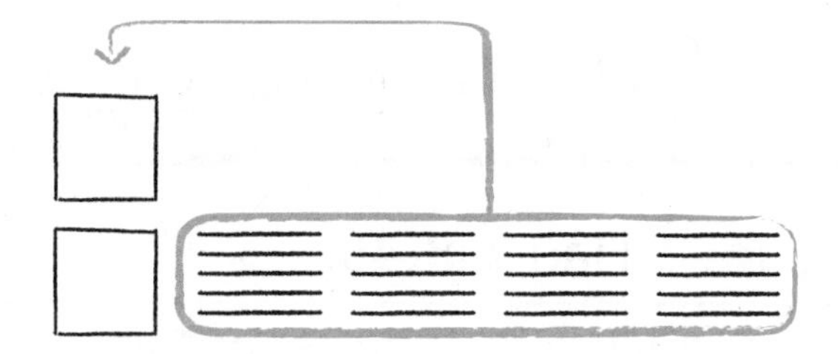

Find the product. Use base-ten blocks or draw a quick picture on your MathBoard.

✓ 2. $7 \times 30 =$ _____   3. _____ $= 2 \times 90$ 4. $8 \times 40 =$ _____   5. _____ $= 4 \times 60$

Find the product.

✓ 6. $\begin{array}{r} 80 \\ \times\ 9 \\ \hline \end{array}$ 7. $\begin{array}{r} 70 \\ \times\ 7 \\ \hline \end{array}$ 8. $\begin{array}{r} 90 \\ \times\ 4 \\ \hline \end{array}$ 9. $\begin{array}{r} 60 \\ \times\ 8 \\ \hline \end{array}$

MATHEMATICAL PRACTICES

Math Talk Explain how you can use place value to solve Exercise 9.

Name ______________________________

On Your Own

Find the product. Use base-ten blocks or draw a quick picture on your MathBoard.

10. $2 \times 70 =$ ____ **11.** $8 \times 50 =$ ____ **12.** ____ $= 3 \times 90$ **13.** $2 \times 80 =$ ____

Find the product.

14. $\begin{array}{r} 80 \\ \times\ 3 \\ \hline \end{array}$ **15.** $\begin{array}{r} 60 \\ \times\ 9 \\ \hline \end{array}$ **16.** $\begin{array}{r} 90 \\ \times\ 8 \\ \hline \end{array}$ **17.** $\begin{array}{r} 80 \\ \times\ 8 \\ \hline \end{array}$

Practice: Copy and Solve **Find the product.**

18. 6×70 **19.** 9×90 **20.** 70×8 **21.** 90×7

Algebra **Find the unknown factor.**

22. $a \times 80 = 480$
$a =$ ____

23. $b \times 30 = 30$
$b =$ ____

24. $7 \times$ ■ $= 420$
■ $=$ ____

25. $50 \times$ ▲ $= 0$
▲ $=$ ____

Problem Solving REAL WORLD

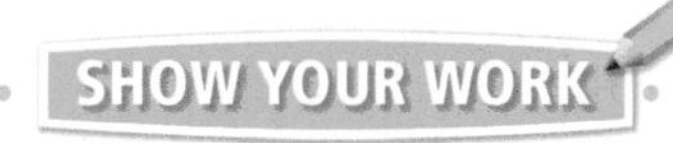

26. Ben writes the product of 7 and 50. Then he subtracts the least 3-digit multiple of 10 from the product. What is the difference?

____

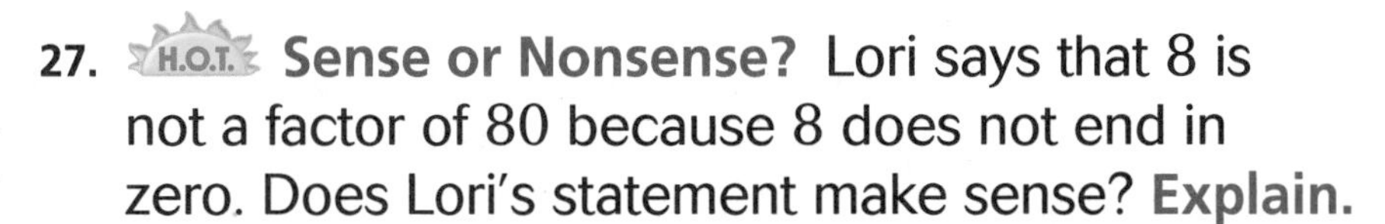

27. H.O.T. **Sense or Nonsense?** Lori says that 8 is not a factor of 80 because 8 does not end in zero. Does Lori's statement make sense? **Explain.**

____

28. H.O.T. The book club members read 200 books in all. Each member read 5 books. Write an equation to find the number of members in the book club. Use a letter to stand for the unknown factor.

____

UNLOCK the Problem REAL WORLD

29. Frank has a 2-digit number on his baseball uniform. The number is a multiple of 10 and has 3 for one of its factors. What three numbers could Frank have on his uniform?

a. What do you need to find?

____

____

b. What information do you need to use?

____

____

c. How can you solve the problem?

d. Complete the sentences.

Frank has a ____________ on his uniform.

The number is a multiple of _____.

One factor of the number is _____.

Frank could have _____, _____, or _____ on his uniform.

30. Write Math Use a property to explain why the product of a multiple of 10 and a number will have a zero in the ones place.

____

____

____

____

____

31. **Test Prep** Wesley jogs 10 miles each week for 2 weeks. Then he jogs 20 miles each week for 3 weeks. How many miles does Wesley jog in 5 weeks?

Ⓐ 20 miles

Ⓑ 30 miles

Ⓒ 60 miles

Ⓓ 80 miles

FOR MORE PRACTICE:
Standards Practice Book, pp. P99–P100

Name ______________________________

Chapter Review/Test

▶ Vocabulary

Choose the best term from the box.

Vocabulary
equation
product

1. An ____________ is a number sentence that uses the equal sign to show that two amounts are equal. (p. 185)

▶ Concepts and Skills

Describe a pattern in the table. Then complete the table.

2.

Tens	1	2	3	4	5
Ones	10	20	30		

____

3.

Jars	2	3	4	5	6
Pickles	16	24		40	

____

Find the unknown factor.

4. $p \times 6 = 42$

$p =$ ______

5. $5 \times s = 15$

$s =$ ______

6. ■ $\times 9 = 81$

■ $=$ ______

Find the product.

7. $6 \times 40 =$ ______

8. $3 \times 80 =$ ______

9. $4 \times 90 =$ ______

10. ______ $= 5 \times 50$

11. 30×6

12. 20×8

13. 60×7

14. 80×9

15. 40×4

16. 70×0

17. 90×9

18. 20×1

GO Online Assessment Options Chapter Test

Fill in the bubble for the correct answer choice.

19. What number makes the equation true?

$$c \times 9 = 27$$

Ⓐ 2

Ⓑ 3

Ⓒ 4

Ⓓ 6

20. The camping club rents 4 rafts. How many people can 4 rafts hold?

Rafts	1	2	3	4
People	8	16	24	■

Ⓐ 20

Ⓑ 26

Ⓒ 32

Ⓓ 40

21. The third-grade locker room has 48 lockers. There are 6 lockers in each row. How many rows of lockers are there in the third-grade locker room?

Ⓐ 9

Ⓑ 8

Ⓒ 7

Ⓓ 6

22. Which equation is an example of the Distributive Property?

Ⓐ $8 \times 20 = 8 \times (10 + 10)$

Ⓑ $8 \times 20 = 20 \times 8$

Ⓒ $20 \times 0 = 0$

Ⓓ $160 = 8 \times 20$

Name ____________________

Fill in the bubble for the correct answer choice.

23. Alya planted 20 trays of flowers. Each tray had 6 flowers in it. How many flowers did she plant?

Ⓐ 80

Ⓑ 120

Ⓒ 160

Ⓓ 180

24. The community center prints a newsletter that uses 4 pieces of paper. How many pieces of paper are needed to print 70 copies of the newsletter?

Ⓐ 110

Ⓑ 140

Ⓒ 210

Ⓓ 280

25. Use the number line to find 5×20. What is the product?

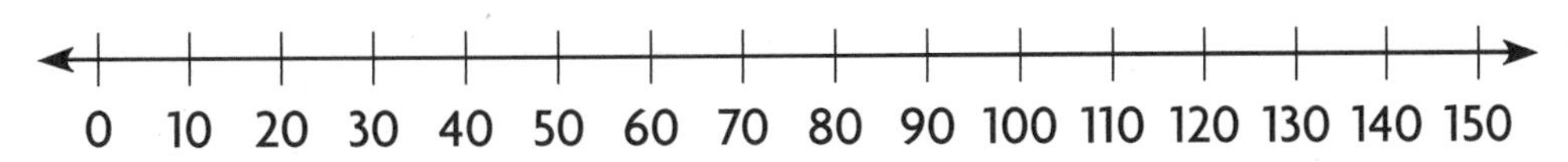

Ⓐ 100

Ⓑ 20

Ⓒ 10

Ⓓ 5

26. Which of the following describes a pattern in the table?

Bags	1	2	3	4	5
Lemons	4	8	12	16	20

Ⓐ Add 3.

Ⓑ Subtract 3.

Ⓒ Multiply by 2.

Ⓓ Multiply by 4.

▶ Constructed Response

27. Devon has 32 books to pack in boxes. She packs 8 books in each box. How many boxes does she need? Write an equation using the letter *n* to stand for the unknown factor. **Explain** how to find the unknown factor.

____

____

28. The bookstore has 6 shelves of poetry books. There are 30 poetry books on each shelf. How many poetry books does the bookstore have? Draw a diagram to show how you can use the Distributive Property to find the number of poetry books.

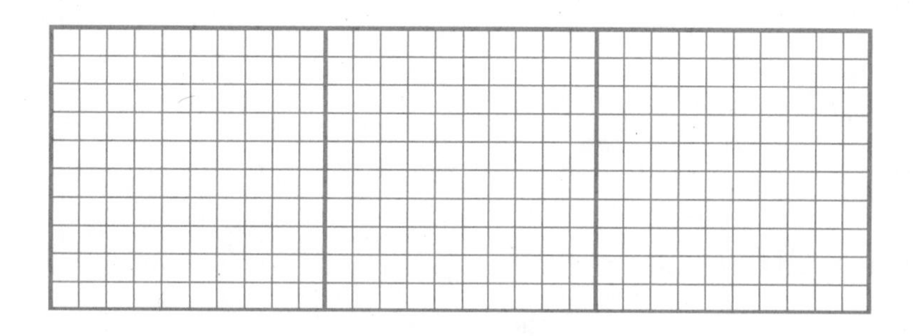

____

▶ Performance Task

29. Ruben is collecting cans for the school recycling contest. He makes two plans to try to collect the most cans.

Plan A: Collect 20 cans a week for 9 weeks.

Plan B: Collect 30 cans a week for 7 weeks.

A Which plan should Ruben choose? ____________

B **Explain** how you made your choice.

____

____

____

____

____

Chapter 6

Understand Division

Show What You Know

Check your understanding of important skills.

Name ____________________________

Count Back to Subtract Use the number line. Write the difference.

1. $8 - 5 =$ _____

2. $9 - 4 =$ _____

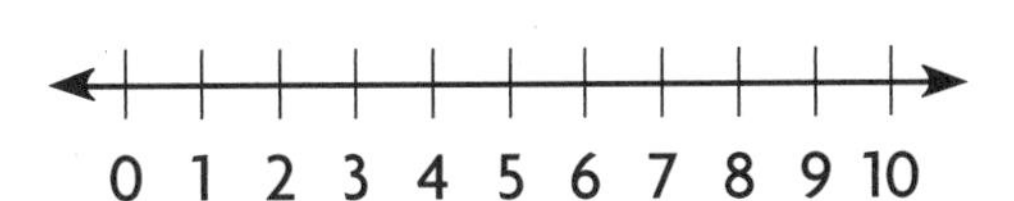

Count Equal Groups Complete.

3.   

_____ groups

_____ in each group

4. 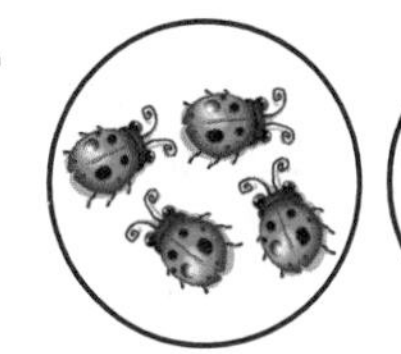 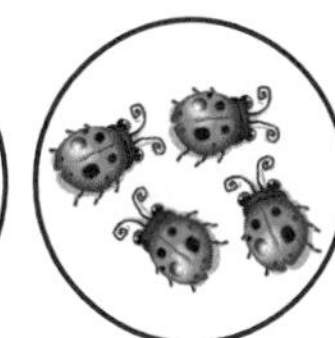

_____ groups

_____ in each group

Multiplication Facts Through 9 Find the product.

5. $8 \times 5 =$ _____

6. _____ $= 7 \times 7$

7. $3 \times 9 =$ _____

The table shows 3 different ways you can score points in basketball. Corina scored 12 points in a basketball game. Be a Math Detective to find the greatest number of field goals she could have scored. Then find the greatest number of 3-pointers she could have scored.

Scoring Points in Basketball	
free throw	1 point
field goal	2 points
3-pointer	3 points

Vocabulary Builder

▶ Visualize It

Complete the bubble map by using the words with a ✓.

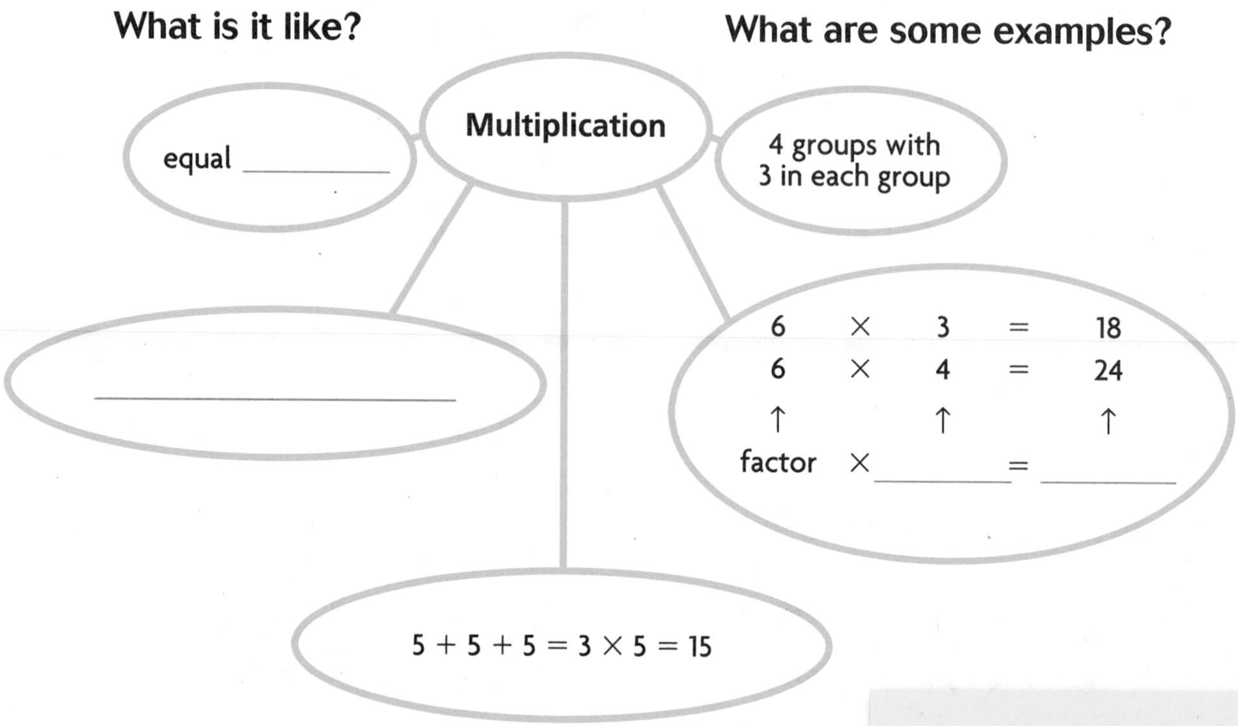

What is it like?

What are some examples?

Multiplication

equal ________

____

4 groups with
3 in each group

6	×	3	=	18
6	×	4	=	24
↑		↑		↑
factor	×	______	=	______

$5 + 5 + 5 = 3 \times 5 = 15$

▶ Understand Vocabulary

Draw a line to match each word or term with its definition.

Preview Words	**Definitions**
1. dividend	A set of related multiplication and division equations
2. related facts	The number that divides the dividend
3. divisor	The number that is to be divided in a division problem

Review Words

- array
- ✓ equal groups
- equation
- ✓ factor
- Identity Property of Multiplication
- ✓ product
- ✓ repeated addition

Preview Words

- divide
- dividend
- divisor
- inverse operations
- quotient
- related facts

GO Online • eStudent Edition • Multimedia eGlossary

Name ______________________________

Problem Solving • Model Division

Essential Question How can you use the strategy *act it out* to solve problems with equal groups?

UNLOCK the Problem REAL WORLD

Stacy has 16 flowers. She puts an equal number of flowers in each of 4 vases. How many flowers does Stacy put in each vase?

Use the graphic organizer below to solve the problem.

Read the Problem

What do I need to find?

I need to find the number of ________ Stacy puts in each ________.

What information do I need to use?

Stacy has ______ flowers. She puts an equal number of flowers in each of ______ vases.

How will I use the information?

I will act out the problem by making equal ________ with counters.

Solve the Problem

Describe how to act out the problem to solve.

First, count out ______ counters.

Next, make ______ equal groups. Place 1 counter at a time in each group until all 16 counters are used.

Last, draw the equal groups by completing the picture below.

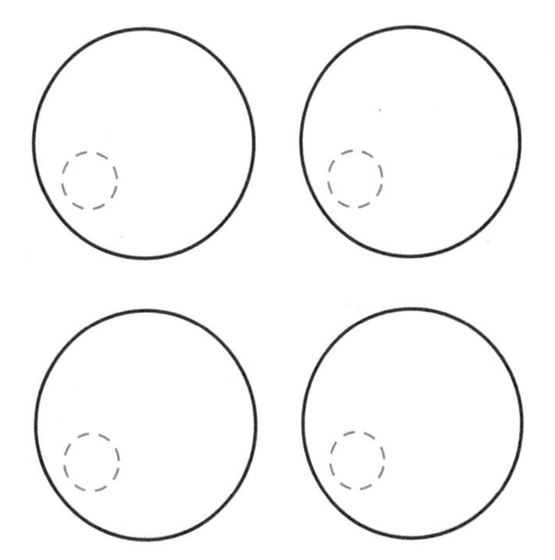

So, Stacy puts ______ flowers in each vase.

Try Another Problem

Hayden is planning a party. He bakes 21 cookies. If he plans to give each person 3 cookies, how many people will be at his party?

Read the Problem	Solve the Problem
What do I need to find?	**Describe how to act out the problem to solve.**
What information do I need to use?	
How will I use the information?	

- How can you check your answer is reasonable? ____

MATHEMATICAL PRACTICES

Math Talk Explain how acting out a problem helps you solve it.

Name ________________________________

Share and Show

Tips

UNLOCK the Problem

- Circle the question.
- Underline important facts.
- Act out the problem with counters.

SHOW YOUR WORK

1. Sue is having a party. She has 16 cups. She puts them in 2 equal stacks. How many cups are in each stack?

   First, decide how to act out the problem.

   You can use counters to represent the __________.

   You can draw __________ to represent the stacks.

   Then, draw to find the number of __________ in each stack.

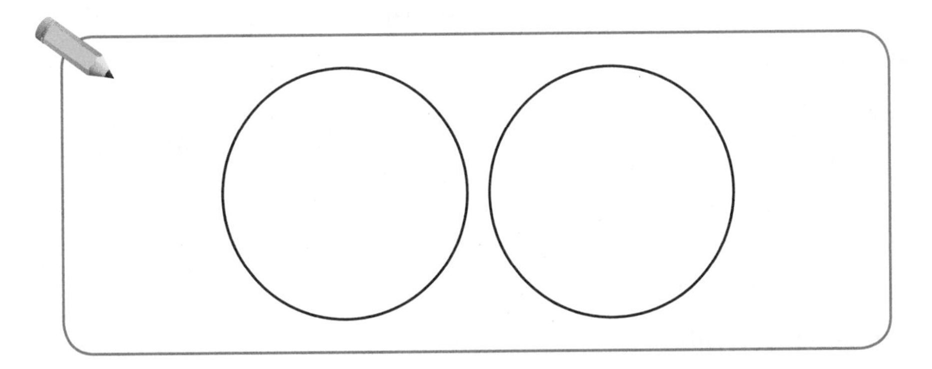

   There are ______ groups. There are ______ counters in each group.

   So, there are ______ cups in each stack.

2. H.O.T. **What if** Sue has 24 cups and puts 4 cups in each stack? If she already made 4 stacks, how many more stacks can she make with the remaining cups?

   ____

3. At Luke's school party, the children get into teams to play a game. If there are 35 children and they make teams of 5, how many teams are there?

   ____

4. Anne put 20 cupcakes on 4 plates. If she put the same number of cupcakes on each one, how many cupcakes did she put on each plate?

   ____

On Your Own

Choose a STRATEGY

- Act It Out
- Draw a Diagram
- Find a Pattern
- Make a Table

Use the table for 5–6.

Sadie's Party Supplies

Item	Number
Plates	30
Napkins	28
Cups	24

5. Sadie's plates came in packages of 5. How many packages of plates did she buy?

6. **Write Math** Sadie bought 4 packages of napkins and 3 packages of cups. Which item had more in each package? How many more? **Explain** how you found your answer.

7. Megan put 3 red balloons and 4 white balloons at each of 4 tables. How many balloons are at the tables altogether?

8. **H.O.T.** There are 12 cookies on plates. List all the ways the cookies could be put equally on the plates.

9. **Test Prep** Miguel bought 18 party favors. He gave 2 party favors to each of the children at his party. How many children were at Miguel's party?

   Ⓐ 20　　Ⓑ 16　　Ⓒ 9　　Ⓓ 8

SHOW YOUR WORK

FOR MORE PRACTICE:
Standards Practice Book, pp. P105–P106

Name ___________________________

Size of Equal Groups

Essential Question How can you model a division problem to find how many in each group?

UNLOCK the Problem REAL WORLD

Hector has 12 rocks from Onondaga Cave State Park. He puts an equal number of his rocks in each of 3 boxes. How many rocks are in each box?

- What do you need to find?

____

- Circle the numbers you need to use.

▲ Onondaga Cave is located in Leasburg, Missouri. Over the years, water has formed flowstones and other colorful deposits.

When you multiply, you put equal groups together. When you **divide**, you separate into equal groups.

You can divide to find the number in each group.

Activity Use counters to model the problem.

Materials ■ counters ■ MathBoard

STEP 1

Use 12 counters.

STEP 2

Draw 3 circles on your MathBoard. Place 1 counter at a time in each circle until all 12 counters are used. Draw the rest of the counters to show your work.

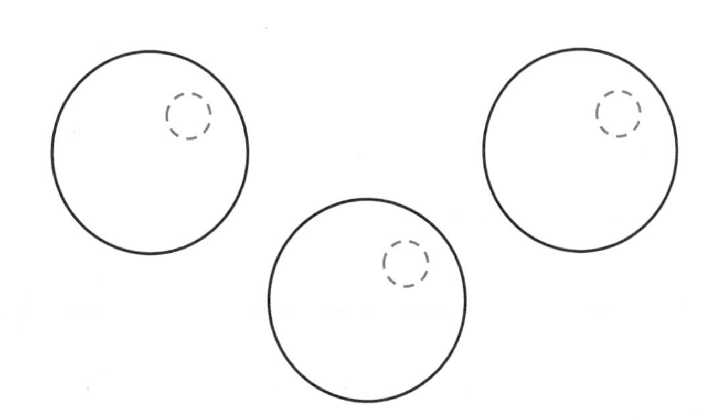

There are ______ counters in each group.

So, there are ______ rocks in each box.

Try This!

Madison has 15 rocks. She puts an equal number of rocks in each of 5 boxes. How many rocks are in each box?

STEP 1

Draw 5 squares to show 5 boxes.

STEP 2

Draw 1 counter in each square to show the rocks. Continue drawing 1 counter at a time in each box until all 15 counters are drawn.

There are _____ counters in each group.

So, there are ______ rocks in each box.

MATHEMATICAL PRACTICES

Math Talk Describe another way to arrange 15 counters to make equal groups.

1. How many counters did you draw? __________
2. How many equal groups did you make? ___________
3. How many counters are in each group? ___________

Name ________________________________

Share and Show 

1. Jon has 8 counters. He makes 4 equal groups. Draw a picture to show the number of counters in each group.

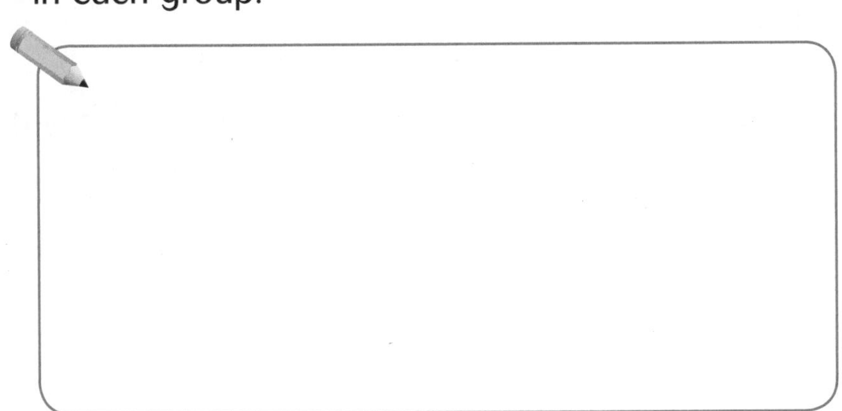

MATHEMATICAL PRACTICES

Math Talk Explain how you made the groups equal.

Use counters or draw a quick picture on your MathBoard. Make equal groups. Complete the table.

	Counters	Number of Equal Groups	Number in Each Group
2.	10	2	
3.	24	6	
4.	12	4	

On Your Own

Use counters or draw a quick picture on your MathBoard. Make equal groups. Complete the table.

	Counters	Number of Equal Groups	Number in Each Group
5.	14	7	
6.	21	3	
7.	20	5	
8.	12	6	
9.	36	9	

Problem Solving REAL WORLD

Use the table for 10–11.

Name	Number of Photos
Madison	28
Joe	25
Ella	15

Photos

10. Madison puts all of her photos in a photo album. She puts an equal number of photos on each of 4 pages in her album. How many photos are on each page?

SHOW YOUR WORK

11. H.O.T. Joe and Ella combine their photos. Then they put an equal number on each page of an 8-page photo album. How many photos are on each page?

12. Write Math Rebekah's family found 30 sand dollars. **Explain** how to share the sand dollars equally among the 6 people in her family.

13. **Test Prep** Zana has 9 rocks from a trip. She puts an equal number of rocks in each of 3 bags. How many rocks are in each bag?

Ⓐ 27 Ⓒ 6

Ⓑ 12 Ⓓ 3

 FOR MORE PRACTICE: Standards Practice Book, pp. P107–P108

Name ______________________________

Number of Equal Groups

Essential Question How can you model a division problem to find how many equal groups?

CONNECT You have learned how to divide to find the number in each group. Now you will learn how to divide to find the number of equal groups.

UNLOCK the Problem REAL WORLD

William has 12 shells and some boxes. He wants to put his shells in groups of 3. How many boxes does he need for his shells?

- Underline what you need to find.
- How many shells does William want in each group? __________

Make equal groups.

- Look at the 12 counters.
- Circle a group of 3 counters.
- Continue circling groups of 3 until all 12 counters are in groups.

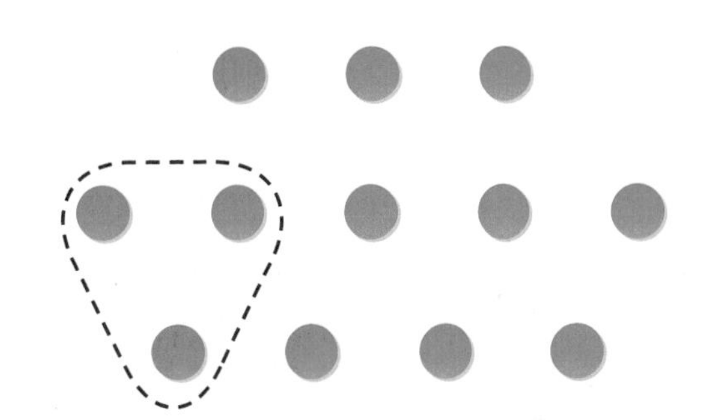

There are _____ groups of counters.

So, William needs _____ boxes for his shells.

MATHEMATICAL PRACTICES

Math Talk Explain how the drawing would change if William wanted to put his shells in groups of 4.

▲ The horse conch can grow to a length of 24 inches!

Try This!

Sarah has 15 shells. She wants to put each group of 5 shells in a box. How many boxes does she need for her shells?

STEP 1

Draw 15 counters.

STEP 2

Make a group of 5 counters by drawing a circle around them. Continue circling groups of 5 until all 15 counters are in groups.

There are _____ groups of 5 counters.

So, Sarah needs _____ boxes for her shells.

• H.O.T. **What if** Sarah puts her 15 shells in groups of 3?

How many boxes does she need? ___________
Draw a quick picture to show your work.

Name ______________________________

Share and Show 

1. Emily has 12 counters. She puts them in groups of 2. Draw a picture to show the number of groups.

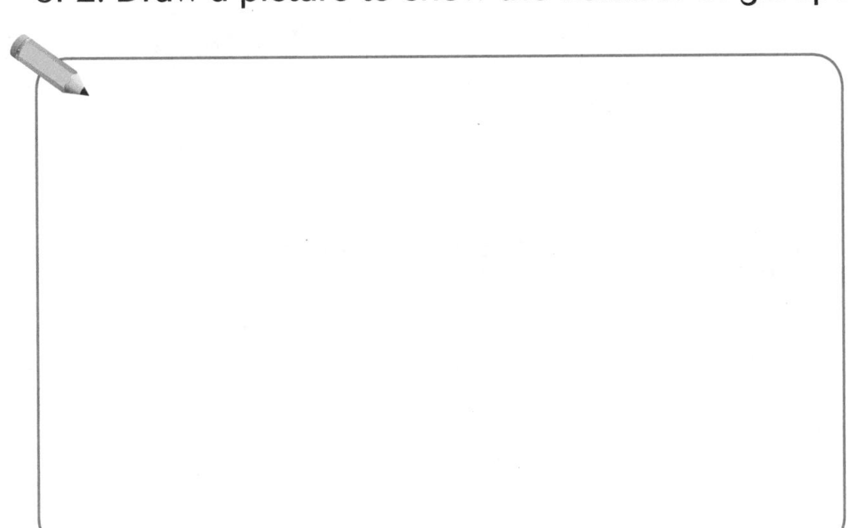

MATHEMATICAL PRACTICES

Math Talk Explain how you find the number of equal groups when you divide.

Draw counters on your MathBoard. Then circle equal groups. Complete the table.

	Counters	Number of Equal Groups	Number in Each Group
✓ 2.	20		4
✓ 3.	24		3
4.	18		2

On Your Own

Draw counters on your MathBoard. Then circle equal groups. Complete the table.

	Counters	Number of Equal Groups	Number in Each Group
5.	16		8
6.	25		5
7.	27		9
8.	32		4

UNLOCK the Problem REAL WORLD

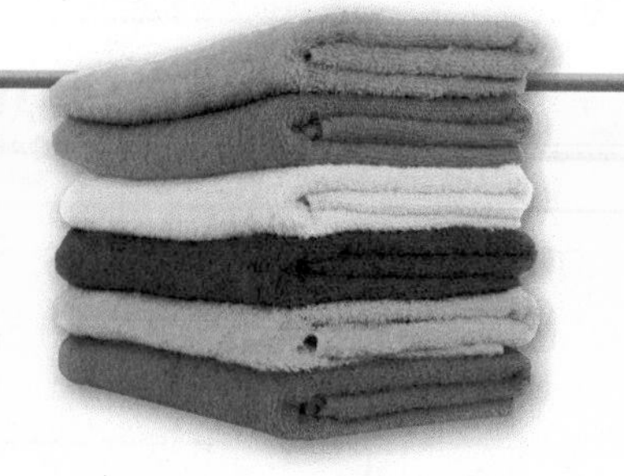

9. A store has 24 beach towels in stacks of 6 towels each. How many stacks of beach towels are at the store?

a. What do you need to find? ______________________

____

b. How will you use what you know about making equal groups to solve the problem? ______________________

____

c. Draw equal groups to find how many stacks of beach towels there are at the store.

d. Complete the sentences.

The store has _____ beach towels.

There are _____ towels in each stack.

So, there are _____ stacks of beach towels at the store.

10. Some friends share 35 beach toys equally. If each friend gets 5 beach toys, how many friends are there?

____

11. **Test Prep** Dan's train is 27 inches long. If each train car is 3 inches long, how many train cars are there?

Ⓐ 9

Ⓑ 8

Ⓒ 7

Ⓓ 6

 FOR MORE PRACTICE:
Standards Practice Book, pp. P109–P110

Name ______________________________

Model with Bar Models

Essential Question How can you use bar models to solve division problems?

UNLOCK the Problem REAL WORLD

A dog trainer has 20 dog treats for 5 dogs in his class. If each dog gets the same number of treats, how many treats will each dog get?

- What do you need to find?

____

Activity 1 Use counters to find how many in each group.

Materials ■ counters ■ MathBoard

- Use 20 counters.
- Draw 5 circles on your MathBoard.
- Place 1 counter at a time in each circle until all 20 counters are used.
- Draw the rest of the counters to show your work.

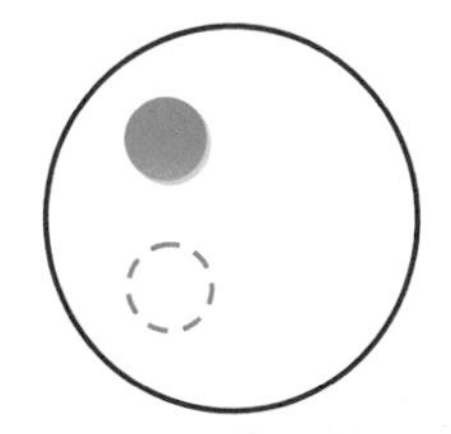
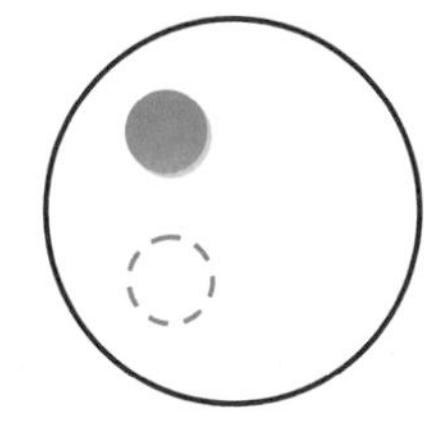
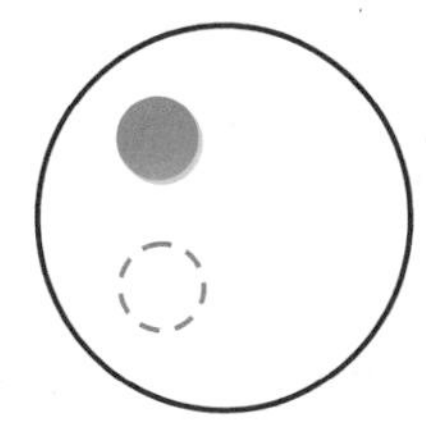
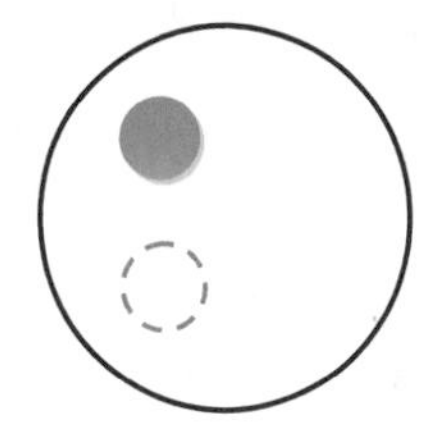
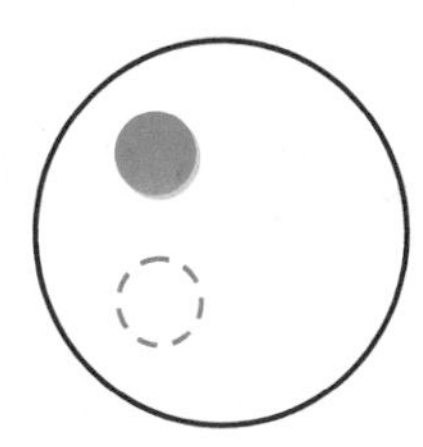

There are _____ counters in each of the 5 groups.

A bar model can show how the parts of a problem are related.

- Complete the bar model to show 20 dog treats divided into 5 equal groups.

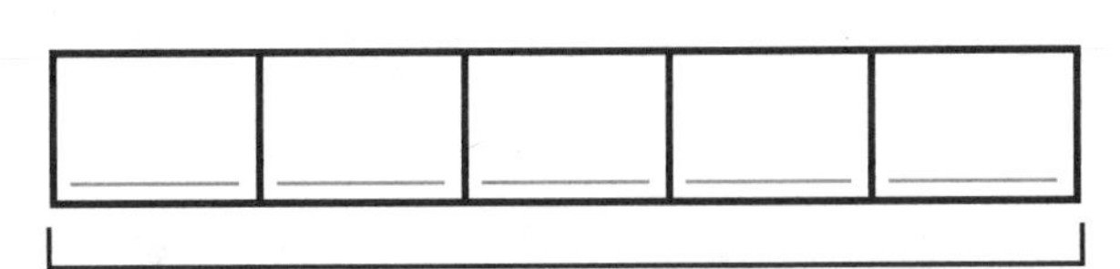

20 dog treats

So, each dog will get _____ treats.

Activity 2 Draw to find how many equal groups.

A dog trainer has 20 dog treats. If he gives 5 treats to each dog in his class, how many dogs are in the class?

- Look at the 20 counters.
- Circle a group of 5 counters.
- Continue circling groups of 5 until all 20 counters are in groups.

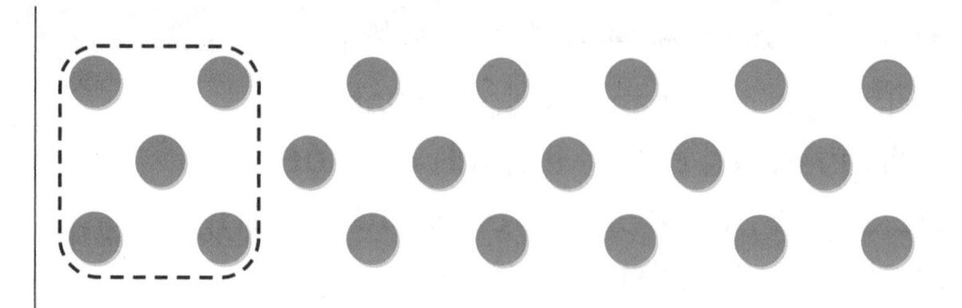

There are _____ groups of 5 counters.

- Complete the bar model to show 20 treats divided into groups of 5 treats.

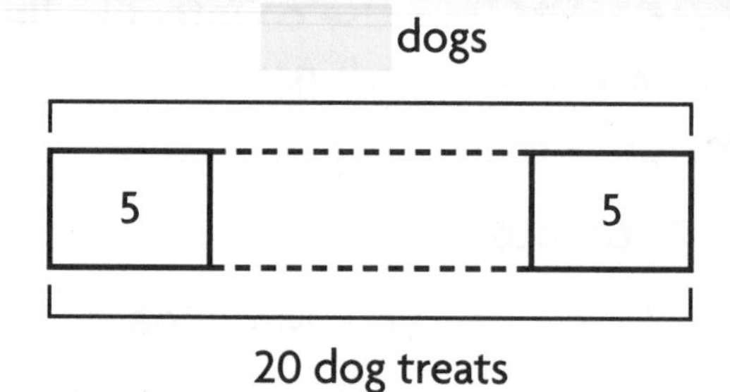

So, there are _____ dogs in the class.

Here are two ways to record division.

Write: $20 \div 5 = 4$

20 ↑ **dividend**; 5 ↑ **divisor**; 4 ↑ **quotient**

$$5\overline{)20}$$

divisor → 5; 4 ← quotient; 20 ↑ dividend

Read: Twenty divided by five equals four.

MATHEMATICAL PRACTICES

Math Talk Describe how you solved the problem. Use the terms *dividend*, *divisor*, and *quotient* in your explanation.

Share and Show

1. Complete the picture to find $12 \div 4$. _____

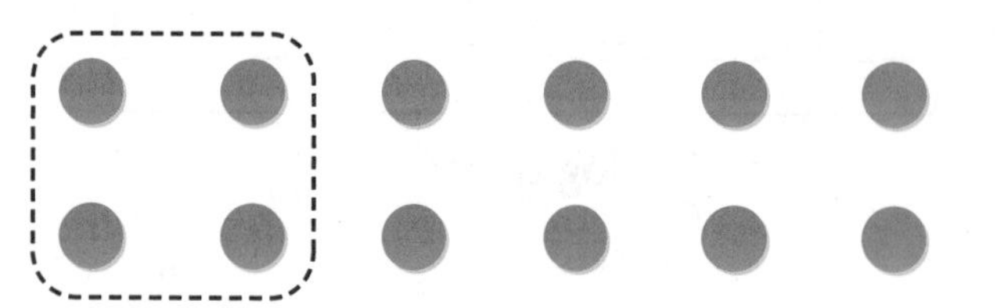

MATHEMATICAL PRACTICES

Math Talk Explain how you know how many groups to make.

Name ______________________

Write a division equation for the picture.

2. 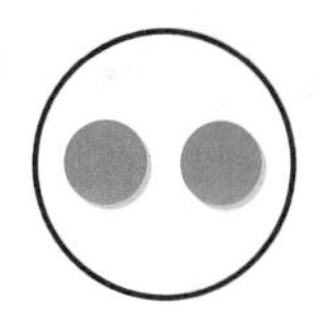 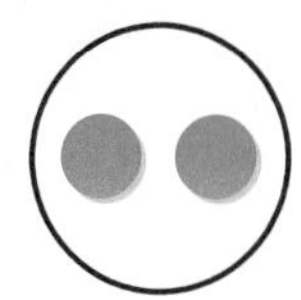 

____

3. 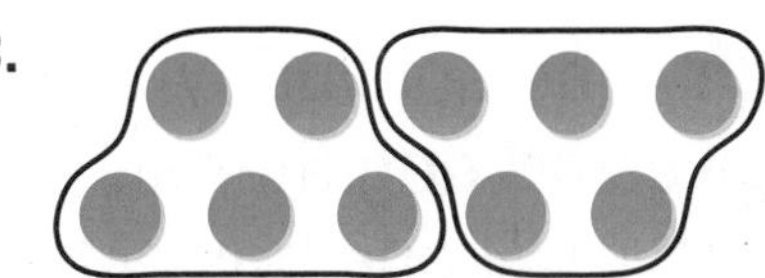

____

On Your Own

Write a division equation for the picture.

4. 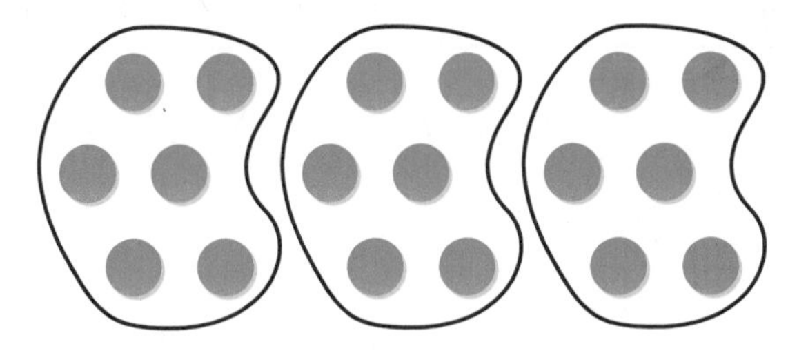

____

5. 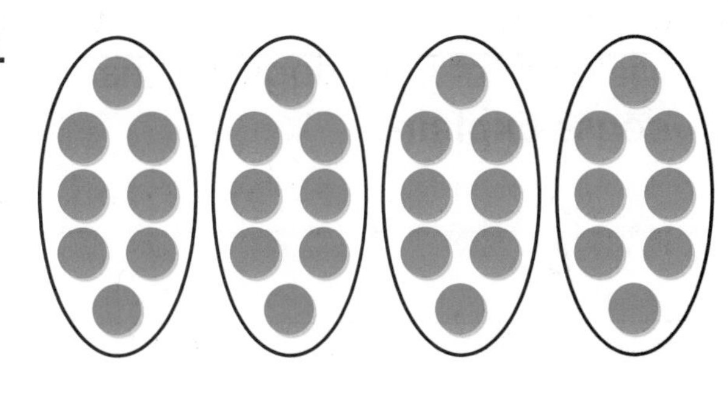

____

Practice: Copy and Solve **Make equal groups to find the quotient. Draw a quick picture to show your work.**

6. $20 \div 2$

7. $27 \div 9$

8. $20 \div 5$

9. $18 \div 3$

Complete the bar model to solve. Then write a division equation for the bar model.

10. There are 24 books in 4 equal stacks. How many books are in each stack?

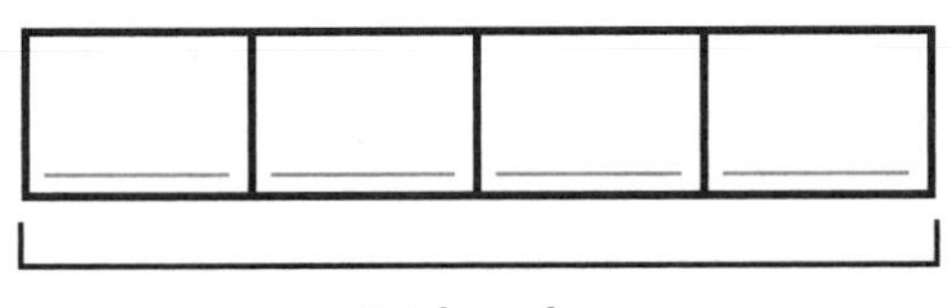

24 books

____

____

11. There are 8 matching socks. How many pairs of socks can you make?

____ pairs

2		2

8 socks

____

____

Problem Solving REAL WORLD

Use the table for 12–13.

Dog Treats	
Type	**Number in Box**
Chew Sticks	14
Chewies	25
Dog Bites	30
Puppy Chips	45

12. Kevin bought a box of Puppy Chips for his dog. If he gives his dog 5 treats each day, for how many days will one box of treats last?

____

13. H.O.T. Pat bought one box of Chew Sticks to share equally between his 2 dogs. Mia bought one box of Chewies to share equally among her 5 dogs. How many more treats will each of Pat's dogs get than each of Mia's dogs? **Explain.**

____

____

____

____

____

14. Write Math **Pose a Problem** Write and solve a problem for $42 \div 7$ in which the quotient is the number of groups.

____

____

____

15. **Test Prep** Ed buys 5 bags of treats. He buys 15 treats in all. How many treats are in each bag?

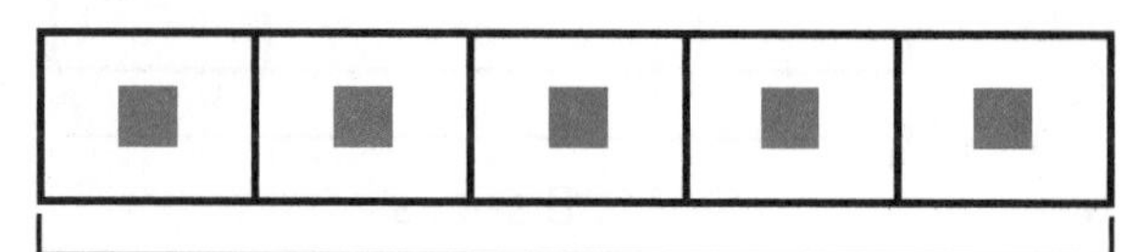

15 treats

Ⓐ 2 Ⓑ 3 Ⓒ 4 Ⓓ 5

SHOW YOUR WORK

FOR MORE PRACTICE:
Standards Practice Book, pp. P111–P112

Name ______________________________

Relate Subtraction and Division

Essential Question How is division related to subtraction?

UNLOCK the Problem REAL WORLD

Serena and Mandy brought a total of 12 newspapers to school for the recycling program. Each girl brought in one newspaper each day. For how many days did the girls bring in newspapers?

- How many newspapers were brought in altogether?

  ____

- How many newspapers did the two girls bring in altogether each day?

  ____

One Way Use repeated subtraction.

- Start with 12.
- Subtract 2 until you reach 0.
- Count the number of times you subtract 2.

	$12 - 2 = 10$	$10 - 2 = 8$	$8 - 2 = $ ___	$6 - 2 = $ ___	$4 - 2 = $ ___	$2 - 2 = $ ___
Number of times you subtract 2:	1	2	3	4	5	6

Since you subtract 2 six times,

there are ______ groups of 2 in 12.

ERROR Alert

Be sure to keep subtracting 2 until you are unable to subtract 2 anymore.

So, Serena and Mandy brought in

newspapers for ______ days.

Write: $12 \div 2 = 6$ or $2\overline{)12}$ with quotient 6

Read: Twelve divided by two equals six.

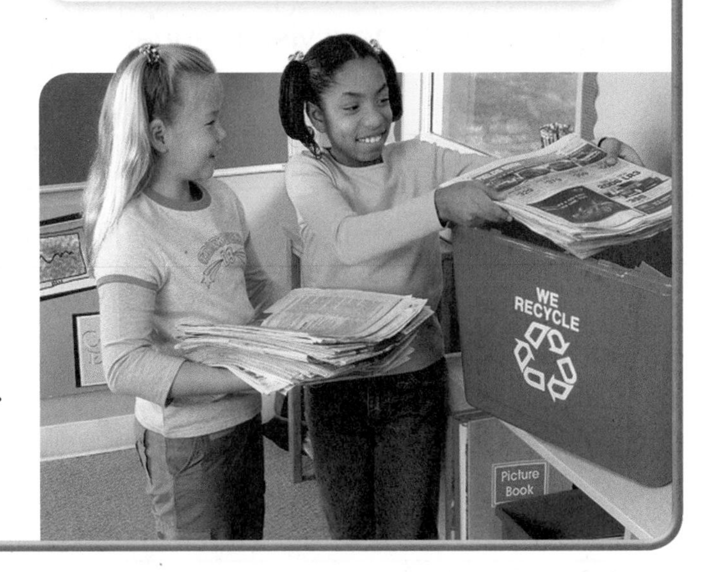

Another Way Count back on a number line.

- Start at 12.
- Count back by 2s as many times as you can. Draw the rest of the jumps on the number line.
- Count the number of times you jumped back 2.

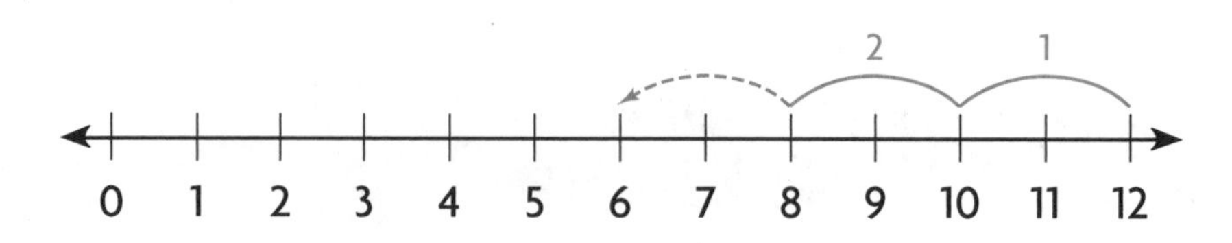

You jumped back by 2 six times.

There are _____ jumps of 2 in 12.

$12 \div 2 =$ _____

Math Talk Explain in your own words how you found the answer.

- What do your jumps of 2 represent? ______________________________

____

Share and Show MATH BOARD

1. Draw the rest of the jumps on the number line to complete the division equation. $12 \div 4 =$ _____

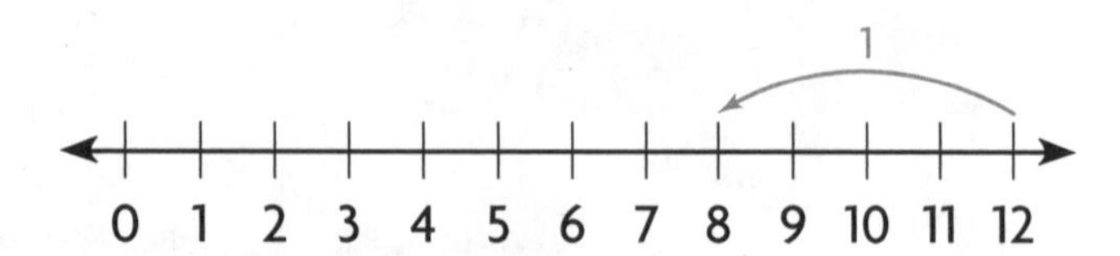

MATHEMATICAL PRACTICES

Math Talk Explain how counting back on a number line is like using repeated subtraction.

Write a division equation.

2.

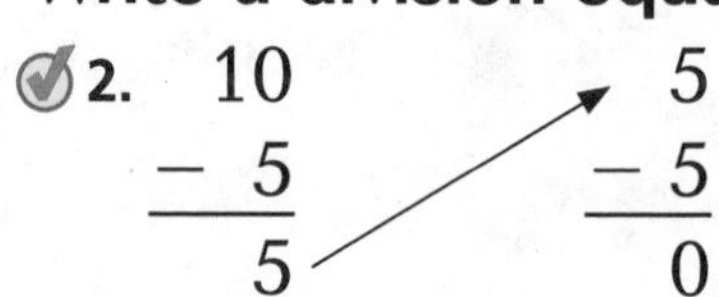

____

3.

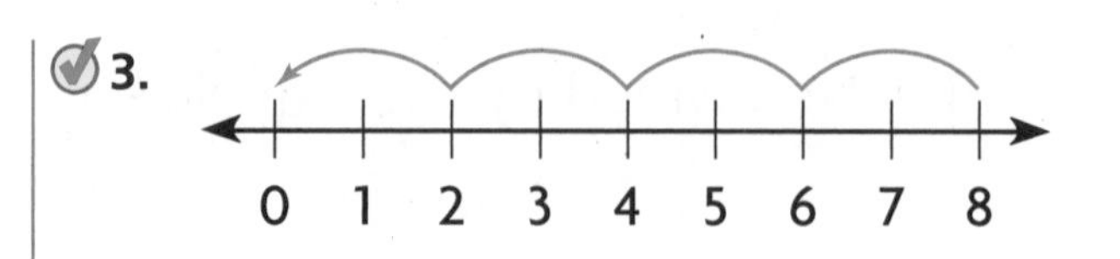

____

Name ______________________________

On Your Own

Write a division equation.

4.

$$\begin{array}{r} 28 \\ -\ 7 \\ \hline 21 \end{array} \quad \begin{array}{r} 21 \\ -\ 7 \\ \hline 14 \end{array} \quad \begin{array}{r} 14 \\ -\ 7 \\ \hline 7 \end{array} \quad \begin{array}{r} 7 \\ -\ 7 \\ \hline 0 \end{array}$$

____

5.

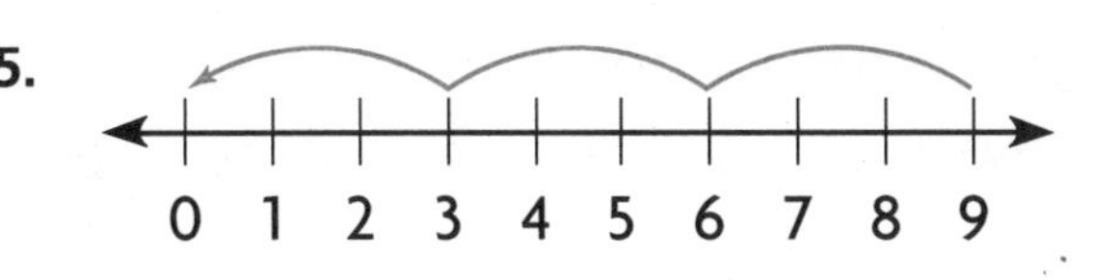

____

6.

$$\begin{array}{r} 24 \\ -\ 8 \\ \hline 16 \end{array} \quad \begin{array}{r} 16 \\ -\ 8 \\ \hline 8 \end{array} \quad \begin{array}{r} 8 \\ -\ 8 \\ \hline 0 \end{array}$$

____

7.

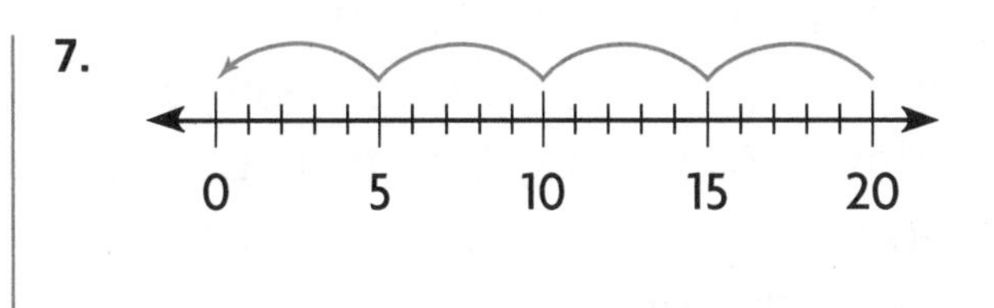

____

8. H.O.T. Write a word problem that can be solved by using one of the division equations above.

____

____

Use repeated subtraction or a number line to solve.

9. $18 \div 6 =$ _____

10. $14 \div 7 =$ _____

11. $9\overline{)27}$

12. $3\overline{)24}$

Problem Solving REAL WORLD

Use the graph for 13–15.

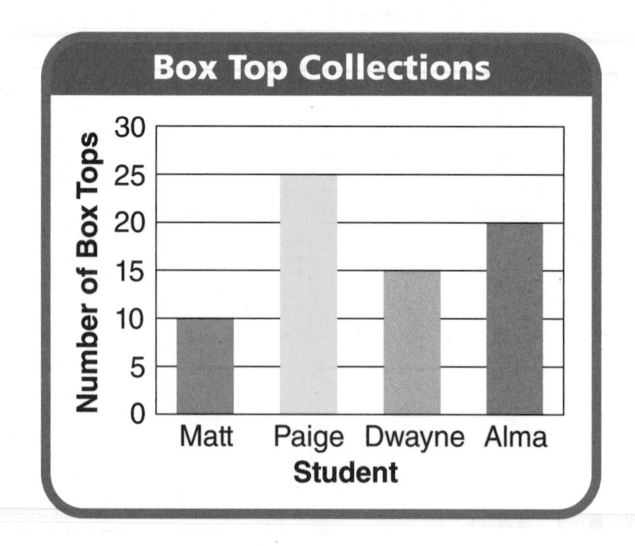

13. Matt puts his box tops in 2 equal piles. How many box tops are in each pile?

____

14. H.O.T. Paige brought an equal number of box tops to school each day for 5 days. Alma also brought an equal number of box tops each day for 5 days. How many box tops did the two students bring in altogether each day? **Explain.**

____

____

____

____

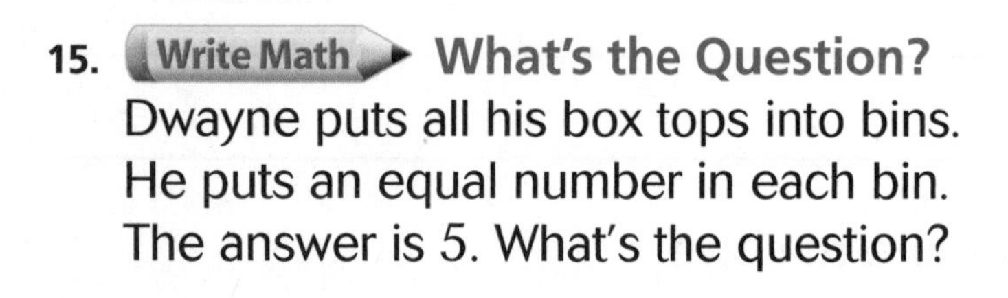

15. Write Math ▸ **What's the Question?** Dwayne puts all his box tops into bins. He puts an equal number in each bin. The answer is 5. What's the question?

____

____

16. **Test Prep** Maya collected 7 box tops each day. She collected 21 box tops in all. For how many days did Maya collect box tops?

Ⓐ 2 days
Ⓑ 3 days
Ⓒ 4 days
Ⓓ 6 days

SHOW YOUR WORK

FOR MORE PRACTICE:
Standards Practice Book, pp. P113–P114

Name ____________________

Mid-Chapter Checkpoint

▶ Vocabulary

Choose the best term from the box to complete the sentence.

Vocabulary
divide
divisor

1. You ________ when you separate into equal groups. (p. 213)

▶ Concepts and Skills

Use counters or draw a quick picture on your MathBoard. Make or circle equal groups. Complete the table.

	Counters	Number of Equal Groups	Number in Each Group
2.	6	2	
3.	30		5
4.	28	7	

Write a division equation for the picture.

5. 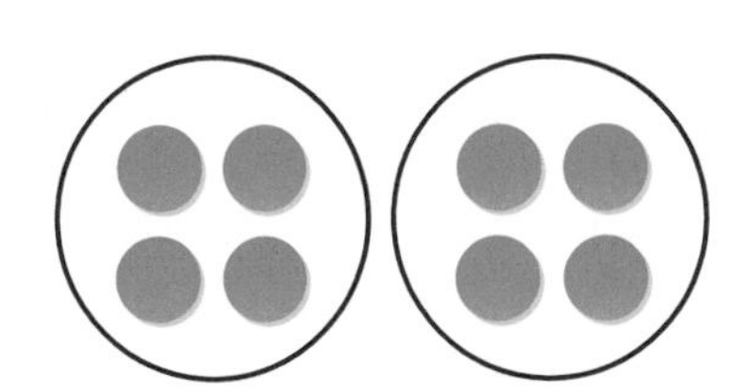

____

6. 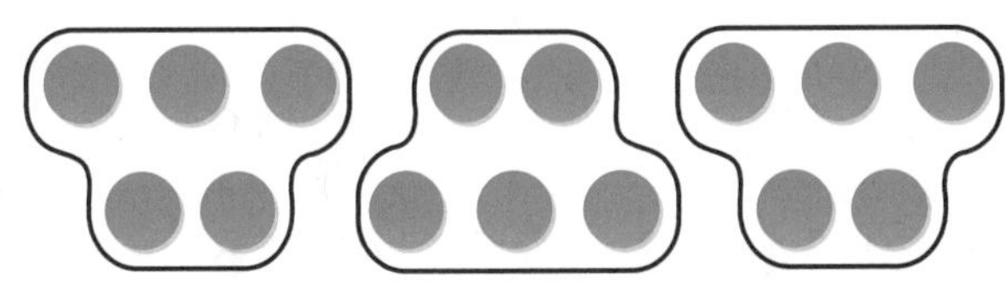

____

Write a division equation.

7.
$$\begin{array}{r} 36 \\ -\ 9 \\ \hline 27 \end{array} \quad \begin{array}{r} 27 \\ -\ 9 \\ \hline 18 \end{array} \quad \begin{array}{r} 18 \\ -\ 9 \\ \hline 9 \end{array} \quad \begin{array}{r} 9 \\ -\ 9 \\ \hline 0 \end{array}$$

____

8. 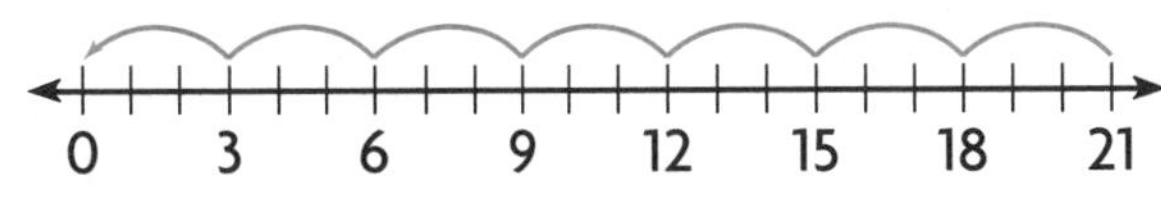

____

Fill in the bubble for the correct answer choice.

9. Adam plants 14 seeds in some flowerpots. If he puts 2 seeds in each pot, how many flowerpots does he use?

   Ⓐ 7 Ⓒ 12
   Ⓑ 8 Ⓓ 16

10. Desiree has 20 stickers. She gives the same number of stickers to each of 5 friends. Which equation can be used to find the number of stickers each friend receives?

    Ⓐ $20 + 5 = \blacksquare$
    Ⓑ $20 - 5 = \blacksquare$
    Ⓒ $20 \times 5 = \blacksquare$
    Ⓓ $20 \div 5 = \blacksquare$

11. Jayden modeled a division equation with some counters. Which division equation matches the model?

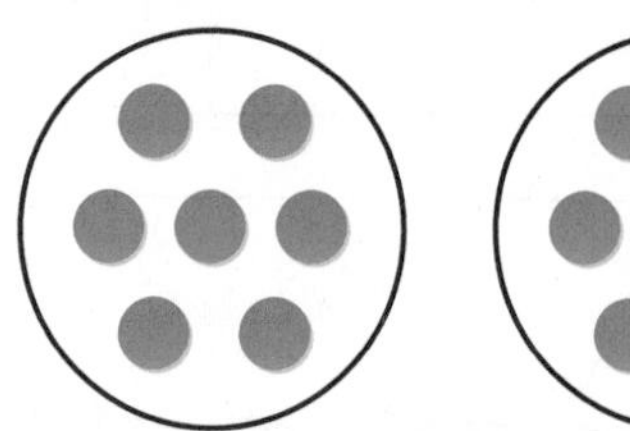
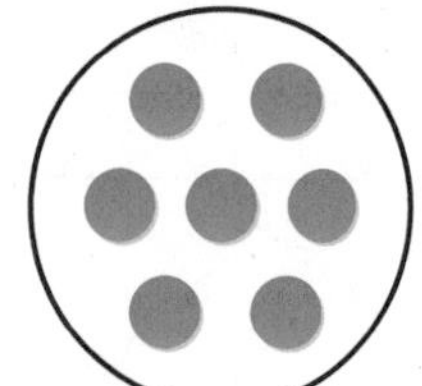

    Ⓐ $12 \div 2 = 6$ Ⓒ $16 \div 2 = 8$
    Ⓑ $14 \div 2 = 7$ Ⓓ $18 \div 2 = 9$

12. Lillian bought 24 cans of cat food. There were 4 cans in each pack. How many packs of cat food did Lillian buy?

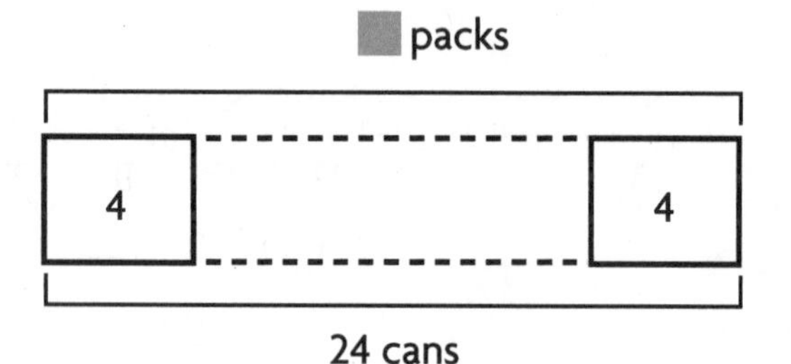

    Ⓐ 5 Ⓑ 6 Ⓒ 7 Ⓓ 8

Name ______________________

Model with Arrays

Essential Question How can you use arrays to solve division problems?

Investigate

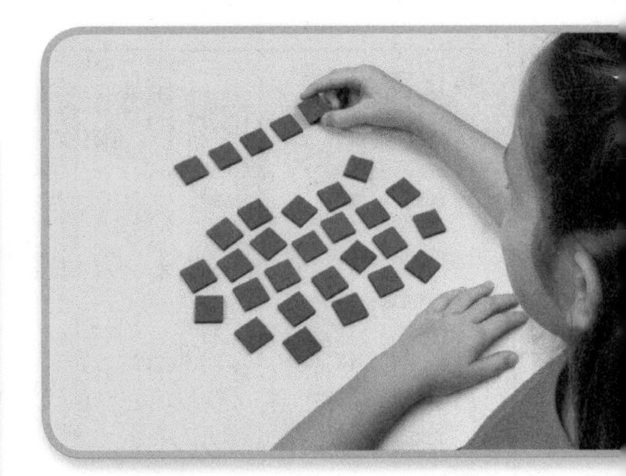

Materials ■ square tiles

You can use arrays to model division and find equal groups.

A. Count out 30 tiles. Make an array to find how many rows of 5 are in 30.

B. Make a row of 5 tiles.

C. Continue to make as many rows of 5 tiles as you can.

How many rows of 5 did you make? __________

Draw Conclusions

1. **Explain** how you used the tiles to find the number of rows of 5 in 30.

____

____

2. What multiplication equation could you write for the array? **Explain**.

____

____

3. H.O.T. **Apply** Tell how to use an array to find how many rows of 6 are in 30.

____

____

Make Connections

You can write a division equation to show how many rows of 5 are in 30. Show the array you made in Investigate by completing the drawing below.

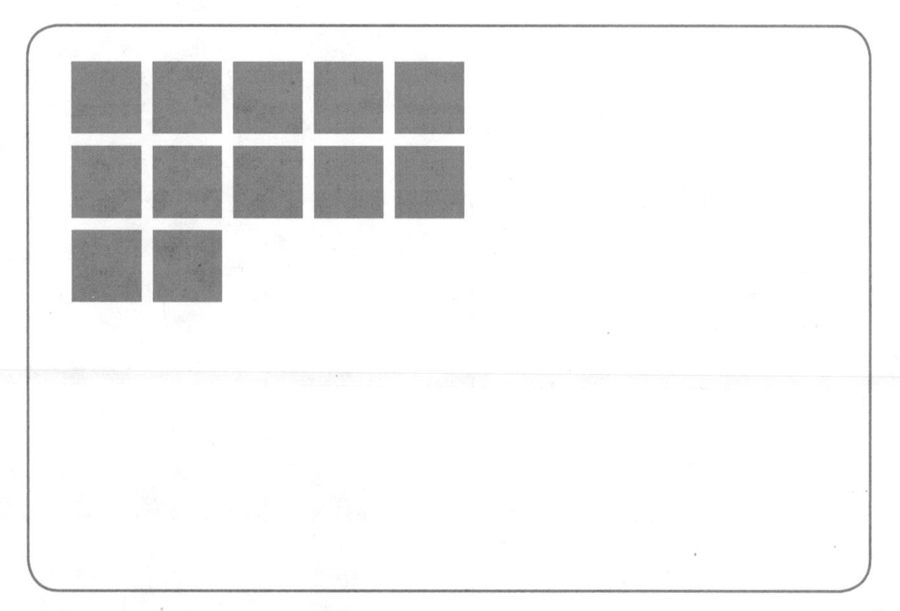

Math Idea

You can divide to find the number of equal rows or to find the number in each row.

$30 \div 5 = \blacksquare$

There are ______ rows of 5 tiles in 30.

So, $30 \div 5 =$ ______.

Try This!

Count out 24 tiles. Make an array with the same number of tiles in 4 rows. Place 1 tile in each of the 4 rows. Then continue placing 1 tile in each row until you use all the tiles. Draw your array below.

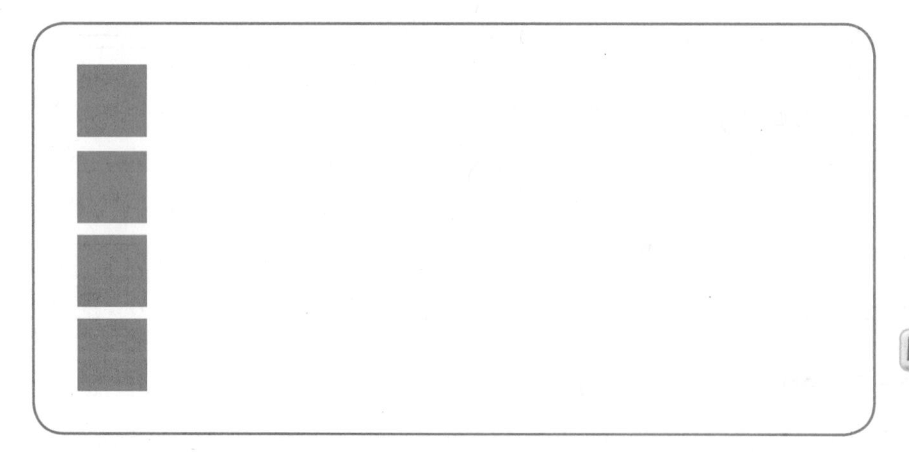

MATHEMATICAL PRACTICES

Math Talk Explain how making an array helps you divide.

- How many tiles are in each row? ______________
- What division equation can you write for your array? ______________

Name ____________________

Share and Show 

Use square tiles to make an array. Solve.

1. How many rows of 3 are in 18? ____________

2. How many rows of 6 are in 12? ____________

3. How many rows of 7 are in 21? ____________

4. How many rows of 8 are in 32? ____________

Make an array. Then write a division equation.

5. 25 tiles in 5 rows ____________

6. 14 tiles in 2 rows ____________

7. 28 tiles in 4 rows ____________

8. 27 tiles in 9 rows ____________

9. How many rows of 3 are in 15? ____________

10. How many rows of 8 are in 24? ____________

MATHEMATICAL PRACTICES

Math Talk **Explain** when you count the number of rows to find the answer and when you count the number of tiles in each row to find the answer.

11. **Write Math** **Show** two ways you could make an array with tiles for $18 \div 6$. Shade squares on the grid to record the arrays.

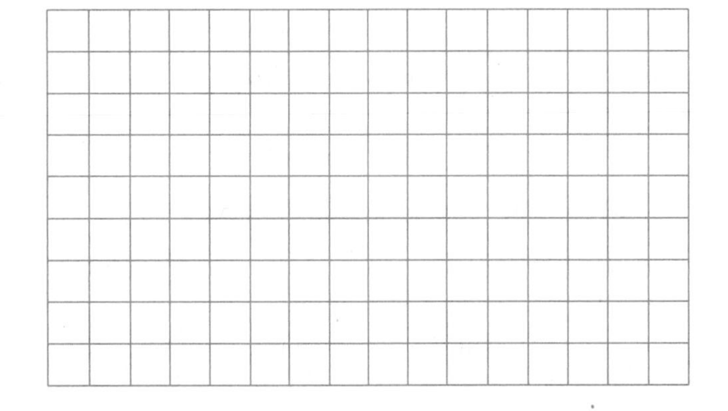

UNLOCK the Problem REAL WORLD

12. Thomas has 28 tomato seedlings to plant in his garden. He wants to plant 4 seedlings in each row. How many rows of tomato seedlings will Thomas plant?

Ⓐ 5 Ⓑ 6 Ⓒ 7 Ⓓ 8

a. What do you need to find? ______________________________

____

b. What operation could you use to solve the problem? ______________

____

c. Draw an array to find the number of rows of tomato seedlings.

d. What is another way you could have solved the problem?

____

____

e. Complete the sentences.

Thomas has ____________ tomato seedlings.

He wants to plant ____________

seedlings in each ____________.

So, Thomas will plant ____________ rows of tomato seedlings.

f. Fill in the bubble for the correct answer choice above.

13. Faith plants 36 flowers in 6 equal rows. How many flowers are in each row?

Ⓐ 42 Ⓑ 30 Ⓒ 7 Ⓓ 6

14. There were 20 plants sold at a store on Saturday. Customers bought 5 plants each. How many customers bought the plants?

Ⓐ 3 Ⓑ 4 Ⓒ 5 Ⓓ 6

 FOR MORE PRACTICE: Standards Practice Book, pp. P115–P116

Name ______________________

Relate Multiplication and Division

Essential Question How can you use multiplication to divide?

UNLOCK the Problem REAL WORLD

Pam went to the fair. She went on the same ride 6 times and used the same number of tickets each time. She used 18 tickets. How many tickets did she use each time she went on the ride?

- What do you need to find?

____

____

- Circle the numbers you need to use.

One Way Use bar models.

You can use bar models to understand how multiplication and division are related.

Complete the bar model to show 18 tickets divided into 6 equal groups.

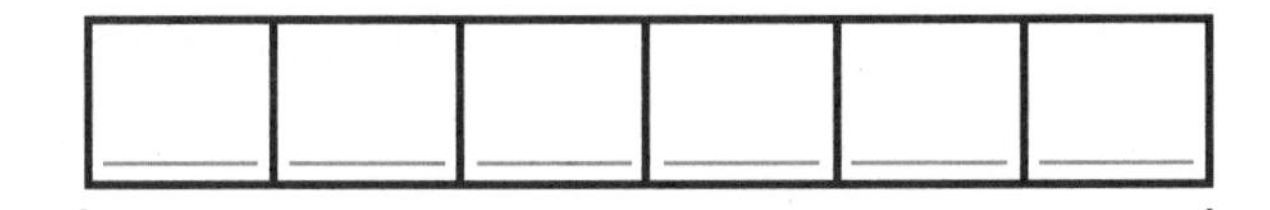

18 tickets

Write: $18 \div 6 =$ ______

So, Pam used ______ tickets each time she went on the ride.

What if the problem said Pam went on the ride 6 times and used 3 tickets each time? How many tickets did Pam use in all?

Complete the bar model to show 6 groups of 3 tickets.

3	3	3	3	3	3

▢ tickets

Write: $6 \times 3 =$ ______

Multiplication and division are opposite operations, or **inverse operations**.

You can think about multiplication to solve a division problem.

To solve $18 \div 6 = \square$, think $6 \times \square = 18$.

Since $6 \times 3 = 18$, then $18 \div 6 = 3$.

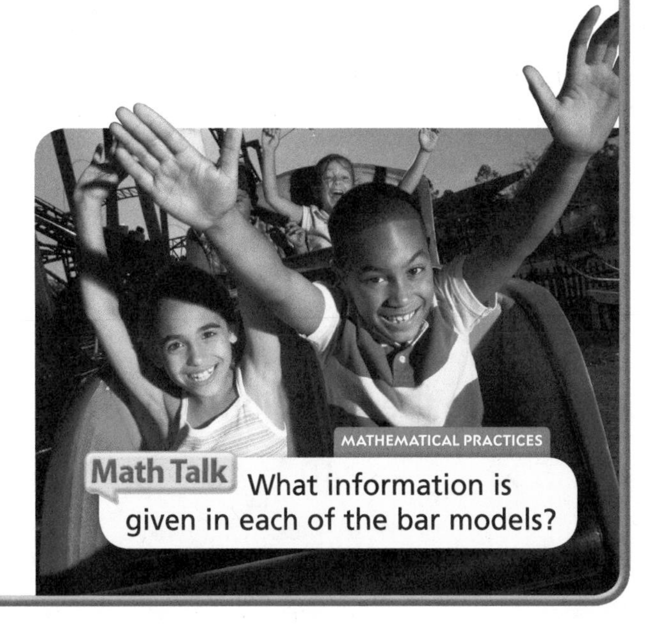

MATHEMATICAL PRACTICES

Math Talk What information is given in each of the bar models?

Another Way Use an array.

You can use an array to see how multiplication and division are related.

Show an array with 18 counters in 3 equal rows by completing the drawing.

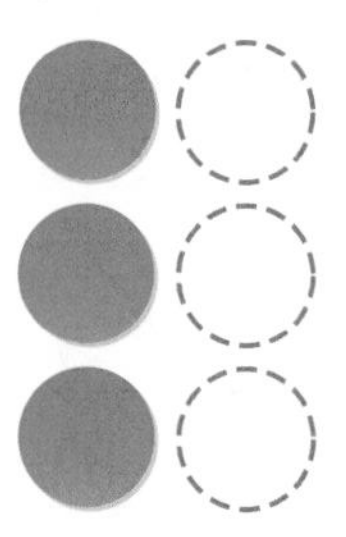

There are _____ counters in each row.

Write: $18 \div 3 =$ _____

The same array can be used to find the total number if you know there are 3 rows with 6 counters in each row.

Write: $3 \times 6 =$ _____

Share and Show MATH BOARD

1. Use the array to complete the equation.

   Think: There are 3 counters in each row.

   $6 \div 2 =$ _____

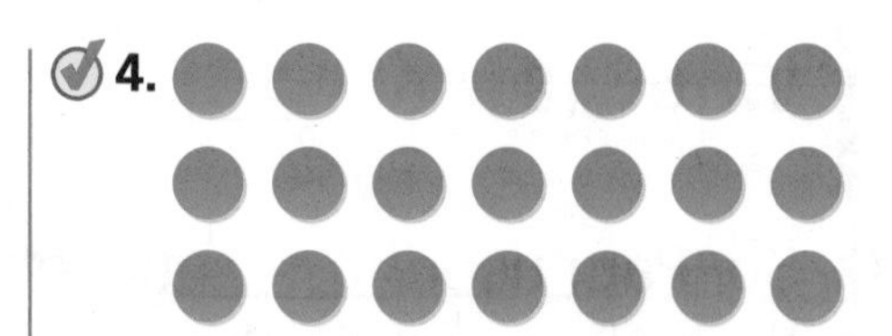

Complete.

2. 3 rows of _____ = 15

   $3 \times$ _____ $= 15$

   $15 \div 3 =$ _____

3. 2 rows of _____ = 12

   $2 \times$ _____ $= 12$

   $12 \div 2 =$ _____

4. 3 rows of _____ = 21

   $3 \times$ _____ $= 21$

   $21 \div 3 =$ _____

Complete the equations.

5. $5 \times$ _____ $= 40$ $40 \div 5 =$ _____

6. $6 \times$ _____ $= 18$ $18 \div 6 =$ _____

Name ____________________

On Your Own

Complete.

7.

5 rows of ____ = 30

5 × ____ = 30

30 ÷ 5 = ____

8.

4 rows of ____ = 20

4 × ____ = 20

20 ÷ 4 = ____

9.

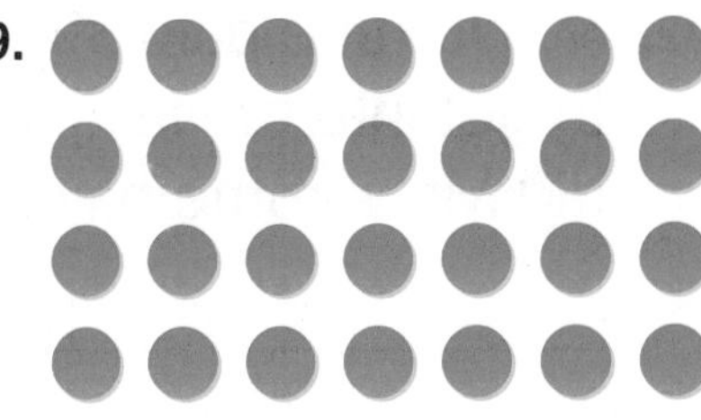

4 rows of ____ = 28

4 × ____ = 28

28 ÷ 4 = ____

Complete the equations.

10. 7 × ____ = 21   21 ÷ 7 = ____

11. 8 × ____ = 16   16 ÷ 8 = ____

12. 4 × ____ = 32   32 ÷ 4 = ____

13. 6 × ____ = 24   24 ÷ 6 = ____

14. 9 × ____ = 18   18 ÷ 9 = ____

15. 5 × ____ = 25   25 ÷ 5 = ____

H.O.T. **Algebra Complete.**

16. 3 × 3 = 27 ÷ ____

17. 16 ÷ 2 = ____ × 2

18. 9 = ____ ÷ 4

19. 5 = ____ ÷ 7

20. 42 ÷ 7 = ____ × 2

21. 30 ÷ ____ = 2 × 3

Problem Solving REAL WORLD

Use the table for 22–23.

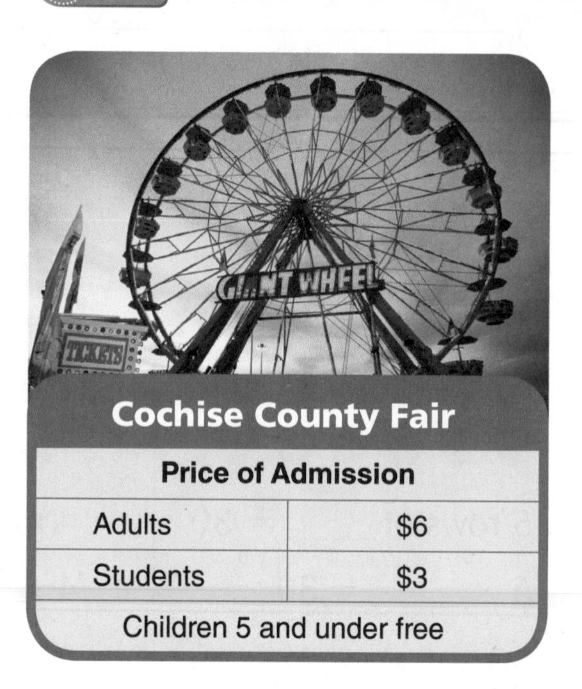

Cochise County Fair

Price of Admission	
Adults	$6
Students	$3
Children 5 and under free	

22. Mr. Jerome paid $24 for some students to get into the fair. How many students did Mr. Jerome pay for?

23. H.O.T. Garrett is 8 years old. He and his family are going to the county fair. What is the price of admission for Garrett, his 2 parents, and baby sister?

24. There are 20 seats on the Wildcat ride. The number of seats in each car is the same. If there are 5 cars on the ride, how many seats are there in each car? Complete the bar model to show the problem. Then answer the question.

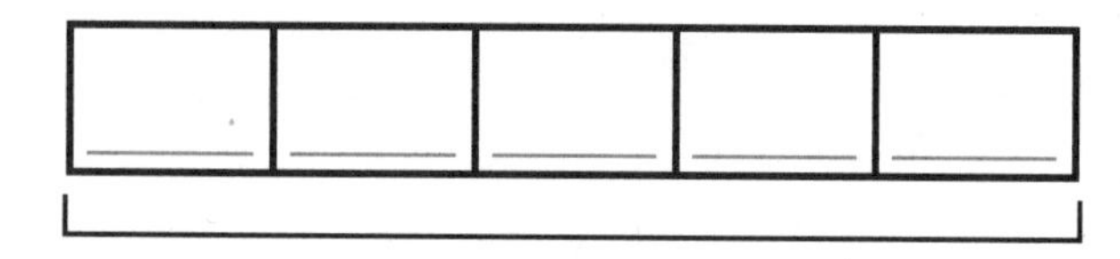

20 seats

25. Write Math ▸ **Pose a Problem** How many days are there in 2 weeks? Write and solve a related word problem to represent the inverse operation.

26. **Test Prep** There are 35 prizes in 5 equal rows. How many prizes are in each row?

Ⓐ 6 Ⓑ 7 Ⓒ 8 Ⓓ 9

SHOW YOUR WORK

 FOR MORE PRACTICE: Standards Practice Book, pp. P117–P118

Name ____________________

Write Related Facts

Essential Question How can you write a set of related multiplication and division facts?

UNLOCK the Problem

Related facts are a set of related multiplication and division equations. What related facts can you write for 2, 4, and 8?

- What model can you use to show how multiplication and division are related?

____

Materials ■ square tiles

STEP 1

Use 8 tiles to make an array with 2 equal rows.

Draw the rest of the tiles.

How many tiles are in each row? ________

Write a division equation for the array using the total number of tiles as the dividend and the number of rows as the divisor.

_____ ÷ _____ = _____

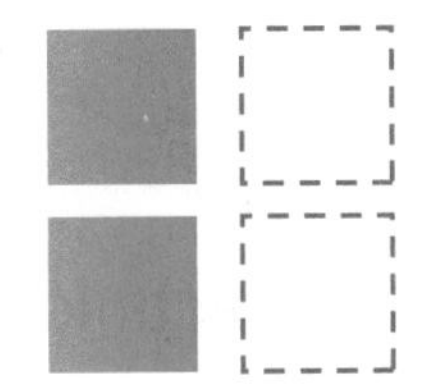

Write a multiplication equation for the array.

_____ × _____ = _____

STEP 2

Now, use 8 tiles to make an array with 4 equal rows.

Draw the rest of the tiles.

How many tiles are in each row? ________

Write a division equation for the array using the total number of tiles as the dividend and the number of rows as the divisor.

_____ ÷ _____ = _____

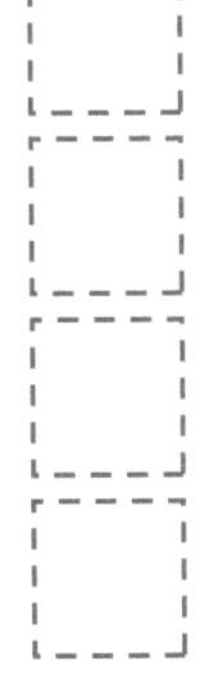

Write a multiplication equation for the array.

_____ × _____ = _____

So, $8 \div 2 =$ _____, $2 \times 4 =$ _____, $8 \div 4 =$ _____,

and $4 \times 2 =$ _____ are related facts.

Try This! **Draw an array with 4 rows of 4 tiles.**

Your array shows the related facts for 4, 4, and 16.

$4 \times 4 =$ ____ $16 \div 4 =$ ____

Since both factors are the same, there are only two equations in this set of related facts.

- Write another set of related facts that has only two equations.

____

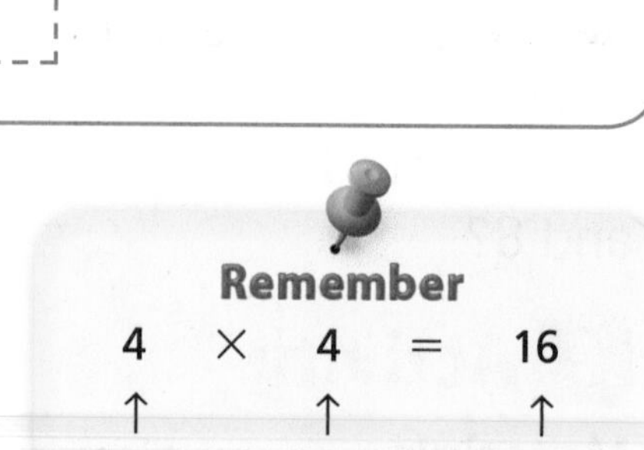

Share and Show

1. Complete the related facts for this array.

$2 \times 8 = 16$ $16 \div 2 = 8$

____ ____

MATHEMATICAL PRACTICES

Math Talk Look at the multiplication and division equations in a set of related facts. What do you notice about the products and dividends? **Explain.**

Write the related facts for the array.

2. 

____

____

____

____

3. 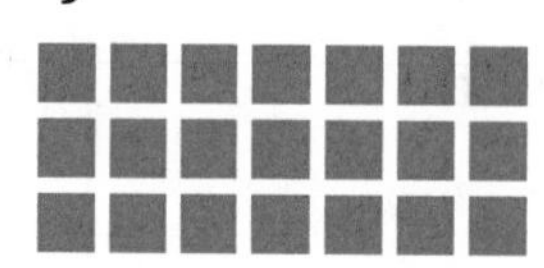

____

____

____

____

4. 

____

____

____

____

5. Why do the related facts for the array in Exercise 2 have only two equations?

____

Name ____________________

On Your Own

Write the related facts for the array.

6.

______
______
______
______

7.

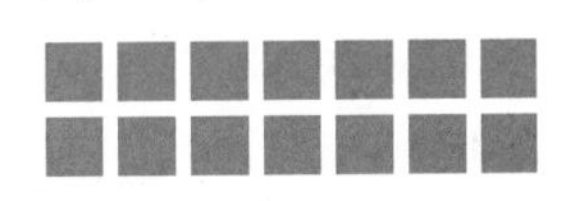

______
______
______
______

8.

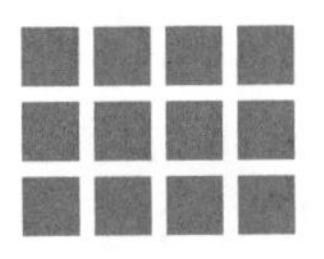

______
______
______
______

Write the related facts for the set of numbers.

9. 2, 5, 10

______
______
______
______

10. 3, 8, 24

______
______
______
______

11. 6, 6, 36

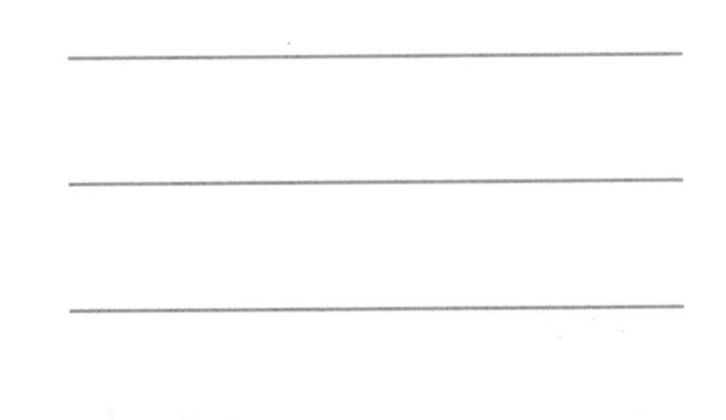

______
______
______
______

Complete the related facts.

12. $4 \times 7 =$ ______

$7 \times$ ______ $= 28$

$28 \div$ ______ $= 4$

$28 \div 4 =$ ______

13. $5 \times$ ______ $= 30$

$6 \times$ ______ $= 30$

$30 \div 6 =$ ______

$30 \div 5 =$ ______

14. ______ $\times 9 = 27$

______ $\times 3 = 27$

______ $\div 9 = 3$

$27 \div$ ______ $= 9$

15. ______ $\times 5 = 20$

______ $\times 4 = 20$

$20 \div$ ______ $= 5$

______ $\div 5 = 4$

16. $2 \times$ ______ $= 12$

$6 \times 2 =$ ______

______ $\div 2 = 6$

$12 \div$ ______ $= 2$

17. $5 \times 8 =$ ______

______ $\times 5 = 40$

$40 \div 5 =$ ______

______ $\div 8 = 5$

Problem Solving REAL WORLD

Use the table for 18–19.

Clay Supplies

Item	Number in Package
Clay	12 sections
Clay tool set	11 tools
Glitter dough	36 sections

18. Mr. Lee divides 1 package of clay and 1 package of glitter dough equally among 4 students. How many more glitter dough sections than clay sections does each student get?

____

SHOW YOUR WORK

19. Write Math ▸ **What's the Error?** Ty has a package of glitter dough. He says he can give 9 friends 5 equal sections. Describe his error.

____

____

____

____

20. H.O.T. **Pose a Problem** Write a word problem that can be solved by using $35 \div 5$. Solve your problem.

____

____

____

____

21. **Test Prep** Which equation is NOT included in the same set of related facts as $9 \times 4 = 36$?

Ⓐ $4 \times 9 = 36$

Ⓑ $36 \div 6 = 6$

Ⓒ $36 \div 4 = 9$

Ⓓ $36 \div 9 = 4$

FOR MORE PRACTICE:
Standards Practice Book, pp. P119–P120

Name ______________________________

Division Rules for 1 and 0

Essential Question What are the rules for dividing with 1 and 0?

UNLOCK the Problem REAL WORLD

What rules for division can help you divide with 1 and 0?

If there is only 1 fishbowl, then all the fish must go in that fishbowl.

$4 \div 1 = 4$

4 ↑ number of fish; 1 ↑ number of bowls; 4 ↑ number in each bowl

Try This! There are 3 fish and 1 fishbowl. Draw a quick picture to show the fish in the fishbowl.

Write the equation your picture shows.

_____ ÷ _____ = _____

Rule A: Any number divided by 1 equals that number.

MATHEMATICAL PRACTICES

Math Talk Explain how Rule A is related to the Identity Property of Multiplication.

If there is the same number of fish and fishbowls, then 1 fish goes in each fishbowl.

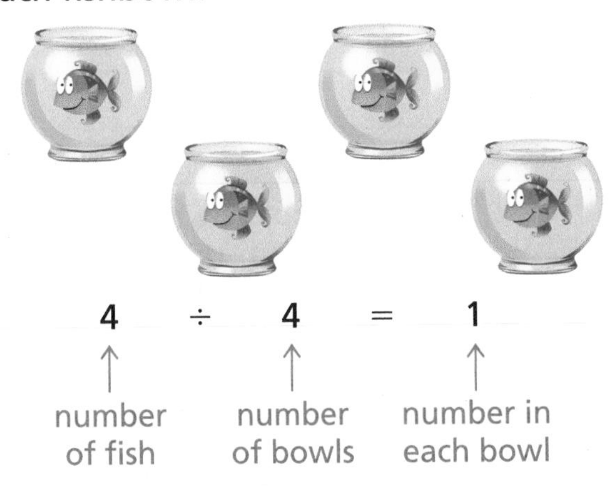

$4 \div 4 = 1$

4 ↑ number of fish; 4 ↑ number of bowls; 1 ↑ number in each bowl

Try This! There are 3 fish and 3 fishbowls. Draw a quick picture to show the fish divided equally among the fishbowls.

Write the equation your picture shows.

_____ ÷ _____ = _____

Rule B: Any number (except 0) divided by itself equals 1.

If there are 0 fish and 4 fishbowls, there will not be any fish in the fishbowls.

$0 \div 4 = 0$

number of fish ↑ number of bowls ↑ number in each bowl ↑

Try This! There are 0 fish and 3 fishbowls. Draw a quick picture to show the fishbowls.

Write the equation your picture shows.

_____ ÷ _____ = _____

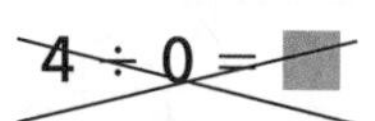

Rule C: Zero divided by any number (except 0) equals 0.

If there are 0 fishbowls, then you cannot separate the fish equally into fishbowls. Dividing by 0 is not possible.

~~$4 \div 0 = \blacksquare$~~

Rule D: You cannot divide by 0.

Share and Show MATH BOARD

1. Use the picture to find $2 \div 2$. _____

MATHEMATICAL PRACTICES

Math Talk Explain what happens when you divide a number (except 0) by itself.

Find the quotient.

2. $7 \div 1 =$ _____

3. $8 \div 8 =$ _____

✓4. $0 \div 5 =$ _____

✓5. $6 \div 6 =$ _____

Name ________________

On Your Own

Find the quotient.

6. $0 \div 8 =$ ____ 7. $5 \div 5 =$ ____ 8. $2 \div 1 =$ ____ 9. $0 \div 7 =$ ____

10. $5\overline{)0}$ 11. $1\overline{)9}$ 12. $7\overline{)7}$ 13. $10\overline{)10}$

Practice: Copy and Solve **Find the quotient.**

14. $6 \div 1$ 15. $25 \div 5$ 16. $0 \div 6$ 17. $18 \div 3$

18. $14 \div 2$ 19. $9 \div 9$ 20. $28 \div 4$ 21. $8 \div 1$

22. $3\overline{)27}$ 23. $5\overline{)10}$ 24. $3\overline{)0}$ 25. $1\overline{)0}$

Problem Solving

SHOW YOUR WORK

26. Claire has 7 parakeets. She puts 4 in a cage. She divides the other parakeets equally among 3 friends to hold. How many parakeets does each friend get to hold?

____

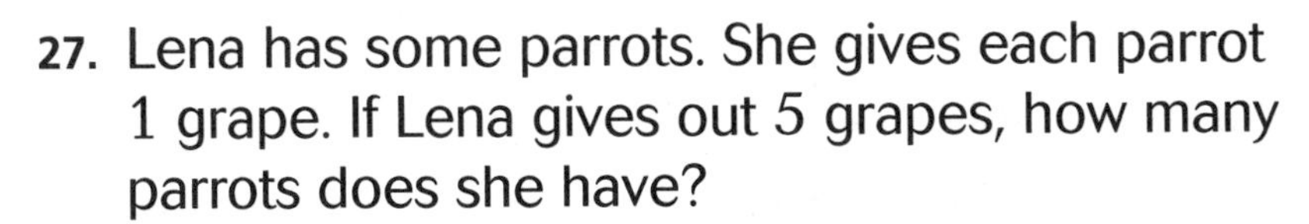

27. Lena has some parrots. She gives each parrot 1 grape. If Lena gives out 5 grapes, how many parrots does she have?

____

28. H.O.T. Write Math Suppose a pet store has 21 birds that are in 21 cages. Use what you know about division rules to find the number of birds in each cage. **Explain** your answer.

____

____

29. **Test Prep** Joe has 4 horses. He puts each horse in its own stall. How many stalls does Joe use?

Ⓐ 0 Ⓑ 1 Ⓒ 2 Ⓓ 4

Connect to Reading

Compare and Contrast

You have learned the rules for division with 1. Compare and contrast them to help you learn how to use the rules to solve problems.

Compare the rules. Think about how they are *alike*.
Contrast the rules. Think about how they are *different*.

Read: Rule A: Any number divided by 1 equals that number.

Rule B: Any number (except 0) divided by itself equals 1.

Compare: How are the rules alike?

- Both are division rules for 1.

Contrast: How are the rules different?

- Rule A is about dividing a number by 1.
  The quotient is that number.
- Rule B is about dividing a number (except 0) by itself.
  The quotient is always 1.

Read the problem. Write an equation. Solve.
Circle *Rule A* or *Rule B* to tell which rule you used.

30. Jamal bought 7 goldfish at the pet store. He put them in 1 fishbowl. How many goldfish did he put in the fishbowl?

____

Rule A Rule B

31. Ava has 6 turtles. She divides them equally among 6 aquariums. How many turtles does she put in each aquarium?

____

Rule A Rule B

FOR MORE PRACTICE:
Standards Practice Book, pp. P121–P122

Name ________________________________

Vocabulary

Choose the best term from the box to complete the sentence.

Vocabulary
dividend
inverse operations
related facts

1. Multiplication and division are opposite operations, or ________________. (p. 235)

2. ________________ are a set of related multiplication and division equations. (p. 239)

Concepts and Skills

Make an array. Then write a division equation.

3. 32 tiles in 4 rows

____

4. 28 tiles in 7 rows

____

Complete the equations.

5. $3 \times$ _____ $= 24$ $24 \div 3 =$ _____

6. $5 \times$ _____ $= 15$ $15 \div 5 =$ _____

7. $7 \times$ _____ $= 49$ $49 \div 7 =$ _____

8. $9 \times$ _____ $= 18$ $18 \div 9 =$ _____

Complete the related facts.

9. $4 \times 8 =$ _____

$8 \times$ _____ $= 32$

$32 \div$ _____ $= 8$

$32 \div 8 =$ _____

10. $3 \times$ _____ $= 9$

_____ $\div 3 = 3$

11. $6 \times$ _____ $= 42$

$7 \times 6 =$ _____

_____ $\div 6 = 7$

$42 \div$ _____ $= 6$

GO Online Assessment Options Chapter Test

Fill in the bubble for the correct answer choice.

12. Caroline has 27 books. She put an equal number of her books on each of 3 shelves. How many books are on each shelf?

    Ⓐ 7

    Ⓑ 8

    Ⓒ 9

    Ⓓ 10

13. Leo has 24 cookies. He wants to put groups of 6 cookies on a plate. How many plates does he need?

    Ⓐ 4

    Ⓑ 5

    Ⓒ 18

    Ⓓ 30

14. A pet shop has 18 hamsters. They sold 10 and put the remaining hamsters into 8 cages. How many hamsters are in each cage?

    Ⓐ 8

    Ⓑ 4

    Ⓒ 2

    Ⓓ 1

15. Aidan bought 14 goldfish and 2 fishbowls. He put an equal number of fish in each fishbowl. Which equation can be used to find how many goldfish are in each fishbowl?

    Ⓐ $14 - 2 = \blacksquare$

    Ⓑ $2 + \blacksquare = 14$

    Ⓒ $14 \div 2 = \blacksquare$

    Ⓓ $2 \times 14 = \blacksquare$

Name ______________________________

Fill in the bubble for the correct answer choice.

16. Which division equation belongs to this set of related facts?

$4 \times 6 = 24$ $6 \times 4 = 24$

Ⓐ $20 \div 4 = 5$
Ⓑ $24 \div 6 = 4$
Ⓒ $18 \div 6 = 3$
Ⓓ $24 \div 3 = 8$

17. Jasmine made some cards for her family. She used 16 stickers and put 4 of them on each card. How many cards did Jasmine make?

Ⓐ 2
Ⓑ 3
Ⓒ 4
Ⓓ 5

18. Which of the following multiplication equations can be used to find $42 \div 7$?

Ⓐ $6 \times 6 = 36$
Ⓑ $5 \times 8 = 40$
Ⓒ $8 \times 6 = 48$
Ⓓ $7 \times 6 = 42$

19. Jill has 6 puppies. The puppies play in 1 crate. How many puppies are in the crate?

Ⓐ 0
Ⓑ 1
Ⓒ 6
Ⓓ 7

▶ Constructed Response

20. Brendan bought 30 baseball cards. They came in packs of 10. How many packs of baseball cards did Brendan buy? **Explain** how you found the answer.

____

____

____

21. Sue drew this array. Write the related facts represented by the array.

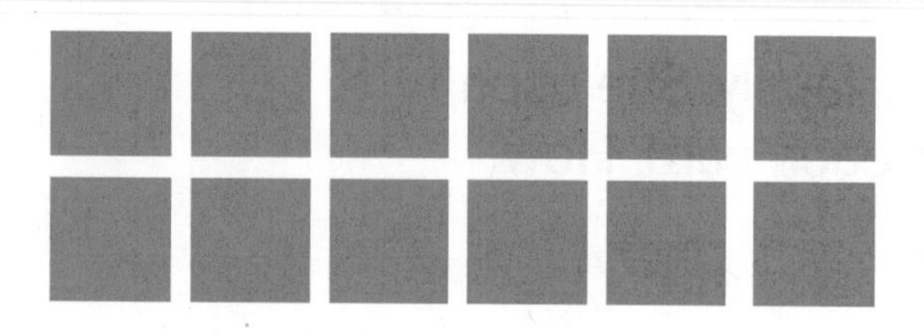

____

____

▶ Performance Task

22. There are 20 students in Mr. Hamilton's class. He wants to put the students in equal groups.

A List as many ways as you can to arrange the students in equal groups.

____

____

____

B Tell how you know you found all the ways.

____

____

____

Chapter 7

Division Facts and Strategies

Show What You Know

Check your understanding of important skills.

Name ______________________

▶ Think Addition to Subtract Write the missing numbers.

1. $10 - 3 = \blacksquare$

   Think: $3 + \blacksquare = 10$

   $3 + \_\_\_\_ = 10$

   So, $10 - 3 = \_\_\_\_$.

2. $12 - 8 = \blacksquare$

   Think: $8 + \blacksquare = 12$

   $8 + \_\_\_\_ = 12$

   So, $12 - 8 = \_\_\_\_$.

▶ Missing Factors Write the missing factor.

3. $2 \times \_\_\_\_ = 10$
4. $42 = \_\_\_\_ \times 7$
5. $\_\_\_\_ \times 6 = 18$

▶ Multiplication Facts Through 9 Find the product.

6. $\_\_\_\_ = 6 \times 9$
7. $3 \times 8 = \_\_\_\_$
8. $4 \times 4 = \_\_\_\_$

On Monday, the students in Mr. Carson's class worked in pairs. On Tuesday, the students worked in groups of 3. On Wednesday, the students worked in groups of 4. Each day the students made equal groups with no student left out of a group. Be a Math Detective to find how many students could be in Mr. Carson's class.

Vocabulary Builder

▶ Visualize It

Sort the review words into the Venn diagram.

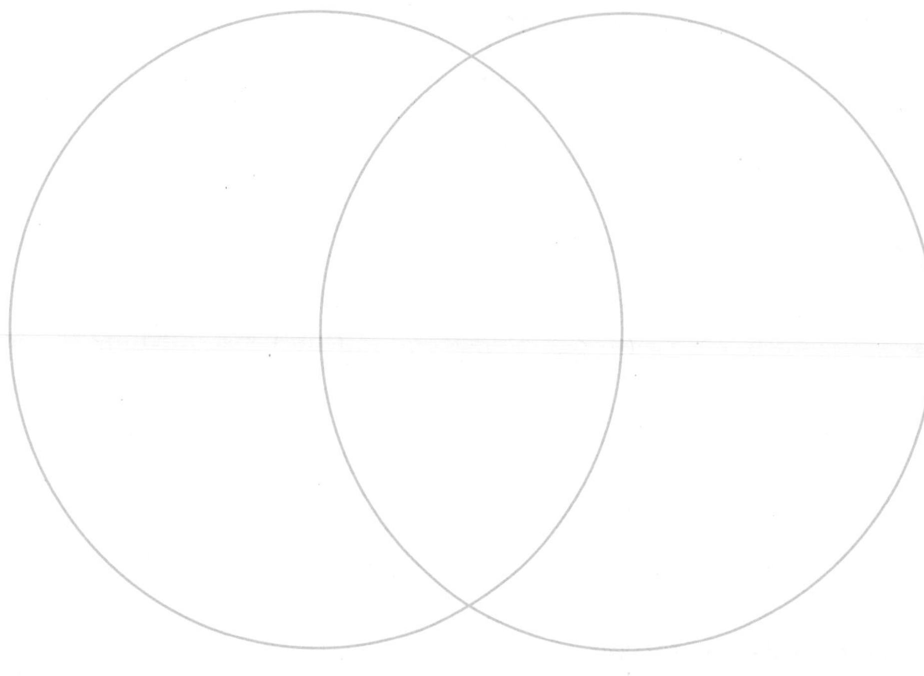

Review Words

- divide
- dividend
- divisor
- equation
- factor
- inverse operations
- multiply
- product
- quotient
- related facts

Preview Word

- order of operations

▶ Understand Vocabulary

Complete the sentences by using the review and preview words.

1. An ________________ is a number sentence that uses the equal sign to show that two amounts are equal.

2. The ________________ is a special set of rules that gives the order in which calculations are done to solve a problem.

3. ________________ are a set of related multiplication and division equations.

Name ______________________

Divide by 2

Essential Question What does dividing by 2 mean?

UNLOCK the Problem REAL WORLD

There are 10 hummingbirds and 2 feeders in Marissa's backyard. If there are an equal number of birds at each feeder, how many birds are at each one?

- What do you need to find?

____

- Circle the numbers you need to use.
- What can you use to help solve the problem? ______________________

Activity 1

Use counters to find how many in each group.

Materials ■ counters ■ MathBoard

MODEL

- Use 10 counters.
- Draw 2 circles on your MathBoard.
- Place 1 counter at a time in each circle until all 10 counters are used.
- Draw the rest of the counters to show your work.

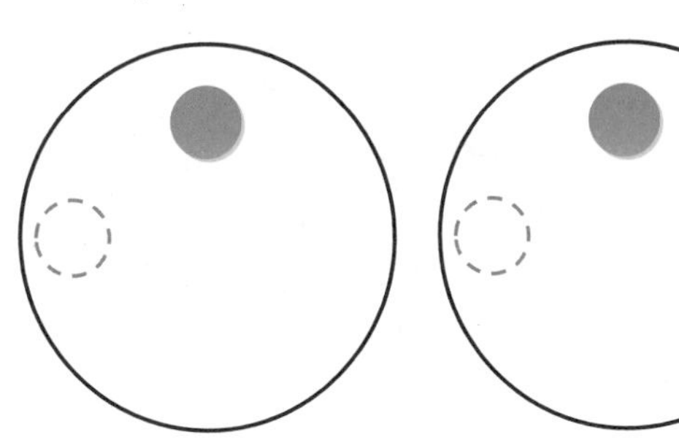

THINK

_____ in all

_____ equal groups

_____ in each group

RECORD

$10 \div 2 = 5$ or $2\overline{)10}$ with quotient 5

Read: Ten divided by two equals five.

There are _____ counters in each of the 2 groups.

So, there are _____ hummingbirds at each feeder.

A hummingbird can fly right, left, up, down, forward, backward, and even upside down! ▶

MATHEMATICAL PRACTICES

Math Talk **Explain** what each number in $10 \div 2 = 5$ represents from the word problem.

Activity 2 Draw to find how many equal groups.

There are 10 hummingbirds in Tyler's backyard. If there are 2 hummingbirds at each feeder, how many feeders are there?

Math Idea

You can divide to find the number in each group or to find the number of equal groups.

MODEL

- Look at the 10 counters.
- Circle a group of 2 counters.
- Continue circling groups of 2 until all 10 counters are in groups.

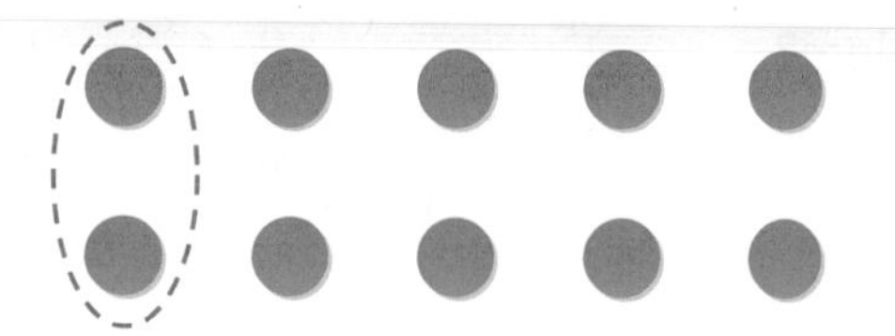

THINK

_____ in all

_____ in each group

_____ equal groups

RECORD

$10 \div 2 = 5$ or $2\overline{)10}$ with quotient 5

Read: Ten divided by two equals five.

There are _____ groups of 2 counters.

So, there are _____ feeders.

Share and Show

1. Complete the picture to find $6 \div 2$. _____

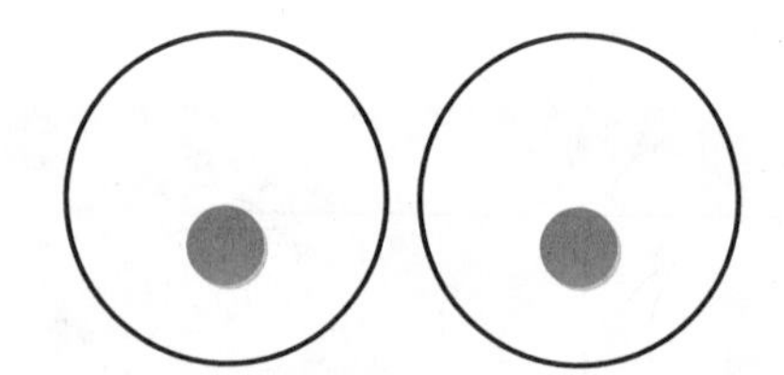

MATHEMATICAL PRACTICES

Math Talk Describe another division equation that could be written for the picture you drew.

Write a division equation for the picture.

2. 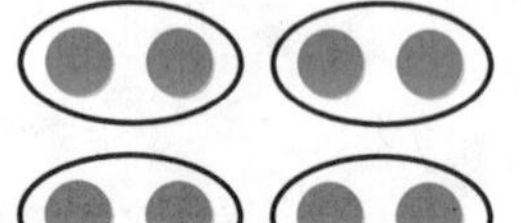

____

✓3. 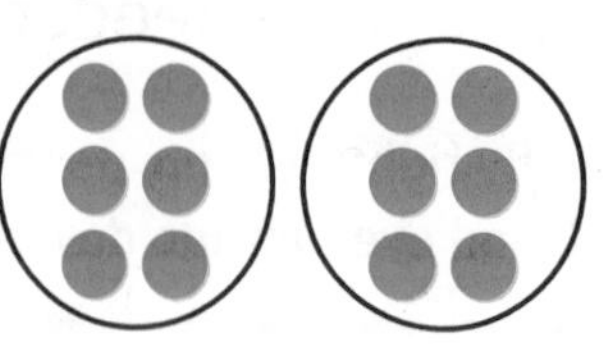

____

✓4. 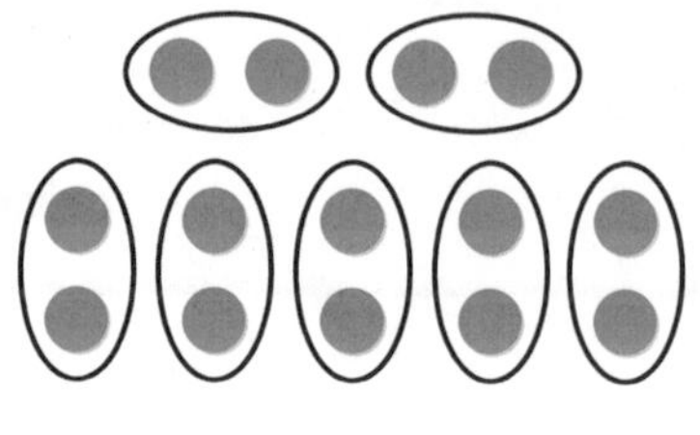

____

Name ______________________________

On Your Own

Write a division equation for the picture.

5. 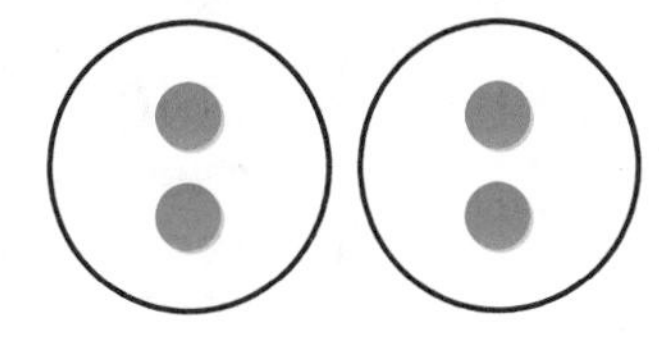

____

6. 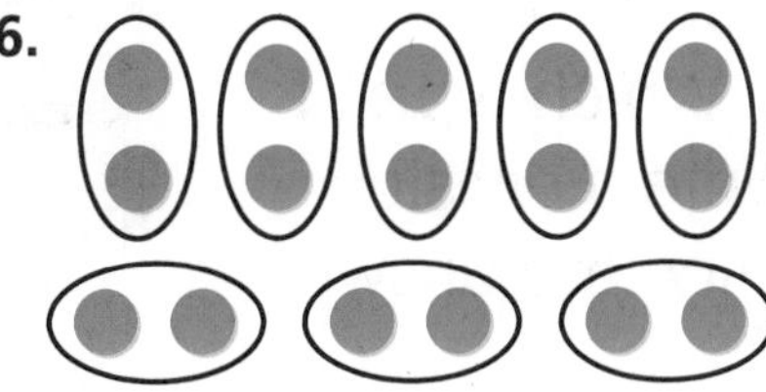

____

7. 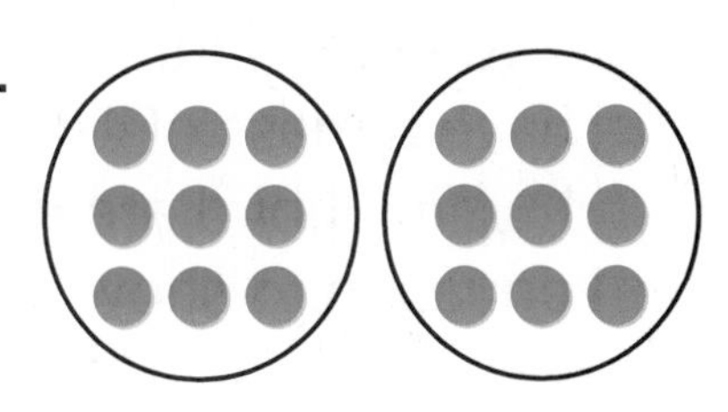

____

Find the quotient. You may want to draw a quick picture to help.

8. $2 \div 2 =$ ____

9. ____ $= 10 \div 2$

10. ____ $= 14 \div 2$

11. ____ $= 18 \div 2$

12. $16 \div 2 =$ ____

13. ____ $= 0 \div 2$

14. $2\overline{)8}$

15. $2\overline{)12}$

16. $2\overline{)20}$

Algebra Find the unknown number.

17. ____ $\div 2 = 5$

18. ____ $\div 2 = 2$

19. ____ $\div 2 = 3$

20. ____ $\div 2 = 8$

Problem Solving REAL WORLD

Use the table for 21–22.

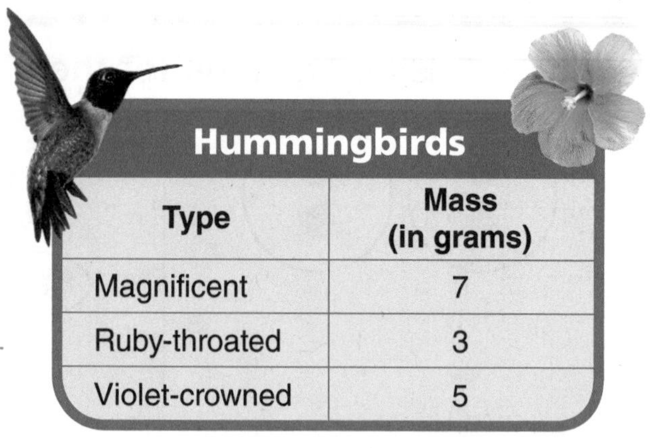

Hummingbirds

Type	Mass (in grams)
Magnificent	7
Ruby-throated	3
Violet-crowned	5

21. Two hummingbirds of the same type have a mass of 10 grams. Which type of hummingbird are they? Write a division equation to show how to find the answer.

____

22. H.O.T. There are 3 ruby-throated hummingbirds and 2 of another type of hummingbird at a feeder. The birds have a mass of 23 grams in all. What other type of hummingbird is at the feeder? **Explain.**

____

____

____

____

____

SHOW YOUR WORK

23. Write Math There are 12 hummingbird eggs in all. If there are 2 eggs in each nest, how many nests are there? **Explain** how you found your answer.

____

____

____

____

24. **Test Prep** Jo sees the same number of birds each hour for 2 hours. She sees 16 birds in all. How many birds does Jo see each hour?

Ⓐ 6 Ⓑ 7 Ⓒ 8 Ⓓ 9

 FOR MORE PRACTICE: Standards Practice Book, pp. P127–P128

Name ______________________________

Divide by 10

Essential Question What strategies can you use to divide by 10?

UNLOCK the Problem REAL WORLD

There are 50 students going on a field trip to the Philadelphia Zoo. The students are separated into equal groups of 10 students each. How many groups of students are there?

- What do you need to find?

  ____

- Circle the numbers you need to use.

One Way Use repeated subtraction.

- Start with 50.
- Subtract 10 until you reach 0.
- Count the number of times you subtract 10.

50 − 10 = 40	40 − 10 = 30	30 − 10 = ___	20 − 10 = ___	10 − 10 = ___
1	2	3	4	5

You subtracted 10 five times. $50 \div 10 =$ _____

So, there are _____ groups of 10 students.

Other Ways

A Use a number line.

- Start at 50 and count back by 10s until you reach 0.
- Count the number of times you jumped back 10.

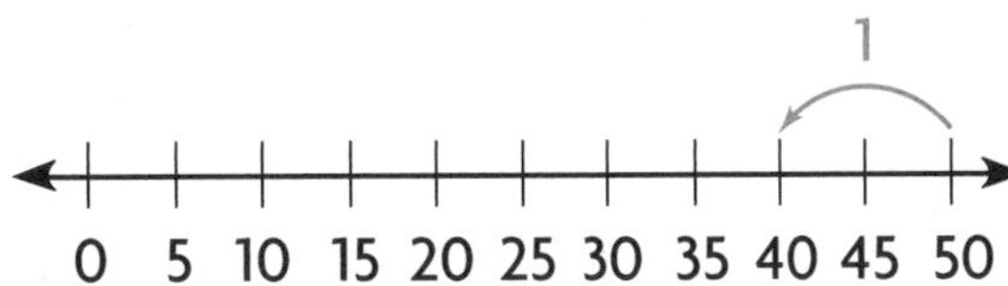

You jumped back by 10 five times.

$50 \div 10 =$ _____

MATHEMATICAL PRACTICES

Math Talk How is counting on a number line to divide by 10 different from counting on a number line to multiply by 10?

B **Use a multiplication table.**

Divide. 50 ÷ 10 = ■

Since division is the opposite of multiplication, you can use a multiplication table to find a quotient.

Think of a related multiplication fact.

■ × 10 = 50

STEP 1 Find the factor, 10, in the top row.

STEP 2 Look down to find the product, 50.

STEP 3 Look left to find the unknown factor, _____.

Since _____ × 10 = 50, then 50 ÷ 10 = _____.

In Step 1, is the divisor or the dividend the given factor in the related multiplication fact?

____

In Step 2, is the divisor or the dividend the product in the related multiplication fact?

____

The quotient is the unknown factor.

×	0	1	2	3	4	5	6	7	8	9	10
0	0	0	0	0	0	0	0	0	0	0	0
1	0	1	2	3	4	5	6	7	8	9	10
2	0	2	4	6	8	10	12	14	16	18	20
3	0	3	6	9	12	15	18	21	24	27	30
4	0	4	8	12	16	20	24	28	32	36	40
5	0	5	10	15	20	25	30	35	40	45	50
6	0	6	12	18	24	30	36	42	48	54	60
7	0	7	14	21	28	35	42	49	56	63	70
8	0	8	16	24	32	40	48	56	64	72	80
9	0	9	18	27	36	45	54	63	72	81	90
10	0	10	20	30	40	50	60	70	80	90	100

Share and Show

1. Use repeated subtraction to find 30 ÷ 10. _____

Think: How many times do you subtract 10?

$$\begin{array}{r}30\\-10\\\hline 20\end{array}\quad\begin{array}{r}20\\-10\\\hline 10\end{array}\quad\begin{array}{r}10\\-10\\\hline \blacksquare\end{array}$$

MATHEMATICAL PRACTICES

Math Talk Describe two other ways to find 30 ÷ 10.

Find the unknown factor and quotient.

2. 10 × _____ = 40   _____ = 40 ÷ 10

✓ 3. 10 × _____ = 60   60 ÷ 10 = _____

Find the quotient.

4. _____ = 20 ÷ 10

5. $10\overline{)50}$

6. $10\overline{)70}$

✓ 7. 90 ÷ 10 = _____

Name ______________________________

On Your Own

Find the unknown factor and quotient.

8. $10 \times \_\_\_\_ = 70$ $\quad 70 \div 10 = \_\_\_\_$

9. $10 \times \_\_\_\_ = 10$ $\quad 10 \div 10 = \_\_\_\_$

10. $10 \times \_\_\_\_ = 80$ $\quad 80 \div 10 = \_\_\_\_$

11. $\_\_\_\_ \times 2 = 12$ $\quad \_\_\_\_ = 12 \div 2$

Find the quotient.

12. $50 \div 10 = \_\_\_\_$

13. $\_\_\_\_ = 60 \div 10$

14. $16 \div 2 = \_\_\_\_$

15. $90 \div 10 = \_\_\_\_$

16. $10 \div 2 = \_\_\_\_$

17. $30 \div 10 = \_\_\_\_$

18. $\_\_\_\_ = 20 \div 2$

19. $\_\_\_\_ = 0 \div 10$

20. $10\overline{)20}$

21. $10\overline{)100}$

22. $10\overline{)40}$

23. $10\overline{)80}$

Algebra Write $<$, $>$, or $=$.

24. $10 \div 1 \bigcirc 4 \times 10$

25. $17 - 6 \bigcirc 18 \div 2$

26. $4 \times 4 \bigcirc 8 + 8$

27. $23 + 14 \bigcirc 5 \times 8$

28. $70 \div 10 \bigcirc 23 - 16$

29. $9 \times 0 \bigcirc 9 + 0$

Problem Solving REAL WORLD

Use the picture graph for 30–32.

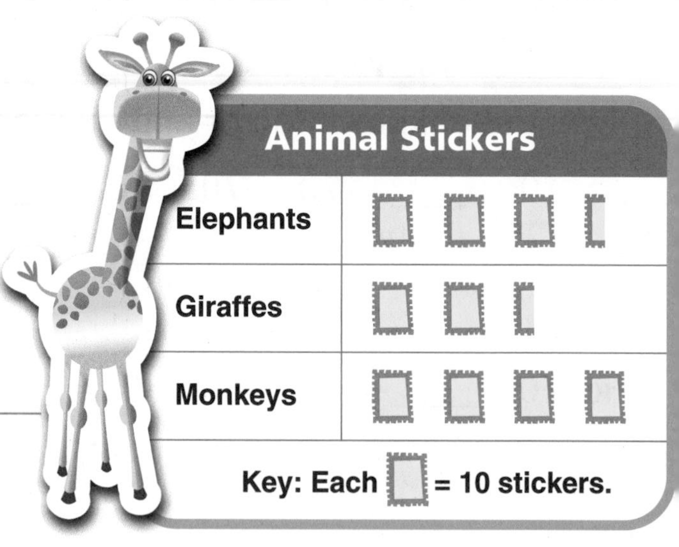

30. Lyle wants to add penguins to the picture graph. There are 30 stickers of penguins. How many symbols should Lyle draw for penguins?

____

31. **Pose a Problem** Write a word problem using information from the picture graph. Then solve your problem.

____

____

____

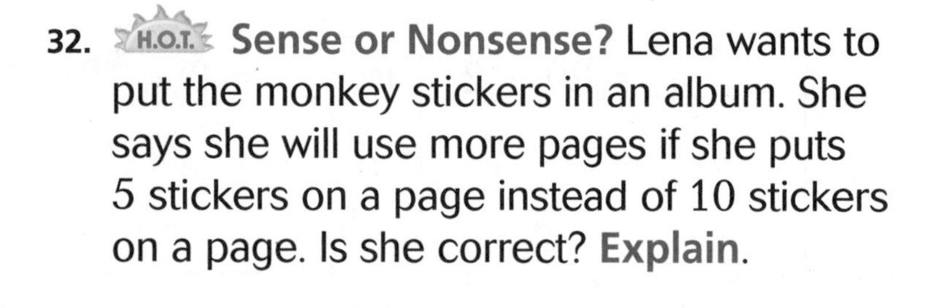

32. H.O.T. **Sense or Nonsense?** Lena wants to put the monkey stickers in an album. She says she will use more pages if she puts 5 stickers on a page instead of 10 stickers on a page. Is she correct? **Explain.**

____

____

____

33. Write Math **Explain** how a division problem is like an unknown factor problem.

____

____

____

34. **Test Prep** Miles has 100¢ in dimes. How many dimes does Miles have?

Ⓐ 1 Ⓑ 4 Ⓒ 10 Ⓓ 100

SHOW YOUR WORK

 FOR MORE PRACTICE: Standards Practice Book, pp. P129–P130

Name ____________________

Divide by 5

Essential Question What does dividing by 5 mean?

UNLOCK the Problem REAL WORLD

Kaley wants to buy a new cage for Coconut, her guinea pig. She has saved 35¢. If she saved a nickel each day, for how many days has she been saving?

- How much is a nickel worth? ____________

One Way Count up by 5s.

- Begin at 0.
- Count up by 5s until you reach 35. 5, 10, ____, ____, ____, ____, ____
  1 2 3 4 5 6 7
- Count the number of times you count up.

You counted up by 5 seven times. $35 \div 5 =$ ____

So, Kaley has been saving for ____ days.

Another Way Count back on a number line.

- Start at 35.
- Count back by 5s until you reach 0. Complete the jumps on the number line.
- Count the number of times you jumped back 5.

You jumped back by 5 ________ times.

$35 \div 5 =$ ____

Math Talk MATHEMATICAL PRACTICES What if Kaley saved 7¢ each day instead of a nickel? What would you do differently to find how many days she has saved?

Strategies for Multiplying and Dividing with 5

You have learned how to use doubles to multiply. Now you will learn how to use doubles to divide by 5.

Use 10s facts, and then take half to multiply with 5.

When one factor is 5, you can use a 10s fact. $5 \times 2 = ■$

First, multiply by 10. $10 \times 2 =$ _____

After you multiply, take half of the product. $20 \div 2 =$ _____

So, $5 \times 2 =$ _____.

Divide by 10, and then double to divide by 5.

When the divisor is 5 and the dividend is even, you can use a 10s fact. $30 \div 5 = ■$

First, divide by 10. $30 \div 10 =$ _____

After you divide, double the quotient. $3 +$ _____ $=$ _____

So, $30 \div 5 =$ _____.

Share and Show

1. Count back on the number line to find $15 \div 5$. _____

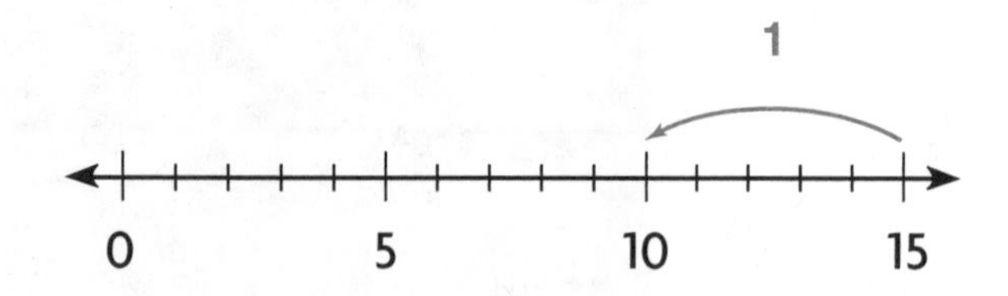

MATHEMATICAL PRACTICES

Math Talk Explain how counting up to solve a division problem is like counting back on a number line.

Use count up or count back on a number line to solve.

2. $10 \div 2 =$ _____

3. $20 \div 5 =$ _____

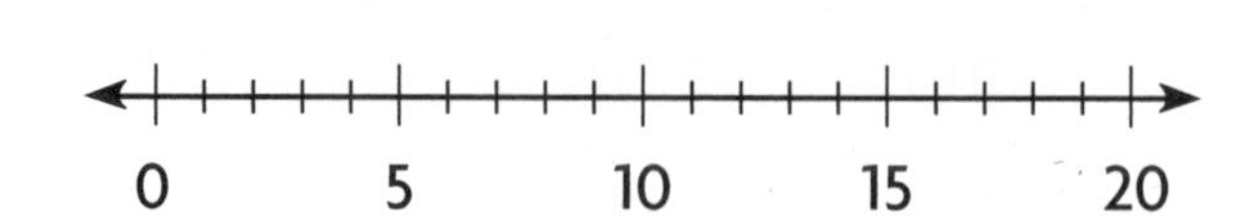

Find the quotient.

4. $50 \div 5 =$ _____

5. $5 \div 5 =$ _____

6. $45 \div 5 =$ _____

Name ______________________________

On Your Own

Use count up or count back on a number line to solve.

7. $30 \div 5 =$ _____

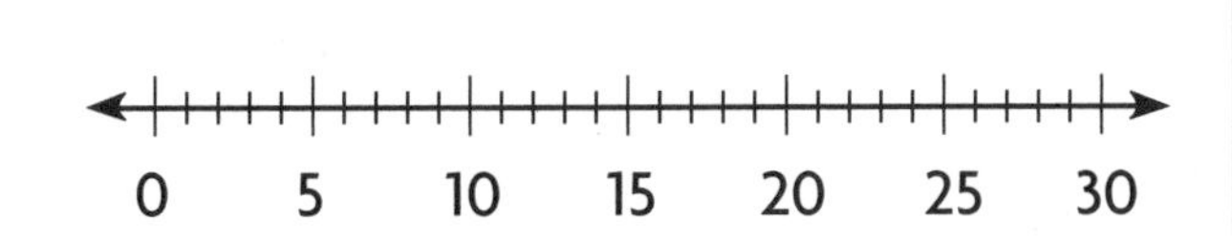

8. $25 \div 5 =$ _____

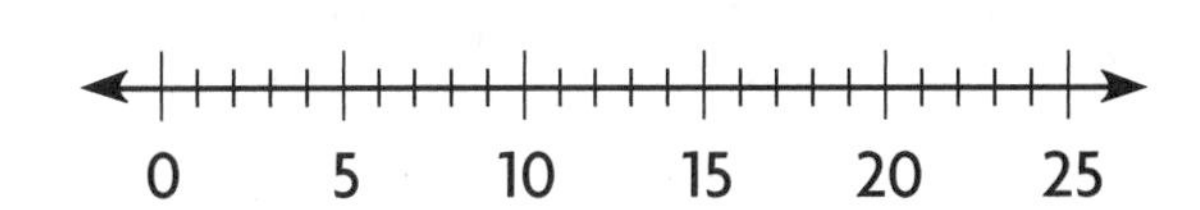

Find the quotient.

9. ____ $= 20 \div 5$

10. $40 \div 5 =$ ____

11. ____ $= 18 \div 2$

12. $0 \div 5 =$ ____

13. $35 \div 5 =$ ____

14. ____ $= 10 \div 5$

15. $40 \div 10 =$ ____

16. ____ $= 4 \div 2$

17. $10\overline{)30}$

18. $2\overline{)16}$

19. $5\overline{)45}$

20. $5\overline{)15}$

Algebra **Complete the table.**

21.

×	1	2	3	4	5
10					
5					

22.

÷	10	20	30	40	50
10					
5					

Problem Solving

23. Guinea pigs eat hay, pellets, and vegetables. If Wonder Hay comes in a 5-pound bag and costs $15, how much does 1 pound of hay cost?

____

24. Guinea pigs sleep about 45 hours every 5 days with their eyes open. About how many hours a day do guinea pigs sleep?

____

25. **Test Prep** The pet store has 20 guinea pigs and plans to put a pair of guinea pigs in each cage. If the pet store has 8 cages, how many more cages will the pet store need?

Ⓐ 1 Ⓑ 2 Ⓒ 4 Ⓓ 5

H.O.T. **Pose a Problem**

26. Maddie went to a veterinary clinic. She saw the vet preparing some carrots for the guinea pigs.

    Write a division problem that can be solved using the picture of carrots.

Pose a problem.	**Solve your problem.**

- Write a similar problem by exchanging the number of equal groups and number in each group. Then solve the problem.

- **Describe** what the quotient in your problem represents.

 FOR MORE PRACTICE:
Standards Practice Book, pp. P131–P132

Name ______________________

Divide by 3

Essential Question What strategies can you use to divide by 3?

UNLOCK the Problem REAL WORLD

For field day, 18 students have signed up for the relay race. Each relay team needs 3 students. How many teams can be made?

- What do you need to find?

____

____

- Circle the numbers you need to use.

One Way Make equal groups.

- Look at the 18 counters below.
- Circle as many groups of 3 as you can.
- Count the number of groups.

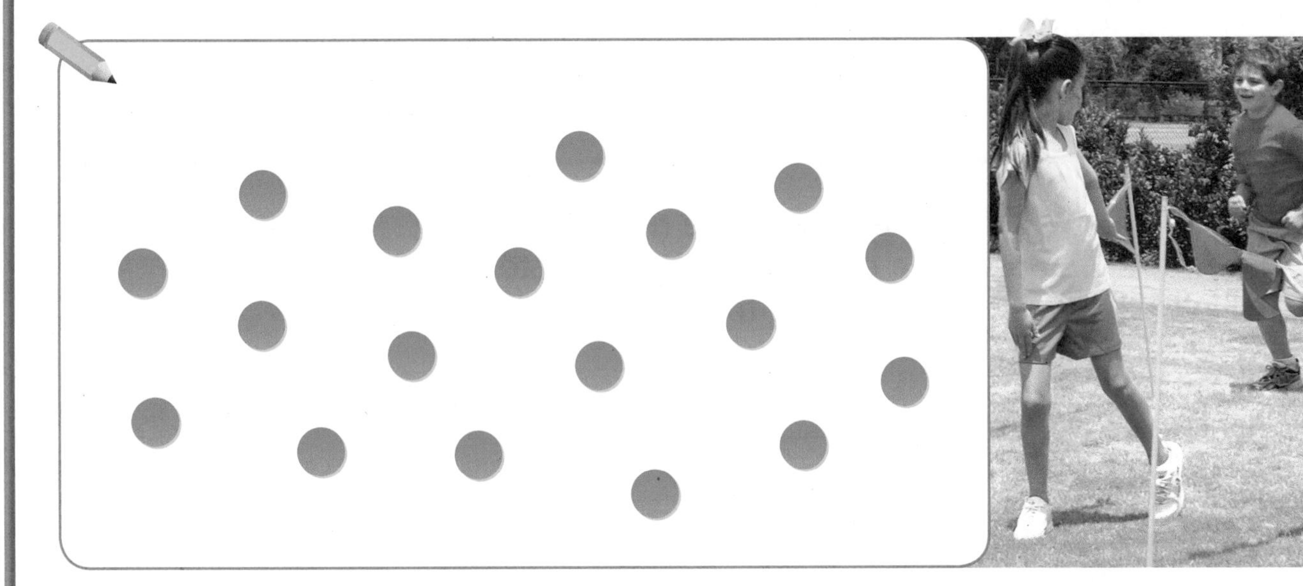

There are ______ groups of 3.

So, ______ teams can be made.

You can write $18 \div 3 =$ ______ or $3\overline{)18}$.

MATHEMATICAL PRACTICES

Math Talk Suppose the question asked how many students would be on 3 equal teams. How would you model 3 equal teams? Would the quotient be the same?

Other Ways

A Count back on a number line.

- Start at 18.
- Count back by 3s as many times as you can. Complete the jumps on the number line.
- Count the number of times you jumped back 3.

ERROR Alert

Be sure to count back the same number of spaces each time you jump back on the number line.

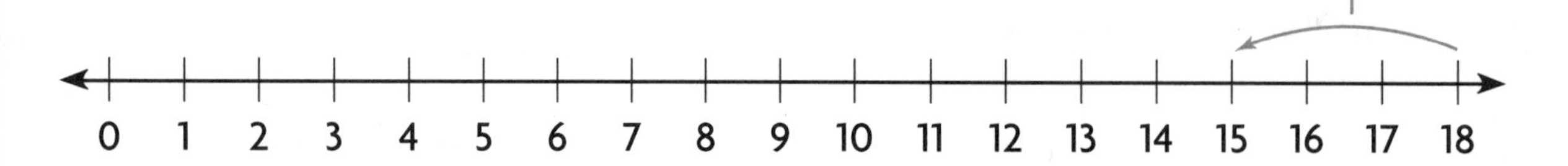

You jumped back by 3 ______ times.

B Use a related multiplication fact.

Since division is the opposite of multiplication, think of a related multiplication fact to find $18 \div 3$.

$\blacksquare \times 3 = 18$

$6 \times 3 = 18$

Think: What number completes the multiplication fact?

So, $18 \div 3 =$ ______ or $3\overline{)18}$.

- **What if** 24 students signed up for the relay race and there were 3 students on each team? What related multiplication fact would you use to find the number of teams?

____

Share and Show

1. Circle groups of 3 to find $12 \div 3$. ______

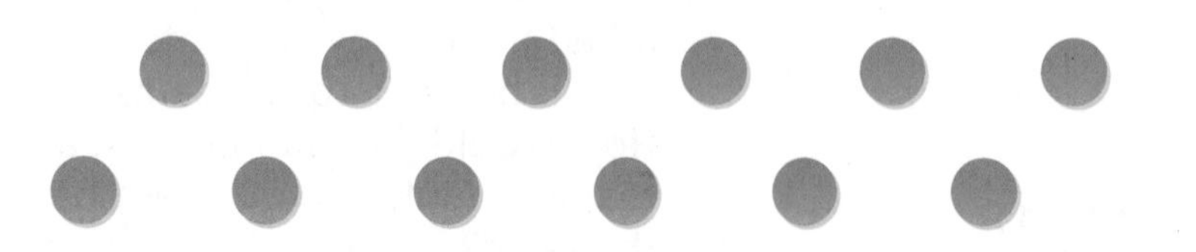

MATHEMATICAL PRACTICES

Math Talk Explain what the number of circles you made tells you.

Find the quotient.

2. $6 \div 3 =$ ____
3. ____ $= 14 \div 2$
4. $21 \div 3 =$ ____
5. ____ $= 30 \div 5$

Name ______________________________

On Your Own

Practice: Copy and Solve Find the quotient. Draw a quick picture to help.

6. $9 \div 3$
7. $10 \div 5$
8. $18 \div 2$
9. $24 \div 3$

Find the quotient.

10. $\_\_\_ = 12 \div 2$
11. $40 \div 5 = \_\_\_$
12. $60 \div 10 = \_\_\_$
13. $\_\_\_ = 20 \div 10$
14. $27 \div 3 = \_\_\_$
15. $\_\_\_ = 0 \div 3$
16. $12 \div 3 = \_\_\_$
17. $\_\_\_ = 8 \div 2$
18. $3\overline{)15}$
19. $2\overline{)4}$
20. $5\overline{)20}$
21. $3\overline{)18}$
22. $2\overline{)16}$
23. $3\overline{)12}$
24. $3\overline{)6}$
25. $5\overline{)35}$
26. $3\overline{)3}$
27. $10\overline{)70}$
28. $3\overline{)30}$
29. $10\overline{)50}$

Algebra Write $+$, $-$, $\times$, or $\div$.

30. $25 \bigcirc 5 = 10 \div 2$
31. $3 \times 3 = 6 \bigcirc 3$
32. $16 \bigcirc 2 = 24 - 16$
33. $13 + 19 = 8 \bigcirc 4$
34. $14 \bigcirc 2 = 6 \times 2$
35. $21 \div 3 = 5 \bigcirc 2$

Problem Solving REAL WORLD

Use the table for 36–37.

Field Day Events

Activity	Number of Students
Relay race	25
Beanbag toss	18
Jump-rope race	27

36. There are 5 equal teams in the relay race. How many students are on each team? Write a division equation that shows the number of students on each team.

____

37. H.O.T. Students doing the jump-rope race and the beanbag toss compete in teams of 3. How many more teams participate in the jump-rope race than in the beanbag toss? **Explain** how you know.

____

____

____

____

SHOW YOUR WORK

38. Write Math **What's the Question?** Michael puts 21 sports cards into stacks of 3. The answer is 7 stacks.

____

____

____

____

39. **Test Prep** Olivia buys 24 beanbags for field day. She buys 3 equal packs. How many beanbags are in each pack?

Ⓐ 6 Ⓒ 8

Ⓑ 7 Ⓓ 9

FOR MORE PRACTICE:
Standards Practice Book, pp. P133–P134

Name ______________________________

Divide by 4

Essential Question What strategies can you use to divide by 4?

UNLOCK the Problem REAL WORLD

A tree farmer plants 12 red maple trees in 4 equal rows. How many trees are in each row?

- What strategy could you use to solve the problem?

____

One Way Make an array.

- Look at the array.
- Continue the array by drawing 1 tile in each of the 4 rows until all 12 tiles are drawn.
- Count the number of tiles in each row.

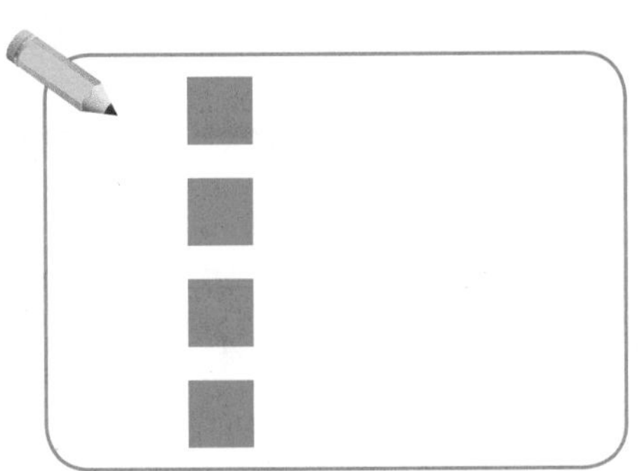

There are ______ tiles in each row.

So, there are ______ trees in each row.

Write: ______ ÷ ______ = ______ or $4\overline{)12}$

Read: Twelve divided by four equals three.

Other Ways

A **Make equal groups.**

- Draw 1 counter in each group.
- Continue drawing 1 counter at a time until all 12 counters are drawn.

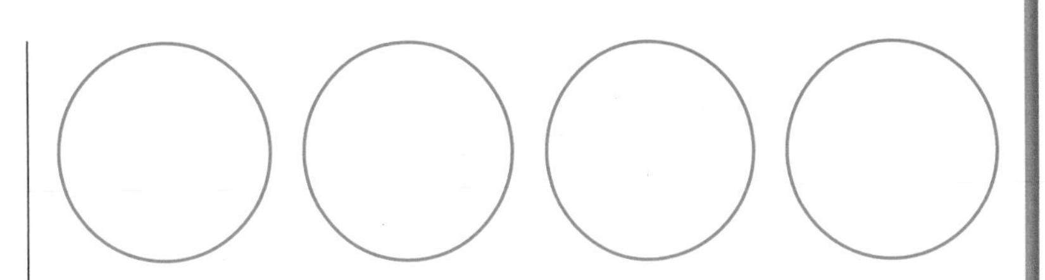

There are ______ counters in each group.

MATHEMATICAL PRACTICES

Math Talk **Explain** how making an array to solve the problem is like making equal groups.

B **Use factors to find 12 ÷ 4.**

The factors of 4 are 2 and 2.

$2 \times 2 = 4$

factors product

To divide by 4, use the factors.

$12 \div 4 = n$

Divide by 2. $12 \div 2 = 6$

Then divide by 2 again. $6 \div 2 = 3$

$12 \div 4 =$ _____

C **Use a related multiplication fact.**

$12 \div 4 = n$

$4 \times n = 12$

$4 \times 3 = 12$

Think: What number completes the multiplication fact?

$12 \div 4 =$ _____ or $4\overline{)12}$

Remember

A letter or symbol, like *n*, can stand for an unknown number.

Try This! **Use factors of 4 to find 16 ÷ 4.**

The factors of 4 are 2 and 2.

$16 \div 4 = ■$

Divide by 2.

$16 \div 2 =$ _____

Then divide by 2 again.

$8 \div 2 =$ _____

So, $16 \div 4 =$ _____.

Think: Dividing by the factors of the divisor is the same as dividing by the divisor.

Share and Show

1. Use the array to find $28 \div 4$. _____

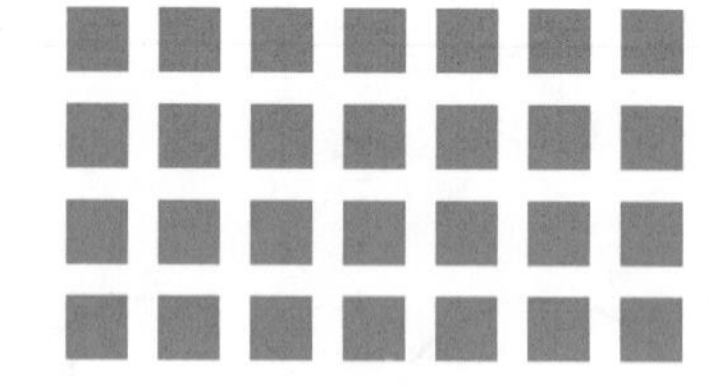

MATHEMATICAL PRACTICES

Math Talk Explain how you used the array to find the quotient.

Find the quotient.

2. _____ $= 21 \div 3$
3. $8 \div 4 =$ _____
4. _____ $= 40 \div 5$
5. $24 \div 4 =$ _____

Find the unknown number.

6. $20 \div 4 = a$

   $a =$ _____

7. $12 \div 2 = p$

   $p =$ _____

8. $27 \div 3 = ▲$

   $▲ =$ _____

9. $12 \div 4 = t$

   $t =$ _____

Name ______________________________

On Your Own

Practice: Copy and Solve Draw tiles to make an array. Find the quotient.

10. $30 \div 10$
11. $15 \div 5$
12. $40 \div 4$
13. $16 \div 2$

Find the quotient.

14. $12 \div 3 =$ ___
15. $20 \div 4 =$ ___
16. ___ $= 0 \div 4$
17. ___ $= 36 \div 4$

18. $4\overline{)28}$
19. $2\overline{)18}$
20. $4\overline{)16}$
21. $5\overline{)25}$

Find the unknown number.

22. $45 \div 5 = b$
    $b =$ ___
23. $20 \div 10 = e$
    $e =$ ___
24. $8 \div 2 = ■$
    $■ =$ ___
25. $24 \div 3 = h$
    $h =$ ___
26. $4 \div 4 = p$
    $p =$ ___
27. $24 \div 4 = t$
    $t =$ ___
28. $16 \div 4 = s$
    $s =$ ___
29. $32 \div 4 = ■$
    $■ =$ ___

Algebra Complete the table.

30.

÷	9	12	15	18
3				

31.

÷	20	24	28	32
4				

H.O.T. **Algebra** Find the unknown number.

32. $14 \div$ ___ $= 7$
33. $30 \div$ ___ $= 6$
34. $8 \div$ ___ $= 2$
35. $24 \div$ ___ $= 8$
36. $36 \div$ ___ $= 9$
37. $40 \div$ ___ $= 4$
38. $3 \div$ ___ $= 1$
39. $35 \div$ ___ $= 7$

Problem Solving REAL WORLD

Use the table for 40–41.

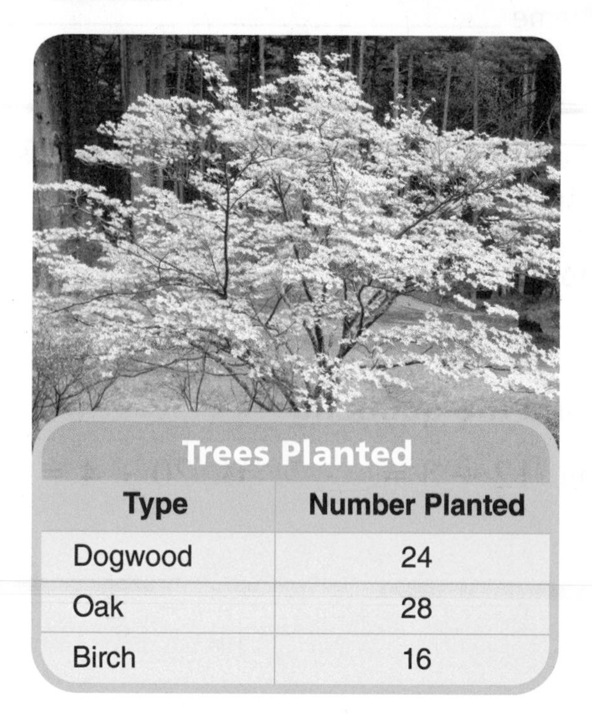

Trees Planted

Type	Number Planted
Dogwood	24
Oak	28
Birch	16

40. Douglas planted the birch trees in 4 equal rows. Then he added 2 more birch trees to each row. How many trees did he plant in each row?

____

41. H.O.T. Mrs. Banks planted the oak trees in 4 equal rows. Mr. Webb planted the dogwood trees in 3 equal rows. Who planted more trees in each row? How many more? **Explain** how you know.

____

____

____

42. Write Math Bryan earns $40 mowing lawns each week. He earns the same amount of money for each lawn. If he mows 4 lawns, how much does Bryan earn for each lawn? **Explain** how you found your answer.

____

____

____

43. **Test Prep** Eric planted 20 flowers in groups of 4. How many groups of flowers did he plant?

Ⓐ 4 Ⓒ 6

Ⓑ 5 Ⓓ 7

SHOW YOUR WORK

Name ____________________

Divide by 6

Essential Question What strategies can you use to divide by 6?

UNLOCK the Problem REAL WORLD

Ms. Sing needs to buy 24 juice boxes for the class picnic. Juice boxes come in packs of 6. How many packs does Ms. Sing need to buy?

- Circle the number that tells you how many juice boxes come in a pack.
- How can you use the information to solve the problem?

____

____

____

One Way Make equal groups.

- Draw 24 counters.
- Circle as many groups of 6 as you can.
- Count the number of groups.

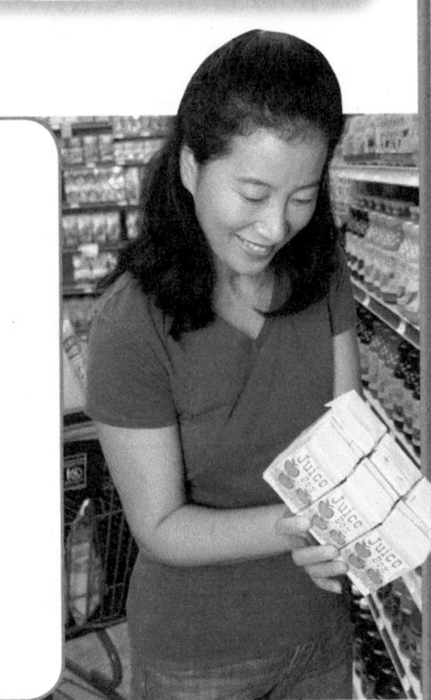

There are ______ groups of 6.

So, Ms. Sing needs to buy ______ packs of juice boxes.

You can write ____ ÷ ____ = ____ or $6\overline{)24}$.

MATHEMATICAL PRACTICES

Math Talk If you divided the 24 counters into groups of 4, how many groups would there be? **Explain** how you know.

Other Ways

A Use a related multiplication fact.

$24 \div 6 = \blacksquare$

$\blacksquare \times 6 = 24$

$4 \times 6 = 24$

Think: What number completes the multiplication fact?

$24 \div 6 = \_\_\_\_$ or $6\overline{)24}$

B Use factors to find 24 ÷ 6.

The factors of 6 are 3 and 2.

$3 \times 2 = 6$

factors — product

To divide by 6, use the factors.

$24 \div 6 = \blacksquare$

Divide by 3. $24 \div 3 = 8$

Then divide by 2. $8 \div 2 = 4$

$24 \div 6 = \_\_\_\_$

- How does knowing $6 \times 9 = 54$ help you find $54 \div 6$?

____

____

Share and Show

1. Continue making equal groups to find $18 \div 6$. ____

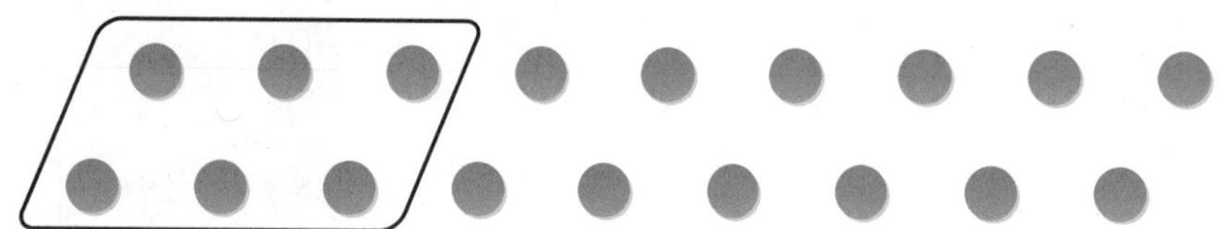

MATHEMATICAL PRACTICES

Math Talk Explain how you could use factors to find $18 \div 6$.

Find the unknown factor and quotient.

2. $\_\_\_ \times 6 = 36$ $36 \div 6 = \_\_\_$

3. $6 \times \_\_\_ = 12$ $12 \div 6 = \_\_\_$

Find the quotient.

4. $\_\_\_ = 0 \div 2$

5. $6 \div 6 = \_\_\_$

6. $\_\_\_ = 28 \div 4$

7. $42 \div 6 = \_\_\_$

Name ______

On Your Own

Find the unknown factor and quotient.

8. $6 \times$ ___ $= 30$ $30 \div 6 =$ ___

9. ___ $\times 6 = 48$ $48 \div 6 =$ ___

10. $2 \times$ ___ $= 16$ ___ $= 16 \div 2$

11. $5 \times$ ___ $= 45$ ___ $= 45 \div 5$

Find the quotient.

12. $12 \div 6 =$ ___

13. ___ $= 6 \div 1$

14. ___ $= 60 \div 6$

15. $27 \div 3 =$ ___

16. $5\overline{)35}$

17. $6\overline{)42}$

18. $6\overline{)6}$

19. $2\overline{)10}$

Find the unknown number.

20. $k = 54 \div 6$
$k =$ ___

21. $20 \div 4 = w$
$w =$ ___

22. $0 \div 6 = s$
$s =$ ___

23. $d = 36 \div 6$
$d =$ ___

24. $24 \div 6 = n$
$n =$ ___

25. $40 \div 5 =$ ▲
▲ $=$ ___

26. $60 \div 10 = m$
$m =$ ___

27. $18 \div 6 =$ ■
■ $=$ ___

H.O.T. **Algebra** **Find the unknown number.**

28. $20 \div$ ___ $= 4$

29. $24 \div$ ___ $= 8$

30. $16 \div$ ___ $= 4$

31. $3 \div$ ___ $= 3$

32. $42 \div$ ___ $= 7$

33. $30 \div$ ___ $= 10$

34. $10 \div$ ___ $= 2$

35. $32 \div$ ___ $= 4$

Problem Solving REAL WORLD

36. Cody baked 12 muffins. He keeps 6 muffins. How many muffins can he give to each of his 6 friends if each friend gets the same number of muffins?

____

SHOW YOUR WORK

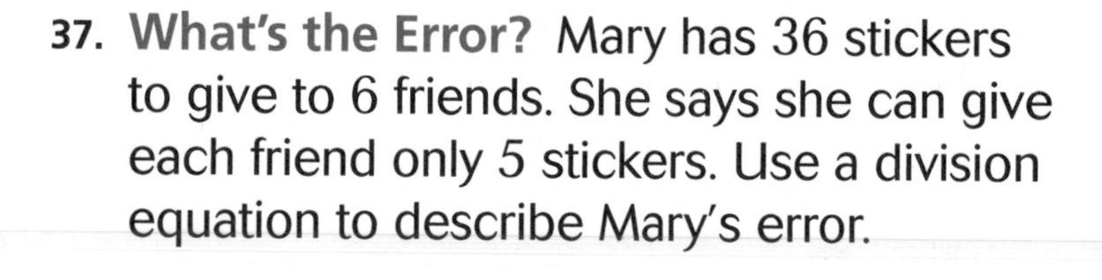

37. **What's the Error?** Mary has 36 stickers to give to 6 friends. She says she can give each friend only 5 stickers. Use a division equation to describe Mary's error.

____

____

____

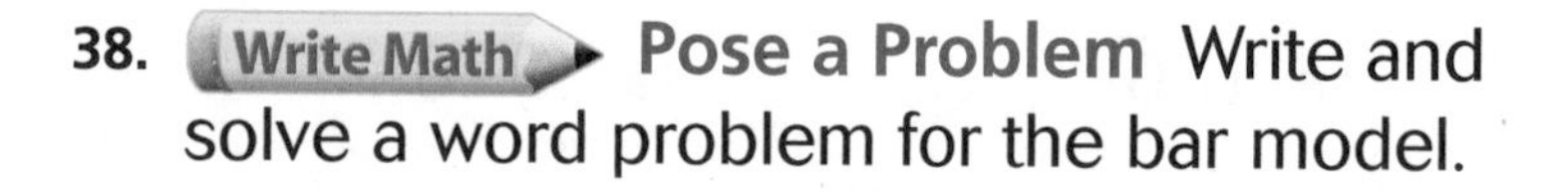

38. Write Math ▸ **Pose a Problem** Write and solve a word problem for the bar model.

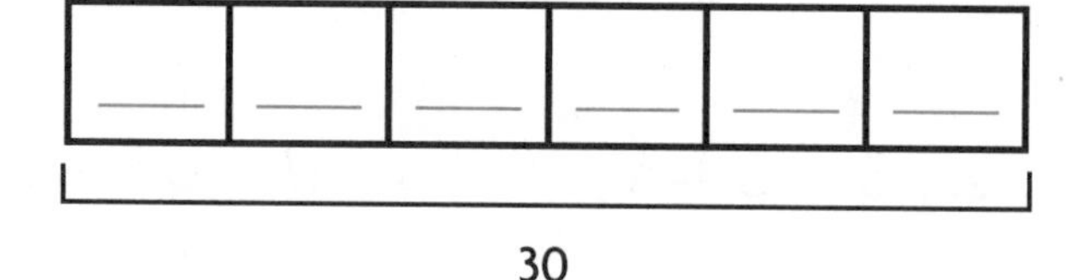

30

____

____

____

____

39. **Test Prep** Each picnic table at a park can seat up to 6 people. How many tables will 48 people fill?

Ⓐ 9 Ⓒ 7

Ⓑ 8 Ⓓ 6

 FOR MORE PRACTICE: Standards Practice Book, pp. P137–P138

Name ____________________

Mid-Chapter Checkpoint

▶ Concepts and Skills

1. **Explain** how to find $20 \div 4$ by making an array.

____

____

____

2. **Explain** how to find $30 \div 6$ by making equal groups.

____

____

____

Find the unknown factor and quotient.

3. $10 \times$ ____ $= 50$ ____ $= 50 \div 10$

4. $2 \times$ ____ $= 16$ ____ $= 16 \div 2$

5. $2 \times$ ____ $= 20$ ____ $= 20 \div 2$

6. $5 \times$ ____ $= 20$ ____ $= 20 \div 5$

Find the quotient.

7. ____ $= 6 \div 6$

8. $21 \div 3 =$ ____

9. ____ $= 0 \div 3$

10. $36 \div 4 =$ ____

11. $5\overline{)35}$

12. $4\overline{)24}$

13. $6\overline{)54}$

14. $3\overline{)9}$

Fill in the bubble for the correct answer choice.

15. Carter has 18 new books. He plans to read 3 of them each week. How many weeks will it take Carter to read all of his new books?

Ⓐ 5 weeks

Ⓑ 6 weeks

Ⓒ 15 weeks

Ⓓ 21 weeks

16. Gabriella made 5 waffles for breakfast. She has 30 berries to put on top of the waffles. She will put an equal number of berries on each waffle. How many berries will Gabriella put on each waffle?

Ⓐ 4

Ⓑ 5

Ⓒ 6

Ⓓ 7

17. There are 60 people at the fair waiting in line for a ride. Each car in the ride can hold 10 people. Which equation could be used to find the number of cars needed to hold all 60 people?

Ⓐ $60 - 10 = \blacksquare$

Ⓑ $\blacksquare + 10 = 60$

Ⓒ $60 \div 10 = \blacksquare$

Ⓓ $60 \times 10 = \blacksquare$

18. Alyssa has 4 cupcakes. She gives 2 cupcakes to each of her cousins. How many cousins does Alyssa have?

Ⓐ 8

Ⓑ 6

Ⓒ 4

Ⓓ 2

Name ______________________________

Divide by 7

Essential Question What strategies can you use to divide by 7?

UNLOCK the Problem REAL WORLD

Erin used 28 large apples to make 7 apple pies. She used the same number of apples for each pie. How many apples did Erin use for each pie?

- Do you need to find the number of equal groups or the number in each group?

____

- What label will your answer have?

____

One Way Make an array.

- Draw 1 tile in each of 7 rows.
- Continue drawing 1 tile in each of the 7 rows until all 28 tiles are drawn.
- Count the number of tiles in each row.

There are _____ tiles in each row.

So, Erin used __________ for each pie.

You can write $28 \div 7 =$ _____ or $7\overline{)28}$.

MATHEMATICAL PRACTICES

Math Talk Why can you use division to solve the problem? **Explain.**

Other Ways

A Use a related multiplication fact.

$28 \div 7 = a$ $\qquad$ $7 \times a = 28$

$7 \times 4 = 28$

Think: What number completes the multiplication fact?

$28 \div 7 =$ _____ or $7\overline{)28}$

B Make equal groups.

- Draw 7 circles to show 7 groups.
- Draw 1 counter in each group.
- Continue drawing 1 counter at a time until all 28 counters are drawn.

There are _____ counters in each group.

Share and Show

1. Use the related multiplication fact to find $42 \div 7$.
$6 \times 7 = 42$

$42 \div 7 =$ _____

MATHEMATICAL PRACTICES

Math Talk Explain why you can use a related multiplication fact to solve a division problem.

Find the unknown factor and quotient.

2. $7 \times$ _____ $= 7$ $\qquad$ $7 \div 7 =$ _____

✓ 3. $7 \times$ _____ $= 35$ $\qquad$ $35 \div 7 =$ _____

Find the quotient.

4. $4 \div 2 =$ _____

5. $56 \div 7 =$ _____

6. _____ $= 20 \div 5$

✓ 7. _____ $= 21 \div 7$

Name ___________________________

On Your Own

Find the unknown factor and quotient.

8. $3 \times$ ____ $= 9$ ____ $= 9 \div 3$

9. $7 \times$ ____ $= 49$ $49 \div 7 =$ ____

10. ____ $\times 7 = 63$ $63 \div 7 =$ ____

11. $4 \times$ ____ $= 32$ ____ $= 32 \div 4$

Find the quotient.

12. ____ $= 14 \div 7$

13. $15 \div 3 =$ ____

14. $0 \div 7 =$ ____

15. ____ $= 25 \div 5$

16. $48 \div 6 =$ ____

17. $7 \div 1 =$ ____

18. ____ $= 42 \div 6$

19. ____ $= 18 \div 2$

20. $7\overline{)56}$

21. $1\overline{)9}$

22. $7\overline{)21}$

23. $2\overline{)8}$

Find the unknown number.

24. $24 \div 4 = x$
$x =$ ____

25. $n = 28 \div 7$
$n =$ ____

26. $r = 54 \div 6$
$r =$ ____

27. $14 \div 7 = j$
$j =$ ____

28. $60 \div 10 =$ ■
■ $=$ ____

29. $70 \div 7 = k$
$k =$ ____

30. $m = 63 \div 9$
$m =$ ____

31. $r = 12 \div 6$
$r =$ ____

Algebra **Complete the table.**

32.

÷	18	30	24	36
6				

33.

÷	56	42	49	35
7				

UNLOCK the Problem REAL WORLD

34. Gavin sold 21 pies to 7 different people. Each person bought the same number of pies. How many pies did Gavin sell to each person?

Ⓐ 3 Ⓑ 4 Ⓒ 14 Ⓓ 28

a. What do you need to find? ______________________

____

b. How can you use a bar model to help you decide which operation to use to solve the problem? ______________________

____

____

c. Complete the bar model to help you find the number of pies Gavin sold to each person.

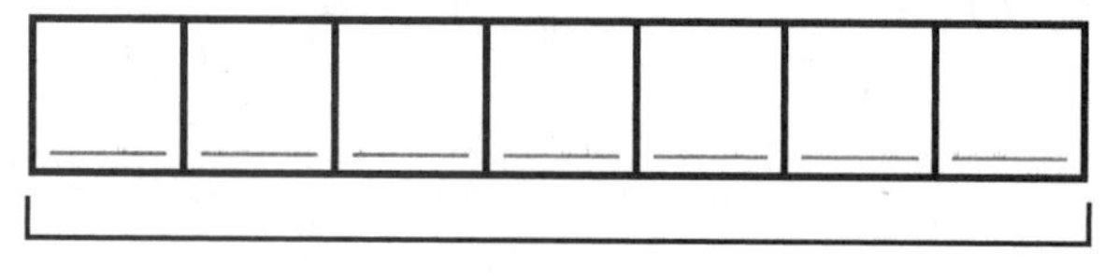

d. What is another way you could have solved the problem?

____

____

e. Complete the sentences.

Gavin sold _____ pies to _____ different people.

Each person bought the same number of __________.

So, Gavin sold _____ pies to each person.

f. Fill in the bubble for the correct answer choice above.

35. Clare bought 35 peaches to make peach pies for a bake sale. She used 7 peaches for each pie. How many pies did Clare make?

Ⓐ 4 Ⓑ 5 Ⓒ 6 Ⓓ 7

36. There are 49 pies in all sitting on 7 shelves at a bakery. If each shelf has the same number of pies, how many pies are on each shelf?

Ⓐ 9 Ⓑ 8 Ⓒ 7 Ⓓ 6

FOR MORE PRACTICE:
Standards Practice Book, pp. P139–P140

Name ______________________________

Divide by 8

Essential Question What strategies can you use to divide by 8?

UNLOCK the Problem REAL WORLD

At Stephen's camping store, firewood is sold in bundles of 8 logs. He has 32 logs to put in bundles. How many bundles of firewood can he make?

- What will Stephen do with the 32 logs?

____

____

One Way Use repeated subtraction.

- Start with 32.
- Subtract 8 until you reach 0.
- Count the number of times you subtract 8.

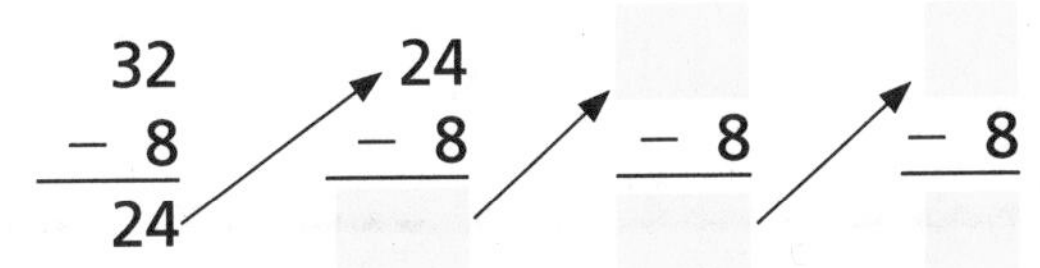

$32 - 8 = 24$ → $24 - 8 = \_\_$ → $\_\_ - 8 = \_\_$ → $\_\_ - 8 = \_\_$

Number of times you subtract 8: 1 2 3 4

You subtracted 8 ______ times.

So, Stephen can make ______ bundles of firewood.

You can write $32 \div 8 =$ ______ or $8\overline{)32}$.

ERROR Alert

Be sure to keep subtracting the divisor until the difference is less than the divisor.

Another Way Use a related multiplication fact.

$32 \div 8 = \blacksquare$

$\blacksquare \times 8 = 32$

$4 \times 8 = 32$

Think: What number completes the multiplication fact?

$32 \div 8 =$ ______ or $8\overline{)32}$

MATHEMATICAL PRACTICES

Math Talk How does knowing $4 \times 8 = 32$ help you find $32 \div 8$?

Example Find the unknown divisor.

Stephen has a log that is 16 feet long. If he cuts the log into pieces that are 2 feet long, how many pieces will Stephen have?

Divide. 16 ÷ ■ = 2

You can also use a multiplication table to find the divisor in a division problem.

Think: ■ × 2 = 16

STEP 1 Find the factor, 2, in the top row.

STEP 2 Look down to find the product, 16.

STEP 3 Look left to find the unknown factor.

×	0	1	2	3	4	5	6	7	8	9	10
0	0	0	0	0	0	0	0	0	0	0	0
1	0	1	2	3	4	5	6	7	8	9	10
2	0	2	4	6	8	10	12	14	16	18	20
3	0	3	6	9	12	15	18	21	24	27	30
4	0	4	8	12	16	20	24	28	32	36	40
5	0	5	10	15	20	25	30	35	40	45	50
6	0	6	12	18	24	30	36	42	48	54	60
7	0	7	14	21	28	35	42	49	56	63	70
8	0	8	16	24	32	40	48	56	64	72	80
9	0	9	18	27	36	45	54	63	72	81	90
10	0	10	20	30	40	50	60	70	80	90	100

The unknown factor is ____.

■ = ____

____ × 2 = 16 Check.

____ = 16 ✓ The equation is true.

So, Stephen will have ____ pieces.

MATHEMATICAL PRACTICES

Math Talk Explain how to use the multiplication table to find the unknown dividend for ■ ÷ 8 = 5.

Share and Show

1. Use repeated subtraction to find 24 ÷ 8. ____

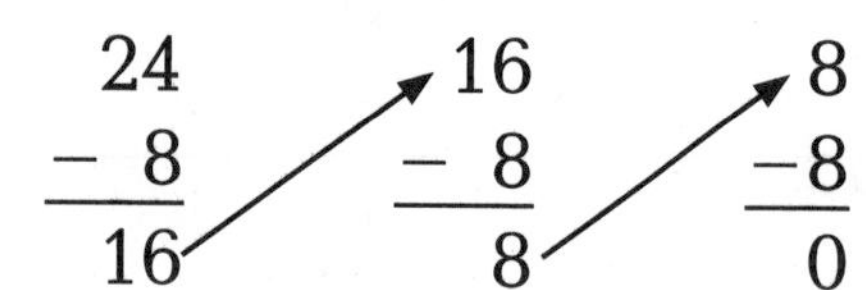

Think: How many times do you subtract 8?

MATHEMATICAL PRACTICES

Math Talk Explain why you subtract 8 from 24 to find 24÷ 8.

Find the unknown factor and quotient.

2. 8 × ____ = 56 56 ÷ 8 = ____

✓3. ____ × 8 = 40 40 ÷ 8 = ____

Find the quotient.

4. 18 ÷ 3 = ____

5. ____ = 48 ÷ 8

6. 56 ÷ 7 = ____

✓7. ____ = 32 ÷ 8

Name ______________________________

On Your Own

Find the unknown factor and quotient.

8. $8 \times$ ____ $= 8$ $\quad 8 \div 8 =$ ____

9. ____ $\times 5 = 35$ $\quad$ ____ $= 35 \div 5$

10. $6 \times$ ____ $= 18$ $\quad 18 \div 6 =$ ____

11. $8 \times$ ____ $= 72$ $\quad$ ____ $= 72 \div 8$

Find the quotient.

12. $28 \div 4 =$ ____

13. $42 \div 7 =$ ____

14. ____ $= 3 \div 3$

15. ____ $= 28 \div 7$

16. $8\overline{)0}$

17. $6\overline{)24}$

18. $8\overline{)64}$

19. $1\overline{)8}$

Find the unknown number.

20. $72 \div ★ = 9$
★ = ____

21. $t \div 8 = 2$
$t =$ ____

22. $64 \div ▲ = 8$
▲ = ____

23. $m \div 8 = 10$
$m =$ ____

24. $▲ \div 2 = 10$
▲ = ____

25. $40 \div ■ = 8$
■ = ____

26. $25 \div k = 5$
$k =$ ____

27. $54 \div n = 9$
$n =$ ____

28. H.O.T. Write a word problem that can be solved by using one of the division facts above.

____

____

Algebra **Write +, −, ×, or ÷.**

29. $6 \times 6 = 32$ ◯ 4

30. 12 ◯ $3 = 19 - 15$

31. $40 \div 8 = 35$ ◯ 7

Problem Solving REAL WORLD

Use the table for 32–33.

Tent Sizes	
Type	**Number of People**
Cabin	10
Vista	8
Trail	4

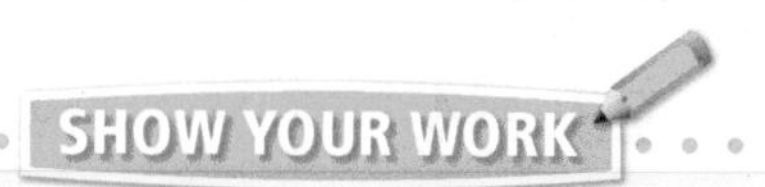

32. H.O.T. There are 36 people camping at Max's family reunion. They have cabin tents and vista tents. How many of each type of tent do they need to sleep exactly 36 people if each tent is filled? **Explain.**

______

______

______

______

SHOW YOUR WORK

33. Write Math ▶ There are 32 people who plan to camp over the weekend. What is the least number of trail tents they need? **Explain.**

______

______

______

34. Josh is dividing 64 marshmallows equally among 8 campers. How many marshmallows will each camper get?

______

35. **Test Prep** Grace set 8 plates at each picnic table so 24 campers could eat dinner. How many picnic tables did Grace have to set?

Ⓐ 3

Ⓑ 4

Ⓒ 5

Ⓓ 6

FOR MORE PRACTICE:
Standards Practice Book, pp. P141–P142

Name ______________________________

Divide by 9

Essential Question What strategies can you use to divide by 9?

UNLOCK the Problem REAL WORLD

Becket's class goes to the aquarium. The 27 students from the class are separated into 9 equal groups. How many students are in each group?

- Do you need to find the number of equal groups or the number in each group?

____

- What label will your answer have?

____

One Way Make equal groups.

- Draw 9 circles to show 9 groups.
- Draw 1 counter in each group.
- Continue drawing 1 counter at a time until all 27 counters are drawn.

There are _____ counters in each group.

So, there are __________ in each group.

You can write $27 \div 9 =$ _____ or $9\overline{)27}$.

Math Talk MATHEMATICAL PRACTICES What is another way you could solve the problem? **Explain.**

Other Ways

A Use factors to find 27 ÷ 9.

The factors of 9 are 3 and 3.

$3 \times 3 = 9$

factors, product

To divide by 9, use the factors.

$27 \div 9 = s$

Divide by 3. $27 \div 3 = 9$

Then divide by 3 again. $9 \div 3 = 3$

$27 \div 9 =$ ______

B Use a related multiplication fact.

$27 \div 9 = s$

$9 \times s = 27$

$9 \times 3 = 27$

Think: What number completes the multiplication fact?

$27 \div 9 =$ ______ or $9\overline{)27}$

- What multiplication fact can you use to find $63 \div 9$? ______

Share and Show

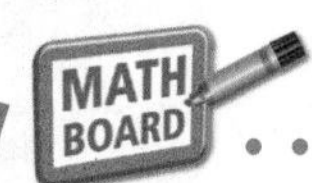

1. Draw counters in the groups to find $18 \div 9$. ______

MATHEMATICAL PRACTICES

Math Talk **Explain** how you would use factors to find $18 \div 9$.

Find the quotient.

2. ______ $= 45 \div 9$
3. $36 \div 6 =$ ______
4. $9 \div 1 =$ ______
5. ______ $= 54 \div 9$

6. $7\overline{)28}$
7. $9\overline{)9}$
8. $5\overline{)40}$
9. $9\overline{)36}$

Name ______________________

On Your Own

Find the quotient.

10. $8 \div 2 =$ ____
11. ____ $= 72 \div 9$
12. $56 \div 8 =$ ____
13. ____ $= 27 \div 9$
14. ____ $= 5 \div 1$
15. ____ $= 36 \div 4$
16. $81 \div 9 =$ ____
17. $30 \div 5 =$ ____
18. ____ $= 0 \div 6$
19. $21 \div 7 =$ ____
20. ____ $= 9 \div 3$
21. $42 \div 6 =$ ____
22. $9\overline{)18}$
23. $3\overline{)27}$
24. $6\overline{)48}$
25. $9\overline{)90}$
26. $4\overline{)12}$
27. $9\overline{)63}$
28. $2\overline{)16}$
29. $5\overline{)25}$

Find the unknown number.

30. $64 \div 8 = e$

    $e =$ ____
31. $0 \div 9 = g$

    $g =$ ____
32. ■ $= 20 \div 4$

    ■ $=$ ____
33. $s = 9 \div 9$

    $s =$ ____
34. $35 \div 5 = \bigstar$

    $\bigstar =$ ____
35. $p = 4 \div 1$

    $p =$ ____
36. $54 \div 9 = w$

    $w =$ ____
37. $a = 14 \div 2$

    $a =$ ____

Algebra **Complete the table.**

38.

÷	24	40	32	48
8				

39.

÷	54	45	72	63
9				

UNLOCK the Problem REAL WORLD

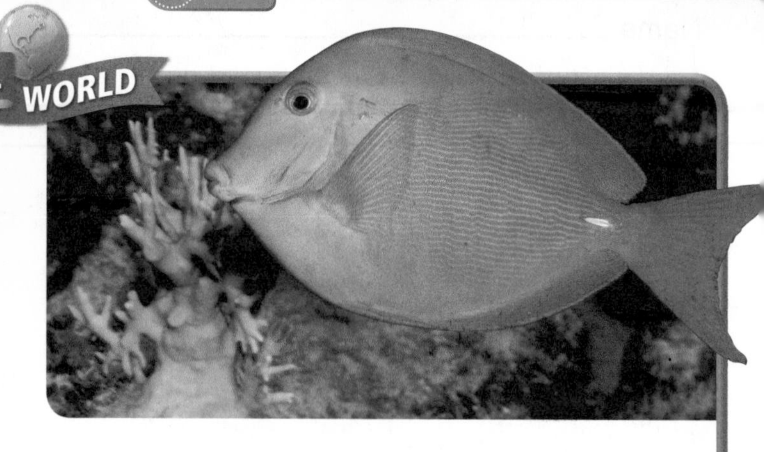

40. Carlos has 28 blue tang fish and 17 yellow tang fish in one large fish tank. He wants to separate the fish so that there are the same number of fish in each of 9 smaller tanks. How many tang fish will Carlos put in each smaller tank?

a. What do you need to find? ______________________

____

b. Why do you need to use two operations to solve the problem? ______________________

____

c. Write the steps to find how many tang fish Carlos will put in each smaller tank.

d. Complete the sentences.

Carlos has _____ blue tang fish

and _____ yellow tang fish in one large fish tank.

He wants to separate the fish so that there are the same number

of fish in each of _____ smaller tanks.

So, Carlos will put _____ fish in each smaller tank.

41. Sophie has a new fish. She feeds it 9 fish pellets each day. If Sophie has fed her fish 72 pellets, for how many days has she had her fish?

____

42. **Test Prep** At the aquarium, Ms. Brady separates 36 students into 9 equal groups. How many students are in each group?

Ⓐ 2
Ⓑ 3
Ⓒ 4
Ⓓ 5

FOR MORE PRACTICE:
Standards Practice Book, pp. P143–P144

Name ______________________________

Problem Solving • Two-Step Problems

Essential Question How can you use the strategy *act it out* to solve two-step problems?

UNLOCK the Problem REAL WORLD

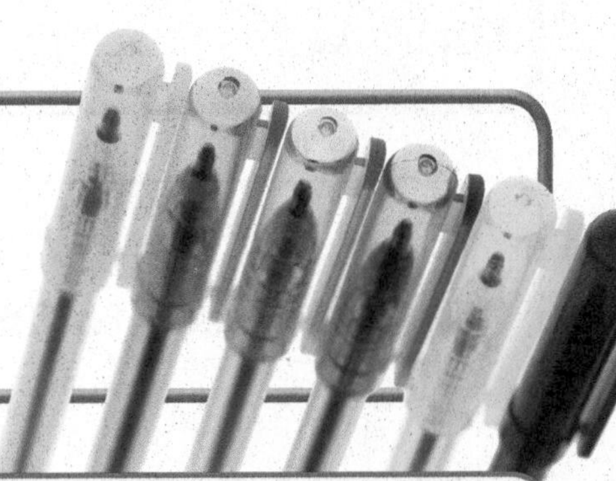

Madilyn bought 2 packs of pens and a notebook for \$11. The notebook cost \$3. Each pack of pens cost the same amount. What is the price of 1 pack of pens?

Read the Problem

What do I need to find?

I need to find the price of 1 pack of ________.

What information do I need to use?

Madilyn spent _____ in all.

She bought _____ packs of pens and _____ notebook.

The notebook cost _____.

How will I use the information?

I will use the information to _____ out the problem.

Solve the Problem

Describe how to act out the problem.

Start with 11 counters. Take away 3 counters.

total cost		cost of notebook		*p*, cost of 2 packs of pens
↓		↓		↓
_____	−	_____	=	*p*
		_____	=	*p*

Now I know that 2 packs of pens cost _____.

Next, make _____ equal groups with the 8 remaining counters.

p, cost of 2 packs of pens		number of packs		*c*, cost of 1 pack of pens
↓		↓		↓
\$8	÷	_____	=	*c*
		_____	=	*c*

So, the price of 1 pack of pens is _____.

MATHEMATICAL PRACTICES

Math Talk Why do you need to use two operations to solve the problem? **Explain.**

Try Another Problem

Chad bought 4 packs of T-shirts. He gave 5 T-shirts to his brother. Now Chad has 19 shirts. How many T-shirts were in each pack?

Read the Problem	Solve the Problem
What do I need to find?	**Describe how to act out the problem.**
What information do I need to use?	
How will I use the information?	

- How can you use multiplication and subtraction to check your answer?

____

____

____

Math Talk MATHEMATICAL PRACTICES Explain another strategy you could use to solve this problem.

Name ________________________________

Share and Show 

Tips
UNLOCK the Problem
- √ Circle the question.
- √ Underline the important facts.
- √ Choose a strategy you know.

1. Mac bought 4 packs of toy cars. Then his friend gave him 9 cars. Now Mac has 21 cars. How many cars were in each pack?

Act out the problem by using counters or the picture and by writing equations.

First, subtract the cars Mac's friend gave him.

total cars ↓		cars given to Mac ↓		c, cars in 4 packs ↓
21	−	_____	=	c
		_____	=	c

Then, divide to find the number of cars in each pack.

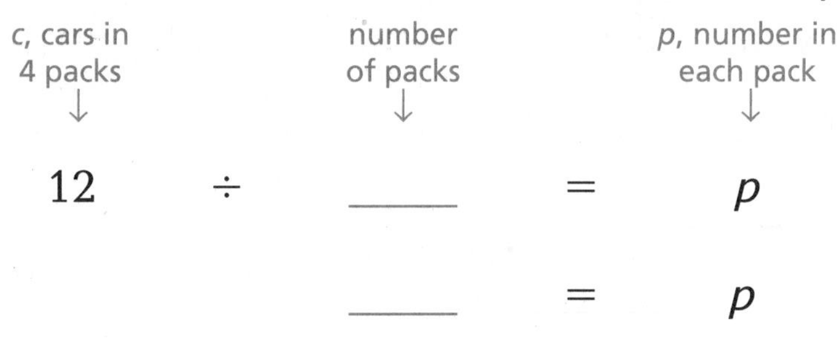

c, cars in 4 packs ↓		number of packs ↓		p, number in each pack ↓
12	÷	_____	=	p
		_____	=	p

So, there were _____ cars in each pack.

2. H.O.T. **What if** Mac bought 8 packs of cars and then he gave his friend 3 cars? If Mac has 13 cars now, how many cars were in each pack?

____

3. Ryan gave 7 of his model cars to a friend. Then he bought 6 more cars. Now Ryan has 13 cars. How many cars did Ryan start with?

____

4. Chloe bought 5 sets of books. She donated 9 of her books to her school. Now she has 26 books. How many books were in each set?

____

SHOW YOUR WORK

On Your Own

Choose a STRATEGY

- Act It Out
- Draw a Diagram
- Find a Pattern
- Make a Table

5. Raul bought 2 packs of erasers. He found 2 erasers in his backpack. Now Raul has 8 erasers. How many erasers were in each pack?

____

6. Blair cuts a ribbon into 2 equal pieces. Then she cuts 4 inches off one piece. That piece is now 5 inches long. What was the length of the original ribbon?

____

SHOW YOUR WORK

7. H.O.T. **Write Math** Rose saw a movie, shopped, and ate at a restaurant. She did not see the movie first. She shopped right after she ate. In what order did Rose do these activities? **Explain** how you know.

____

____

____

____

____

____

8. **Test Prep** Mr. Acosta bought 3 packs of markers. Each pack had the same number of markers. He gave 4 markers to his daughter. Now he has 14 markers. How many markers were in each pack?

Ⓐ 6

Ⓑ 11

Ⓒ 14

Ⓓ 26

FOR MORE PRACTICE:
Standards Practice Book, pp. P145–P146

Name ______________________

Order of Operations

Essential Question Why are there rules such as the order of operations?

Investigate

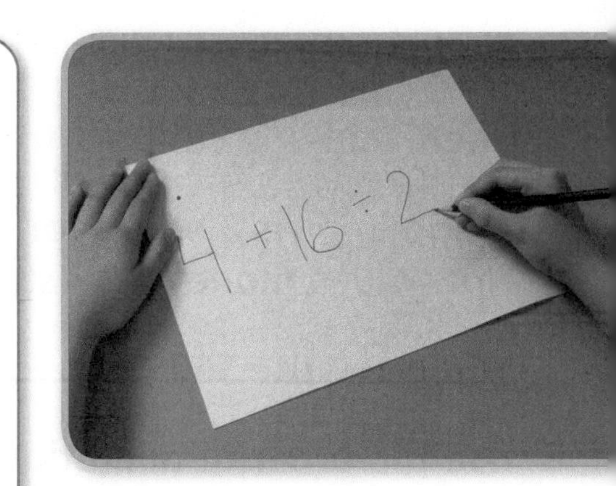

CONNECT You can use what you know about acting out a two-step problem to write one equation to describe and solve a two-step problem.

- If you solved a two-step problem in a different order, what do you think might happen?

____

Use different orders to find $4 + 16 \div 2$.

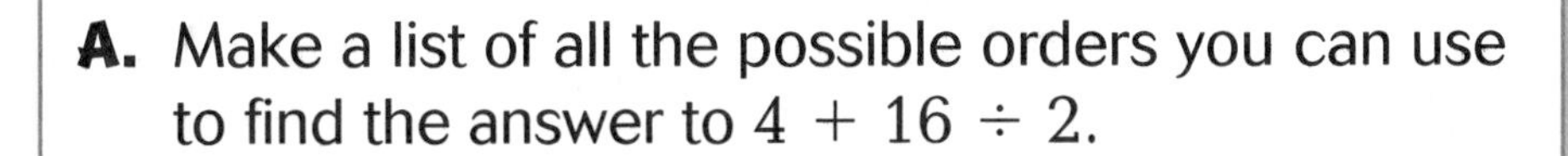

A. Make a list of all the possible orders you can use to find the answer to $4 + 16 \div 2$.

____

B. Use each order in your list to find the answer. Show the steps you used.

____

____

Draw Conclusions

1. Did following different orders change the answer? ______________________

2. If a problem has more than one type of operation, how does the order in which you perform the operations affect the answer?

____

3. H.O.T. **Explain** the need for setting an order of operations that everyone follows.

____

____

Make Connections

When solving problems with more than one type of operation, you need to know which operation to do first. A special set of rules, called the **order of operations**, gives the order in which calculations are done in a problem.

First, multiply and divide from left to right.

Then, add and subtract from left to right.

Meghan buys 2 books for $4 each. She pays with a $10 bill. How much money does she have left?

You can write $\$10 - 2 \times \$4 = c$ to describe and solve the problem.

Use the order of operations to solve $\$10 - 2 \times \$4 = c$.

STEP 1

Multiply from left to right.

$\$10 - 2 \times \$4 = c$
$\$10 - \$8 = c$

STEP 2

Subtract from left to right.

$\$10 - \$8 = c$
$\$2 = c$

So, Meghan has ______ left.

- Does your answer make sense? **Explain.**

____

____

____

MATHEMATICAL PRACTICES

Math Talk What operation should you do first to find: $12 - 6 \div 2$ and $12 \div 6 - 2$? What is the answer to each problem?

Share and Show 

Write *correct* if the operations are listed in the correct order. If not correct, write the correct order of operations.

1. $4 + 5 \times 2$ multiply, add

____

2. $8 \div 4 \times 2$ multiply, divide

____

3. $12 + 16 \div 4$ add, divide

____

4. $9 + 2 \times 3$ add, multiply

____

5. $4 + 6 \div 3$ divide, add

____

6. $36 - 7 \times 3$ multiply, subtract

____

Name ________________________________

Follow the order of operations to find the unknown number. Use your MathBoard.

7. $63 \div 9 - 2 = f$

$f =$ ______

8. $7 - 5 + 8 = y$

$y =$ ______

9. $3 \times 6 - 2 = h$

$h =$ ______

10. $80 - 64 \div 8 = n$

$n =$ ______

11. $3 \times 4 + 6 = a$

$a =$ ______

12. $2 \times 7 \div 7 = c$

$c =$ ______

13. $20 \div 5 + 5 = g$

$g =$ ______

14. $42 \div 6 - 4 = b$

$b =$ ______

15. $20 + 12 - 5 = t$

$t =$ ______

H.O.T. **Algebra** **Use the numbers listed to make the equation true.**

16. 2, 6, and 5

______ + ______ × ______ = 16

17. 4, 12, and 18

______ − ______ ÷ ______ = 15

18. 8, 9, and 7

______ × ______ − ______ = 47

19. 2, 4, and 9

______ ÷ ______ + ______ = 11

20. H.O.T. **Pose a Problem** Write a word problem that can be solved by using $2 \times 5 \div 5$. Solve your problem.

____

____

____

____

21. **Write Math** Is $4 + 8 \times 3$ equal to $4 + 3 \times 8$? **Explain** how you know without finding the answers.

____

____

____

Connect to Social Studies

Picture Book Art

The Eric Carle Museum of Picture Book Art in Amherst, Massachusetts, is the first museum in the United States that is devoted to picture book art. Picture books introduce literature to young readers.

The museum has 3 galleries, a reading library, a café, an art studio, an auditorium, and a museum shop. The exhibits change every 3 to 6 months, depending on the length of time the picture art is on loan and how fragile it is.

The table shows prices for some souvenirs in the bookstore in the museum.

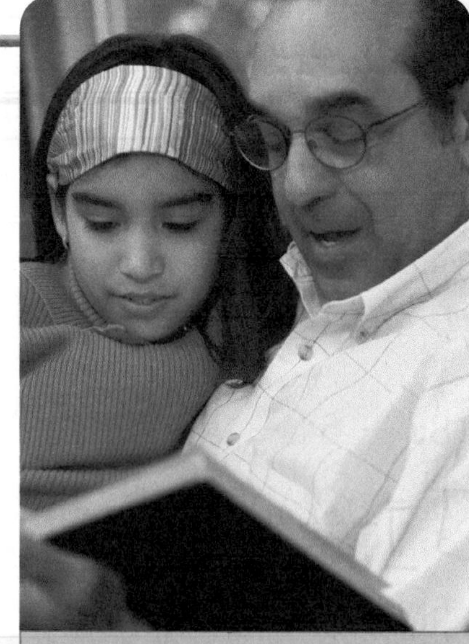

Souvenir Prices

Souvenir	Price
Firefly Picture Frame	$25
Exhibition Posters	$10
Caterpillar Note Cards	$8
Caterpillar Pens	$4
Sun Note Pads	$3

22. Kallon bought 3 Caterpillar note cards and 1 Caterpillar pen. How much did he spend on souvenirs?

____

23. Raya and 4 friends bought their teacher 1 Firefly picture frame. They shared the cost equally. Then Raya bought an Exhibition poster. How much money did Raya spend in all? **Explain.**

____

____

____

24. In Jaylene's class, 7 people bought Caterpillar pens and 1 person bought a poster. What was the total amount spent?

____

FOR MORE PRACTICE:
Standards Practice Book, pp. P147–P148

Name ____________________

Vocabulary

Choose the best term from the box.

Vocabulary
equation
order of operations

1. A special set of rules that gives the order in which calculations are done in a problem is called the ______________. (p. 296)

Concepts and Skills

Find the quotient.

2. $12 \div 4 =$ ____
3. $35 \div 7 =$ ____
4. ____ $= 0 \div 2$
5. ____ $= 27 \div 3$
6. ____ $= 48 \div 6$
7. ____ $= 9 \div 9$
8. $64 \div 8 =$ ____
9. $45 \div 5 =$ ____
10. $7\overline{)49}$
11. $3\overline{)9}$
12. $10\overline{)70}$
13. $9\overline{)72}$
14. $2\overline{)8}$
15. $5\overline{)40}$
16. $6\overline{)18}$
17. $8\overline{)32}$

Find the unknown factor and quotient.

18. $6 \times$ ____ $= 42$ $42 \div 6 =$ ____
19. $7 \times$ ____ $= 28$ ____ $= 28 \div 7$
20. ____ $\times 8 = 16$ $16 \div 8 =$ ____
21. $6 \times$ ____ $= 60$ ____ $= 60 \div 6$

GO Online Assessment Options Chapter Test

Fill in the bubble for the correct answer choice.

22. Maria has 54 flower seeds and 9 pots. If she wants to put an equal number of seeds in each pot, how many seeds will Maria put in each pot?

    Ⓐ 6

    Ⓑ 7

    Ⓒ 8

    Ⓓ 9

23. There are 56 students at basketball camp. The coaches put 8 players on each team. How many teams did they make?

    Ⓐ 6

    Ⓑ 7

    Ⓒ 48

    Ⓓ 64

24. Tristan bought 21 tickets to go on his favorite ride at the fair. The ride costs 3 tickets. How many times can Tristan go on his favorite ride?

    Ⓐ 24

    Ⓑ 18

    Ⓒ 9

    Ⓓ 7

25. Liza bought 63 inches of ribbon to make bows. If she uses 9 inches of ribbon for each bow, how many bows can she make?

    Ⓐ 7

    Ⓑ 8

    Ⓒ 54

    Ⓓ 72

Name ____

Fill in the bubble for the correct answer choice.

26. There are 42 students going on a field trip. They are riding in 7 vans. Each van has the same number of students. How many students are riding in each van?

Ⓐ 8

Ⓑ 7

Ⓒ 6

Ⓓ 5

27. There are 15 boys going camping. They brought 5 tents. An equal number of boys sleep in each tent. How many boys will sleep in each tent?

Ⓐ 2

Ⓑ 3

Ⓒ 10

Ⓓ 25

28. Kendall baked 30 muffins. Each of her baking trays holds 6 muffins. How many trays of muffins did Kendall bake?

Ⓐ 4

Ⓑ 5

Ⓒ 24

Ⓓ 36

▶ Constructed Response

29. Ginger bought 3 boxes of markers. She gave 6 markers to her sister. Now Ginger has 18 markers. How many markers were in each box? **Explain** how you found the answer.

____

____

____

30. Jordan had 30 minutes left to play 3 innings of his baseball game. If 2 innings lasted 10 minutes each, how much time is left for the last inning? **Explain**.

____

____

____

▶ Performance Task

31. There are 54 students in the marching band. They form equal rows, so that there are no fewer than 5 rows and no more than 10 students in each row.

A How many rows can they form and how many students will be in each row? **Explain**.

____

____

____

B One day, only 47 students came to practice. There were 5 students marching in the first row. How could the rest of the students be arranged to make equal rows? **Explain**.

____

____

____

Fractions

Developing understanding of fractions, especially unit fractions (fractions with numerator 1)

The Missouri quarter shows explorers Lewis and Clark traveling down the Missouri River. The Gateway Arch is in the background.

Project

Coins in the U.S.

Many years ago, a coin called a *piece of eight* was sometimes cut into 8 equal parts. Each part was equal to one eighth ($\frac{1}{8}$) of the whole. Now, U.S. coin values are based on the dollar. Four quarters are equal in value to 1 dollar. So, 1 quarter is equal to one fourth ($\frac{1}{4}$) of a dollar.

Get Started

Work with a partner. In which year were the Missouri state quarters minted? Use the Important Facts to help you. Then write fractions to answer these questions:

1. 2 quarters are equal to what part of a dollar?
2. 1 nickel is equal to what part of a dime?
3. 2 nickels are equal to what part of a dime?

Important Facts

- The U.S. government minted state quarters every year from 1999 to 2008 in the order that the states became part of the United States.
- 1999—Delaware, Pennsylvania, New Jersey, Georgia, Connecticut
- 2000—Massachusetts, Maryland, South Carolina, New Hampshire, Virginia
- 2001—New York, North Carolina, Rhode Island, Vermont, Kentucky
- 2002—Tennessee, Ohio, Louisiana, Indiana, Mississippi
- 2003—Illinois, Alabama, Maine, Missouri, Arkansas
- 2004—Michigan, Florida, Texas, Iowa, Wisconsin
- 2005—California, Minnesota, Oregon, Kansas, West Virginia
- 2006—Nevada, Nebraska, Colorado, North Dakota, South Dakota
- 2007—Montana, Washington, Idaho, Wyoming, Utah
- 2008—Oklahoma, New Mexico, Arizona, Alaska, Hawaii

Completed by ______________________

Chapter 8

Understand Fractions

Show What You Know

Check your understanding of important skills.

Name ______________________

▶ **Equal Parts** Circle the shape that has equal parts.

1. 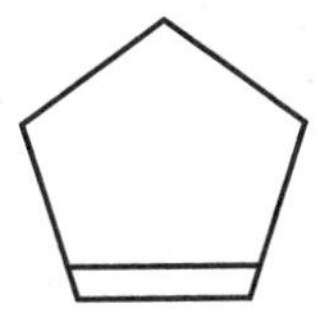 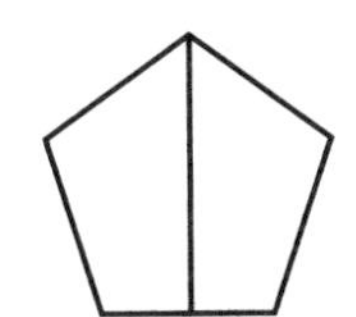

2. 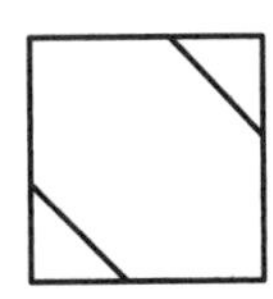 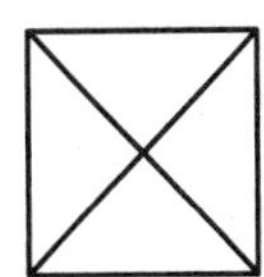 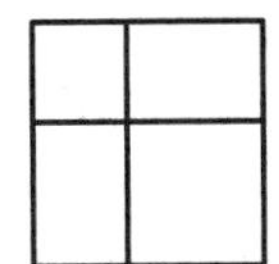

▶ **Combine Plane Shapes** Write the number of ▲ needed to cover the shape.

3. 

_____ triangles

4. 

_____ triangles

5. 

_____ triangles

▶ **Count Equal Groups** Complete.

6.    

_____ groups

_____ in each group

7.   

_____ groups

_____ in each group

Casey shared a pizza with some friends. They each ate $\frac{1}{3}$ of the pizza. Be a Math Detective to find how many people shared the pizza.

GO Online Assessment Options: Soar to Success Math

Vocabulary Builder

▶ Visualize It

Complete the bubble map by using the words with a ✓.

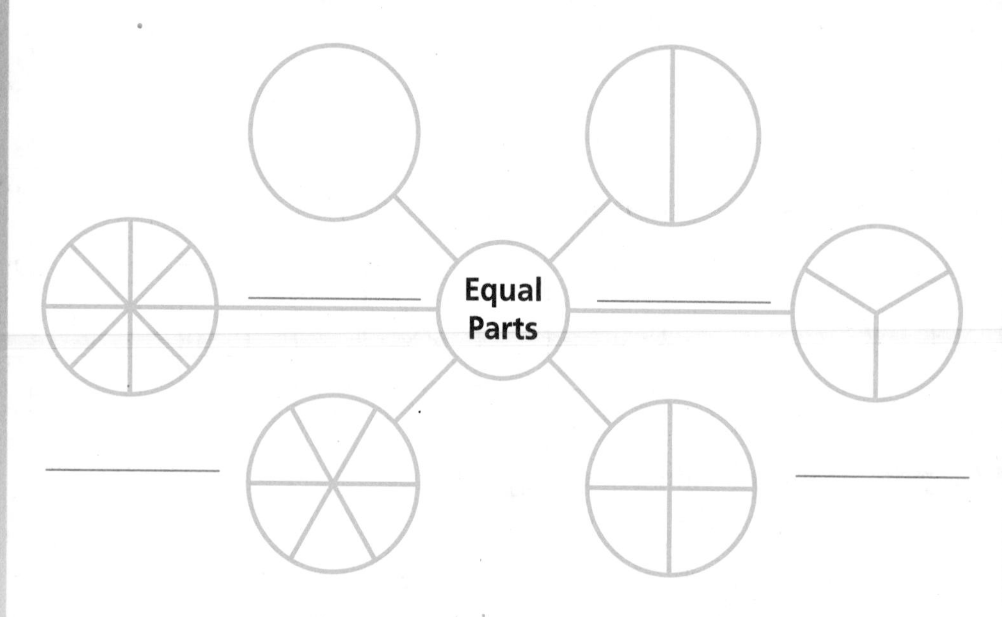

Preview Words
denominator
✓ eighths
equal parts
✓ fourths
fraction
fraction greater than 1
✓ halves
numerator
✓ sixths
✓ thirds
unit fraction
✓ whole

▶ Understand Vocabulary

Read the description. Write the preview word.

1. It is a number that names part of a whole or part of a group. ________________

2. It is the part of a fraction above the line, which tells how many parts are being counted.

   ____

3. It is the part of a fraction below the line, which tells how many equal parts there are in the whole or in the group. ________________

4. It is a number that names 1 equal part of a whole and has 1 as its numerator. ________________

GO Online • eStudent Edition • Multimedia eGlossary

Name ______________________________

Equal Parts of a Whole

Essential Question What are equal parts of a whole?

UNLOCK the Problem REAL WORLD

Lauren shares a sandwich with her brother. They each get an equal part. How many equal parts are there?

- What do you need to find?

____

____

- How many people share the sandwich? ________

Each whole shape below is divided into equal parts. A **whole** is all of the parts of one shape or group. **Equal parts** are exactly the same size.

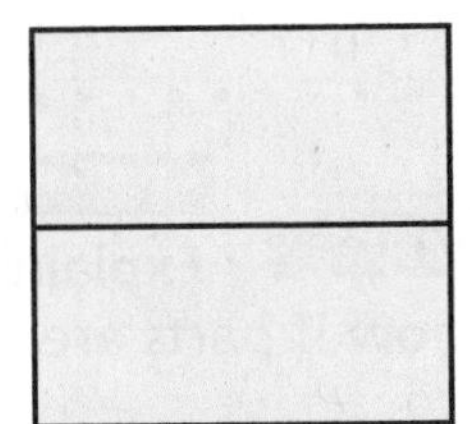

2 **halves**

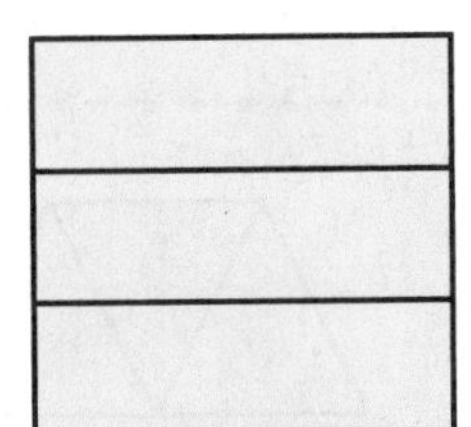

3 **thirds**

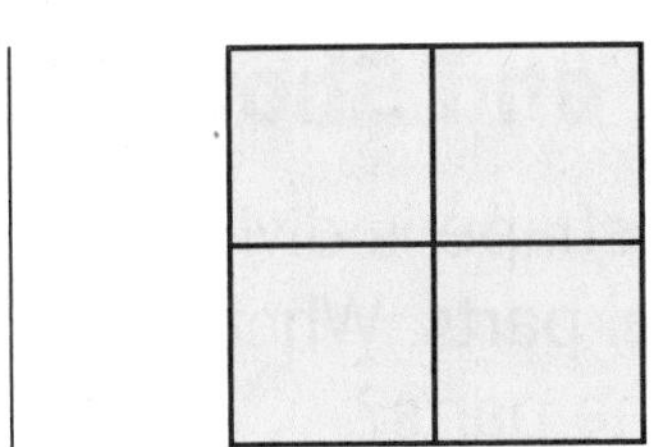

4 **fourths**

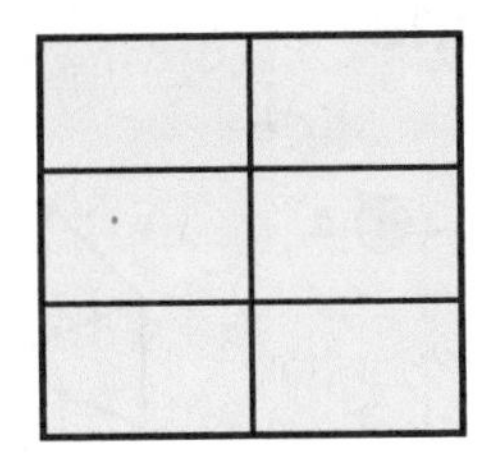

_____ **sixths**

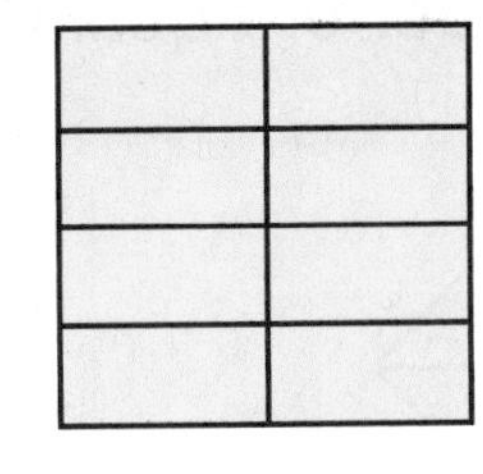

_____ **eighths**

Lauren's sandwich is divided into halves.

So, there are _____ equal parts.

- Draw a picture to show a different way Lauren's sandwich could have been divided into halves.

MATHEMATICAL PRACTICES

Math Talk Are your halves the same shape as your classmates' halves? **Explain** why both halves represent the same size.

Try This! Write whether the shape is divided into *equal* parts or *unequal* parts.

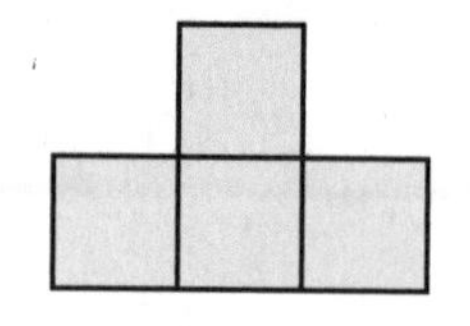

4 ______ parts
fourths

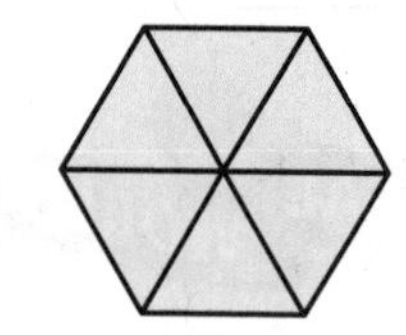

6 ______ parts
sixths

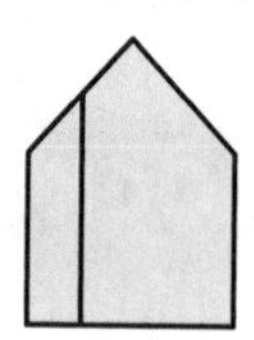

2 ______ parts
These are not halves.

ERROR Alert

Be sure the parts are equal in size.

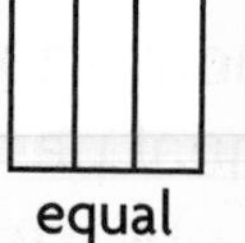

Share and Show

1. This shape is divided into 3 equal parts. What is the name for the parts?

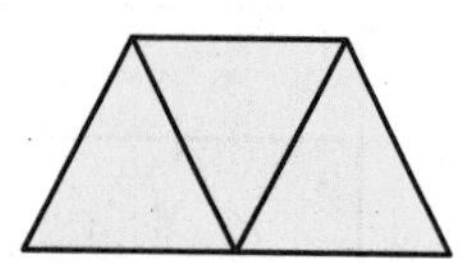

______

MATHEMATICAL PRACTICES

Math Talk Explain how you know if parts are equal.

Write the number of equal parts. Then write the name for the parts.

2.

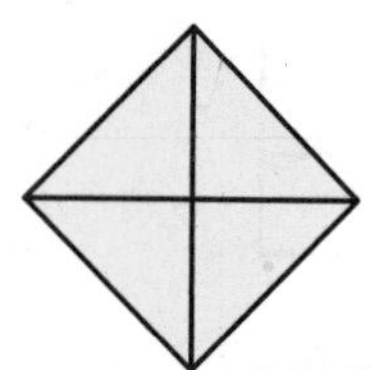

4 equal parts
fourths

3.

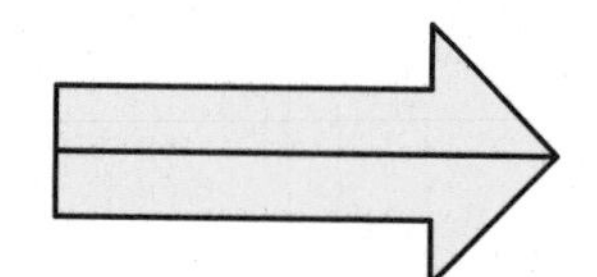

2 equal parts

hales

4.

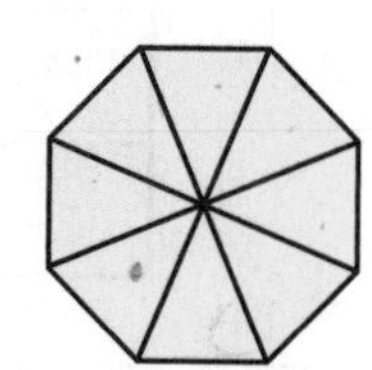

8 equal parts
eght

Write whether the shape is divided into *equal* parts or *unequal* parts.

5.

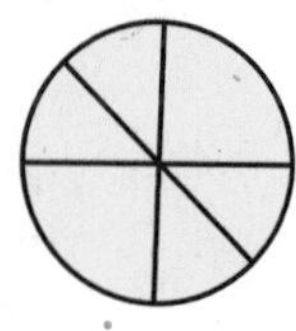

unequal parts

6.

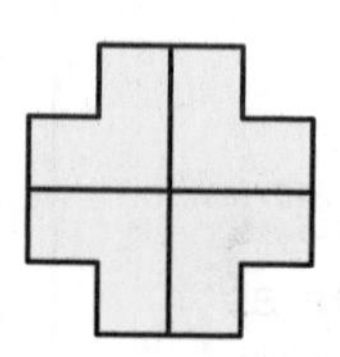

equal parts

7.

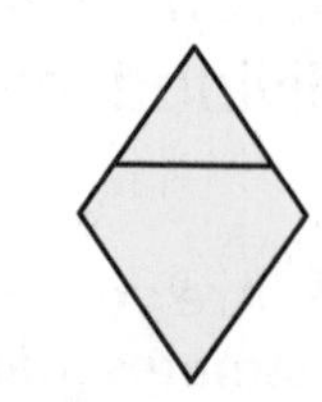

unequal parts

Name ______________________

On Your Own

Write the number of equal parts. Then write the name for the parts.

8.

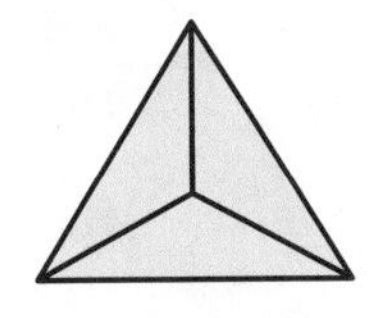

_____ equal parts

____

9.

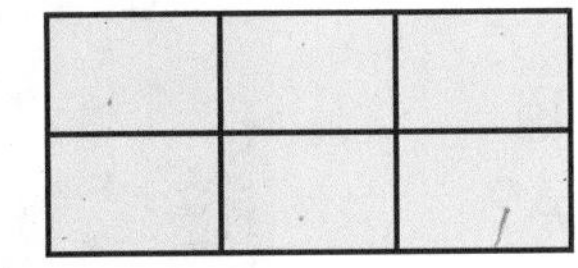

_____ equal parts

____

10.

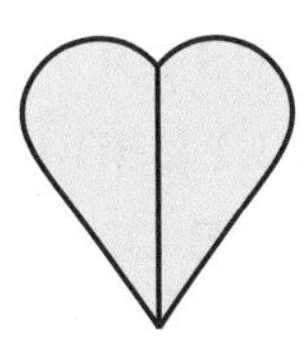

_____ equal parts

____

11.

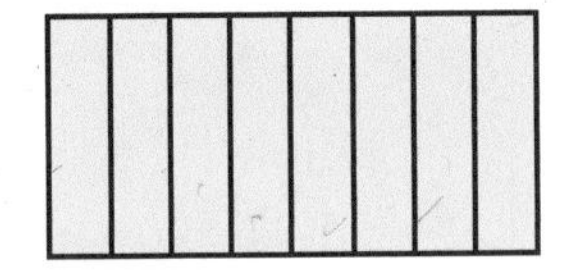

_____ equal parts

____

12.

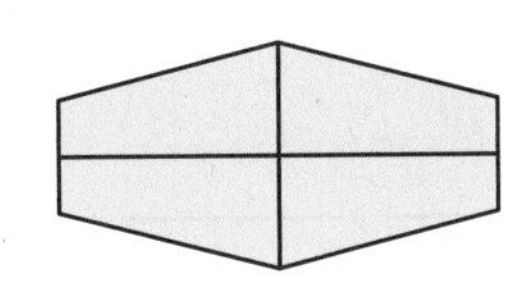

_____ equal parts

____

13.

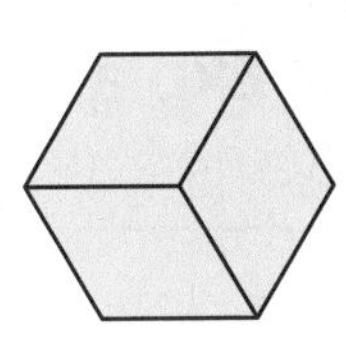

_____ equal parts

____

Write whether the shape is divided into *equal* parts or *unequal* parts.

14.

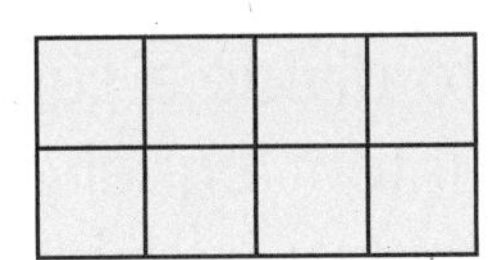

__________ parts

15.

__________ parts

16.

__________ parts

Draw lines to divide the circles into equal parts.

17. 2 halves

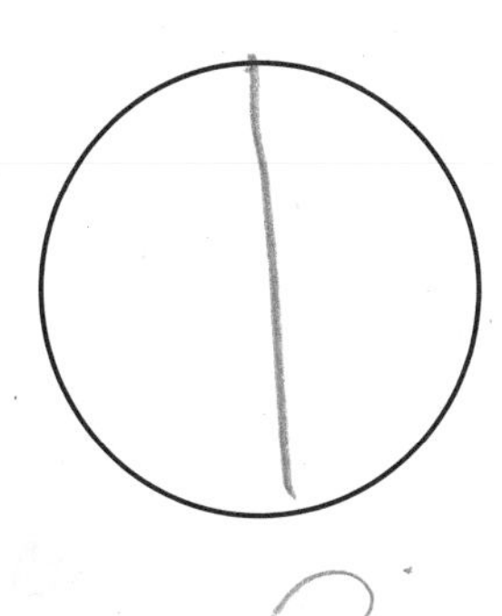

2

18. 4 fourths

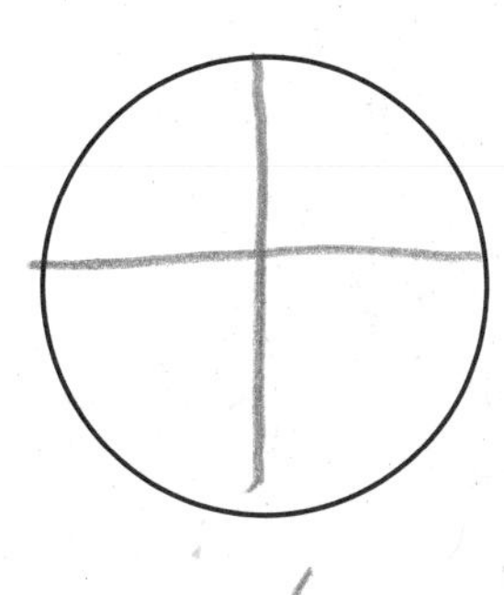

4

19. 8 eighths

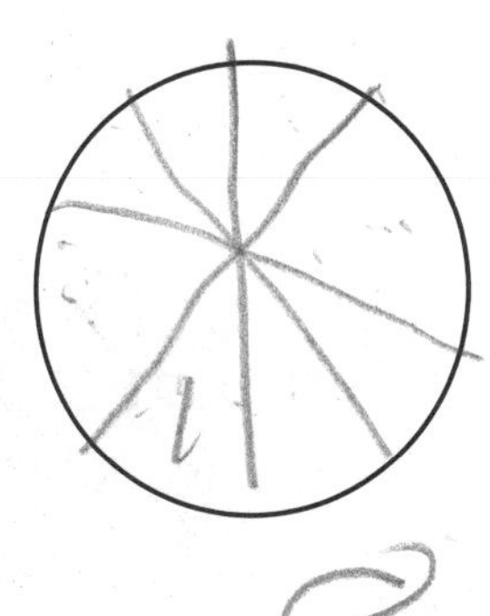

8

Problem Solving REAL WORLD

Use the pictures for 20–21.

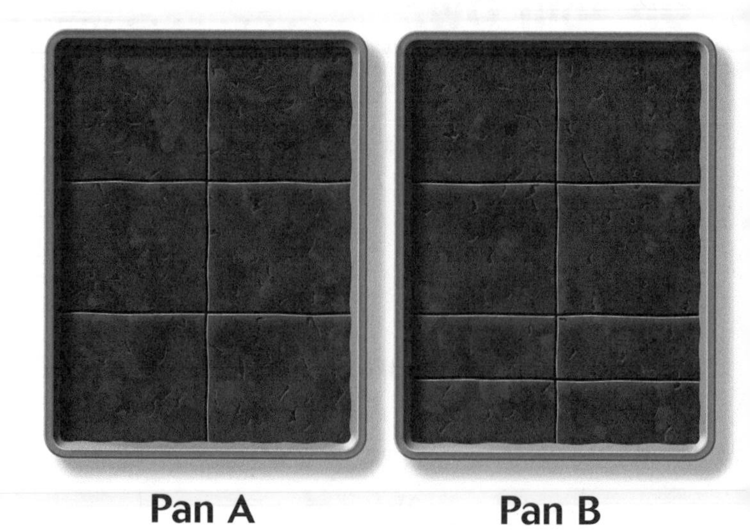
Pan A Pan B

20. Mrs. Rivera made 2 pans of brownies for Alex's party. She cut each pan into parts. What is the name of the parts for Pan A?

____

21. H.O.T. **Sense or Nonsense?** Alex said his mom divided Pan B into eighths. Does his statement make sense? **Explain.**

____

____

22. H.O.T. Write Math **Explain** why the rectangle is divided into 4 equal parts.

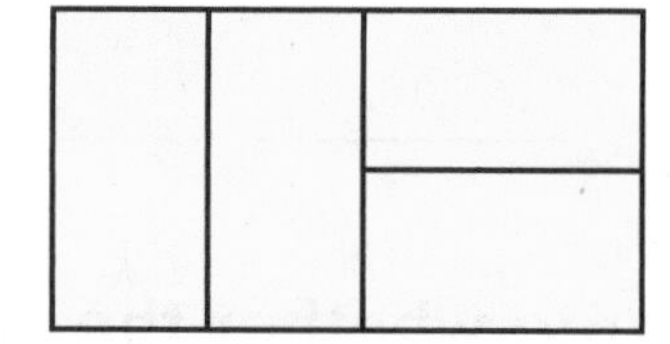

____

____

23. Shelby cut a triangle out of paper. She wants to divide the triangle into 2 equal parts. Draw a quick picture to show what her triangle could look like.

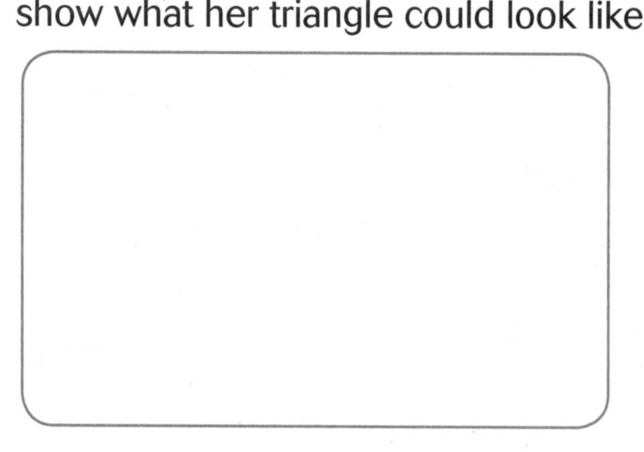

24. Andrew wants to divide a square piece of paper into 4 equal parts. Draw two different quick pictures to show what his paper could look like.

25. **Test Prep** Parker divides a fruit bar into thirds. How many equal parts are there?

Ⓐ 2
Ⓑ 3
Ⓒ 4
Ⓓ 8

Name ____________________________

Equal Shares

Essential Question Why do you need to know how to make equal shares?

UNLOCK the Problem REAL WORLD

Four friends share 2 small pizzas equally. What are two ways the pizza could be divided equally? How much pizza will each friend get?

- How might the two ways be different?

____

____

Draw to model the problem.

Draw 2 circles to show the pizzas.

One Way

There are _____ friends.

So, cut each pizza into 4 slices.

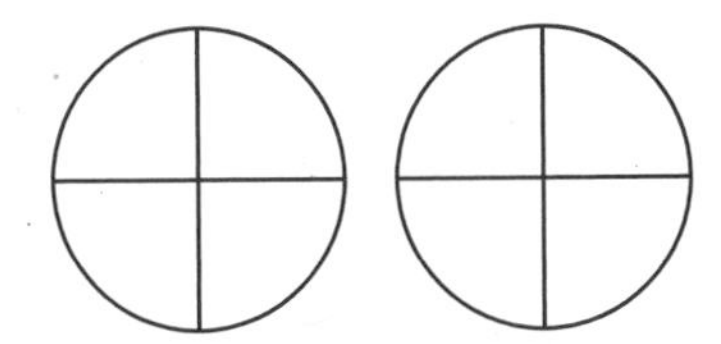

There are _____ equal parts.

Each friend can have 2 equal parts. Each friend will get 2 eighths of all the pizza.

Another Way

There are _____ friends.

So, cut all the pizza into 4 slices.

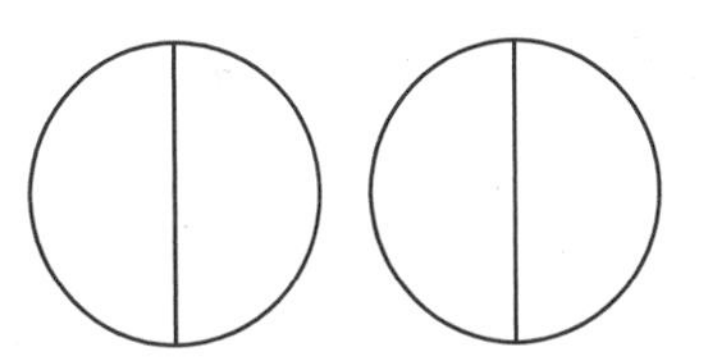

There are _____ equal parts.

Each friend can have 1 equal part. Each friend will get 1 half of a pizza.

MATHEMATICAL PRACTICES

Math Talk Explain why both ways let the friends have an equal share.

Try This! Four girls share 3 oranges equally. Draw a quick picture to find out how much each girl gets.

- Draw 3 circles to show the oranges.
- Draw lines to divide the circles equally.
- Shade the part 1 girl gets.
- Describe what part of an orange each girl gets.

____

Example

Melissa and Kyle are planning to share one pan of lasagna with 6 friends. They do not agree on the way to cut the pan into equal parts. Will each friend get an equal share using Melissa's way? Using Kyle's way?

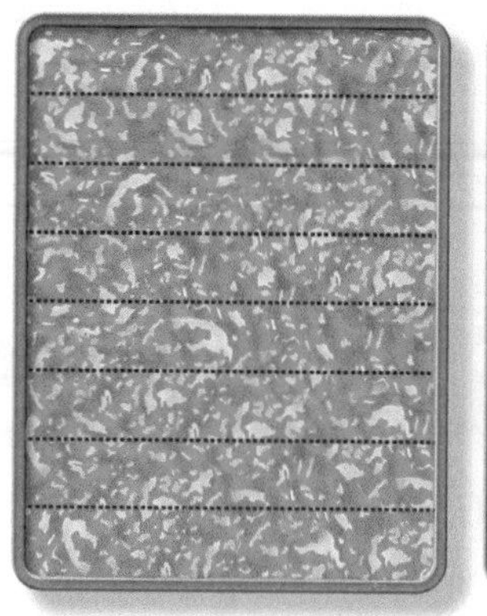
Melissa's Way

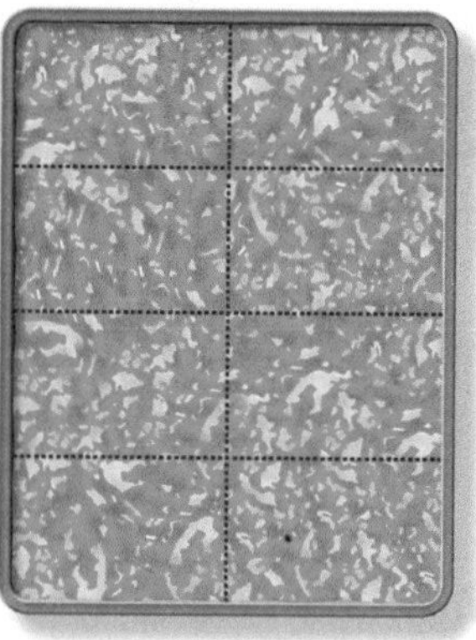
Kyle's Way

- Will Melissa's shares and Kyle's shares have the same shape? _____
- Will their shares using either way be the same size? _____

So, each friend will get an _____ share using either way.

- H.O.T. **Explain** why both ways let the friends have the same amount.

____

____

Share and Show

1. Two friends share 4 cookies equally. Use the picture to find how much each friend gets.

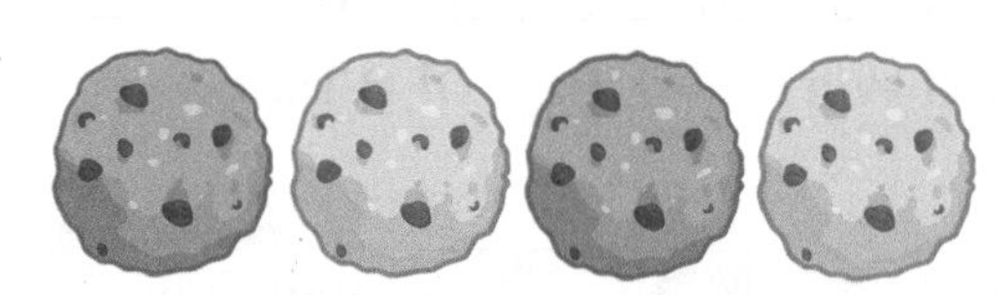

Think: There are more cookies than friends.

____

MATHEMATICAL PRACTICES

Math Talk **Explain** another way the cookies could have been divided. Tell how much each friend will get.

Draw lines to show how much each person gets. Write the answer.

2. 8 sisters share 3 brownies equally.

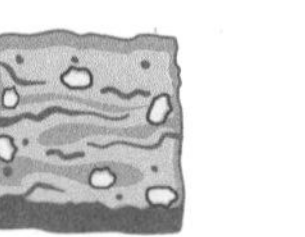

____

3. 6 neighbors share 4 pies equally.

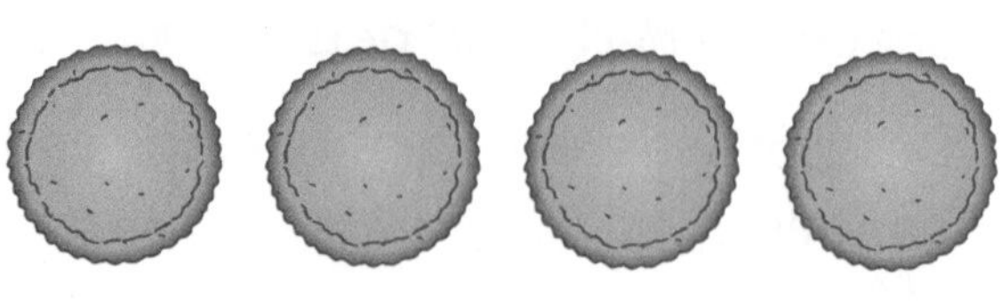

____

Name ________________________

On Your Own

Draw lines to show how much each person gets. Write the answer.

4. 3 classmates share 2 granola bars equally.

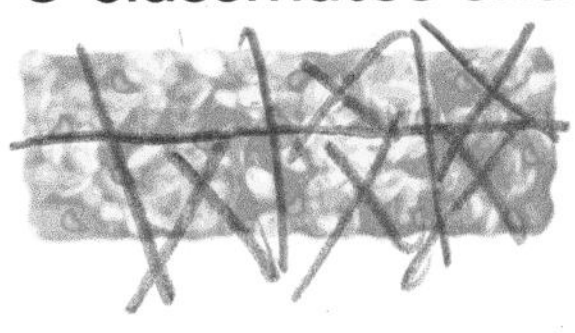

____

5. 4 brothers share 2 sandwiches equally.

____

Draw to show how much each person gets. Shade the amount that one person gets. Write the answer.

6. 8 friends share 4 sheets of construction paper equally.

____

7. 4 sisters share 3 muffins equally.

____

8. 6 students share 5 small pizzas equally.

____

UNLOCK the Problem REAL WORLD

9. Julia has 4 adults and 3 children coming over for dessert. She is going to serve 2 small blueberry pies. If she plans to give each person, including herself, an equal share of pie, how much pie will each person get?

a. What do you need to find? ______________________

____

b. How will you use what you know about drawing equal shares to solve the problem? ______________________

____

c. Draw a quick picture to find the share of pie each person will get.

d. Complete the sentences.

Julia has _____ adults and _____ children coming over.

In all, there will be _____ people having pie.

She is going to serve _____ blueberry pies.

So, each person will get __________ of a blueberry pie.

10. Abby baked 5 cherry pies. She wants to share them equally among 8 of her neighbors. How much of a pie will each neighbor get?

____

11. **Test Prep** Six friends share 2 pizzas equally. How much of a pizza does each friend get?

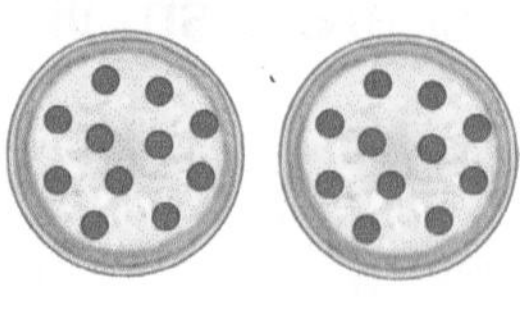

Ⓐ 2 sixths
Ⓑ 3 sixths
Ⓒ 6 halves
Ⓓ 6 thirds

FOR MORE PRACTICE:
Standards Practice Book, pp. P155–P156

Name ___________________________

Unit Fractions of a Whole

Essential Question What do the top and bottom numbers of a fraction tell?

A **fraction** is a number that names part of a whole or part of a group.

In a fraction, the top number tells how many equal parts are being counted. → 1

The bottom number tells how many equal parts are in the whole or in the group. → 6

A **unit fraction** names 1 equal part of a whole. It has 1 as its top number. $\frac{1}{6}$ is a unit fraction.

UNLOCK the Problem REAL WORLD

Luke's family picked strawberries. They used them to make a strawberry pie. They cut it into 6 equal pieces. Luke ate 1 piece. What fraction of the strawberry pie did he eat?

Find part of a whole.

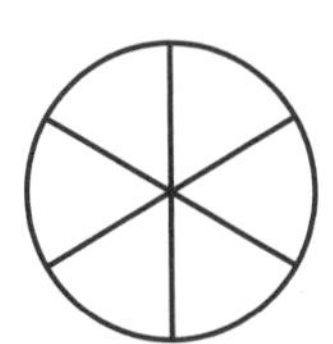

Shade 1 of the 6 equal parts.

Read: one sixth **Write:** $\frac{1}{6}$

So, Luke ate ______ of the strawberry pie.

Use a fraction to find a whole.

This shape  is $\frac{1}{4}$ of the whole. Here are examples of what the whole could look like.

MATHEMATICAL PRACTICES

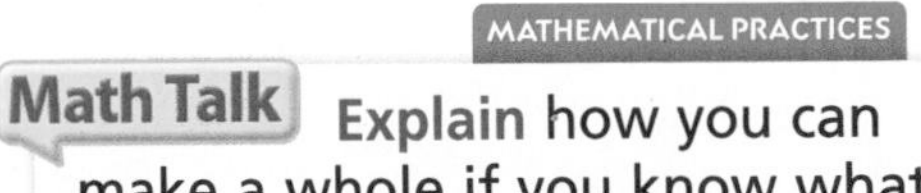

Math Talk Explain how you can make a whole if you know what one equal part looks like.

A

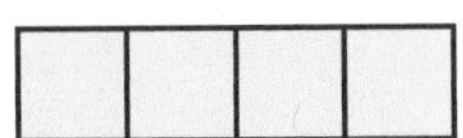

B

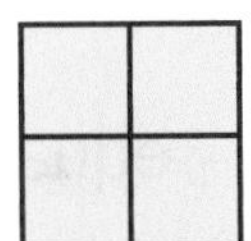

C

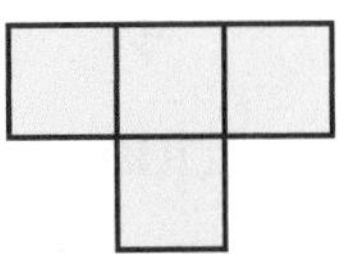

Try This! Look again at the examples at the bottom of page 315. Draw two other pictures of how the whole might look.

Share and Show 

1. What fraction names the shaded part? ____

Think: 1 out of 3 equal parts is shaded.

MATHEMATICAL PRACTICES

Math Talk Explain how you knew what number to write as the bottom number of the fraction in Exercise 1.

Write the number of equal parts in the whole. Then write the fraction that names the shaded part.

2. 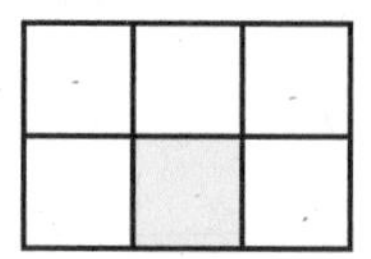

____ equal parts

3. 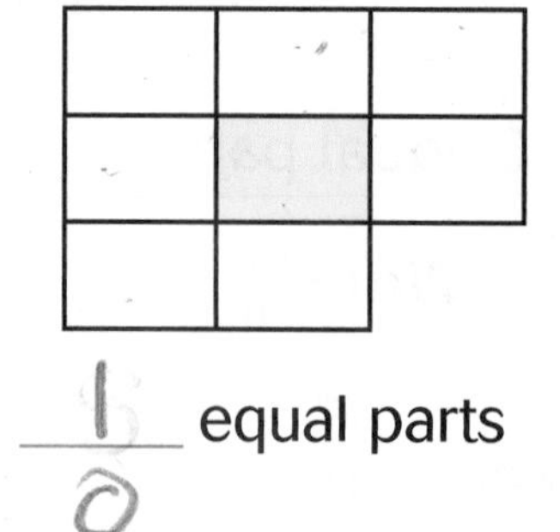

____ equal parts

____

4. 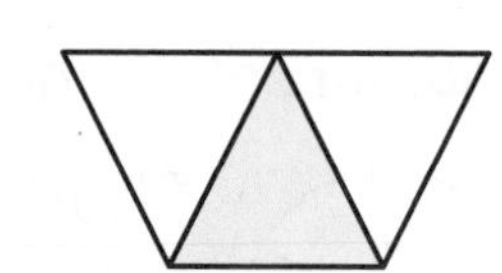

____ equal parts

5. 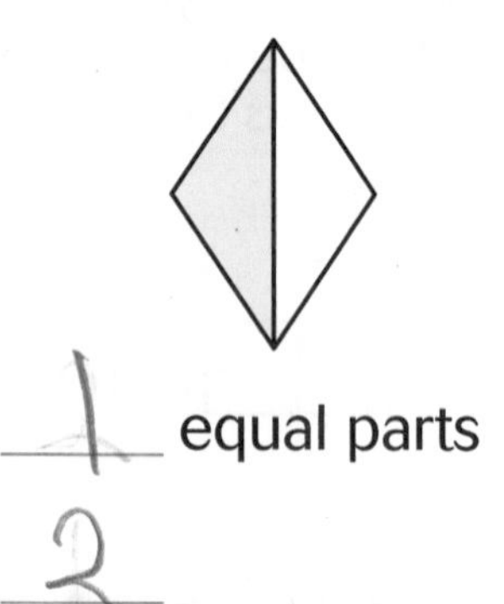

____ equal parts

____

6. 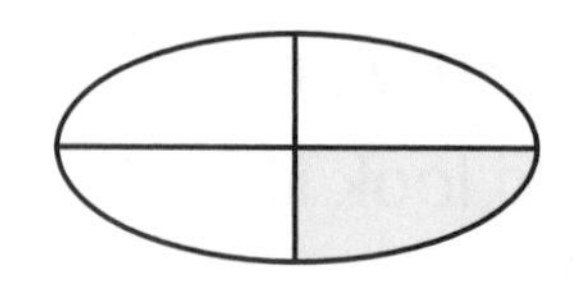

____ equal parts

7. 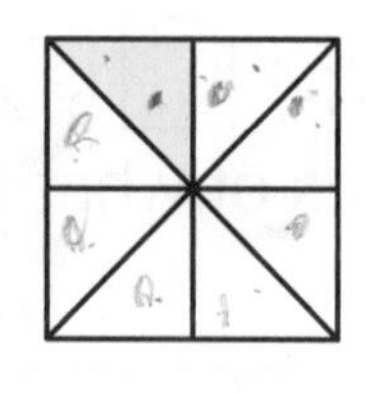

____ equal parts

Name ______________________________

On Your Own

Write the number of equal parts in the whole.
Then write the fraction that names the shaded part.

8.

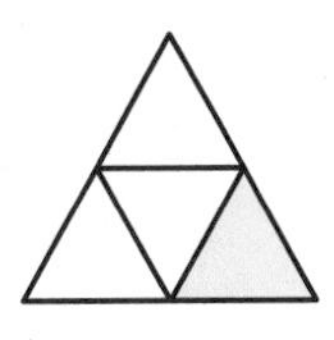

_____ equal parts

_____

9.

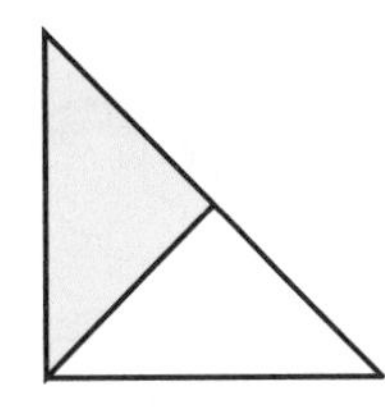

_____ equal parts

_____

10.

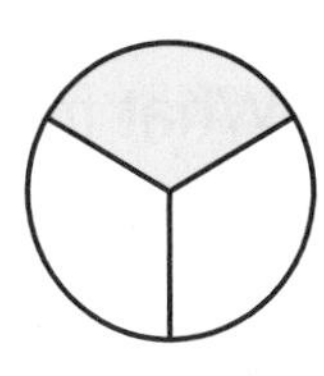

_____ equal parts

_____

11.

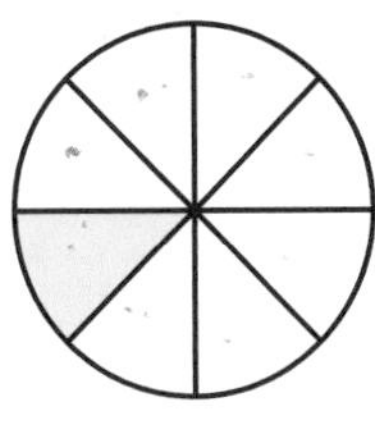

1 equal parts

12.

1 equal parts

13. H.O.T.

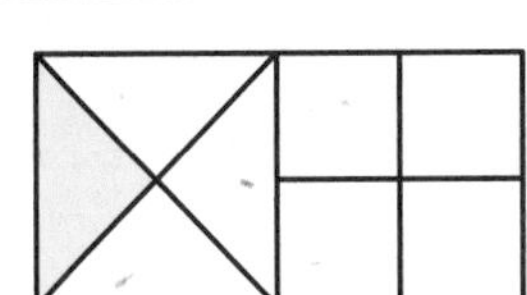

_____ equal parts

_____

Draw a picture of the whole.

14. $\frac{1}{2}$ is

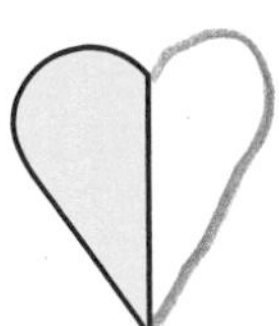

15. $\frac{1}{3}$ is

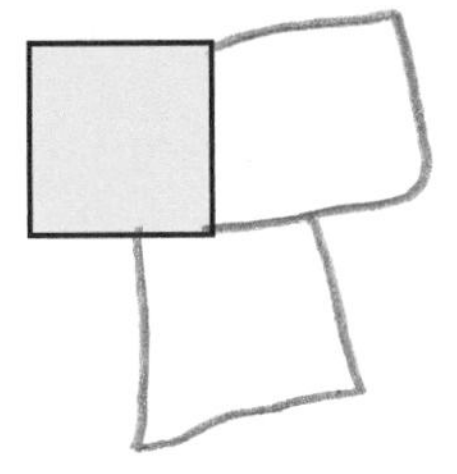

16. $\frac{1}{6}$ is

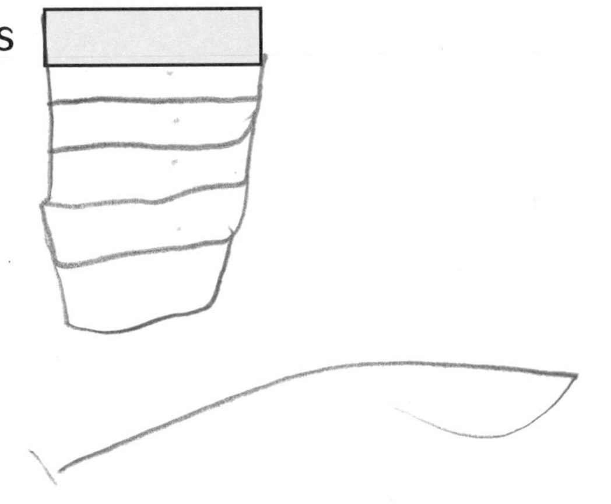

17. $\frac{1}{4}$ is

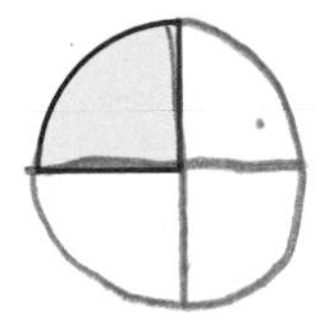

Problem Solving REAL WORLD

Use the pictures for 18–20.

Kylie's Lunch	Dylan's Lunch
sandwich	pizza
cookie	fruit bar

18. The missing parts of the pictures show what Kylie and Dylan ate for lunch. What fraction of the pizza did Dylan eat?

____

19. What fraction of the cookie did Kylie eat? Write the fraction in numbers and in words.

____ ______________

20. Write Math ▸ **What's the Question?** The answer is $\frac{1}{4}$.

____

____

21. Diego drew lines to divide the square into 6 pieces as shown. Then he shaded part of the square. Diego says he shaded $\frac{1}{6}$ of the square. Is he correct? **Explain** how you know.

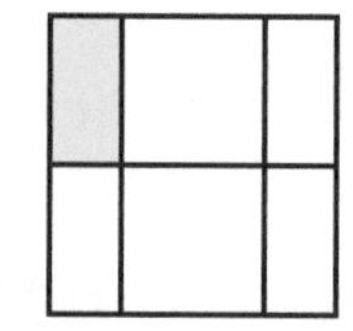

____

____

22. H.O.T. Riley's granola bar is broken into equal pieces. She ate one piece, which was $\frac{1}{4}$ of the bar. How many more pieces does Riley need to eat to finish the granola bar? Draw a picture.

____

23. **Test Prep** Mary shaded part of a rectangle. What fraction names the part she shaded?

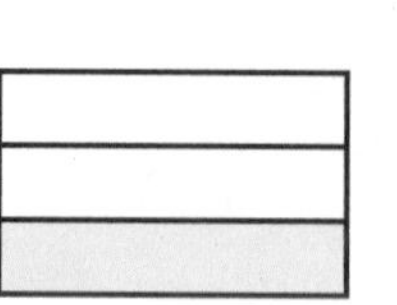

Ⓐ $\frac{1}{2}$ Ⓒ $\frac{1}{4}$

Ⓑ $\frac{1}{3}$ Ⓓ $\frac{3}{1}$

Name ________________________________

Fractions of a Whole

Essential Question How does a fraction name part of a whole?

UNLOCK the Problem REAL WORLD

The first pizzeria in America opened in New York in 1905. The pizza recipe came from Italy. Look at Italy's flag. What fraction of the flag is not red?

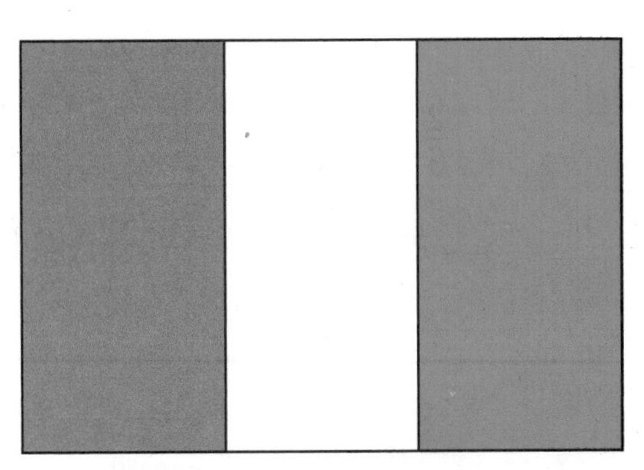

▲ Italy's flag has three equal parts.

 **Name equal parts of a whole.**

A fraction can name more than 1 equal part of a whole.

The flag is divided into 3 equal parts, and 2 parts are not red.

2 parts not red → 2 ← numerator
3 equal parts in all → 3 ← denominator

Read: two thirds or two parts out of three equal parts

Write: $\frac{2}{3}$

So, _____ of the flag is not red.

Math Idea

When all the parts are shaded, one whole shape is equal to all of its parts. It represents the whole number 1.

$\frac{3}{3} = 1$

The **numerator** tells how many parts are being counted.

The **denominator** tells how many equal parts are in the whole or in the group.

You can count equal parts, such as sixths, to make a whole.

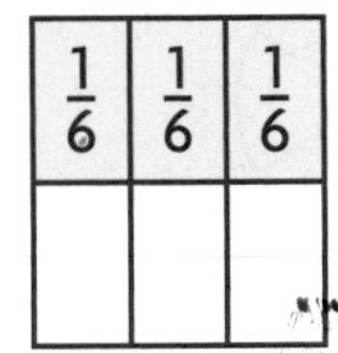

One $\frac{1}{6}$ part	Two $\frac{1}{6}$ parts	Three $\frac{1}{6}$ parts	Four $\frac{1}{6}$ parts	Five $\frac{1}{6}$ parts	Six $\frac{1}{6}$ parts
$\frac{1}{6}$	$\frac{2}{6}$	$\frac{3}{6}$	$\frac{4}{6}$	$\frac{5}{6}$	$\frac{6}{6}$

For example, $\frac{6}{6}$ = one whole, or 1.

Try This! **Write the missing word or number to name the shaded part.**

A

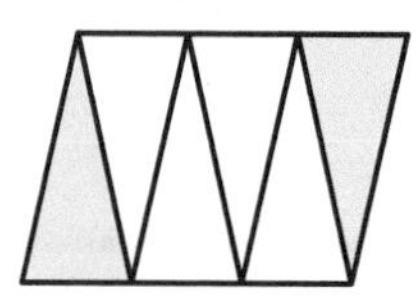

$\frac{2}{6}$

two sixths

B

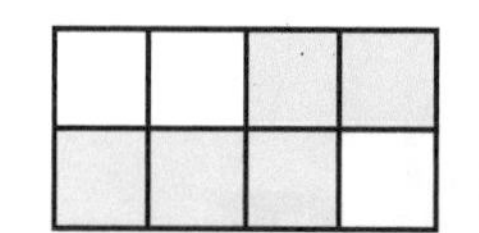

$\frac{5}{8}$

five eighths

C

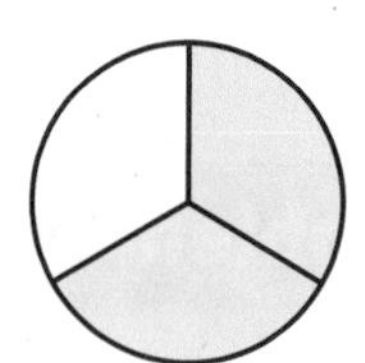

$\frac{2}{3}$

two thirds

D

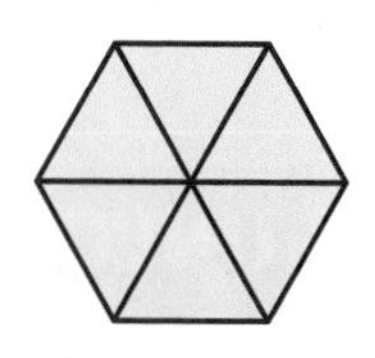

$\frac{6}{6}$, or 1

six sixths, or one whole

Share and Show

MATHEMATICAL PRACTICES

Math Talk **Explain** what the numerator and denominator of a fraction tell you.

1. Shade two parts out of eight equal parts. Write a fraction in words and in numbers to name the shaded part.

Think: Each part is $\frac{1}{8}$.

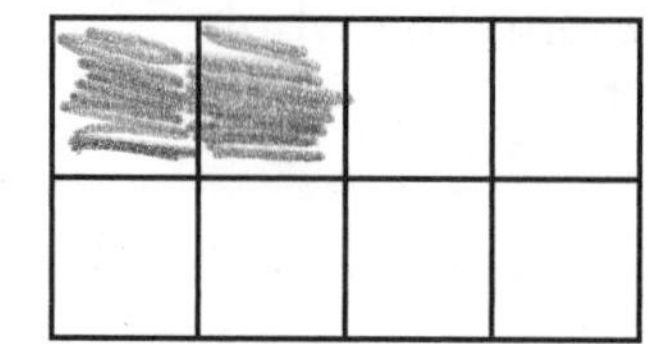

Read: two eighths **Write:**

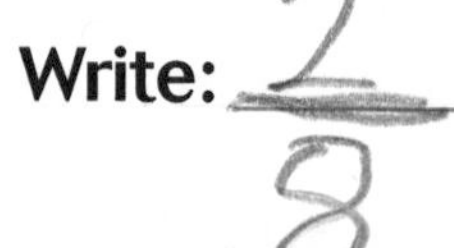

$\frac{2}{8}$

Write the fraction that names each part. Write a fraction in words and in numbers to name the shaded part.

2.

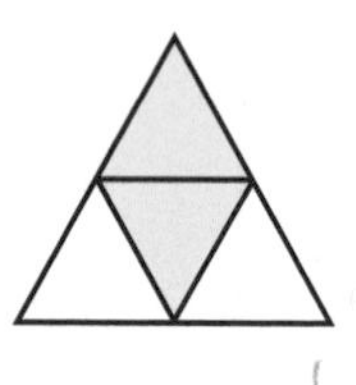

Each part is $\frac{1}{4}$.

2 fourths

4

3.

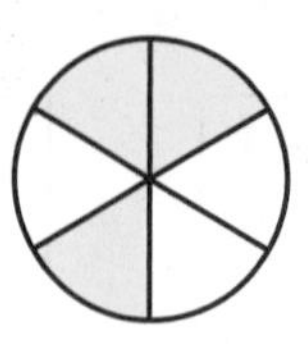

Each part is _____.

3 sixths

6

4.

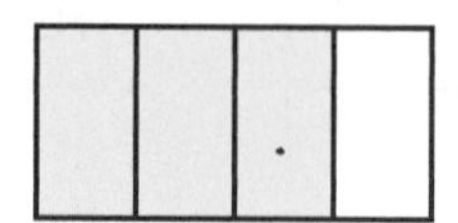

Each part is $\frac{1}{4}$.

1 fourths

____

Name ______________________________

On Your Own

Write the fraction that names each part. Write a fraction in words and in numbers to name the shaded part.

5.

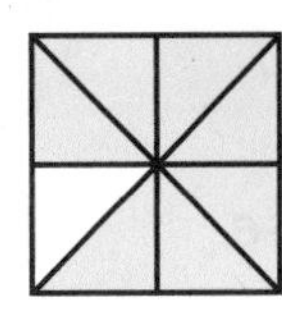

Each part is _____.

_________ eighths

____

6.

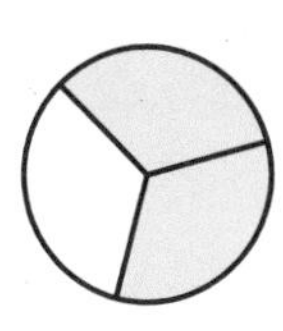

Each part is _____.

_________ thirds

____

7.

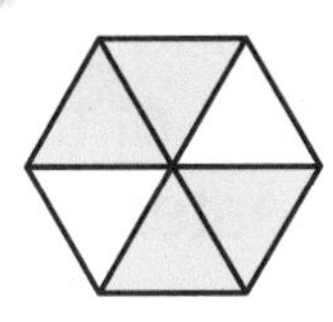

Each part is _____.

_________ sixths

____

8.

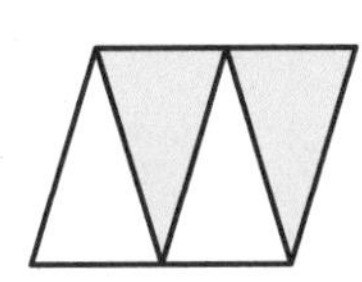

Each part is _____.

_________ fourths

____

9.

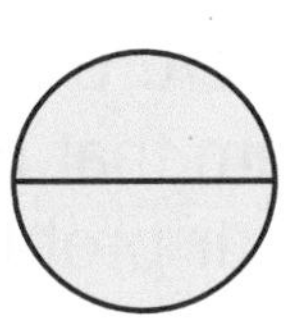

Each part is _____.

_________ halves

____

10.

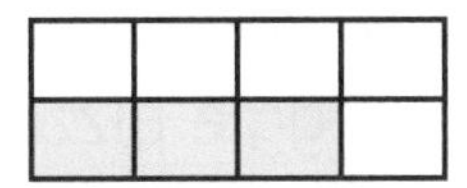

Each part is _____.

_________ eighths

____

Shade the fraction circle to model the fraction. Then write the fraction in numbers.

11. six out of eight

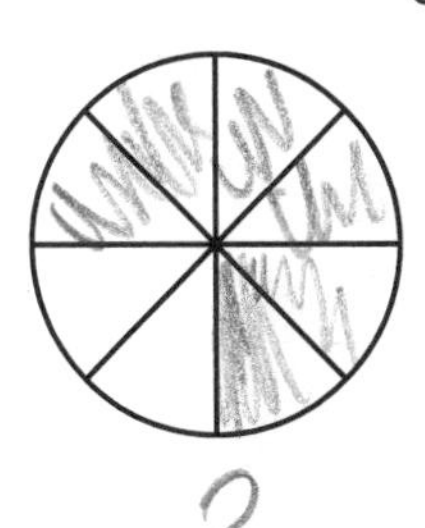

12. three fourths

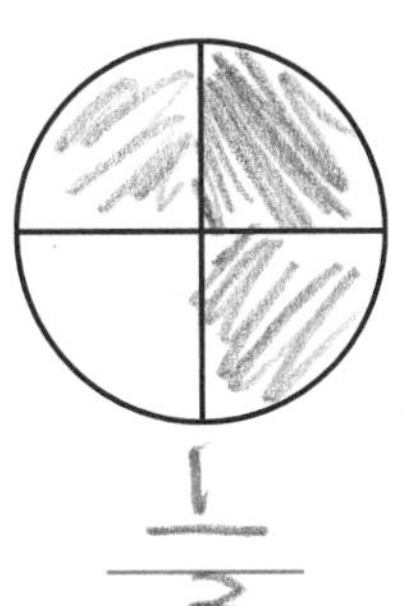

13. three out of three

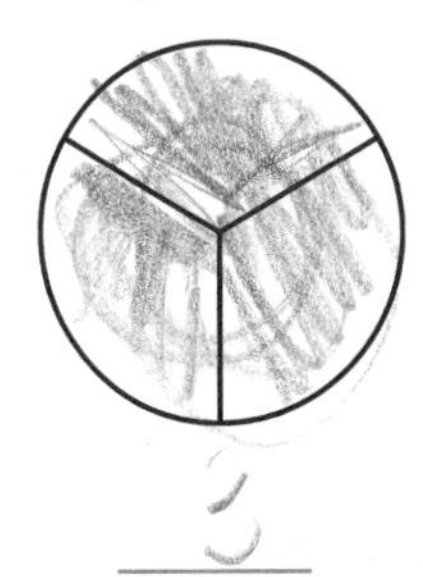

14. one out of two

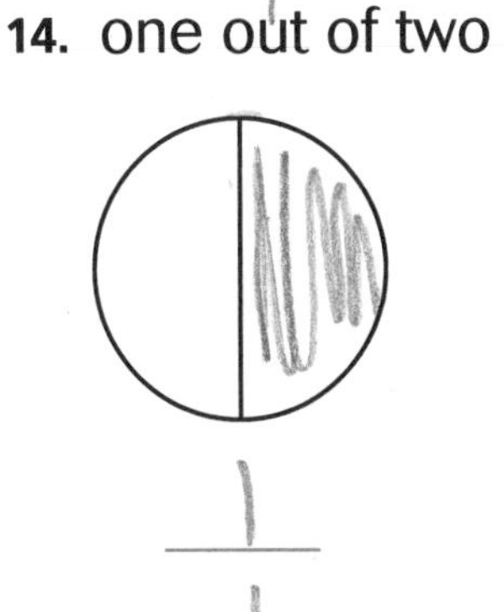

15. five sixths

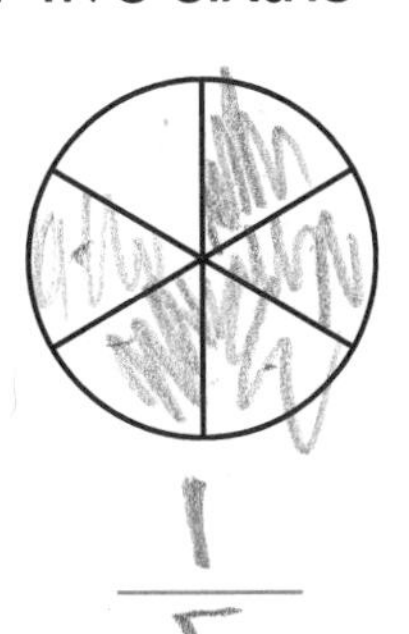

16. one out of four

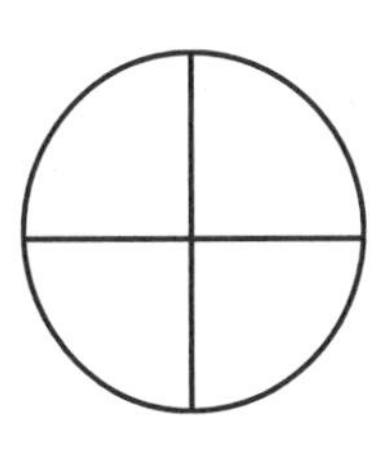

Problem Solving REAL WORLD

Use the pictures for 17–19.

Pepperoni

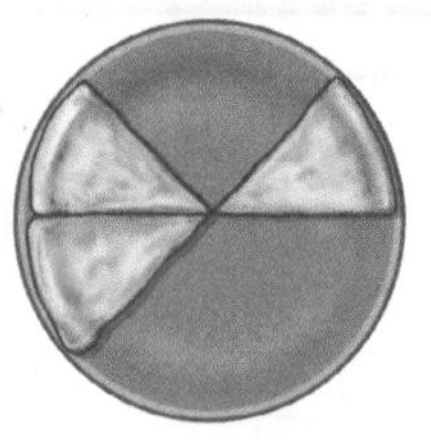
Cheese

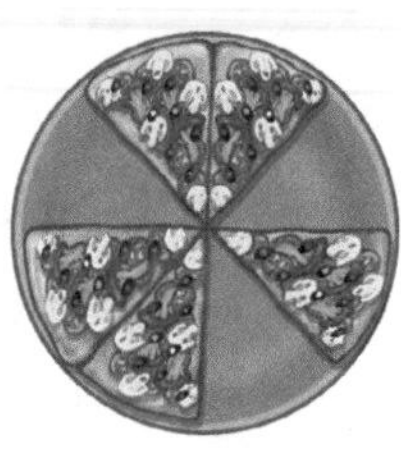
Veggie

17. Mrs. Ormond ordered pizza. Each pizza had 8 equal slices. What fraction of the pepperoni pizza was eaten?

____

18. What fraction of the cheese pizza is left?

____

19. H.O.T. **Pose a Problem** Use the picture of the veggie pizza to write a problem that has a fraction as the answer. Solve your problem.

____

____

____

20. H.O.T. **What's the Error?** Kate says that $\frac{2}{4}$ of the rectangle is shaded. Describe her error. Write the correct fraction for the shaded part.

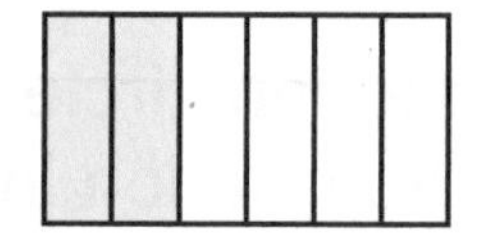

____

____

____

21. **Test Prep** What fraction names the shaded part of the shape?

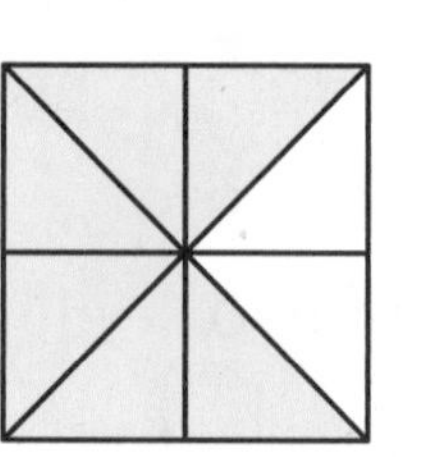

Ⓐ 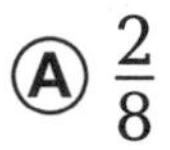 $\frac{2}{8}$

Ⓑ 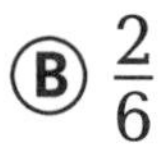 $\frac{2}{6}$

Ⓒ 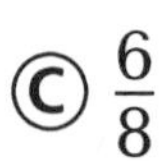 $\frac{6}{8}$

Ⓓ $\frac{8}{6}$

Name ______________________

Fractions on a Number Line

Essential Question How can you represent and locate fractions on a number line?

UNLOCK the Problem REAL WORLD

Billy's family is traveling from his house to his grandma's house. They stop at gas stations when they are $\frac{1}{4}$ and $\frac{3}{4}$ of the way there. How can he represent those distances on a number line?

You can use a number line to show fractions. The length from one whole number to the next whole number represents one whole. The line can be divided into any number of equal parts, or lengths.

Math Idea

A point on a number line shows the endpoint of a length, or distance, from zero. A number or fraction can name the distance.

Activity Locate fractions on a number line.

Materials ■ fraction strips

Billy's House — Grandma's House

0 — 1

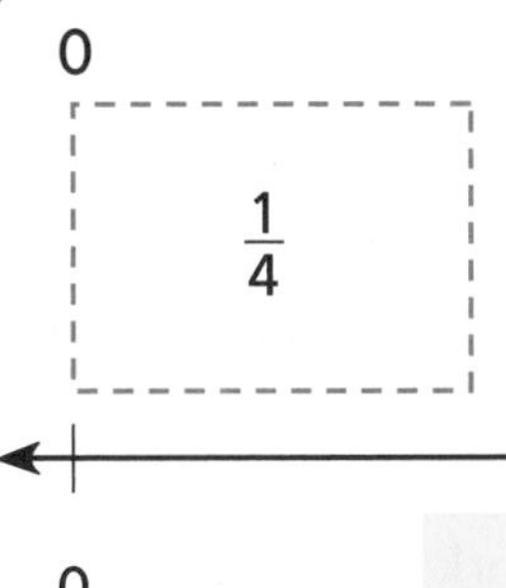

$\frac{0}{4}$ ___ ___ ___ $\frac{4}{4}$

STEP 1 Divide the line into four equal lengths, or fourths. Place four $\frac{1}{4}$-fraction strips end-to-end above the line to help.

STEP 2 At the end of each strip, draw a mark on the line.

STEP 3 Count the fourths from zero to 1 to label the distances from zero.

STEP 4 Think: $\frac{1}{4}$ is 1 out of 4 equal lengths.
Draw a point at $\frac{1}{4}$ to represent the distance from 0 to $\frac{1}{4}$.
Label the point *G1*.

STEP 5 Think: $\frac{3}{4}$ is 3 out of 4 equal lengths.
Draw a point at $\frac{3}{4}$ to represent the distance from 0 to $\frac{3}{4}$.
Label the point *G2*.

Example Complete the number line to name the point.

Materials ■ color pencils

Write the fraction that names the point on the number line.

Think: This number line is divided into six equal lengths, or sixths.

The length of one equal part is ______.

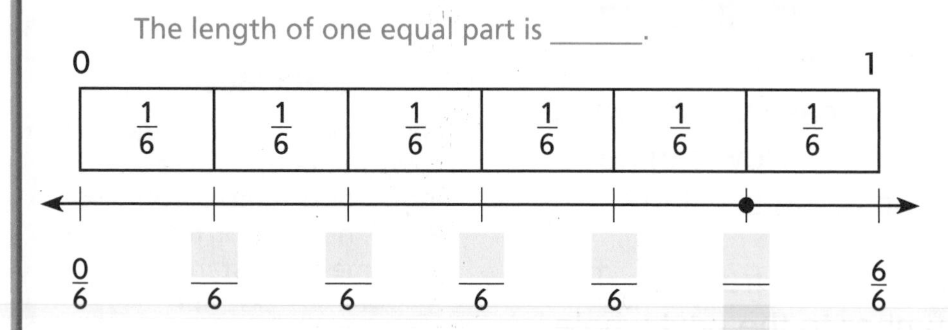

Shade the fraction strips to show the location of the point.

There are ______ out of ______ equal lengths shaded.
The shaded length shows $\frac{5}{6}$.

So, ______ names the point.

Share and Show MATH BOARD

1. Complete the number line. Draw a point to show $\frac{2}{3}$.

0 1

$\frac{1}{3}$ $\frac{1}{3}$ $\frac{1}{3}$

$\frac{0}{\ }$

MATHEMATICAL PRACTICES

Math Talk Explain what the length between each mark on this number line represents.

Write the fraction that names the point.

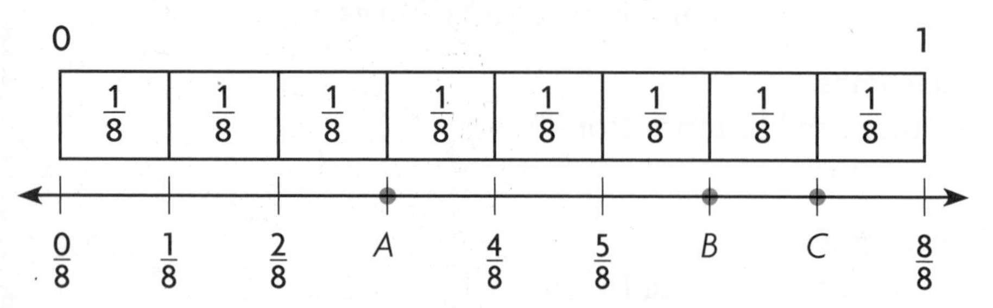

2. point *A* ______

✓ 3. point *B* ______

✓ 4. point *C* ______

Name ______________________________

On Your Own

Use fraction strips to help you complete the number line. Then locate and draw a point for the fraction.

5. $\frac{2}{6}$

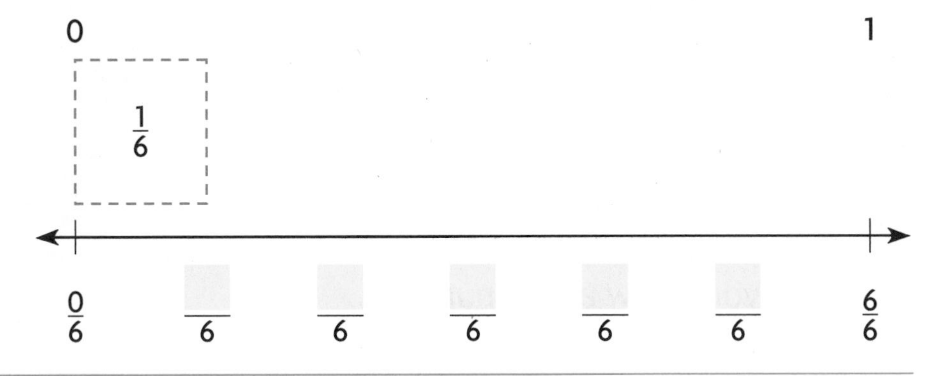

6. $\frac{1}{2}$

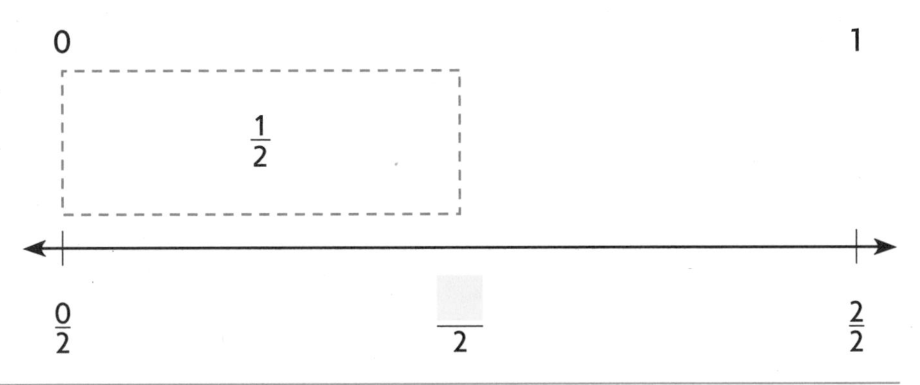

7. $\frac{2}{3}$

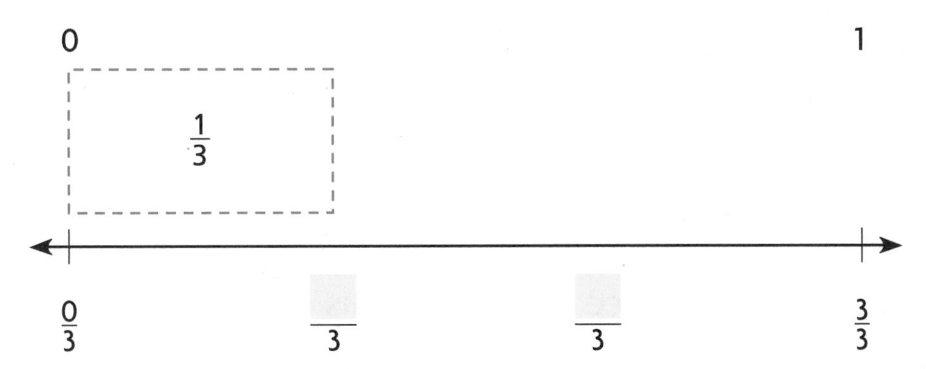

Write the fraction that names the point.

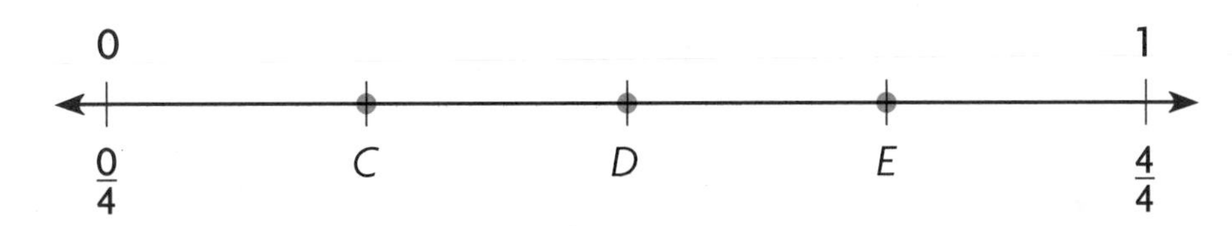

8. point C ______ **9.** point D ______ **10.** point E ______

UNLOCK the Problem REAL WORLD

11. Javia ran 8 laps around a track to run a total of 1 mile on Monday. How many laps will she need to run on Tuesday to run $\frac{3}{8}$ of a mile?

Ⓐ 1 Ⓑ 3 Ⓒ 6 Ⓓ 8

a. What do you need to find?

____

b. How will you use what you know about number lines to help you solve the problem?

____

____

c. H.O.T. Make a model to solve the problem.

d. Complete the sentences.

There are _____ laps in 1 mile.

Each lap represents _____ of a mile.

_____ laps represent the distance of three eighths of a mile.

So, Javia will need to run _____ laps to run $\frac{3}{8}$ of a mile.

e. Fill in the bubble for the correct answer choice above.

12. Which fraction names point *F* on the number line?

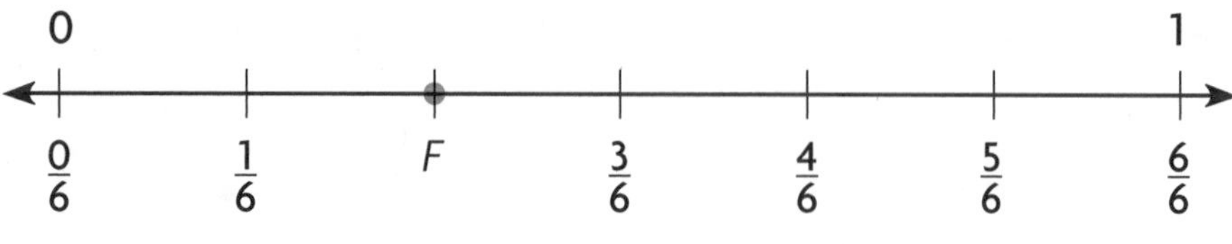

Ⓐ $\frac{1}{6}$ Ⓑ $\frac{2}{6}$ Ⓒ $\frac{6}{2}$ Ⓓ $\frac{6}{6}$

 FOR MORE PRACTICE: Standards Practice Book, pp. P161–P162

Name ______________________________

Mid-Chapter Checkpoint

▶ Vocabulary

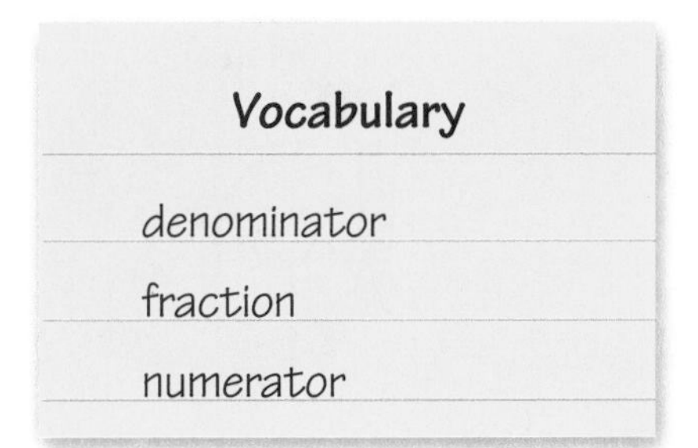

Choose the best term from the box to complete the sentence.

1. A ________________ is a number that names part of a whole or part of a group. (p. 315)

2. The ________________ tells how many equal parts are in the whole or in the group. (p. 319)

▶ Concepts and Skills

Write the number of equal parts. Then write the name for the parts.

3. 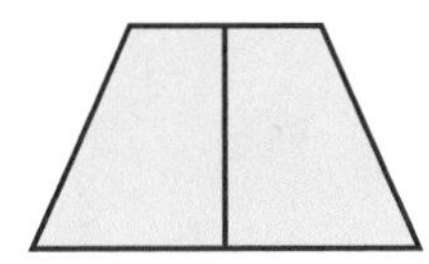

_____ equal parts

____

4. 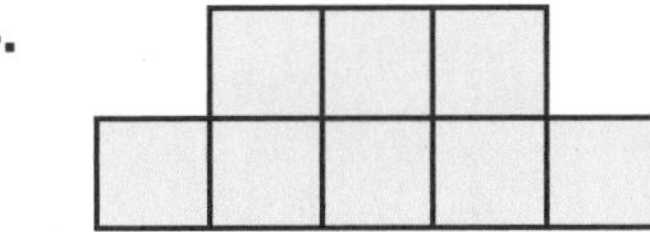

_____ equal parts

____

5. 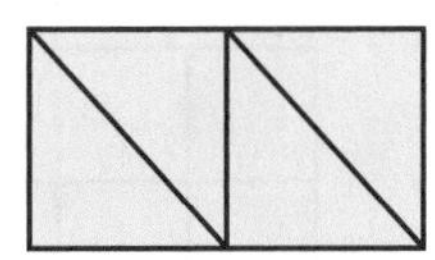

_____ equal parts

____

Write the number of equal parts in the whole. Then write the fraction in numbers that names the shaded part.

6. 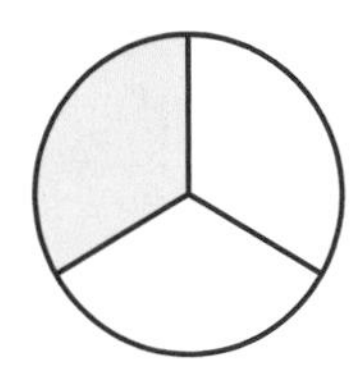

_____ equal parts

_____

7. 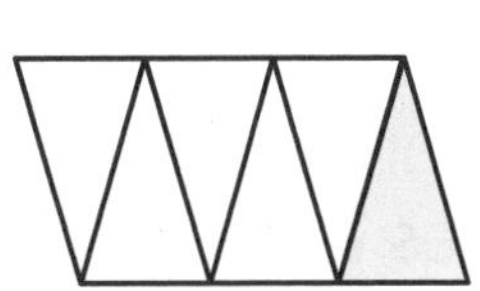

_____ equal parts

_____

8. 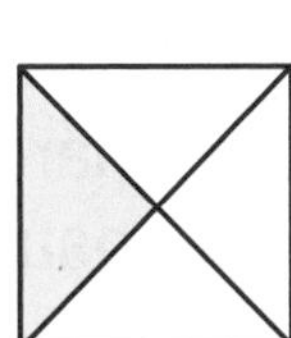

_____ equal parts

_____

Write the fraction that names the point.

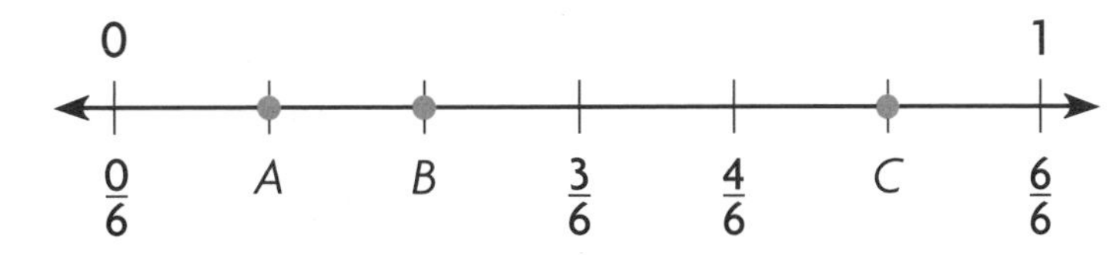

9. point *A* _____

10. point *B* _____

11. point *C* _____

Fill in the bubble for the correct answer choice.

12. Jessica ordered a pizza. What fraction of the pizza has mushrooms?

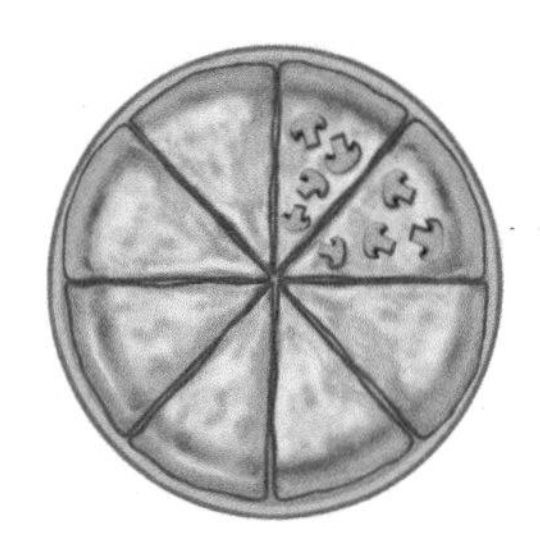

Ⓐ $\frac{2}{6}$

Ⓑ $\frac{2}{8}$

Ⓒ $\frac{6}{8}$

Ⓓ $\frac{8}{6}$

13. Which fraction names the shaded part?

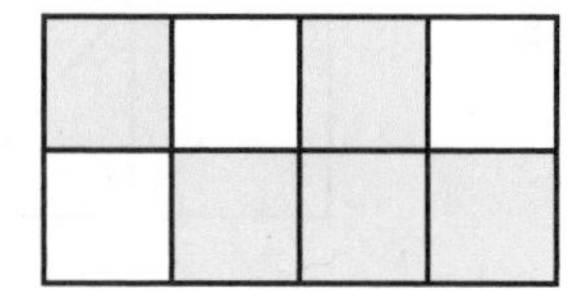

Ⓐ $\frac{8}{8}$

Ⓑ $\frac{5}{8}$

Ⓒ $\frac{3}{8}$

Ⓓ $\frac{1}{8}$

14. Six friends share 3 oatmeal squares equally. How much of an oatmeal square does each friend get?

Ⓐ 1 half

Ⓑ 1 third

Ⓒ 1 fourth

Ⓓ 1 sixth

Name ___________________________________

Relate Fractions and Whole Numbers

Essential Question When might you use a fraction greater than 1 or a whole number?

UNLOCK the Problem REAL WORLD

Steve ran 1 mile and Jenna ran $\frac{4}{4}$ of a mile. Did Steve and Jenna run the same distance?

Math Idea
If two numbers are located at the same point on a number line, then they are equal and represent the same distance.

Locate 1 and $\frac{4}{4}$ on a number line.

- Shade 4 lengths of $\frac{1}{4}$ and label the number line.
- Draw a point at 1 and $\frac{4}{4}$.

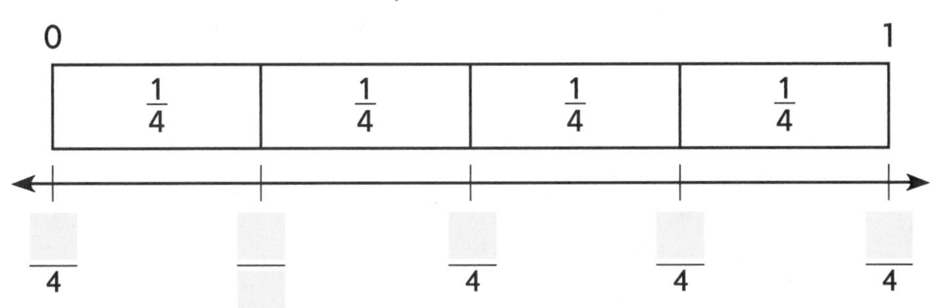

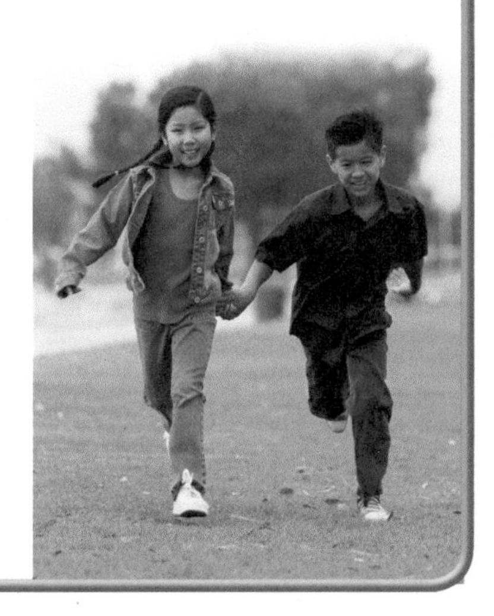

Since the distance _____ and _____ end at the same point, they are equal.

So, Steve and Jenna ran the _____ distance.

Try This! Complete the number line. Locate and draw points at $\frac{3}{6}$, $\frac{6}{6}$, and 1.

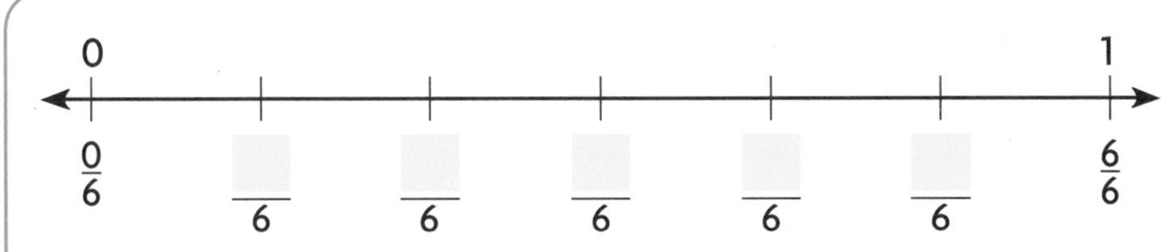

A Are $\frac{3}{6}$ and 1 equal? **Explain.**

Think: Do the distances end at the same point?

____

____

So, $\frac{3}{6}$ and 1 are __________.

B Are $\frac{6}{6}$ and 1 equal? **Explain.**

Think: Do the distances end at the same point?

____

____

So, $\frac{6}{6}$ and 1 are __________.

CONNECT The number of equal parts the whole is divided into is the denominator of a fraction. The number of parts being counted is the numerator. A **fraction greater than 1** has a numerator greater than its denominator.

Examples

Each shape is 1 whole. Write a whole number and a fraction greater than 1 for the parts that are shaded.

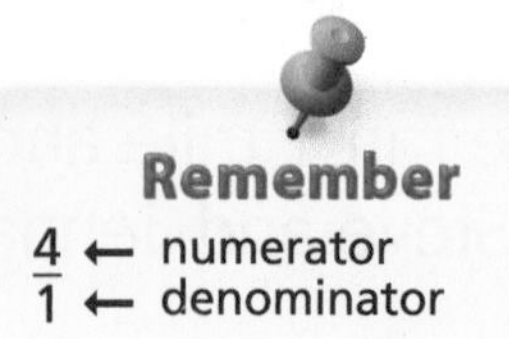

Remember

$\frac{4}{1}$ ← numerator
← denominator

A

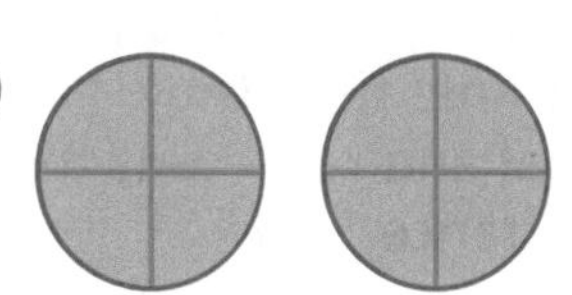

There are 2 wholes.

Each whole is divided into 4 equal parts, or fourths. $2 = \frac{8}{4}$

There are _____ equal parts shaded.

B

There are 3 wholes.

Each whole is divided into 1 equal part. $3 = \frac{3}{1}$

There are _____ equal parts shaded.

1. **Explain** what *each whole is divided into 1 equal part* means in Example B.

____

____

Read Math

Read $\frac{3}{1}$ as *three ones*.

2. How do you divide a whole into 1 equal part?

____

____

Try This!

Each shape is 1 whole. Write a whole number and a fraction greater than 1 for the parts that are shaded.

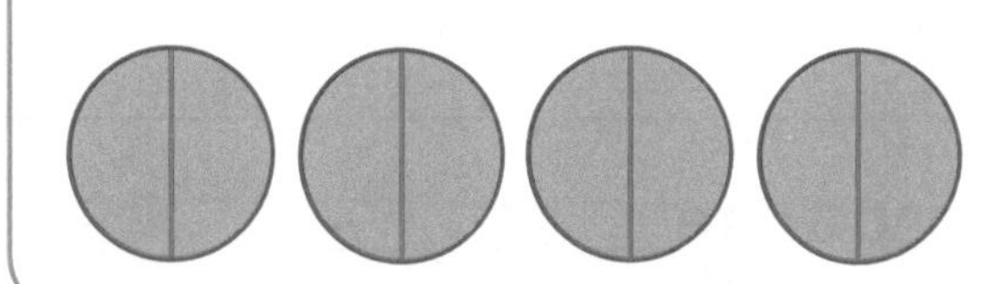

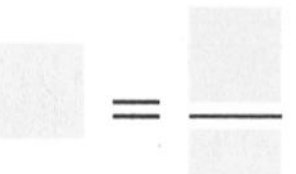

Name ______________________________

Share and Show

1. Each shape is 1 whole. Write a whole number and a fraction greater than 1 for the parts that are shaded.

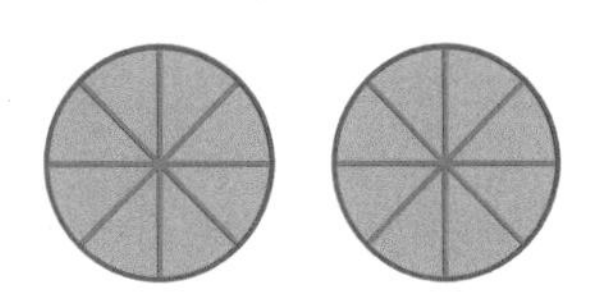

There are _____ wholes.

Each whole is divided into _____ equal parts.

There are _____ equal parts shaded.

$\square = \frac{\square}{\square}$

Use the number line to find whether the two numbers are equal. Write *equal* or *not equal.*

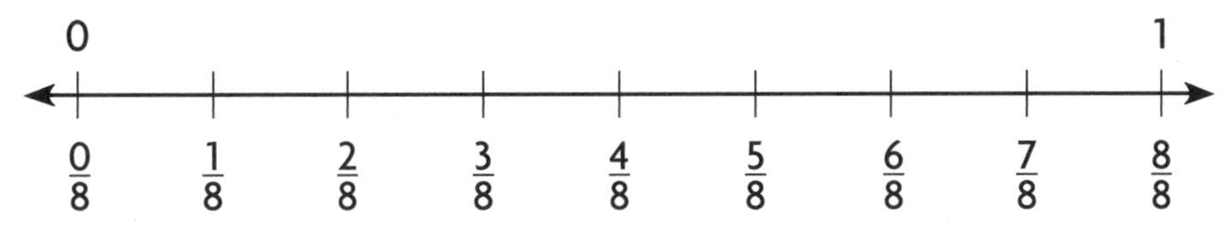

2. $\frac{1}{8}$ and $\frac{8}{8}$ __________

3. $\frac{8}{8}$ and 1 __________

4. 1 and $\frac{4}{8}$ __________

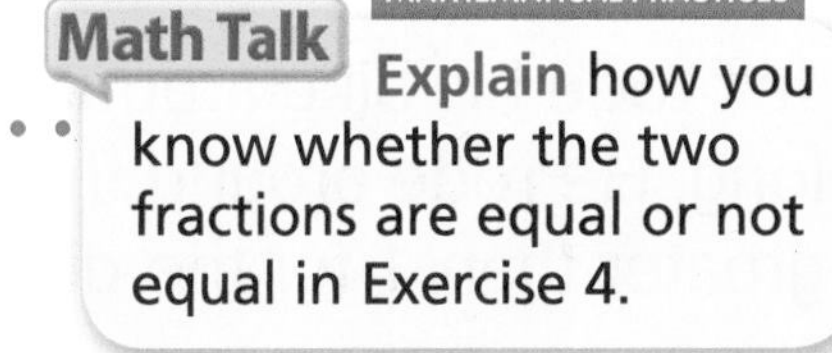

On Your Own

Use the number line to find whether the two numbers are equal. Write *equal* or *not equal.*

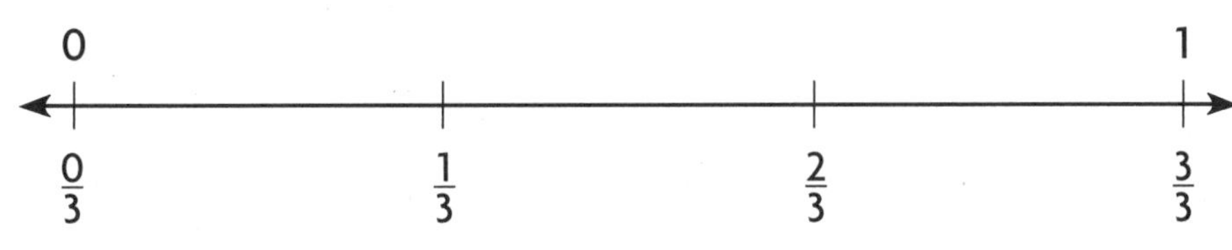

5. $\frac{0}{3}$ and 1 __________

6. 1 and $\frac{2}{3}$ __________

7. $\frac{3}{3}$ and 1 __________

Each shape is 1 whole. Write a fraction greater than 1 for the parts that are shaded.

8. 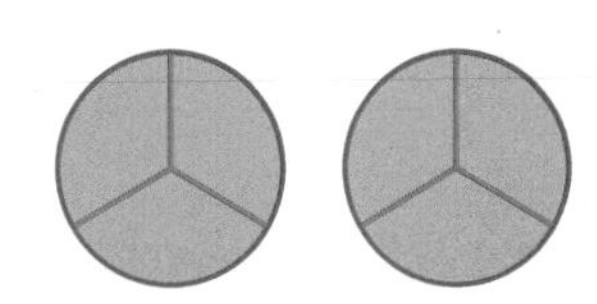

$2 =$ _____

9. 

$1 =$ _____

10. 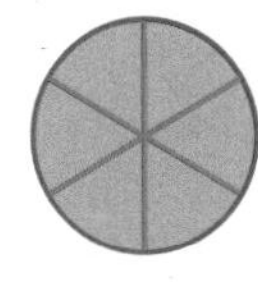 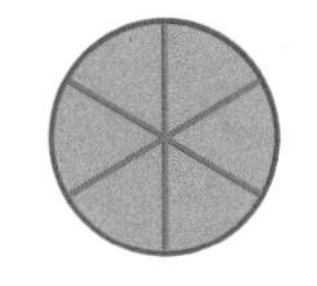

$3 =$ _____

11.  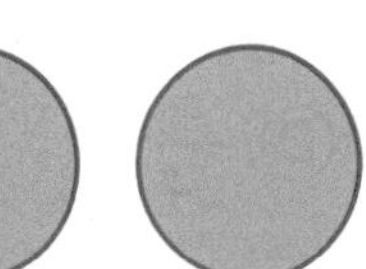

$2 =$ _____

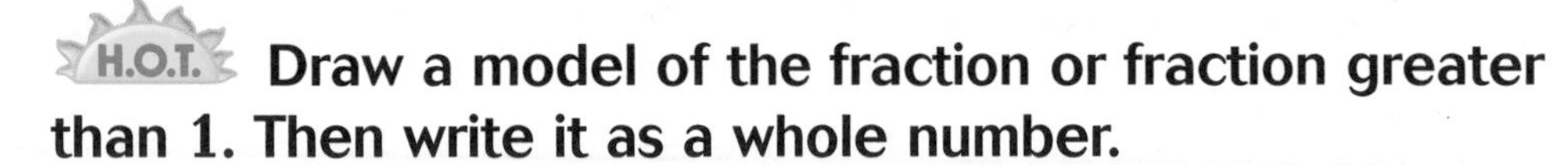

H.O.T. **Draw a model of the fraction or fraction greater than 1. Then write it as a whole number.**

12. $\frac{8}{4} =$ ______

13. $\frac{6}{6} =$ ______

14. $\frac{5}{1} =$ ______

Problem Solving REAL WORLD

15. Jeff rode his bike around a bike trail that was $\frac{1}{3}$ of a mile long. He rode around the trail 9 times. Write a fraction greater than 1 for the distance. How many miles did Jeff ride?

____

16. H.O.T. Write Math **What's the Error?** Andrea drew the number line below. She said that $\frac{9}{8}$ and 1 were equal. **Explain** her error.

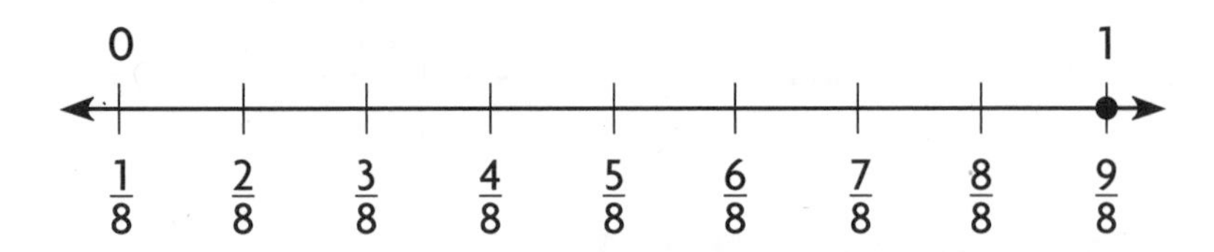

____

17. **Test Prep** Each shape is 1 whole. Which fraction greater than 1 names the parts that are shaded?

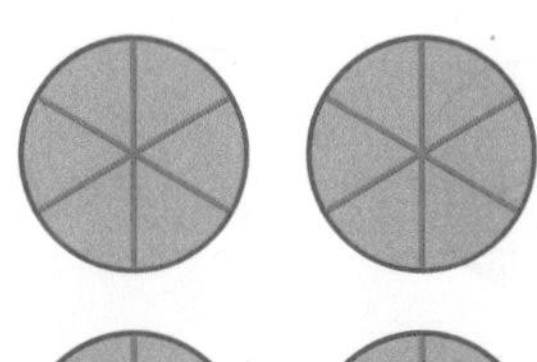

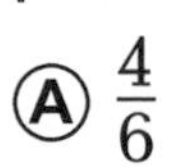

Ⓐ $\frac{4}{6}$

Ⓑ $\frac{24}{6}$

Ⓒ $\frac{24}{4}$

Ⓓ $\frac{4}{24}$

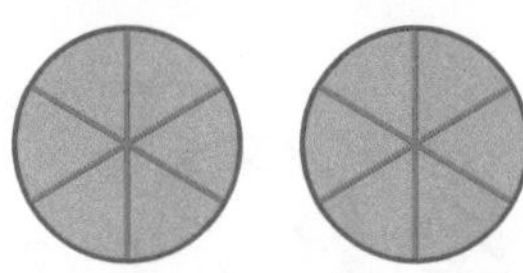

Name ______________________

Fractions of a Group

Essential Question How can a fraction name part of a group?

UNLOCK the Problem REAL WORLD

Jake and Emma each have a collection of marbles. What fraction of each collection is blue?

You can use a fraction to name part of a group.

Jake's Marbles

number of blue marbles → ☐ ← numerator

total number of marbles → 8 ← denominator

Read: three eighths, or three out of eight

Write: $\frac{3}{8}$

So, _____ of Jake's marbles are blue.

Emma's Marbles

bags of blue marbles → ☐ ← numerator

total number of bags → 4 ← denominator

Read: one fourth, or one out of four

Write: $\frac{1}{4}$

So, _____ of Emma's marbles are blue.

Try This! **Name part of a group.**

Draw 2 red counters and 6 yellow counters.

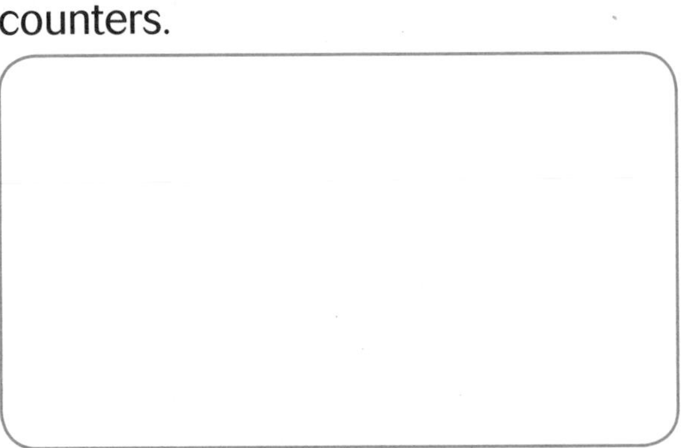

Write the fraction of counters that are red.

☐ ← number of red counters

☐ ← total number of counters

Write the fraction of counters that are not red.

☐ ← number of yellow counters

☐ ← total number of counters

So, _____ of the counters are red and _____ are not red.

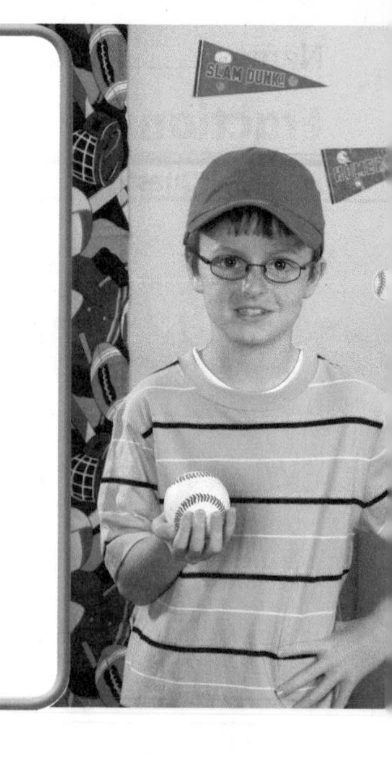

Fractions Greater Than 1

Sometimes a fraction can name more than a whole group.

Daniel collects baseballs. He has collected 8 so far. He puts them in cases that hold 4 baseballs each. What part of the baseball cases has Daniel filled?

Think: 1 case = 1

Daniel has two full cases of 4 baseballs each.

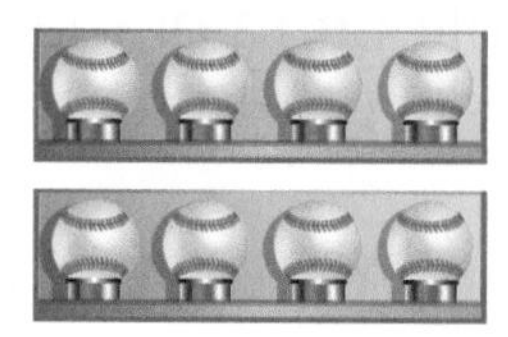

So, 2, or $\frac{8}{4}$, baseball cases are filled.

Try This! **Complete the whole number and the fraction greater than 1 to name the part filled.**

  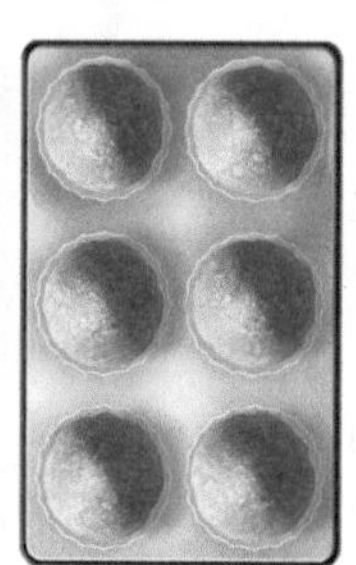

Think: 1 pan = 1

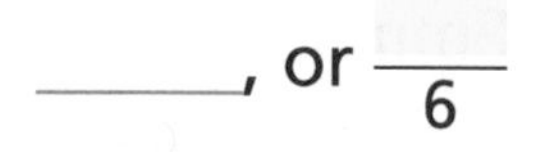

______, or $\frac{\quad}{6}$

B

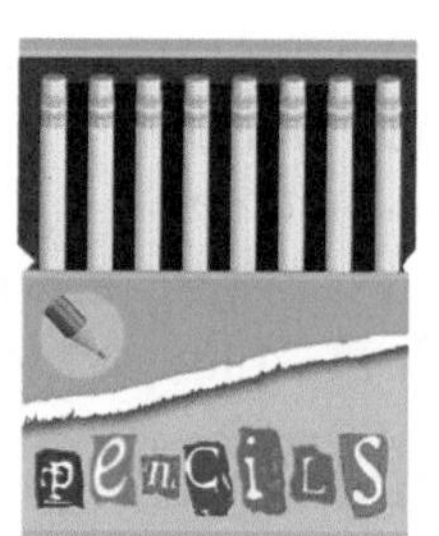

Think: 1 box = 1

______, or $\frac{\quad}{8}$

Share and Show

1. What fraction of the counters are red? ______

Think: How many red counters are there? How many counters are there in all?

MATHEMATICAL PRACTICES

Math Talk Explain another way to name the fraction for Exercise 3.

Write a fraction to name the red part of each group.

2.    ______

3. 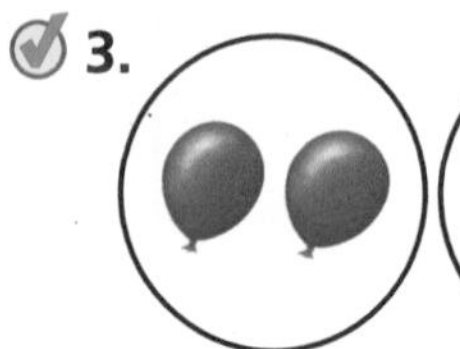  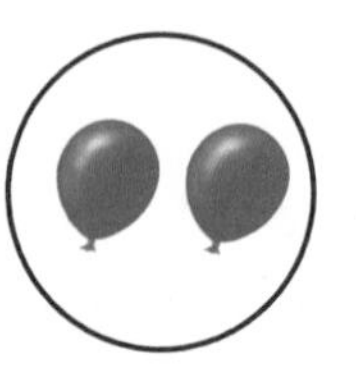 ______

Name ________________________________

Write a whole number and a fraction greater than 1 to name the part filled.

4.   

Think: 1 carton = 1

______ ______

5.  

Think: 1 container = 1

______ ______

On Your Own

Write a fraction to name the blue part of each group.

6.  ______

7.  ______

8.  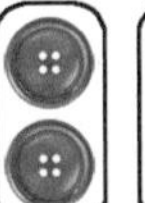 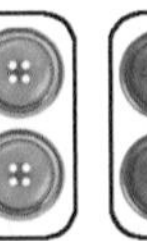 ______

9. 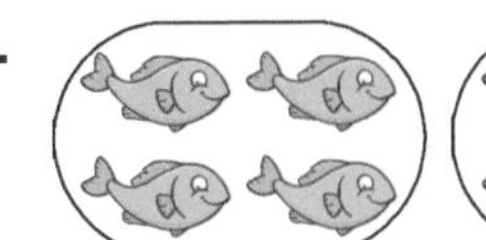 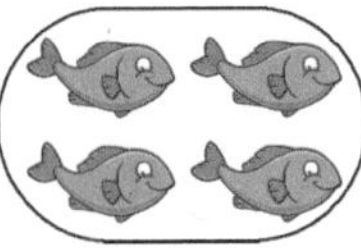 ______

Write a whole number and a fraction greater than 1 to name the part filled.

10. 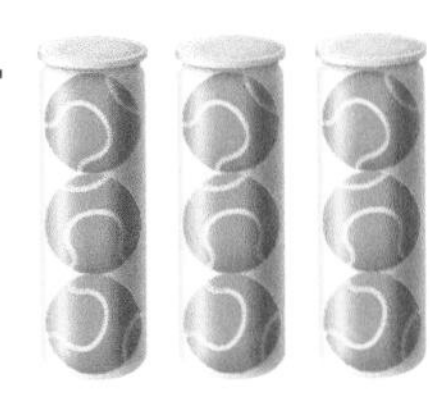

Think: 1 container = 1

______ ______

11. H.O.T. 

Think: 1 carton = 1

______ ______

Draw a quick picture on your MathBoard. Then write a fraction to name the shaded part of the group.

12. Draw 8 circles. Shade 8 circles.

______

13. Draw 8 triangles. Make 4 groups. Shade 1 group.

______

14. Draw 4 rectangles. Shade 2 rectangles.

______

Problem Solving REAL WORLD

Use the graph for 15–16.

15. The bar graph shows the winners of the Smith Elementary School Marble Tournament. How many games were played? What fraction of the games did Scott win?

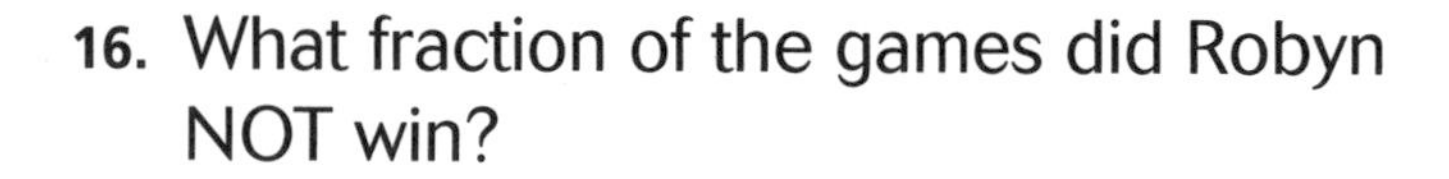

16. What fraction of the games did Robyn NOT win?

17. H.O.T. Liam has 6 marbles. Of them, $\frac{1}{3}$ are blue. The rest are red. Draw a picture to show Liam's marbles.

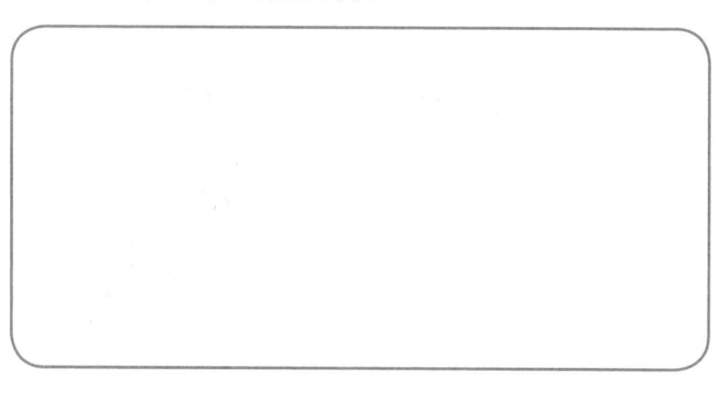

SHOW YOUR WORK

18. Write Math ▸ **What's the Question?** A bag has 2 yellow cubes, 3 blue cubes, and 1 white cube. The answer is $\frac{1}{6}$.

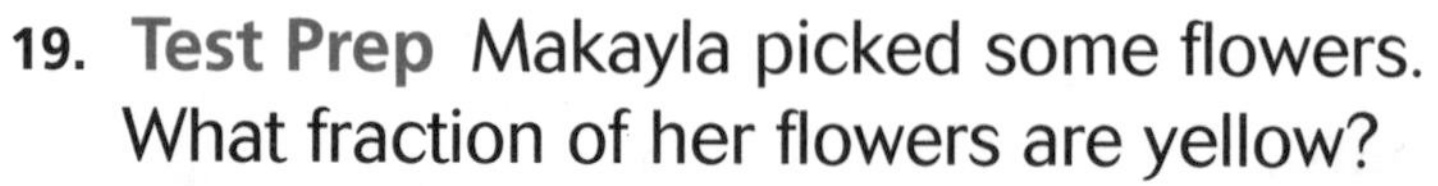

19. **Test Prep** Makayla picked some flowers. What fraction of her flowers are yellow?

Ⓐ $\frac{1}{8}$ Ⓑ $\frac{2}{8}$ Ⓒ $\frac{3}{8}$ Ⓓ $\frac{4}{8}$

Name ____________________

Find Part of a Group Using Unit Fractions

Essential Question How can a fraction tell how many are in part of a group?

UNLOCK the Problem REAL WORLD

Audrey buys a bouquet of 12 flowers. One third of them are red. How many of the flowers are red?

- How many flowers does Audrey buy in all? ____________
- What fraction of the flowers are red? ____________

Activity

Materials ■ two-color counters ■ MathBoard

- Put 12 counters on your MathBoard.
- Since you want to find $\frac{1}{3}$ of the group, there should be ______ equal groups. Draw the counters below.

- Circle one of the groups to show ______.
  Then count the number of counters in that group.

There are ______ counters in 1 group. $\frac{1}{3}$ of $12 =$ ______

So, ______ of the flowers are red.

- **What if** Audrey buys a bouquet of 9 flowers and one third of them are yellow? Use your MathBoard and counters to find how many of the flowers are yellow.

____

MATHEMATICAL PRACTICES

Math Talk Explain how you can use the numerator and denominator in a fraction to find part of a group.

Try This! **Find part of a group.**

Joseph picks 20 flowers from his mother's garden. One fourth of them are purple. How many of the flowers are purple?

STEP 1 Draw a row of 4 counters.

Think: To find $\frac{1}{4}$, make 4 equal groups.

STEP 2 Continue to draw as many rows of 4 counters as you can until you have 20 counters.

STEP 3 Then circle _____ equal groups.

Think: Each group represents $\frac{1}{4}$ of the flowers.

$\frac{1}{4}$ $\frac{1}{4}$ $\frac{1}{4}$ $\frac{1}{4}$

There are _____ counters in 1 group.

$\frac{1}{4}$ of 20 = _____

So, _____ of the flowers are purple.

Share and Show

1. Use the model to find $\frac{1}{2}$ of 8. _____

Think: How many counters are in 1 of the 2 equal groups?

MATHEMATICAL PRACTICES

Math Talk Explain why you count the number of counters in just one of the groups in Exercise 1.

Circle equal groups to solve. Count the number of flowers in 1 group.

2. $\frac{1}{4}$ of 8 = _____

3. $\frac{1}{3}$ of 6 = _____

4. $\frac{1}{6}$ of 12 = _____

Name ______________________

On Your Own

Circle equal groups to solve. Count the number of flowers in 1 group.

5. $\frac{1}{4}$ of $12 =$ _____

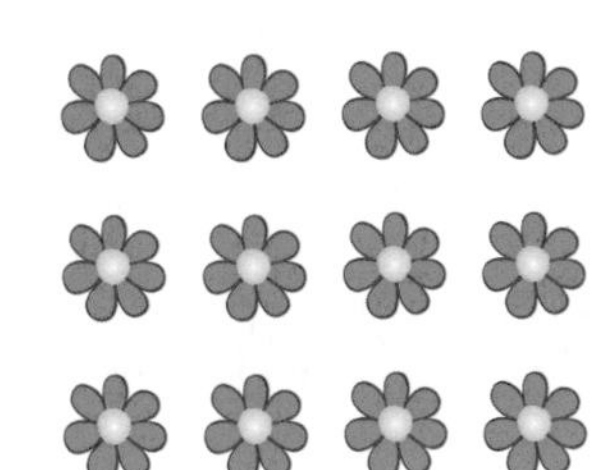

6. $\frac{1}{3}$ of $15 =$ _____

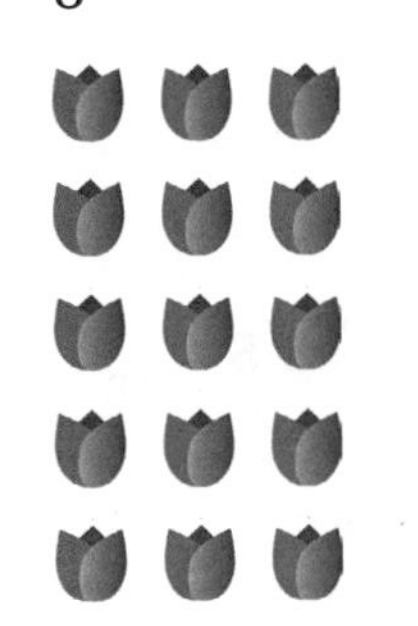

7. $\frac{1}{4}$ of $16 =$ _____

8. $\frac{1}{3}$ of $9 =$ _____

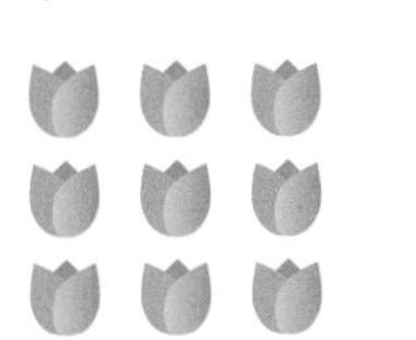

9. $\frac{1}{6}$ of $18 =$ _____

10. $\frac{1}{8}$ of $8 =$ _____

11. $\frac{1}{6}$ of $30 =$ _____

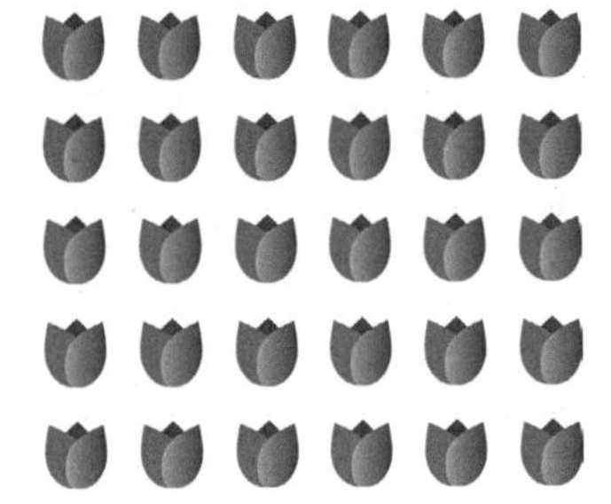

12. $\frac{1}{3}$ of $12 =$ _____

13. H.O.T. $\frac{1}{2}$ of $6 =$ _____

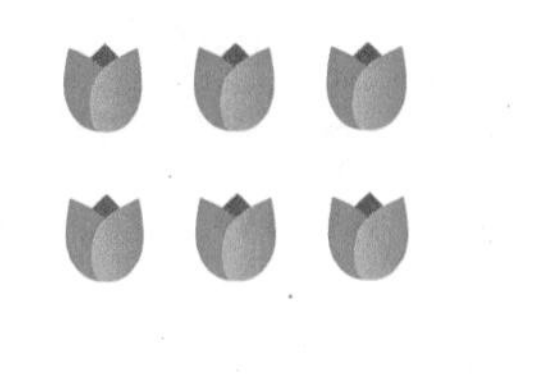

Draw counters. Then circle equal groups to solve.

14. $\frac{1}{8}$ of $16 =$ _____

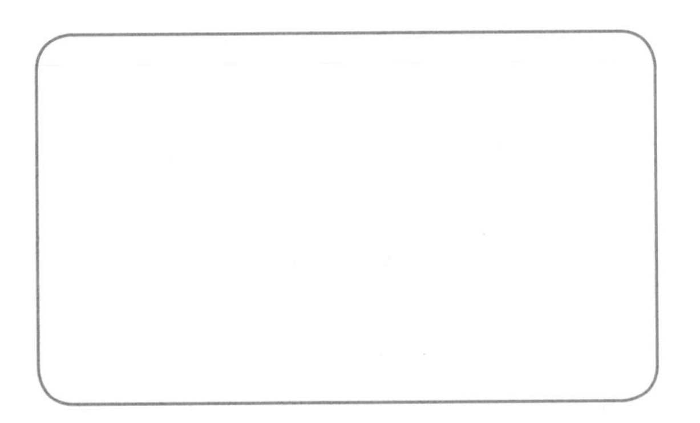

15. $\frac{1}{6}$ of $24 =$ _____

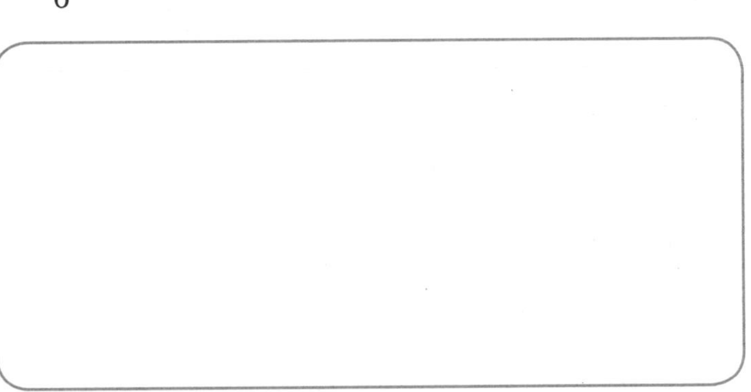

Problem Solving REAL WORLD

Use the table for 16–17.

Flower Seeds Bought	
Name	**Number of Packs**
Ryan	8
Brooke	12
Cole	20

16. One fourth of the seed packs Ryan bought are violet seeds. How many packs of violet seeds did Ryan buy? Draw counters to solve.

____

17. H.O.T. Write Math One third of Brooke's seed packs and one fourth of Cole's seed packs are daisy seeds. How many packs of daisy seeds did they buy altogether? **Explain** how you know.

____

____

____

18. **Sense or Nonsense?** Sophia bought 12 pots. One sixth of them are green. Sophia said she bought 2 green pots. Does her answer make sense? **Explain** how you know.

____

____

____

____

19. **Test Prep** Bailey picked 15 flowers. One third of them are yellow. How many yellow flowers did Bailey pick?

Ⓐ 3　　Ⓒ 10

Ⓑ 5　　Ⓓ 15

SHOW YOUR WORK

FOR MORE PRACTICE:
Standards Practice Book, pp. P167–P168

Name ______________________________

Problem Solving • Find the Whole Group Using Unit Fractions

Essential Question How can you use the strategy *draw a diagram* to solve fraction problems?

UNLOCK the Problem REAL WORLD

Cameron has 4 clown fish in his fish tank. One third of the fish in the tank are clown fish. How many fish does Cameron have in his tank?

Use the graphic organizer to help you solve the problem.

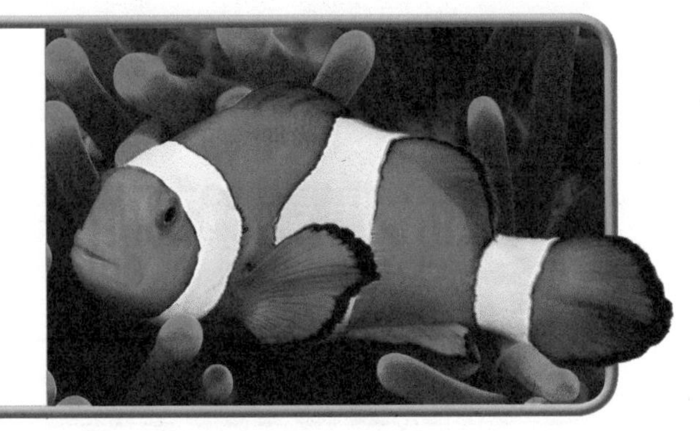

Read the Problem

What do I need to find?

I need to find ______________ are in Cameron's fish tank.

What information do I need to use?

Cameron has _____ clown fish.

__________ of the fish in the tank are clown fish.

How will I use the information?

I will use the information in the problem to draw a __________.

Solve the Problem

Describe how to draw a diagram to solve.

The denominator in $\frac{1}{3}$ tells you that there are _____ equal parts in the whole group. Draw 3 circles to show _____ equal parts.

Since 4 fish are $\frac{1}{3}$ of the whole group, draw _____ counters in the first circle.

Since there are _____ counters in the first circle, draw _____ counters in each of the remaining circles. Then find the total number of counters.

So, Cameron has _____ fish in his tank.

Try Another Problem

A pet store has 2 gray rabbits. One eighth of the rabbits at the pet store are gray. How many rabbits does the pet store have?

Read the Problem	Solve the Problem
What do I need to find?	
What information do I need to use?	
How will I use the information?	

1. How do you know that your answer is reasonable?

____

____

2. How did your diagram help you solve the problem? ______________________

____

MATHEMATICAL PRACTICES

Math Talk Suppose $\frac{1}{2}$ of the rabbits are gray. **Explain** how you can find the number of rabbits at the pet store.

Name ______________________________

Share and Show

MATH BOARD

1. Lily has 3 dog toys that are red. One fourth of all her dog toys are red. How many dog toys does Lily have?

   First, draw _____ circles to show _____ equal parts.

   Next, draw _____ toys in _____ circle since _____ circle represents the number of red toys.

   Last, draw _____ toys in each of the remaining circles. Find the total number of toys.

   So, Lily has _____ dog toys.

2. H.O.T. **What if** Lily had 4 toys that are red? How many dog toys would she have?

   ____

3. The pet store sells bags of pet food. There are 4 bags of cat food. One sixth of the bags of food are bags of cat food. How many bags of pet food does the pet store have?

   ____

4. Rachel owns 2 parakeets. One fourth of all her birds are parakeets. How many birds does Rachel own?

   ____

Tips

UNLOCK the Problem

- √ Circle the question.
- √ Underline important facts.
- √ Put the problem in your own words.
- √ Choose a strategy you know.

SHOW YOUR WORK

On Your Own

Choose a STRATEGY

Act It Out
Draw a Diagram
Find a Pattern
Make a Table

Draw a quick picture to solve for 5–6.

5. There are 18 bottlenose dolphins at an aquarium. One sixth of them are not adults. How many of the bottlenose dolphins are not adults?

____

6. H.O.T. Six friends share 5 small pizzas. Each friend first eats half of a pizza. How much more pizza does each friend need to eat to finish all the pizzas and share them equally?

____

7. Write Math Braden bought 4 packs of dog treats. He gave 4 treats to his neighbor's dog. Now Braden has 24 treats left for his dog. How many dog treats were in each pack? **Explain** how you know.

____

____

____

____

____

8. **Test Prep** Two hats are $\frac{1}{3}$ of the group. How many hats are in the whole group?

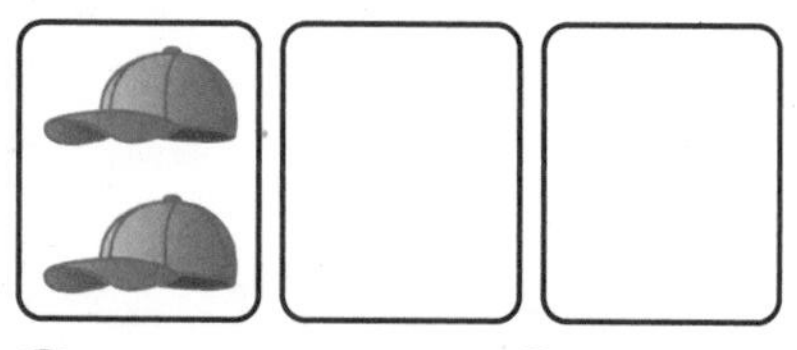

Ⓐ 4   Ⓒ 6

Ⓑ 5   Ⓓ 8

SHOW YOUR WORK

FOR MORE PRACTICE:
Standards Practice Book, pp. P169–P170

Name ____________________

Chapter Review/Test

▶ Vocabulary

Choose the best term from the box to complete the sentence.

Vocabulary
denominator
fraction greater than 1
numerator
unit fraction

1. The ____________________ tells how many parts are being counted. (p. 319)

2. A ____________________ has a numerator greater than its denominator. (p. 330)

3. A ____________________ has 1 as its numerator. (p. 315)

▶ Concepts and Skills

Each shape is 1 whole. Write a fraction greater than 1 for the parts that are shaded.

4. 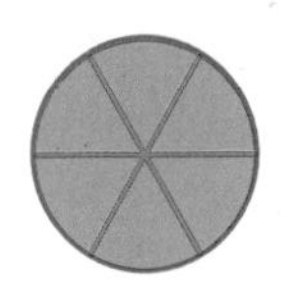 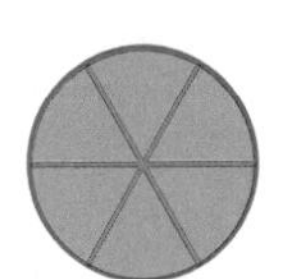 $2 =$ ______

5. 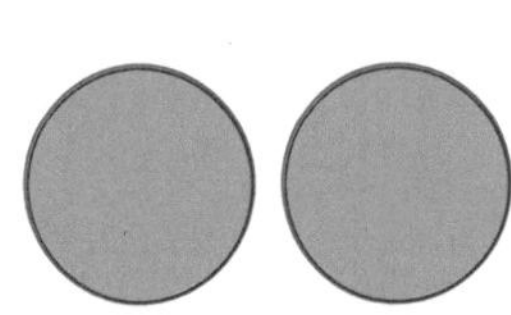 $2 =$ ______

Write a fraction to name the blue part of each group.

6.     ______

7. 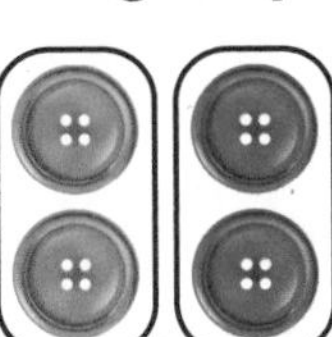 ______

Circle equal groups to solve. Count the number of flowers in 1 group.

8. $\frac{1}{3}$ of $6 =$ ______

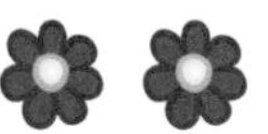

9. $\frac{1}{4}$ of $12 =$ ______

10. $\frac{1}{3}$ of $9 =$ ______

 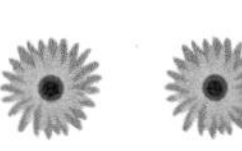
 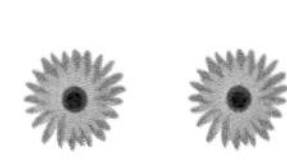

GO Online Assessment Options Chapter Test

Fill in the bubble for the correct answer choice.

11. Mason's mom bought a pumpkin pie. Mason wants to eat 1 slice for dessert. What fraction of the pie will he eat?

Ⓐ $\frac{1}{4}$ Ⓒ $\frac{1}{8}$

Ⓑ $\frac{1}{6}$ Ⓓ $\frac{1}{10}$

12. Joshua walks $\frac{7}{8}$ of a mile to the park. Which point represents the distance that Joshua walked?

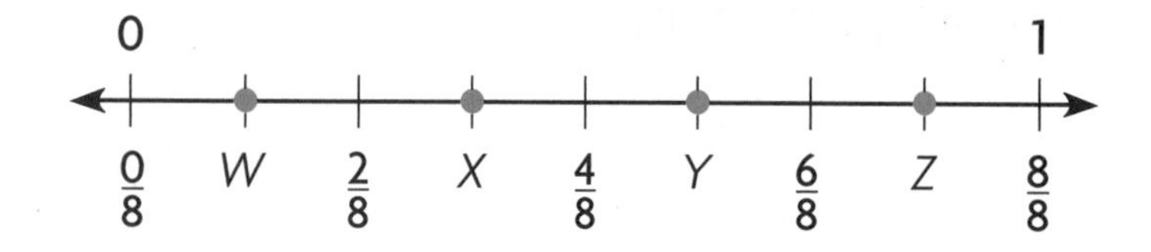

Ⓐ point W Ⓒ point Y

Ⓑ point X Ⓓ point Z

13. Kennedy shares 2 oranges equally with two friends. How much of an orange does each friend get?

Ⓐ 6 whole oranges

Ⓑ 2 whole oranges

Ⓒ 2 thirds of an orange

Ⓓ 1 half and 1 third of an orange

Name ________________________________

Fill in the bubble for the correct answer choice.

14. Jessica has these pieces of fruit in a bowl.

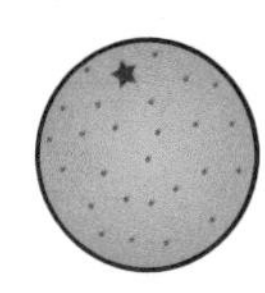

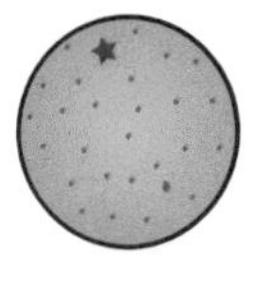

What fraction of the fruit are oranges?

Ⓐ $\frac{1}{6}$

Ⓑ $\frac{2}{6}$

Ⓒ $\frac{3}{6}$

Ⓓ $\frac{2}{4}$

15. Max drew the shape below and divided it into equal parts. What is the name for the parts?

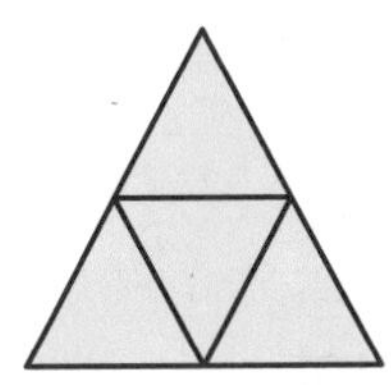

Ⓐ halves

Ⓑ thirds

Ⓒ fourths

Ⓓ sixths

16. Avery has 2 blue marbles. One sixth of Avery's marbles are blue. How many marbles does Avery have?

Ⓐ 2

Ⓑ 6

Ⓒ 8

Ⓓ 12

▶ Constructed Response

17. Destiny has 6 library books on her desk. One third of them are about animals. How many of Destiny's library books are about animals? Draw a diagram to show your work.

______

18. Julian made a flag for his clubhouse. What fraction of his flag is green? **Explain** how you know.

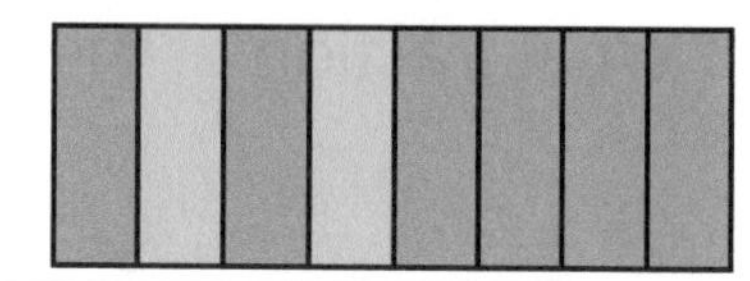

______
______
______

▶ Performance Task

19. Taylor baked a cake to serve after dinner.

A Suppose she cut the cake into 8 equal size pieces and 6 people ate all the pieces. **Explain** how they could have divided the pieces so that everyone ate the same amount of cake. You may show your work in a drawing.

______
______
______
______

B Suppose there are 2 pieces of the cake left over, but you don't know how many pieces were in the whole cake. **Explain** how you could find the number of pieces in the whole cake if Taylor told you $\frac{1}{6}$ of the cake was left. You may show your work in a drawing.

______
______
______

Chapter 9

Compare Fractions

Show What You Know

Check your understanding of important skills.

Name ______________________

▶ Halves and Fourths

1. Find the shape that is divided into 2 equal parts. Color $\frac{1}{2}$.

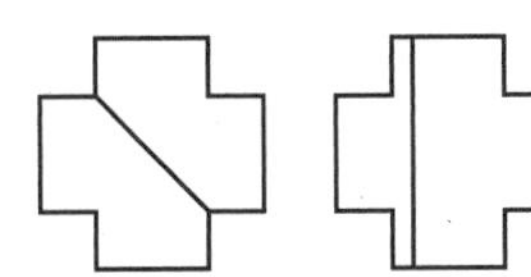

2. Find the shape that is divided into 4 equal parts. Color $\frac{1}{4}$.

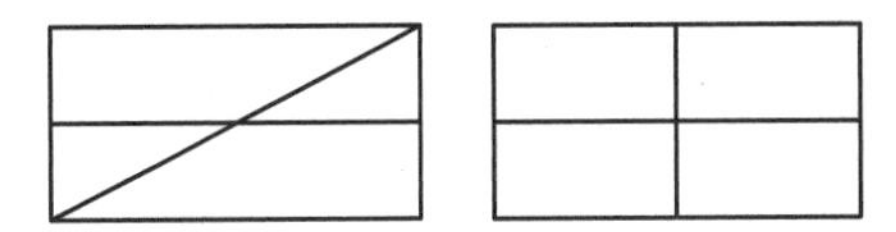

▶ Parts of a Whole

Write the number of shaded parts and the number of equal parts.

3. _____ shaded parts

_____ equal parts

4. _____ shaded parts

_____ equal parts

▶ Fractions of a Whole

Write the fraction that names the shaded part of each shape.

5. 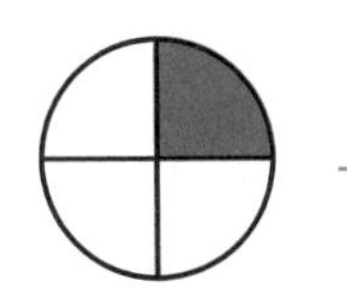 _____

6. 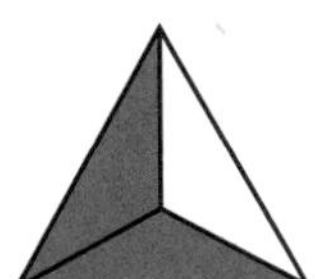 _____

7. 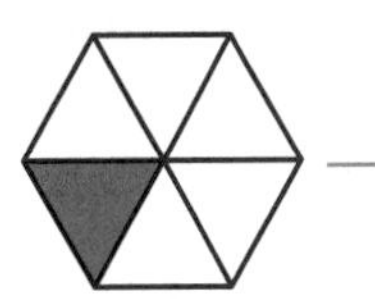 _____

Hannah keeps her marbles in bags with 4 marbles in each bag. She writes $\frac{3}{4}$ to show the number of red marbles in each bag. Be a Math Detective to find another fraction to name the number of red marbles in 2 bags.

Vocabulary Builder

▶ Visualize It

Complete the flow map by using the words with a ✓.

Fractions and Whole Numbers

What is it?		What are some examples?
______________	→	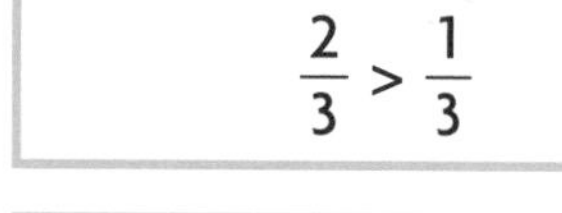 $\frac{2}{3} > \frac{1}{3}$
______________	→	$\frac{1}{4} < \frac{2}{4}$
______________	→	$\frac{1}{2} = \frac{2}{4}$
______________	→	$\frac{1}{3}, \frac{1}{4}$
______________	→	$\frac{2}{2}, \frac{4}{2}$

Review Words

- compare
- denominator
- eighths
- equal parts
- equal to (=)
- fourths
- fraction
- ✓ greater than (>)
- halves
- ✓ less than (<)
- numerator
- order
- sixths
- thirds
- ✓ unit fractions
- ✓ whole numbers

Preview Word

- ✓ equivalent fractions

▶ Understand Vocabulary

Write the review word or preview word that answers the riddle.

1. We are two fractions that name the same amount.

   ____

2. I am the part of a fraction above the line. I tell how many parts are being counted.

   ____

3. I am the part of a fraction below the line. I tell how many equal parts are in the whole or in the group.

   ____

GO Online • eStudent Edition • Multimedia eGlossary

Name ______________________________

Problem Solving • Compare Fractions

Essential Question How can you use the strategy *act it out* to solve comparison problems?

UNLOCK the Problem REAL WORLD

Mary and Vincent climbed up a rock wall at the park. Mary climbed $\frac{3}{4}$ of the way up the wall. Vincent climbed $\frac{3}{8}$ of the way up the wall. Who climbed higher?

You can act out the problem by using manipulatives to help you compare fractions.

Remember

< is less than

> is greater than

= is equal to

Read the Problem	Solve the Problem
What do I need to find? ______________________	**Record the steps you used to solve the problem.** 1 $\frac{1}{4}$ $\frac{1}{4}$ $\frac{1}{4}$ $\frac{1}{8}$ $\frac{1}{8}$ $\frac{1}{8}$ Compare the lengths. _____ ◯ _____ The length of the $\frac{3}{4}$ model is _____ than the length of the $\frac{3}{8}$ model. So, __________ climbed higher on the rock wall.
What information do I need to use? Mary climbed _____ of the way. Vincent climbed _____ of the way.	
How will I use the information? I will use ______________ and ______________ the lengths of the models to find who climbed __________.	

MATHEMATICAL PRACTICES

Math Talk How do you know who climbed higher?

Try Another Problem

The camp leader is making waffles for breakfast. Tracy ate $\frac{3}{6}$ of her waffle. Kim ate $\frac{5}{6}$ of her waffle. Who ate more of her waffle?

Read the Problem	Solve the Problem
What do I need to find?	**Record the steps you used to solve the problem.**
What information do I need to use?	
How will I use the information?	

Math Talk MATHEMATICAL PRACTICES **Explain** how you know that $\frac{5}{6}$ is greater than $\frac{3}{6}$ without using models.

1. How did your model help you solve the problem? ____________________

____

2. Tracy and Kim each had a carton of milk with lunch. Tracy drank $\frac{5}{8}$ of her milk. Kim drank $\frac{7}{8}$ of her milk. Who drank more of her milk? **Explain**.

____

____

Name ______________________

Share and Show

1. At the park, people can climb a rope ladder to its top. Rosa climbed $\frac{2}{8}$ of the way up the ladder. Justin climbed $\frac{2}{6}$ of the way up the ladder. Who climbed higher on the rope ladder?

   First, what are you asked to find?

   ____

   Then, model and compare the fractions.

   Think: Compare $\frac{2}{8}$ and $\frac{2}{6}$.

   Last, find the greater fraction.

   ____ ○ ____

   So, __________ climbed higher on the rope ladder.

2. H.O.T. **What if** Cara also tried the rope ladder and climbed $\frac{2}{4}$ of the way up? Who climbed highest on the rope ladder: Rosa, Justin, or Cara? **Explain** how you know.

   ____

   ____

   ____

   ____

3. Ted walked $\frac{2}{3}$ mile to his soccer game. Then he walked $\frac{1}{3}$ mile to his friend's house. Which distance is shorter? **Explain** how you know.

   ____

   ____

Tips

√ Circle the question.

√ Underline important facts.

√ Act out the problem using manipulatives.

SHOW YOUR WORK

On Your Own

Choose a STRATEGY

- Act It Out
- Draw a Diagram
- Find a Pattern
- Make a Table

Use the table for 4–5.

Suri's Cupcakes

Frosting Flavor	Fraction of Cupcakes
Vanilla	$\frac{3}{8}$
Chocolate	$\frac{4}{8}$
Strawberry	$\frac{1}{8}$

4. Suri is frosting 8 cupcakes for her party. The table shows the fraction of cupcakes frosted with each frosting flavor. Which flavor did Suri use on the most cupcakes?

Hint: Use 8 counters to model the cupcakes.

____

5. Write Math **What's the Question?** The answer is strawberry.

____

____

6. H.O.T. Suppose Suri had also used peanut butter frosting on the cupcakes. She frosted $\frac{1}{2}$ of the cupcakes with vanilla, $\frac{1}{4}$ with chocolate, $\frac{1}{8}$ with strawberry, and $\frac{1}{8}$ with peanut butter. Which flavor of frosting did Suri use on the most cupcakes?

____

7. Ms. Gordon has many cookie recipes. One recipe uses $\frac{1}{3}$ cup oatmeal and $\frac{1}{2}$ cup flour. Will Ms. Gordon use more oatmeal or more flour? **Explain.**

____

8. **Test Prep** Rick lives $\frac{4}{6}$ mile from school. Noah lives $\frac{3}{6}$ mile from school. Which of the following correctly compares the fractions?

Ⓐ $\frac{4}{6} = \frac{3}{6}$

Ⓑ $\frac{4}{6} < \frac{3}{6}$

Ⓒ $\frac{4}{6} > \frac{3}{6}$

Ⓓ $\frac{3}{6} > \frac{4}{6}$

SHOW YOUR WORK

FOR MORE PRACTICE:
Standards Practice Book, pp. P175–P176

Name ______________________________

Compare Fractions with the Same Denominator

Essential Question How can you compare fractions with the same denominator?

UNLOCK the Problem REAL WORLD

Jeremy and Christina are each making quilt blocks. Both blocks are the same size and both are made of 4 equal-size squares. $\frac{2}{4}$ of Jeremy's squares are green. $\frac{1}{4}$ of Christina's squares are green. Whose quilt block has more green squares?

- Circle the two fractions you need to compare.
- How are the two fractions alike?

____

Compare fractions of a whole.

- Shade $\frac{2}{4}$ of Jeremy's quilt block.
- Shade $\frac{1}{4}$ of Christina's quilt block.
- Compare $\frac{2}{4}$ and $\frac{1}{4}$.

Jeremy's Quilt Block

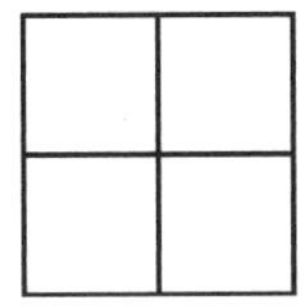

Christina's Quilt Block

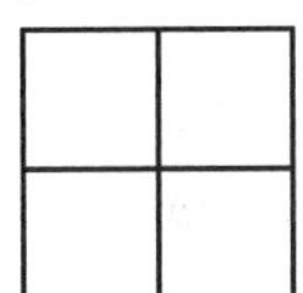

The greater fraction will have the larger amount of the whole shaded.

$\frac{2}{4} \bigcirc \frac{1}{4}$

So, __________ quilt block has more green squares.

Math Idea

You can compare two fractions when they refer to the same whole or to groups that are the same size.

Compare fractions of a group.

Jen and Maggie each have 6 buttons.

- Shade 3 of Jen's buttons to show the number of buttons that are red. Shade 5 of Maggie's buttons to show the number that are red.
- Write a fraction to show the number of red buttons in each group. Compare the fractions.

Jen's Buttons

______

Maggie's Buttons

◯◯◯◯◯◯ ______

There are the same number of buttons in each group, so you can count the number of red buttons to compare the fractions.

$3 <$ ______, so $\frac{\_\_}{6} < \frac{\_\_}{6}$.

So, __________ has a greater fraction of red buttons.

Use fraction strips and a number line.

At the craft store, one piece of ribbon is $\frac{2}{8}$ yard long. Another piece of ribbon is $\frac{7}{8}$ yard long. If Sean wants to buy the longer piece of ribbon, which piece should he buy?

- On a number line, a fraction farther to the right is greater than a fraction to its left.
- On a number line, a fraction farther to the left is ______ a fraction to its right.

Compare $\frac{2}{8}$ and $\frac{7}{8}$.

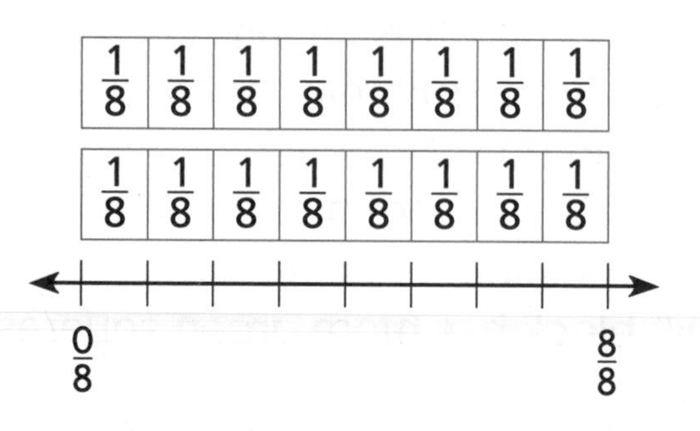

- Shade the fraction strips to show the locations of $\frac{2}{8}$ and $\frac{7}{8}$.
- Draw and label points on the number line to represent the distances $\frac{2}{8}$ and $\frac{7}{8}$.
- Compare the lengths.

  $\frac{2}{8}$ is to the left of $\frac{7}{8}$. It is closer to $\frac{0}{8}$, or ______.

  $\frac{7}{8}$ is to the ______ of $\frac{2}{8}$. It is closer to $\frac{\square}{\square}$, or ______.

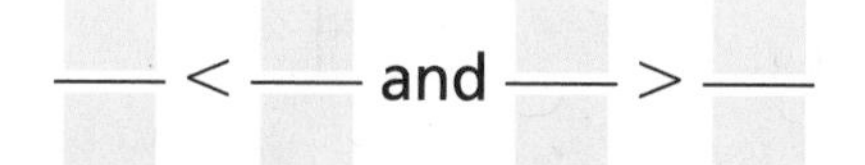

$\frac{\square}{\square} < \frac{\square}{\square}$ and $\frac{\square}{\square} > \frac{\square}{\square}$

So, Sean should buy the piece of ribbon that is $\frac{\square}{\square}$ yard long.

Use reasoning.

Molly and Omar are decorating same-size bookmarks. Molly covers $\frac{3}{3}$ of her bookmark with glitter. Omar covers $\frac{1}{3}$ of his bookmark with glitter. Whose bookmark is covered with more glitter?

Compare $\frac{3}{3}$ and $\frac{1}{3}$.

- When the denominators are the same, the whole is divided into same-size pieces. You can look at the ______ to compare the number of pieces.
- Both fractions involve third-size pieces. ______ pieces are more than ______ piece. 3 > ______, so $\frac{\square}{\square} > \frac{\square}{\square}$.

So, ______ bookmark is covered with more glitter.

Math Talk MATHEMATICAL PRACTICES

Explain how you can use reasoning to compare fractions with the same denominator.

Name ____________________________________

Share and Show 

1. Draw points on the number line to show $\frac{1}{6}$ and $\frac{5}{6}$. Then compare the fractions.

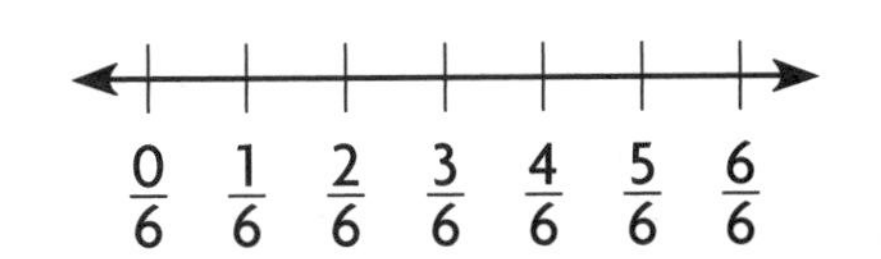

Think: $\frac{1}{6}$ is to the left of $\frac{5}{6}$ on the number line.

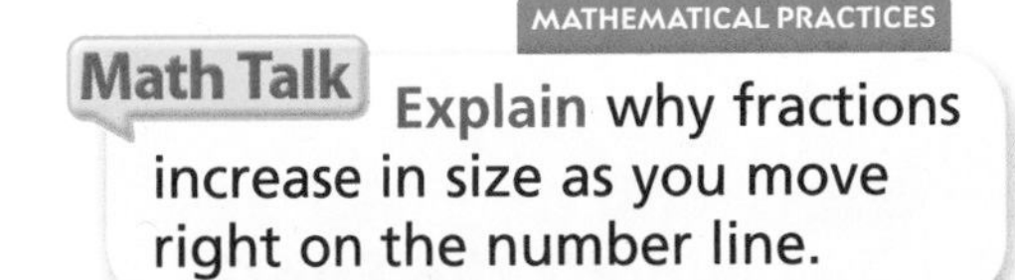

$\frac{1}{6}$ ○ $\frac{5}{6}$

Compare. Write <, >, or =.

2. $\frac{4}{8}$ ○ $\frac{3}{8}$

✓3. $\frac{1}{4}$ ○ $\frac{4}{4}$

4. $\frac{1}{2}$ ○ $\frac{1}{2}$

✓5. $\frac{3}{6}$ ○ $\frac{2}{6}$

On Your Own

Compare. Write <, >, or =.

6. $\frac{2}{4}$ ○ $\frac{3}{4}$

7. $\frac{2}{3}$ ○ $\frac{2}{3}$

8. $\frac{4}{6}$ ○ $\frac{2}{6}$

9. $\frac{0}{8}$ ○ $\frac{2}{8}$

H.O.T. **Write a fraction less than, greater than, or equal to the given fraction.**

10. $\frac{1}{2} < \frac{\square}{\square}$

11. $\frac{\square}{\square} < \frac{12}{6}$

12. $\frac{8}{8} = \frac{\square}{\square}$

13. $\frac{\square}{\square} > \frac{2}{4}$

Problem Solving 

14. Carlos finished $\frac{5}{8}$ of his art project on Monday. Tyler finished $\frac{7}{8}$ of his art project on Monday. Who finished more of his art project on Monday?

____

15. Ms. Endo made two pies that are the same size. Her family ate $\frac{1}{4}$ of the apple pie and $\frac{3}{4}$ of the cherry pie. Which pie had less left over?

____

16. Test Prep Todd and Lisa are comparing fraction strips. Which statement is NOT correct?

Ⓐ $\frac{1}{4} < \frac{4}{4}$ Ⓑ $\frac{5}{6} < \frac{4}{6}$ Ⓒ $\frac{2}{3} > \frac{1}{3}$ Ⓓ $\frac{5}{8} > \frac{4}{8}$

What's the Error?

17. Gary and Vanessa are comparing fractions. Vanessa models $\frac{2}{4}$ and Gary models $\frac{3}{4}$. Vanessa writes $\frac{3}{4} < \frac{2}{4}$. Look at Gary's model and Vanessa's model and describe her error.

Vanessa's Model

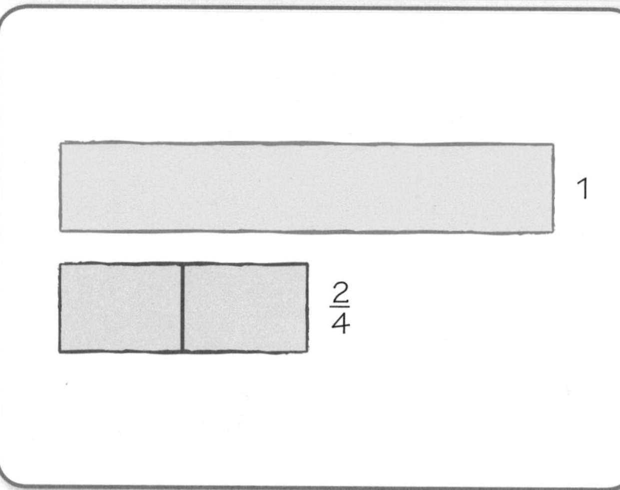

Gary's Model

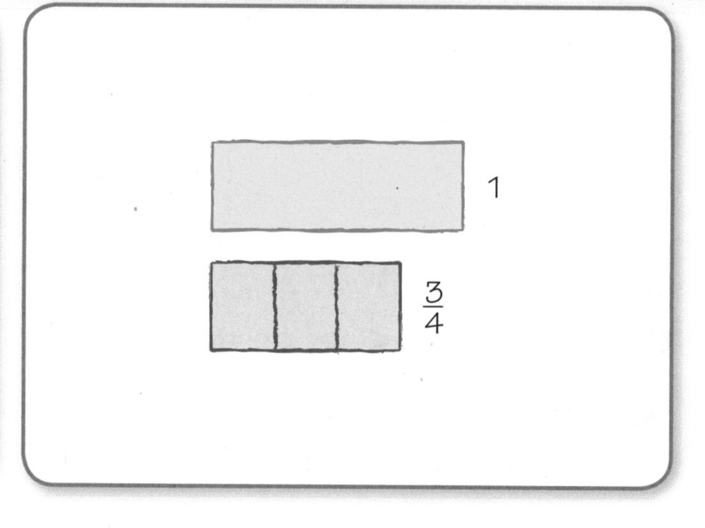

- Describe Vanessa's error.

____

____

- **Explain** how to correct Vanessa's error. Then show the correct model.

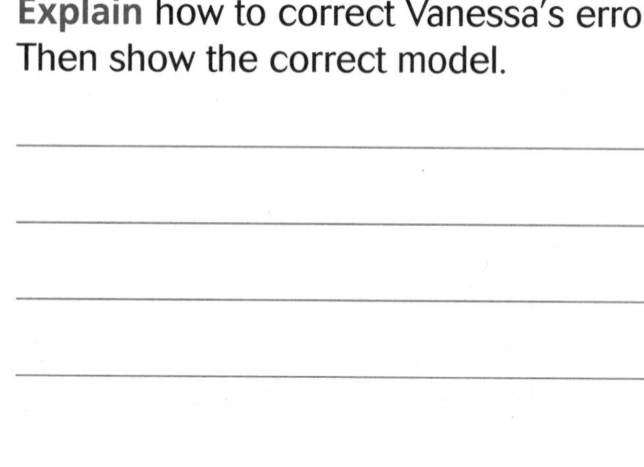

____

____

____

____

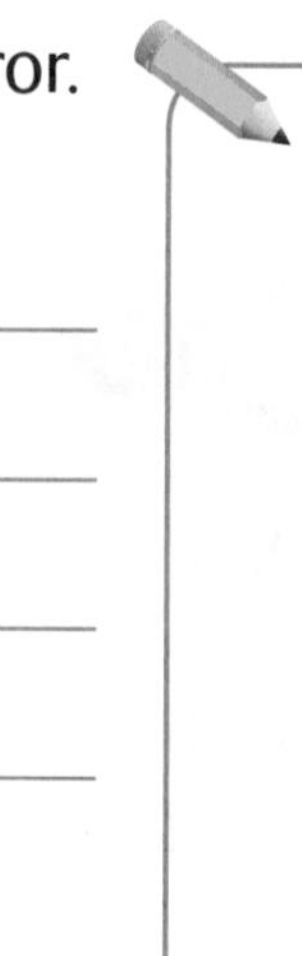

Name ______________________

Compare Fractions with the Same Numerator

Essential Question How can you compare fractions with the same numerator?

UNLOCK the Problem REAL WORLD

Josh is at Enzo's Pizza Palace. He can sit at a table with 5 of his friends or at a different table with 7 of his friends. The same-size pizza is shared equally among the people at each table. At which table should Josh sit to get more pizza?

- Including Josh, how many friends will be sharing pizza at each table?

____

- What will you compare?

____

____

Model the problem.

There will be 6 friends sharing Pizza A or 8 friends sharing Pizza B.

So, Josh will get either $\frac{1}{6}$ or $\frac{1}{8}$ of a pizza.

- Shade $\frac{1}{6}$ of Pizza A.
- Shade $\frac{1}{8}$ of Pizza B.
- Which piece of pizza is larger?
- Compare $\frac{1}{6}$ and $\frac{1}{8}$.

$\frac{1}{6} \bigcirc \frac{1}{8}$

Pizza A

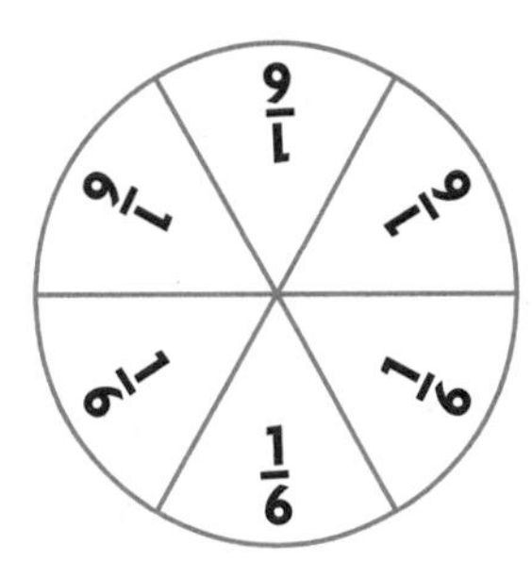

Pizza B

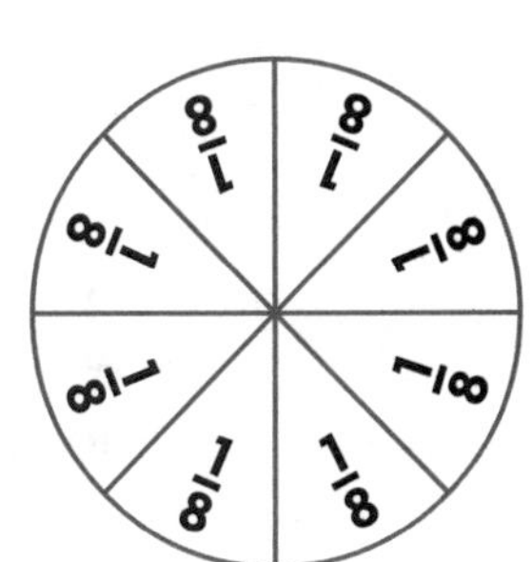

So, Josh should sit at the table with _____ friends to get more pizza.

1. Which pizza has more pieces? _____
The *more* pieces a whole is divided into,

the ________________ the pieces are.

2. Which pizza has fewer pieces? _____
The *fewer* pieces a whole is divided into,

the ________________ the pieces are.

MATHEMATICAL PRACTICES

Math Talk Suppose Josh wants two pieces of one of the pizzas above. Is $\frac{2}{6}$ or $\frac{2}{8}$ of the pizza a greater amount? **Explain** how you know.

 **Use fraction strips.**

On Saturday, the campers paddled $\frac{2}{8}$ of their planned route down the river. On Sunday, they paddled $\frac{2}{3}$ of their route down the river. On which day did the campers paddle farther?

Compare $\frac{2}{8}$ and $\frac{2}{3}$.

1							
$\frac{1}{8}$	$\frac{1}{8}$	$\frac{1}{8}$	$\frac{1}{8}$	$\frac{1}{8}$	$\frac{1}{8}$	$\frac{1}{8}$	$\frac{1}{8}$

$\frac{1}{3}$	$\frac{1}{3}$	$\frac{1}{3}$

- Place a ✓ next to the fraction strips that show more parts in the whole.
- Shade $\frac{2}{8}$. Then shade $\frac{2}{3}$. Compare the shaded parts.
- $\frac{2}{8}$ ◯ $\frac{2}{3}$

Think: $\frac{1}{8}$ is less than $\frac{1}{3}$, so $\frac{2}{8}$ is less than $\frac{2}{3}$.

So, the campers paddled farther on ______________.

 **Use reasoning.**

For her class party, Becky baked two cakes that were the same size. After the party, she had $\frac{3}{4}$ of the carrot cake and $\frac{3}{6}$ of the apple cake left over. Was more carrot cake or more apple cake left over?

Compare $\frac{3}{4}$ and $\frac{3}{6}$.

- Since the numerators are the same, look at the denominators to compare the size of the pieces.

$\frac{3}{4}$ ● $\frac{3}{6}$

- The *more* pieces a whole is divided into, the ______________ the pieces are.
- The *fewer* pieces a whole is divided into, the ______________ the pieces are.

ERROR Alert

When comparing fractions with the same numerator, be sure the symbol shows that the fraction with fewer pieces in the whole is the greater fraction.

- $\frac{1}{4}$ is ______________ than $\frac{1}{6}$ because there are ______________ pieces.

- $\frac{3}{4}$ ◯ $\frac{3}{6}$

So, there was more ______________ cake left over.

Name ______________________________

Share and Show 

1. Shade the models to show $\frac{1}{6}$ and $\frac{1}{4}$.

   Then compare the fractions.

   $\frac{1}{6}$ ◯ $\frac{1}{4}$

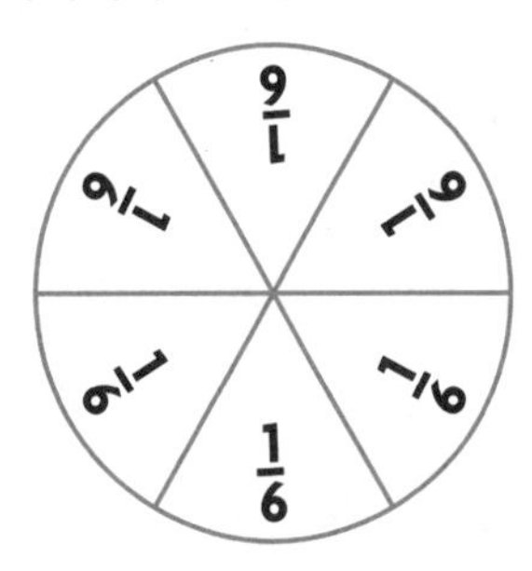

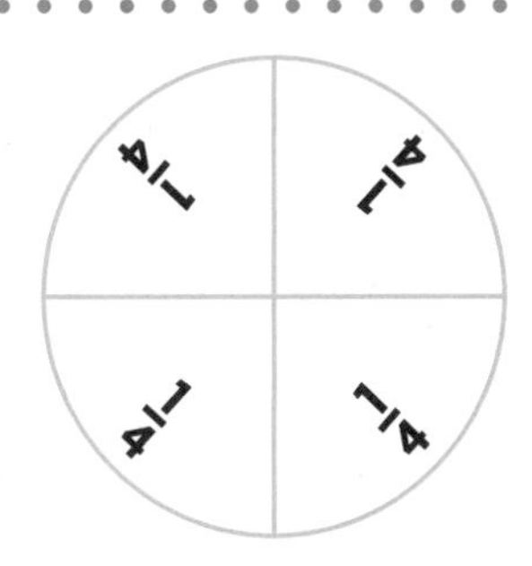

Compare. Write <, >, or =.

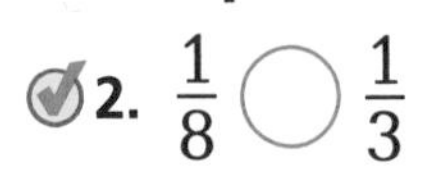

2. $\frac{1}{8}$ ◯ $\frac{1}{3}$

3. $\frac{3}{4}$ ◯ $\frac{3}{8}$

4. $\frac{2}{6}$ ◯ $\frac{2}{3}$

5. $\frac{4}{8}$ ◯ $\frac{4}{4}$

6. $\frac{3}{6}$ ◯ $\frac{3}{6}$

7. $\frac{8}{4}$ ◯ $\frac{8}{8}$

Math Talk MATHEMATICAL PRACTICES
Explain why $\frac{1}{2}$ is greater than $\frac{1}{4}$.

On Your Own

Compare. Write <, >, or =.

8. $\frac{1}{3}$ ◯ $\frac{1}{4}$

9. $\frac{2}{3}$ ◯ $\frac{2}{6}$

10. $\frac{4}{8}$ ◯ $\frac{4}{2}$

11. $\frac{6}{8}$ ◯ $\frac{6}{6}$

12. $\frac{1}{6}$ ◯ $\frac{1}{2}$

13. $\frac{7}{8}$ ◯ $\frac{7}{8}$

14. H.O.T. **Sense or Nonsense?** James ate $\frac{3}{4}$ of his pancake. David ate $\frac{2}{3}$ of his pancake. Both pancakes are the same size. Who ate more of his pancake?

    James said he knows he ate more because he looked at the amounts left. Does his answer make sense? Shade the models. **Explain.**

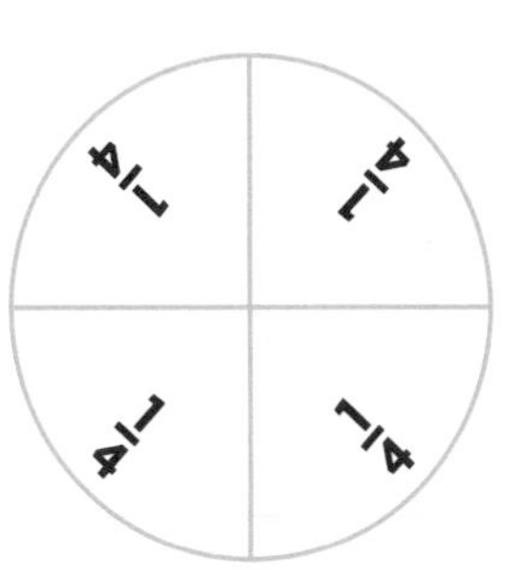

James

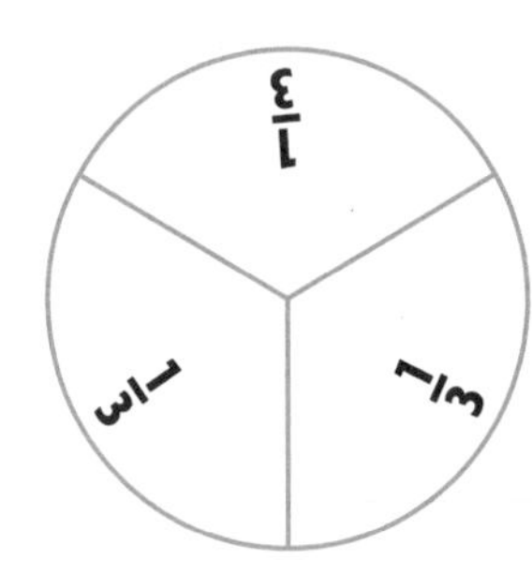

David

____

____

UNLOCK the Problem REAL WORLD

15. Quinton and Hunter are biking on trails in Katy Trail State Park. They biked $\frac{5}{6}$ mile in the morning and $\frac{5}{8}$ mile in the afternoon. Did they bike a greater distance in the morning or in the afternoon?

a. What do you need to know? ______________________

b. The numerator is 5 in both fractions, so compare $\frac{1}{6}$ and $\frac{1}{8}$. **Explain**.

____

c. How can you solve the problem?

d. Complete the sentences.

In the morning, the boys biked

__________ mile.

In the afternoon, they biked

__________ mile.

The boys biked a greater distance

in the __________. $\frac{5}{6}$ ◯ $\frac{5}{8}$

16. H.O.T. **Write Math** Zach has a piece of pie that is $\frac{1}{4}$ of a pie. Max has a piece of pie that is $\frac{1}{2}$ of a pie. Max's piece is smaller than Zach's piece. **Explain** how this could happen. Draw a picture to show your answer.

____

____

____

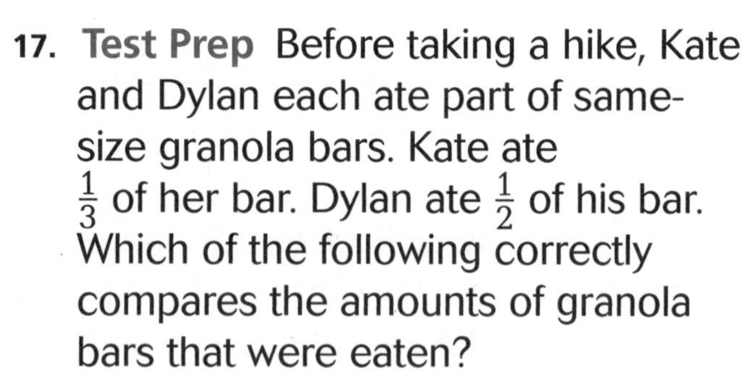

17. **Test Prep** Before taking a hike, Kate and Dylan each ate part of same-size granola bars. Kate ate $\frac{1}{3}$ of her bar. Dylan ate $\frac{1}{2}$ of his bar. Which of the following correctly compares the amounts of granola bars that were eaten?

Ⓐ $\frac{1}{3} > \frac{1}{2}$

Ⓑ $\frac{1}{2} < \frac{1}{3}$

Ⓒ $\frac{1}{2} > \frac{1}{3}$

Ⓓ $\frac{1}{3} = \frac{1}{2}$

FOR MORE PRACTICE:
Standards Practice Book, pp. P179–P180

Name ____________________

Compare Fractions

Essential Question What strategies can you use to compare fractions?

UNLOCK the Problem REAL WORLD

Luka and Ann are eating the same-size small pizzas. One plate has $\frac{3}{4}$ of Luka's cheese pizza. Another plate has $\frac{5}{6}$ of Ann's sausage pizza. Whose plate has more pizza?

- Circle the numbers you need to compare.
- How many pieces make up each whole pizza?

____

Compare $\frac{3}{4}$ and $\frac{5}{6}$.

Missing Pieces Strategy

- You can compare fractions by comparing pieces missing from a whole.

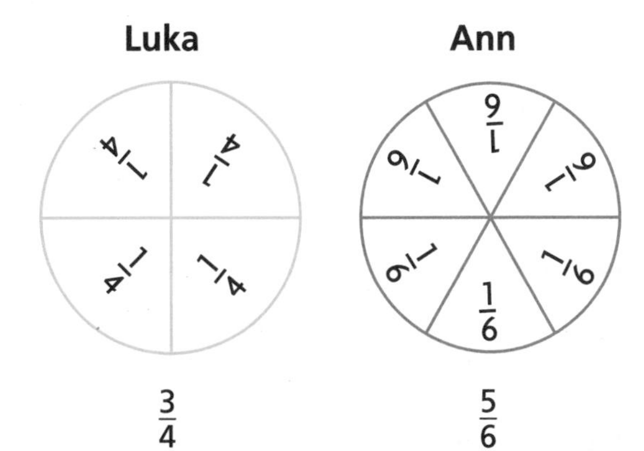

- Shade $\frac{3}{4}$ of Luka's pizza and $\frac{5}{6}$ of Ann's pizza. Each fraction represents a whole that is missing one piece.
- Since $\frac{1}{6}$ ◯ $\frac{1}{4}$, a smaller piece is missing from Ann's pizza.
- If a smaller piece is missing from Ann's pizza, she must have more pizza.

So, __________ plate has more pizza.

MATHEMATICAL PRACTICES

Math Talk Explain how knowing that $\frac{1}{4}$ is less than $\frac{1}{3}$ helps you compare $\frac{3}{4}$ and $\frac{2}{3}$.

Morgan ran $\frac{2}{3}$ mile. Alexa ran $\frac{1}{3}$ mile. Who ran farther?

Compare $\frac{2}{3}$ and $\frac{1}{3}$.

$\frac{\square}{3} > \frac{\square}{3}$

Same Denominator Strategy

- When the denominators are the same, you can compare only the number of pieces, or the numerators.

So, ____________ ran farther.

Ms. Davis is making a fruit salad with $\frac{3}{4}$ pound of cherries and $\frac{3}{8}$ pound of strawberries. Which weighs less, the cherries or the strawberries?

Compare $\frac{3}{4}$ and $\frac{3}{8}$.

Same Numerator Strategy

- When the numerators are the same, look at the denominators to compare the size of the pieces.

$\frac{3}{\_\_} < \frac{3}{\_\_}$

So, the ______________ weigh less.

Share and Show

1. Compare $\frac{7}{8}$ and $\frac{5}{6}$.

   Think: What is missing from each whole?

   Write <, >, or =. $\frac{7}{8}$ ◯ $\frac{5}{6}$

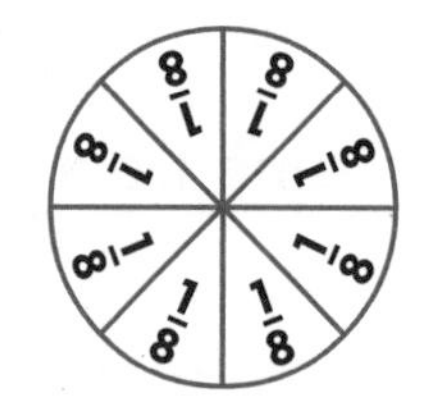

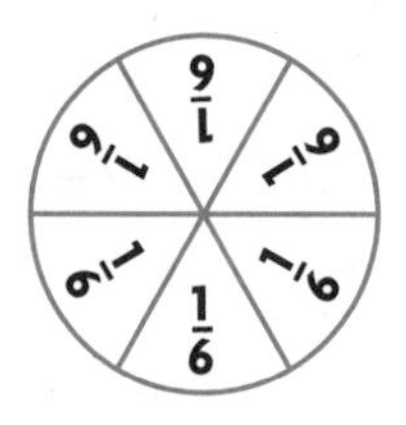

Compare. Write <, >, or =. Write the strategy you used.

2. $\frac{1}{2}$ ◯ $\frac{2}{3}$

   ____

✓ 3. $\frac{3}{4}$ ◯ $\frac{2}{4}$

   ____

✓ 4. $\frac{3}{8}$ ◯ $\frac{3}{6}$

   ____

5. $\frac{3}{4}$ ◯ $\frac{7}{8}$

   ____

MATHEMATICAL PRACTICES

Math Talk **Explain** how the missing pieces in Exercise 1 help you compare $\frac{7}{8}$ and $\frac{5}{6}$.

Name ______________________

On Your Own

Compare. Write <, >, or =. Write the strategy you used.

6. $\frac{1}{2} \bigcirc \frac{2}{2}$

____

7. $\frac{1}{3} \bigcirc \frac{1}{4}$

____

8. $\frac{2}{3} \bigcirc \frac{5}{6}$

____

9. $\frac{2}{8} \bigcirc \frac{0}{8}$

____

10. $\frac{5}{6} \bigcirc \frac{5}{6}$

____

11. $\frac{4}{6} \bigcirc \frac{4}{2}$

____

Name a fraction that is less than or greater than the given fraction. Draw to justify your answer.

12. greater than $\frac{2}{3}$ ______

13. less than $\frac{5}{6}$ ______

14. less than $\frac{2}{4}$ ______

15. greater than $\frac{3}{8}$ ______

16. H.O.T. **What's the Error?** Jack says that $\frac{5}{8}$ is greater than $\frac{5}{6}$ because the denominator 8 is greater than the denominator 6. Describe Jack's error.

____

____

UNLOCK the Problem REAL WORLD

17. Tracy is making a blueberry cake. She is using $\frac{4}{4}$ cup of sugar and $\frac{4}{2}$ cups of flour. Which of the following correctly compares the fractions?

Ⓐ $\frac{4}{4} = \frac{4}{2}$ Ⓒ $\frac{4}{2} > \frac{4}{4}$

Ⓑ $\frac{4}{4} > \frac{4}{2}$ Ⓓ $\frac{4}{2} < \frac{4}{4}$

a. What do you need to know?

____

b. What strategy will you use to compare the fractions?

____

c. Show the steps you used to solve the problem.

d. Complete the comparison.

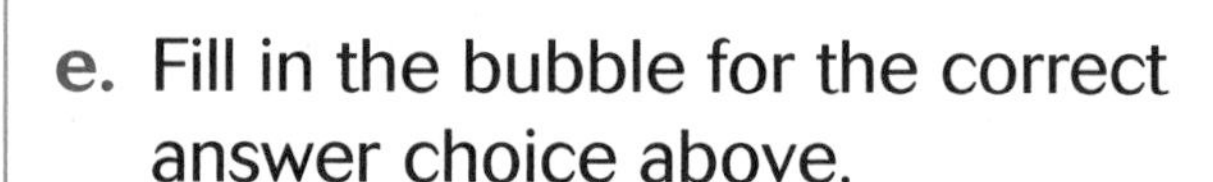

$\underline{\quad} > \underline{\quad}$

e. Fill in the bubble for the correct answer choice above.

18. Luke and Seth have empty glasses. Mr. Gabel pours $\frac{3}{4}$ cup of orange juice in Seth's glass. Then he pours $\frac{1}{4}$ cup of orange juice in Luke's glass. Which statement is correct?

Ⓐ Luke has less orange juice than Seth.

Ⓑ Luke has more orange juice than Seth.

Ⓒ Seth has less orange juice than Luke.

Ⓓ Luke and Seth have the same amount of orange juice.

 FOR MORE PRACTICE: Standards Practice Book, pp. P181–P182

Name ___________________________________

Mid-Chapter Checkpoint

▶ Concepts and Skills

1. When two fractions refer to the same whole, **explain** why the fraction with a lesser denominator has larger pieces than the fraction with a greater denominator.

____

____

____

2. When two fractions refer to the same whole and have the same denominators, **explain** why you can compare only the numerators.

____

____

____

Compare. Write <, >, or =.

3. $\frac{1}{6}$ ◯ $\frac{1}{4}$

4. $\frac{1}{8}$ ◯ $\frac{1}{8}$

5. $\frac{2}{8}$ ◯ $\frac{2}{3}$

6. $\frac{4}{2}$ ◯ $\frac{1}{2}$

7. $\frac{7}{8}$ ◯ $\frac{3}{8}$

8. $\frac{5}{6}$ ◯ $\frac{2}{3}$

9. $\frac{2}{4}$ ◯ $\frac{3}{4}$

10. $\frac{6}{6}$ ◯ $\frac{6}{8}$

11. $\frac{3}{4}$ ◯ $\frac{7}{8}$

Name a fraction that is less than or greater than the given fraction. Draw to justify your answer.

12. greater than $\frac{2}{6}$ ______

13. less than $\frac{2}{3}$ ______

Fill in the bubble for the correct answer choice.

14. Two walls in Tiffany's room are the same size. Tiffany paints $\frac{1}{4}$ of one wall. Jake paints $\frac{1}{8}$ of the other wall. Which of the following correctly compares the fractions?

Ⓐ $\frac{1}{4} < \frac{1}{8}$

Ⓑ $\frac{1}{8} = \frac{1}{4}$

Ⓒ $\frac{1}{4} > \frac{1}{8}$

Ⓓ $\frac{1}{8} > \frac{1}{4}$

15. Matthew ran $\frac{5}{8}$ mile during track practice. Paul ran $\frac{5}{6}$ mile. Which of the following correctly compares the fractions?

Ⓐ $\frac{5}{6} = \frac{5}{8}$

Ⓑ $\frac{5}{8} < \frac{5}{6}$

Ⓒ $\frac{5}{8} > \frac{5}{6}$

Ⓓ $\frac{5}{6} < \frac{5}{8}$

16. Mallory bought 6 roses for her mother. Two-sixths of the roses are red and $\frac{4}{6}$ are yellow. Which of the following correctly compares the fractions?

Ⓐ $\frac{4}{6} < \frac{2}{6}$

Ⓑ $\frac{2}{6} < \frac{4}{6}$

Ⓒ $\frac{4}{6} = \frac{2}{6}$

Ⓓ $\frac{2}{6} > \frac{4}{6}$

17. Lani used $\frac{2}{3}$ cup of raisins and $\frac{3}{4}$ cup of oatmeal to bake cookies. Which statement is correct?

Ⓐ Lani used the same amount of raisins and oatmeal.

Ⓑ Lani used more raisins than oatmeal.

Ⓒ Lani used less oatmeal than raisins.

Ⓓ Lani used less raisins than oatmeal.

Name ______________________________

Compare and Order Fractions

Essential Question How can you compare and order fractions?

UNLOCK the Problem REAL WORLD

Harrison, Tad, and Dale ride their bikes to school. Harrison rides $\frac{3}{4}$ mile, Tad rides $\frac{3}{8}$ mile, and Dale rides $\frac{3}{6}$ mile. Compare and order the distances the boys ride from least to greatest.

- Circle the fractions you need to use.
- Underline the sentence that tells you what you need to do.

Activity 1 Order fractions with the same numerator.

Materials ■ color pencil

You can order fractions by reasoning about the size of unit fractions.

1			
$\frac{1}{4}$	$\frac{1}{4}$	$\frac{1}{4}$	$\frac{1}{4}$

$\frac{1}{8}$	$\frac{1}{8}$	$\frac{1}{8}$	$\frac{1}{8}$	$\frac{1}{8}$	$\frac{1}{8}$	$\frac{1}{8}$	$\frac{1}{8}$

$\frac{1}{6}$	$\frac{1}{6}$	$\frac{1}{6}$	$\frac{1}{6}$	$\frac{1}{6}$	$\frac{1}{6}$

Remember

- The *more* pieces a whole is divided into, the smaller the pieces are.
- The *fewer* pieces a whole is divided into, the larger the pieces are.

STEP 1 Shade one unit fraction for each fraction strip.

______ is the longest unit fraction.

______ is the shortest unit fraction.

STEP 2 Shade one more unit fraction for each fraction strip.

Are the shaded fourths still the longest? ______

Are the shaded eighths still the shortest? ______

STEP 3 Continue shading the fraction strips so that three unit fractions are shaded for each strip.

Are the shaded fourths still the longest? ______

Are the shaded eighths still the shortest? ______

$\frac{3}{4}$ mile is the __________ distance. $\frac{3}{8}$ mile is the __________ distance. $\frac{3}{6}$ mile is *between* the other two distances.

So, the distances in order from least to greatest are

______ mile, ______ mile, ______ mile.

Try This! **Order $\frac{2}{6}$, $\frac{2}{3}$, and $\frac{2}{4}$ from greatest to least.**

Order the fractions $\frac{2}{6}$, $\frac{2}{3}$, and $\frac{2}{4}$ by thinking about the length of the unit fraction strip. Then label the fractions *shortest, between,* or *longest.*

Fraction	Unit Fraction	Length
$\frac{2}{6}$		
$\frac{2}{3}$		
$\frac{2}{4}$		

MATHEMATICAL PRACTICES

Math Talk When ordering three fractions, what do you know about the third fraction when you know which fraction is shortest and which fraction is longest? **Explain** your answer.

- When the numerators are the same, think about the ______ of the pieces to compare and order fractions.

So, the order from greatest to least is ______, ______, ______.

Activity 2 Order fractions with the same denominator.

Materials ■ color pencil

Shade fraction strips to order $\frac{5}{8}$, $\frac{8}{8}$, and $\frac{3}{8}$ from least to greatest.

1								
$\frac{1}{8}$	$\frac{1}{8}$	$\frac{1}{8}$	$\frac{1}{8}$	$\frac{1}{8}$	$\frac{1}{8}$	$\frac{1}{8}$	$\frac{1}{8}$	Shade $\frac{5}{8}$.
$\frac{1}{8}$	$\frac{1}{8}$	$\frac{1}{8}$	$\frac{1}{8}$	$\frac{1}{8}$	$\frac{1}{8}$	$\frac{1}{8}$	$\frac{1}{8}$	Shade $\frac{8}{8}$.
$\frac{1}{8}$	$\frac{1}{8}$	$\frac{1}{8}$	$\frac{1}{8}$	$\frac{1}{8}$	$\frac{1}{8}$	$\frac{1}{8}$	$\frac{1}{8}$	Shade $\frac{3}{8}$.

- When the denominators are the same, the size of the pieces is the ______.

So, think about the ______ of pieces to compare and order fractions.

______ is the shortest. ______ is the longest.

______ is between the other two fractions.

So, the order from least to greatest is ______, ______, ______.

Name ______________________________

Share and Show 

1. Shade the fraction strips to order $\frac{4}{6}$, $\frac{4}{4}$, and $\frac{4}{8}$ from least to greatest.

1							
$\frac{1}{6}$	$\frac{1}{6}$	$\frac{1}{6}$	$\frac{1}{6}$	$\frac{1}{6}$	$\frac{1}{6}$		
$\frac{1}{4}$	$\frac{1}{4}$	$\frac{1}{4}$	$\frac{1}{4}$				
$\frac{1}{8}$	$\frac{1}{8}$	$\frac{1}{8}$	$\frac{1}{8}$	$\frac{1}{8}$	$\frac{1}{8}$	$\frac{1}{8}$	$\frac{1}{8}$

MATHEMATICAL PRACTICES

Math Talk Explain how you would order the fractions $\frac{2}{3}$, $\frac{1}{3}$, and $\frac{3}{3}$ from greatest to least.

______ is the shortest. ______ is the longest.

______ is between the other two lengths. ______, ______, ______

Write the fractions in order from least to greatest.

2. $\frac{1}{2}, \frac{0}{2}, \frac{2}{2}$ ______, ______, ______

3. $\frac{1}{6}, \frac{1}{2}, \frac{1}{3}$ ______, ______, ______

On Your Own

Write the fractions in order from greatest to least.

4. $\frac{6}{6}, \frac{2}{6}, \frac{5}{6}$ ______, ______, ______

5. $\frac{1}{8}, \frac{1}{4}, \frac{1}{2}$ ______, ______, ______

6. $\frac{3}{6}, \frac{3}{4}, \frac{3}{1}$ ______, ______, ______

7. $\frac{3}{8}, \frac{1}{8}, \frac{5}{8}$ ______, ______, ______

Write the fractions in order from least to greatest.

8. $\frac{0}{4}, \frac{3}{4}, \frac{2}{4}$ ______, ______, ______

9. $\frac{2}{4}, \frac{2}{6}, \frac{2}{8}$ ______, ______, ______

10. H.O.T. $\frac{6}{3}, \frac{6}{2}, \frac{6}{8}$ ______, ______, ______

11. H.O.T. $\frac{4}{2}, \frac{2}{2}, \frac{8}{2}$ ______, ______, ______

Problem Solving REAL WORLD

12. In fifteen minutes, Greg's sailboat went $\frac{3}{6}$ mile, Gina's sailboat went $\frac{6}{6}$ mile, and Stuart's sailboat went $\frac{4}{6}$ mile. Whose sailboat went the longest distance in fifteen minutes?

____

Whose sailboat went the shortest distance?

____

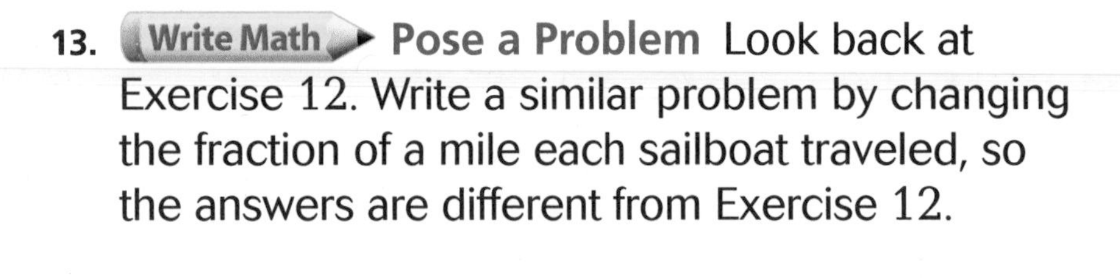

13. **Write Math** ▶ **Pose a Problem** Look back at Exercise 12. Write a similar problem by changing the fraction of a mile each sailboat traveled, so the answers are different from Exercise 12.

____

____

____

____

____

SHOW YOUR WORK

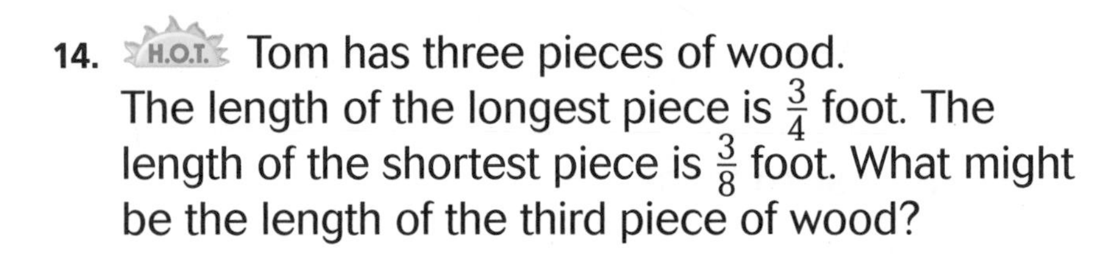

14. H.O.T. Tom has three pieces of wood. The length of the longest piece is $\frac{3}{4}$ foot. The length of the shortest piece is $\frac{3}{8}$ foot. What might be the length of the third piece of wood?

____

15. **Test Prep** Jesse ran $\frac{2}{4}$ mile on Monday, $\frac{2}{3}$ mile on Tuesday, and $\frac{2}{8}$ mile on Wednesday. Which list orders the fractions from least to greatest?

Ⓐ $\frac{2}{4}, \frac{2}{3}, \frac{2}{8}$

Ⓑ $\frac{2}{3}, \frac{2}{4}, \frac{2}{8}$

Ⓒ $\frac{2}{8}, \frac{2}{4}, \frac{2}{3}$

Ⓓ $\frac{2}{8}, \frac{2}{3}, \frac{2}{4}$

Name ______________________

Model Equivalent Fractions

Essential Question How can you use models to find equivalent fractions?

Investigate

Materials ■ sheet of paper ■ crayon or color pencil

Two or more fractions that name the same amount are called **equivalent fractions**. You can use a sheet of paper to model fractions equivalent to $\frac{1}{2}$.

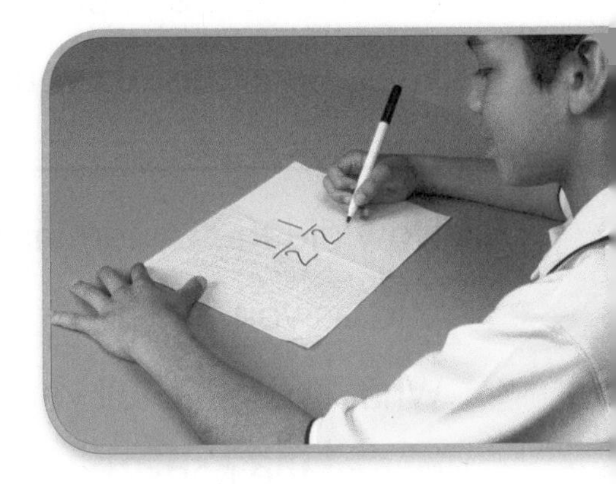

A. First, fold a sheet of paper into two equal parts. Open the paper and count the parts.

There are _____ equal parts. Each part is _____ of the paper.

Shade one of the halves. Write $\frac{1}{2}$ on each of the halves.

B. Next, fold the paper in half two times. Open the paper.

Now there are _____ equal parts. Each part is _____ of the paper.

Write $\frac{1}{4}$ on each of the fourths.

Look at the shaded parts. $\frac{1}{2} = \frac{\square}{4}$

C. Last, fold the paper in half three times.

Now there are _____ equal parts. Each part is _____ of the paper.

Write $\frac{1}{8}$ on each of the eighths.

Find the fractions equivalent to $\frac{1}{2}$ on your paper.

So, $\frac{1}{2}$, $\frac{\square}{\square}$, and $\frac{\square}{\square}$ are equivalent.

Draw Conclusions

1. **Explain** how many $\frac{1}{8}$ parts are equivalent to one $\frac{1}{4}$ part on your paper.

____

____

> **Math Idea**
> Two or more numbers that have the same value or name the same amount are *equivalent*.

2. H.O.T. **Analyze** What do you notice about how the numerators changed for the shaded part as you folded the paper? ____________________

   What does this tell you about the change in the number of parts? ____________________

   How did the denominators change for the shaded part as you folded? ____________________

   What does this tell you about the change in the size of the parts? ____________________

Make Connections

You can use a number line to find equivalent fractions.

Find a fraction equivalent to $\frac{2}{3}$.

Materials ■ fraction strips

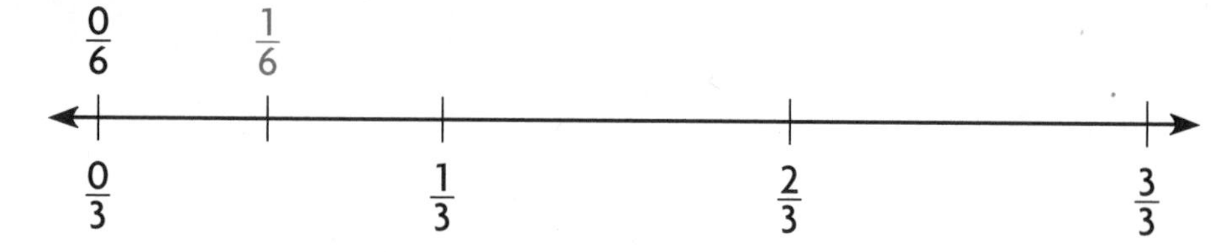

> MATHEMATICAL PRACTICES
> **Math Talk** Explain how the number of sixths in a distance on the number line is related to the number of thirds in the same distance.

STEP 1 Draw a point on the number line to represent the distance $\frac{2}{3}$.

STEP 2 Use fraction strips to divide the number line into sixths. At the end of each strip, draw a mark on the number line and label the marks to show sixths.

STEP 3 Identify the fraction that names the same point as $\frac{2}{3}$. ______

So, $\frac{2}{3} = \frac{\square}{6}$.

Name ______________________________

Share and Show 

Shade the model. Then divide the pieces to find the equivalent fraction.

1. 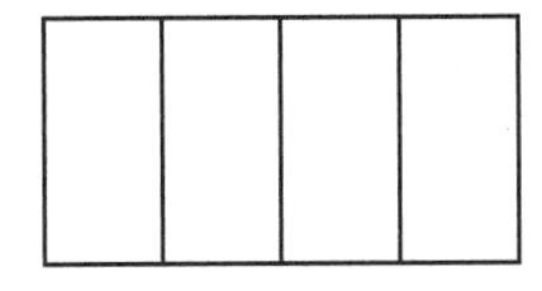

$\frac{1}{4} = \frac{\square}{8}$

✓ 2. 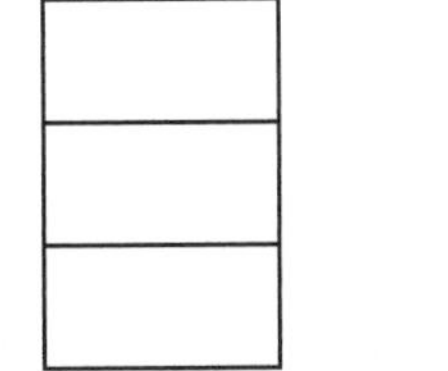

$\frac{2}{3} = \frac{\square}{6}$

MATHEMATICAL PRACTICES

Math Talk Explain how you decided to divide the pieces in Exercise 2.

Use the number line to find the equivalent fraction.

3. 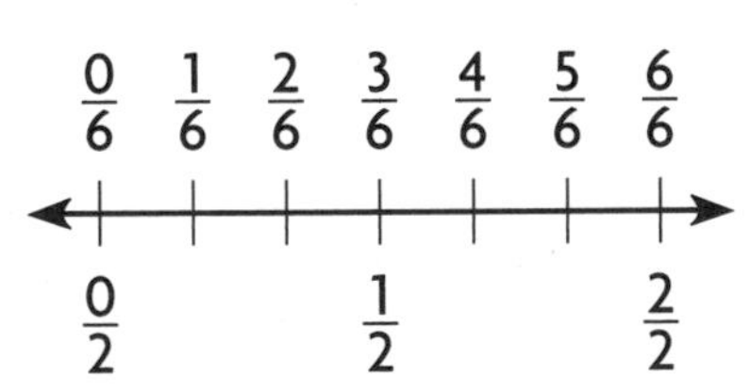

$\frac{1}{2} = \frac{\square}{6}$

✓ 4. 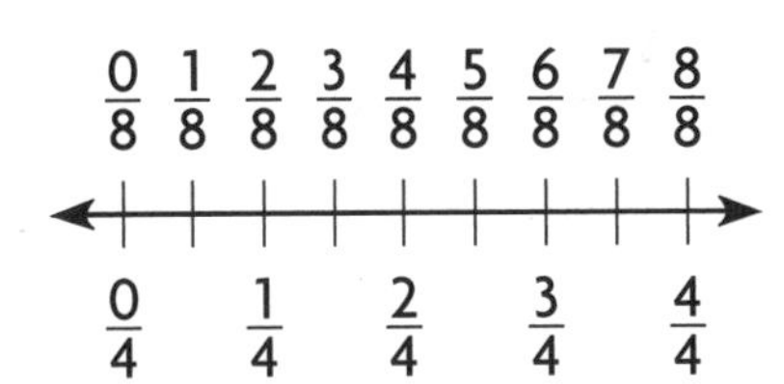

$\frac{3}{4} = \frac{\square}{8}$

5. 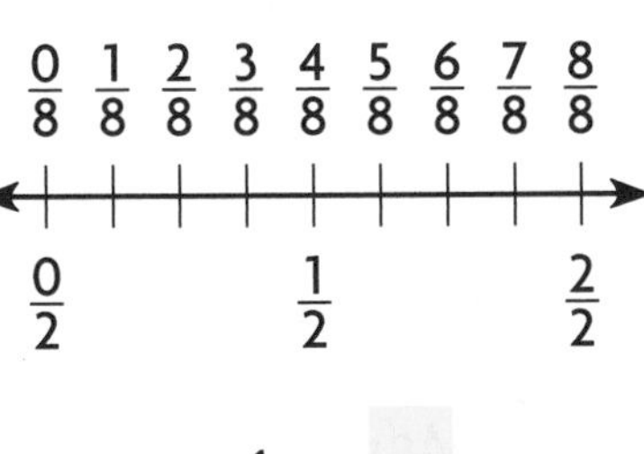

$\frac{1}{2} = \frac{\square}{8}$

6. 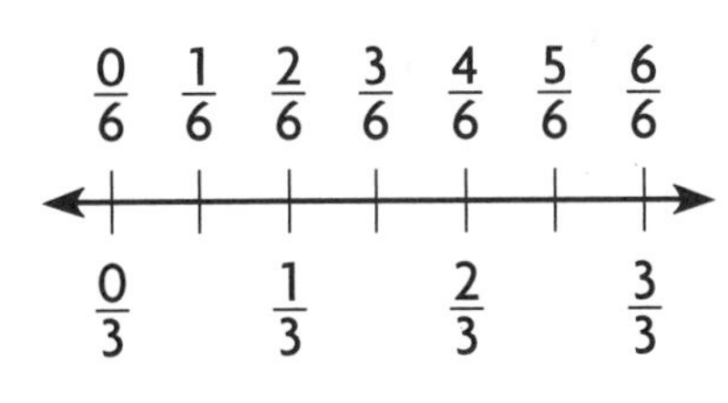

$\frac{3}{3} = \frac{\square}{6}$

7. **Write Math** Explain why $\frac{2}{2} = 1$. Write another fraction that is equal to 1. Draw to justify your answer.

____

____

____

Connect to Reading

Summarize

You can *summarize* the information in a problem by underlining it or writing the information needed to answer a question.

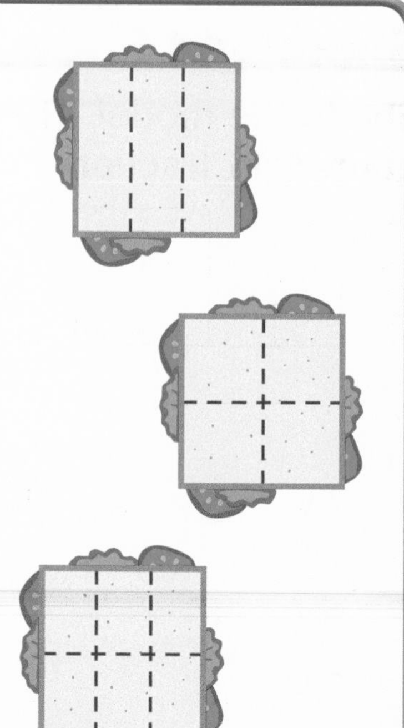

Read the problem. Underline the important information.

Mrs. Akers bought three sandwiches that were the same size. She cut the first one into thirds. She cut the second one into fourths and the third one into sixths. Marian ate 2 pieces of the first sandwich. Jason ate 2 pieces of the second sandwich. Marcos ate 3 pieces of the third sandwich. Which children ate the same amount of a sandwich? **Explain.**

The first sandwich was cut into ________.	The second sandwich was cut into ________.	The third sandwich was cut into ________.
Marian ate ______ pieces of the sandwich. Shade the part Marian ate.	Jason ate ______ pieces of the sandwich. Shade the part Jason ate.	Marcos ate ______ pieces of the sandwich. Shade the part Marcos ate.
Marian ate $\frac{\square}{\square}$ of the first sandwich.	Jason ate $\frac{\square}{\square}$ of the second sandwich.	Marcos ate $\frac{\square}{\square}$ of the third sandwich.

Are all the fractions equivalent? ______

Which fractions are equivalent? $\frac{\square}{\square} = \frac{\square}{\square}$

So, ____________ and ____________ ate the same amount of a sandwich.

FOR MORE PRACTICE:
Standards Practice Book, pp. P185–P186

Name ____________________________

Equivalent Fractions

Essential Question How can you use models to name equivalent fractions?

UNLOCK the Problem REAL WORLD

Cole brought a submarine sandwich to the picnic. He shared the sandwich equally with 3 friends. The sandwich was cut into eighths. What are two ways to describe the part of the sandwich each friend ate?

- How many people shared the sandwich?

  ____

Cole grouped the smaller pieces into twos. Draw circles to show equal groups of two pieces to show what each friend ate.

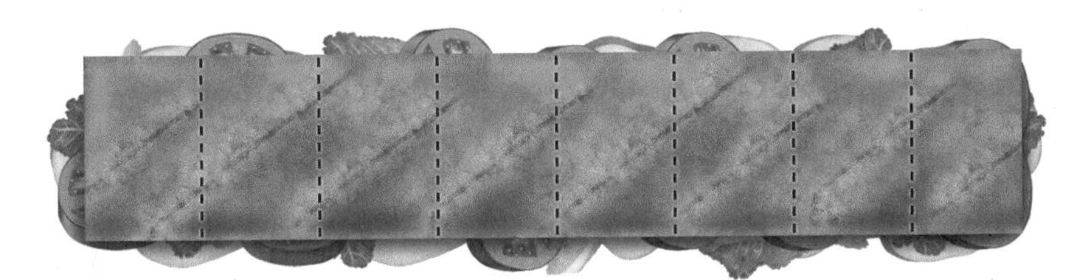

There are 4 equal groups. Each group is $\frac{1}{4}$ of the whole sandwich. So, each friend ate $\frac{1}{4}$ of the whole sandwich.

How many eighths did each friend eat? ______

$\frac{1}{4}$ and ______ are equivalent fractions since they both name the ______ amount of the sandwich.

So, $\frac{1}{4}$ and ______ of the sandwich are two ways to describe the part of the sandwich each friend ate.

Try This! **Circle equal groups. Write an equivalent fraction for the shaded part of the whole.**

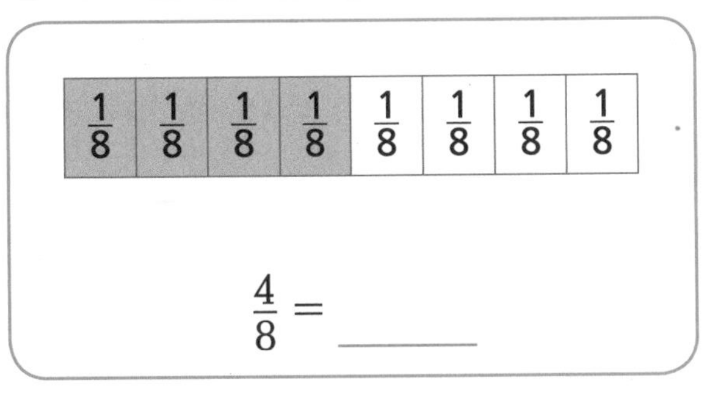

$\frac{4}{8}$ = ______

MATHEMATICAL PRACTICES

Math Talk Explain a different way you could have circled the equal groups.

Example Model the problem.

Heidi ate $\frac{3}{6}$ of her fruit bar. Molly ate $\frac{4}{8}$ of her fruit bar, which is the same size. Which girl ate more of her fruit bar?

Heidi

$\frac{1}{6}$	$\frac{1}{6}$	$\frac{1}{6}$
$\frac{1}{6}$	$\frac{1}{6}$	$\frac{1}{6}$

Molly

$\frac{1}{8}$	$\frac{1}{8}$	$\frac{1}{8}$	$\frac{1}{8}$
$\frac{1}{8}$	$\frac{1}{8}$	$\frac{1}{8}$	$\frac{1}{8}$

Shade $\frac{3}{6}$ of Heidi's fruit bar and $\frac{4}{8}$ of Molly's fruit bar.

- Is $\frac{3}{6}$ greater than, less than, or equal to $\frac{4}{8}$? ________

So, both girls ate the ________ amount.

Try This! **Each shape is 1 whole. Write an equivalent fraction for the shaded part of the models.**

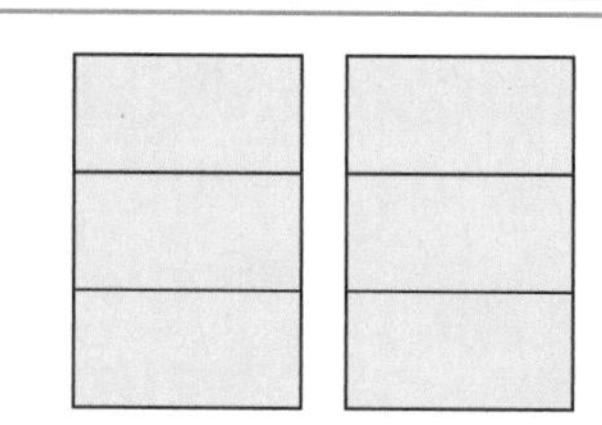
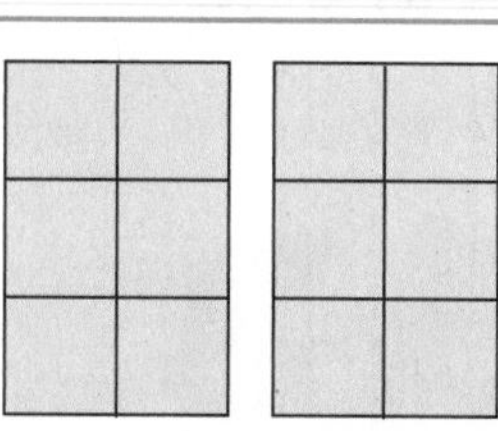

$$\frac{6}{3} = \frac{\square}{6}$$

Share and Show

1. Each shape is 1 whole. Use the model to find the equivalent fraction.

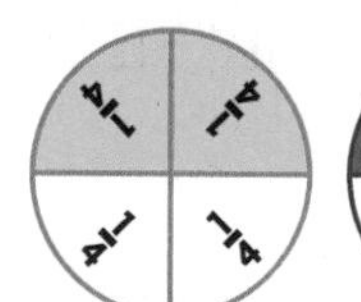

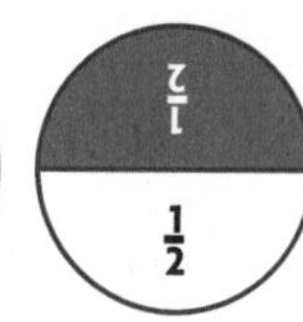

$$\frac{2}{4} = \frac{\square}{2}$$

MATHEMATICAL PRACTICES
Math Talk Explain why both fractions name the same amount.

Each shape is 1 whole. Shade the model to find the equivalent fraction.

2. 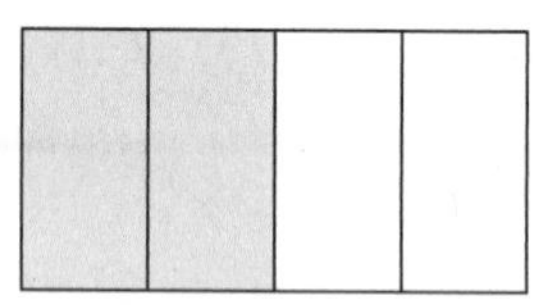 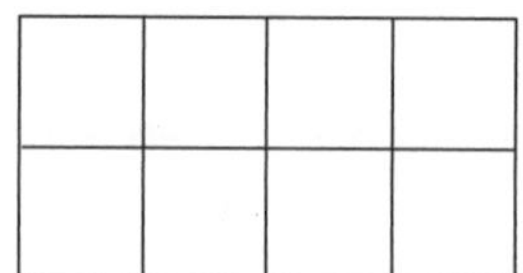

$$\frac{2}{4} = \frac{\square}{8}$$

3. 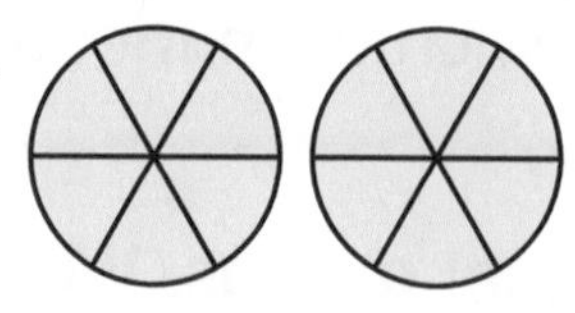 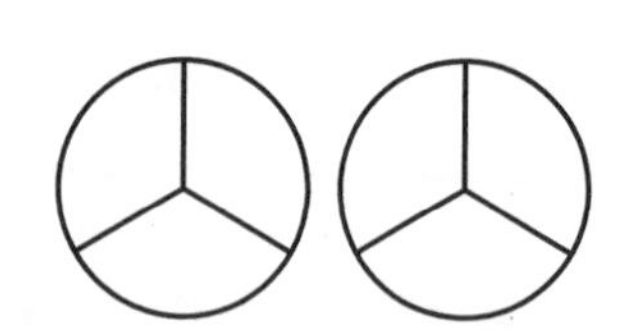

$$\frac{12}{6} = \frac{\square}{3}$$

4. Andy swam $\frac{8}{8}$ mile in a race. Use the number line to find a fraction that is equivalent to $\frac{8}{8}$.

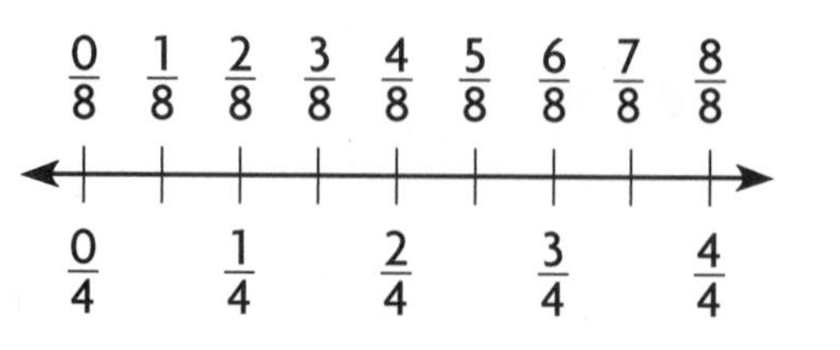

$$\frac{8}{8} = \frac{\square}{\square}$$

Name ______________________________

Circle equal groups to find the equivalent fraction.

5.

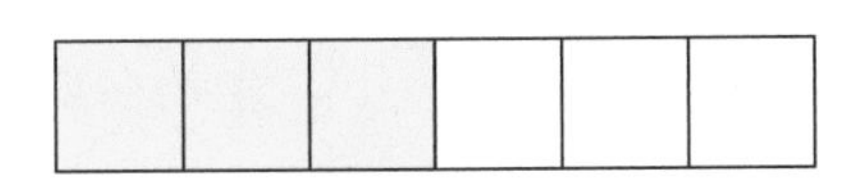

$\frac{3}{6} = \frac{\square}{2}$

6.

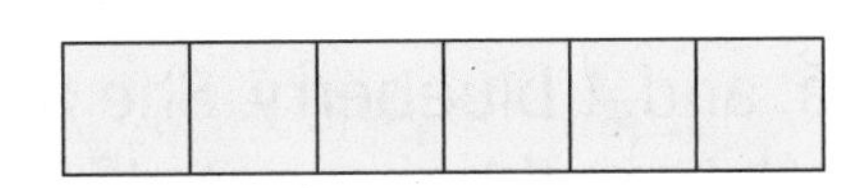

$\frac{6}{6} = \frac{\square}{3}$

On Your Own

Each shape is 1 whole. Shade the model to find the equivalent fraction.

7.

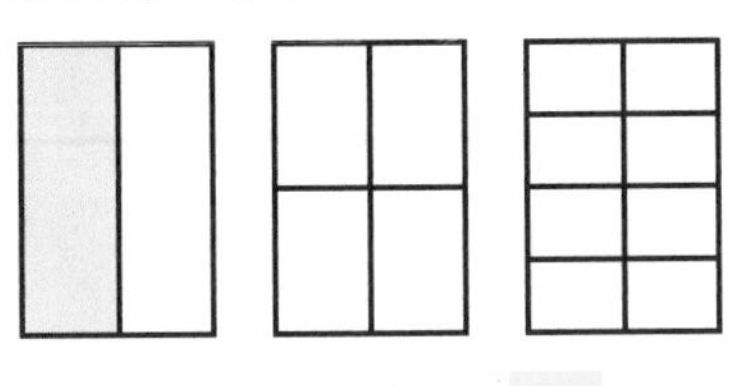

$\frac{1}{2} = \frac{2}{\square} = \frac{\square}{8}$

8.

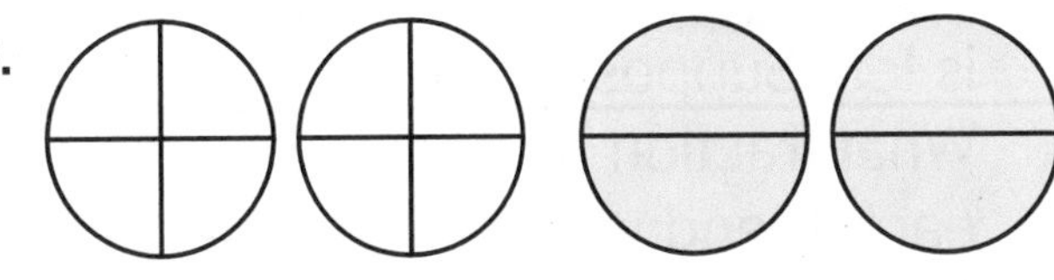

$\frac{8}{\square} = \frac{4}{2}$

Circle equal groups to find the equivalent fraction.

9.

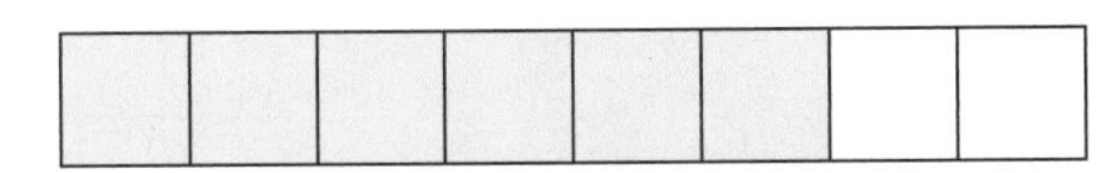

$\frac{6}{8} = \frac{\square}{4}$

10.

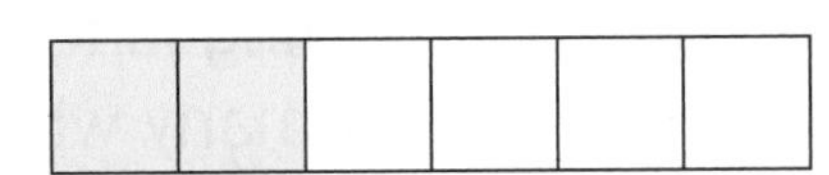

$\frac{2}{6} = \frac{\square}{3}$

11. Write the fraction that names the shaded part of each circle.

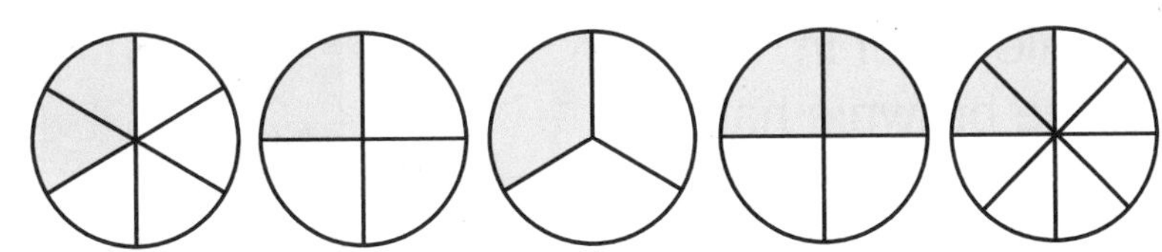

_____ _____ _____ _____ _____

Which pairs of fractions are equivalent? ______________________________

12. Matt cut his small pizza into 6 equal pieces and ate 4 of them. Josh cut his small pizza, which is the same size, into 3 equal pieces and ate 2 of them. Write fractions for the amount they each ate. Are the fractions equivalent? **Draw** to explain.

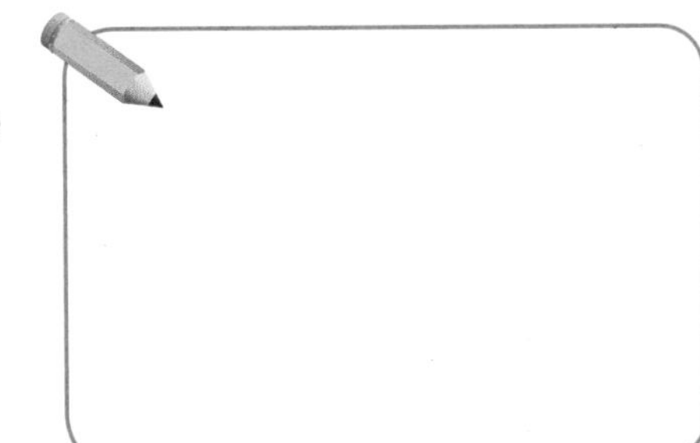

____

Problem Solving REAL WORLD

13. Christy bought 8 muffins. She chose 2 chocolate, 2 banana, and 4 blueberry. She and her family ate the chocolate and banana muffins for breakfast. What fraction of the muffins did they eat? Write an equivalent fraction. Draw a picture.

____

____

14. H.O.T. Write Math After dessert, $\frac{2}{3}$ of a cherry pie is left. Suppose 4 friends want to share it equally. What fraction names how much of the whole pie each friend will get? Use the model on the right. **Explain** your answer.

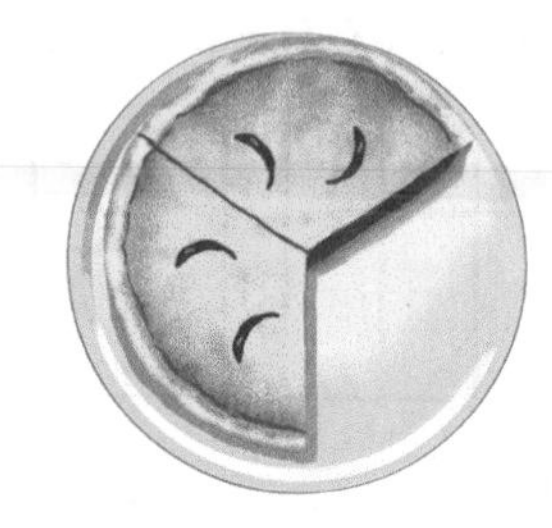

____

____

15. There are 16 people having lunch. Each person wants $\frac{1}{4}$ of a pizza. How many whole pizzas are needed? Draw a picture to show your answer.

____

16. Lucy has 5 brownies. Each brownie is cut in half. What fraction names all of the brownie halves? $\frac{\square}{2}$

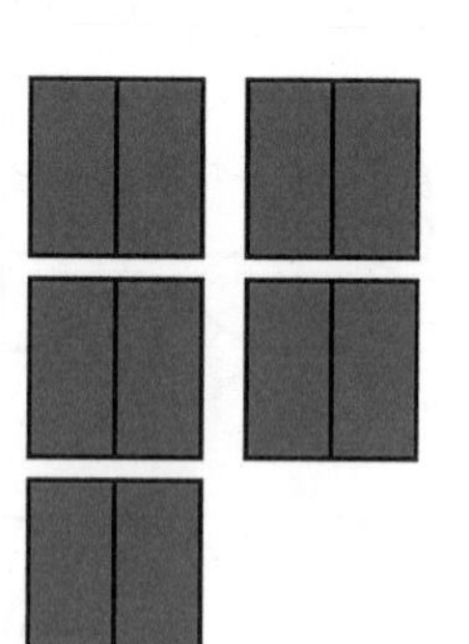

What if Lucy cuts each part of the brownie into 2 equal pieces to share with friends? What fraction names all of the brownie pieces now? $\frac{\square}{4}$

$\frac{\square}{2}$ and $\frac{\square}{4}$ are equivalent fractions.

17. **Test Prep** Mr. Peters made an apple pie. There is $\frac{6}{8}$ of the pie left over. Which fraction is equal to the part of the pie that is left over?

Ⓐ $\frac{1}{4}$ Ⓑ $\frac{2}{4}$ Ⓒ $\frac{1}{2}$ Ⓓ $\frac{3}{4}$

FOR MORE PRACTICE: Standards Practice Book, pp. P187–P188

Name ___________________________

Vocabulary

Choose the best term from the box to complete the sentence.

Vocabulary
equivalent fractions
unit fractions

1. ______________ are two or more fractions that name the same amount. (p. 373)

Concepts and Skills

Compare. Write <, >, or =.

2. $\frac{2}{6}$ ◯ $\frac{2}{8}$

3. $\frac{1}{4}$ ◯ $\frac{1}{8}$

4. $\frac{3}{8}$ ◯ $\frac{7}{8}$

5. $\frac{2}{4}$ ◯ $\frac{2}{4}$

6. $\frac{3}{3}$ ◯ $\frac{2}{3}$

7. $\frac{4}{6}$ ◯ $\frac{4}{2}$

Write the fractions in order from least to greatest.

8. $\frac{1}{6}, \frac{4}{6}, \frac{3}{6}$ _____, _____, _____

9. $\frac{1}{8}, \frac{1}{2}, \frac{1}{3}$ _____, _____, _____

10. $\frac{3}{3}, \frac{1}{3}, \frac{0}{3}$ _____, _____, _____

11. $\frac{4}{8}, \frac{4}{4}, \frac{4}{6}$ _____, _____, _____

Use the model or number line to find the equivalent fraction.

12. 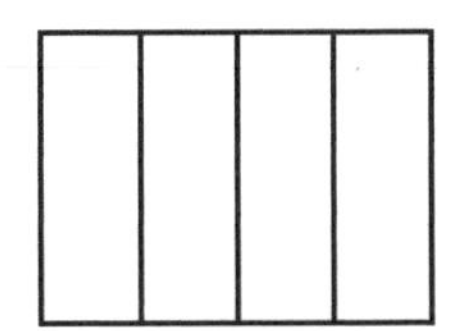

$\frac{3}{4} = \frac{\square}{8}$

13. 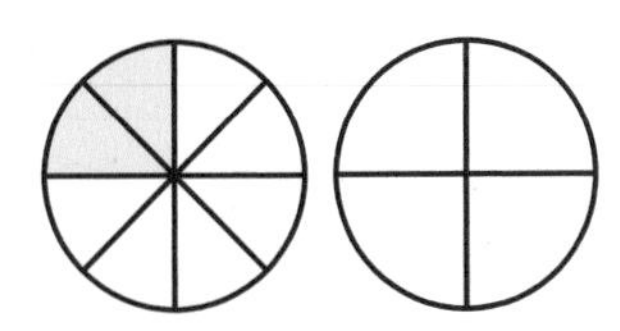

$\frac{2}{8} = \frac{\square}{4}$

14. 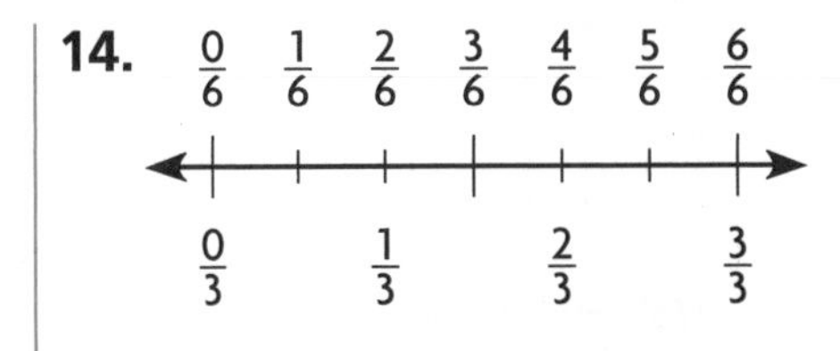

$\frac{1}{3} = \frac{\square}{6}$

GO Online Assessment Options Chapter Test

Fill in the bubble for the correct answer choice.

15. Tali ran $\frac{3}{8}$ mile at the park. Which distance is less than $\frac{3}{8}$ mile?

Ⓐ $\frac{1}{8}$ mile Ⓒ $\frac{4}{8}$ mile

Ⓑ $\frac{7}{8}$ mile Ⓓ $\frac{3}{8}$ mile

16. Mrs. Harrison is ordering a dozen flowers for a party. She says that $\frac{1}{6}$ of the flowers must be red and that $\frac{1}{3}$ of the flowers must be pink. Which of the following correctly compares the two fractions?

Ⓐ $\frac{1}{6} > \frac{1}{3}$

Ⓑ $\frac{1}{6} = \frac{1}{3}$

Ⓒ $\frac{1}{3} < \frac{1}{6}$

Ⓓ $\frac{1}{6} < \frac{1}{3}$

17. David, Maria, and Simone are shading same-size index cards for a science project. David shaded $\frac{2}{4}$ of his index card. Maria shaded $\frac{2}{8}$ of her card, and Simone shaded $\frac{2}{6}$ of her card. Which shows the shaded parts of the index cards in order from greatest to least?

Ⓐ $\frac{2}{4}, \frac{2}{8}, \frac{2}{6}$

Ⓑ $\frac{2}{4}, \frac{2}{6}, \frac{2}{8}$

Ⓒ $\frac{2}{8}, \frac{2}{6}, \frac{2}{4}$

Ⓓ $\frac{2}{6}, \frac{2}{8}, \frac{2}{4}$

Name ______________________________

Fill in the bubble for the correct answer choice.

18. Marissa is decorating a paper plate. She is shading $\frac{1}{4}$ of the plate blue. Which model shows $\frac{1}{4}$ shaded?

Ⓐ 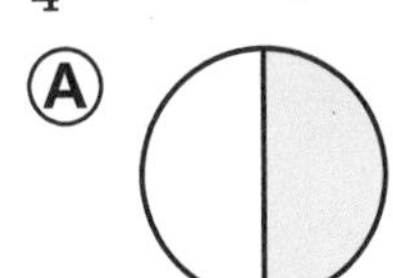

Ⓒ 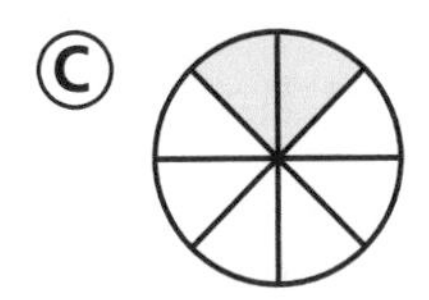

Ⓑ 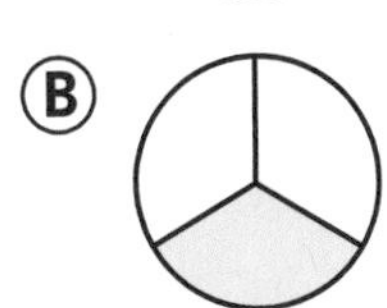

Ⓓ 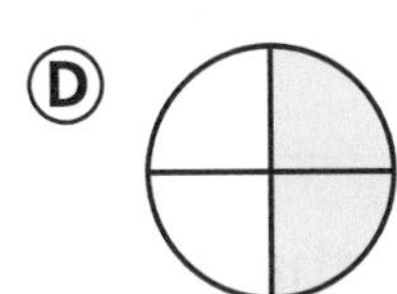

19. Olivia is painting two same-size tiles for art class. She paints $\frac{7}{8}$ of one tile green. Then Olivia paints $\frac{5}{8}$ of the other tile yellow. Which of the following correctly compares the painted parts of the two tiles?

Ⓐ $\frac{5}{8} > \frac{7}{8}$

Ⓑ $\frac{7}{8} > \frac{5}{8}$

Ⓒ $\frac{5}{8} = \frac{7}{8}$

Ⓓ $\frac{7}{8} < \frac{5}{8}$

20. Nicholas ran $\frac{5}{6}$ mile on Monday, $\frac{2}{6}$ mile on Tuesday, and $\frac{6}{6}$ mile on Wednesday. Which list orders the fractions from least to greatest?

Ⓐ $\frac{2}{6}, \frac{5}{6}, \frac{6}{6}$

Ⓑ $\frac{2}{6}, \frac{6}{6}, \frac{5}{6}$

Ⓒ $\frac{6}{6}, \frac{5}{6}, \frac{2}{6}$

Ⓓ $\frac{5}{6}, \frac{2}{6}, \frac{6}{6}$

▶ Constructed Response

21. Alexa and Rachael each ordered a medium pizza. Alexa's pizza was cut into 8 equal slices. Rachael's pizza was cut into 6 equal slices. Each pizza had only 3 slices with mushrooms on top.

Write a fraction for each girl's pizza to tell what part had mushrooms. Then tell whose pizza had a greater part with mushrooms. **Explain.**

____

22. Bryan and Casey's mother made two small pizzas that were the same size. She cut one pizza into eighths and the other pizza into sixths. She placed $\frac{7}{8}$ of the first pizza in Bryan's lunch and $\frac{5}{6}$ of the second pizza in Casey's lunch.

Which boy had less pizza in his lunch? **Explain** the strategy you used to solve the problem.

____

____

____

▶ Performance Task

23. Mrs. Reed baked four cakes for the party. Use the rectangles to represent the cakes.

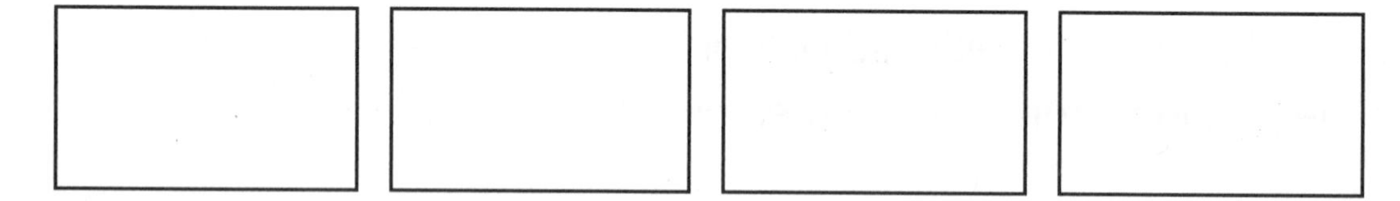

A Draw lines to show how Mrs. Reed could cut one cake into thirds, one into fourths, one into sixths, and one into eighths.

B At the end of the party, equivalent amounts of two pairs of cakes were left over. Shade equivalent amounts on pairs of rectangles to show the cake that might have been left over. Then write two pairs of equivalent fractions.

__________ __________

Measurement

Developing understanding of the structure of rectangular arrays and of area

Measurement tools and data are used to design and build a safe and enjoyable playground.

Project

Plan a Playground

Is there a playground at your school, in your neighborhood, or in a nearby park? Playgrounds provide a fun and safe outdoor space for you to climb, swing, slide, and play.

Important Facts

Playground Features

- Bench
- Jungle Gym
- Playhouse
- Sandbox
- Seesaw
- Slide
- Swing Set
- Water Fountain

Get Started

Suppose you want to help plan a playground for a block in your neighborhood.

- Draw a large rectangle on the grid paper to show a fence around your playground. Find the distance around your playground by counting the number of units on each side. Record the distance.
- Use the Important Facts to help you decide on features to have in your playground. Shade parts of your playground to show each feature's location. Then find the number of unit squares the feature covers and record it on your plan.

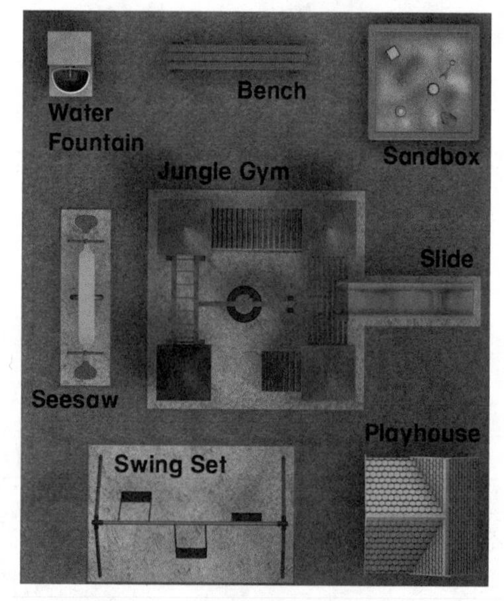

▲ This drawing shows a plan for a playground.

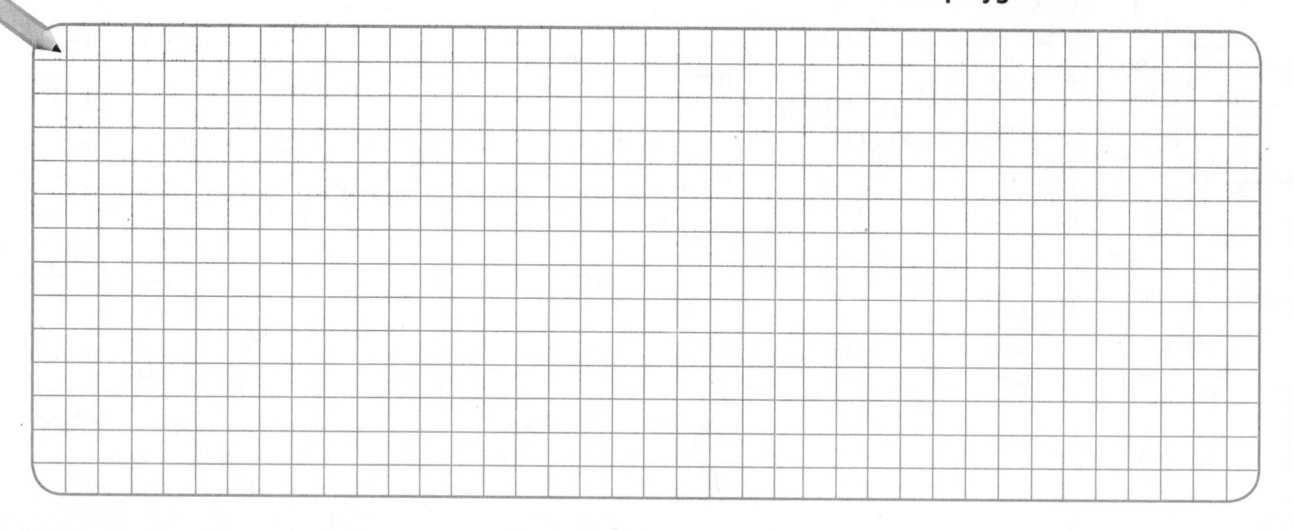

Completed by ____________________

Chapter 10

Time, Length, Liquid Volume, and Mass

Show What You Know

Check your understanding of important skills.

Name ______________________________

▶ Time to the Half Hour Read the clock. Write the time.

1. 

____

2. 

____

▶ Skip Count by Fives

Skip count by fives. Write the missing numbers.

3. 5, 10, 15, ____, 25, ____, 35

4. 55, 60, ____, 70, ____, ____, 85

▶ Inches Use a ruler to measure the length to the nearest inch.

5. 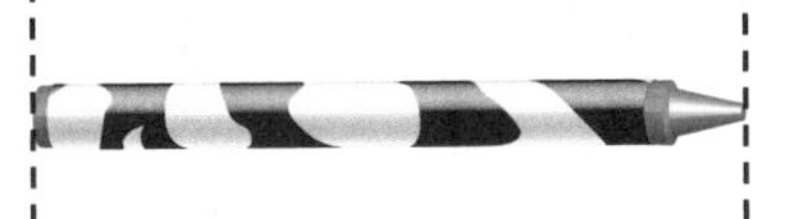

about _____ inches

6. 

about _____ inch

You can look at the time the sun rises and sets to find the amount of daylight each day. The table shows the time the sun rose and set from January 10 to January 14 in Philadelphia, Pennsylvania. Be a Math Detective to find which day had the least daylight and which day had the most daylight.

Sunrise and Sunset Times

Date	Sunrise	Sunset
Jan 10	7:22 A.M.	4:55 P.M.
Jan 11	7:22 A.M.	4:56 P.M.
Jan 12	7:22 A.M.	4:57 P.M.
Jan 13	7:21 A.M.	4:58 P.M.
Jan 14	7:21 A.M.	4:59 P.M.

Vocabulary Builder

▶ Visualize It

Complete the graphic organizer by using the words with a ✓. Write the words in order from the greatest to the least length of time.

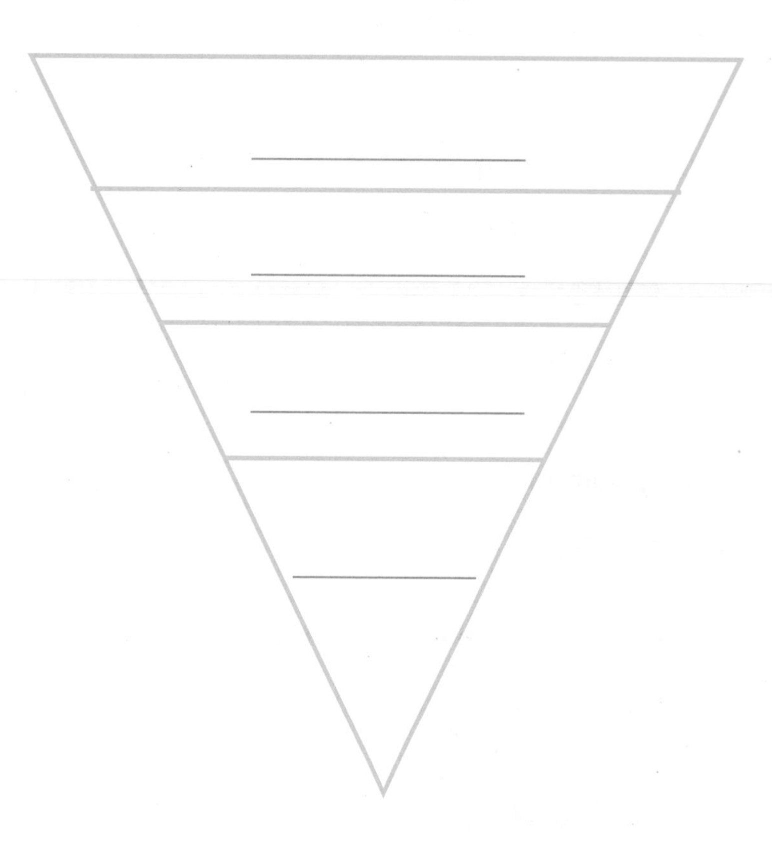

Review Words

- analog clock
- digital clock
- fourth
- half
- ✓ half hour
- ✓ hour (hr)
- inch (in.)
- ✓ quarter hour

Preview Words

- A.M.
- elapsed time
- gram (g)
- kilogram (kg)
- liquid volume
- liter (L)
- mass
- midnight
- ✓ minute (min)
- noon
- P.M.

▶ Understand Vocabulary

Write the word that answers the riddle.

1. I am written with times after midnight and before noon. ____________
2. I am the time when it is 12:00 in the daytime. ____________
3. I am the amount of liquid in a container. ____________
4. I am the time that passes from the start of an activity to the end of that activity. ____________
5. I am the amount of matter in an object. ____________

GO Online • eStudent Edition • Multimedia eGlossary

Name ______________________

Time to the Minute

Essential Question How can you tell time to the nearest minute?

UNLOCK the Problem REAL WORLD

Groundhog Day is February 2. People say that if a groundhog can see its shadow on that morning, winter will last another 6 weeks. The clock shows the time when the groundhog saw its shadow. What time was it?

- Underline the question.
- Where will you look to find the time?

____

Example

Look at the time on this clock face.

- What does the hour hand tell you?

____

- What does the minute hand tell you?

____

In 1 **minute**, the minute hand moves from one mark to the next on a clock. It takes 5 minutes for the minute hand to move from one number to the next on a clock.

You can count on by fives to tell time to five minutes. Count zero at the 12.

0, 5, 10, 15, ____, ____, ____, ____

Write: 7:35

Read:

- seven ______
- thirty-five minutes after ______

So, the groundhog saw its shadow at ______.

MATHEMATICAL PRACTICES

Math Talk How does skip counting by fives help you tell the time when the minute hand points to a number?

- Is 7:35 a reasonable answer? **Explain.** ______________________

____

Time to the Minute

Count by fives and ones to help you.

One Way Find minutes after the hour.

Look at the time on this clock face.

- What does the hour hand tell you?

____

- What does the minute hand tell you?

____

Count on by fives and ones from the 12 on the clock to where the minute hand is pointing. Write the missing counting numbers next to the clock.

When a clock shows 30 or fewer minutes after the hour, you can read the time as a number of minutes *after* the hour.

Write: _______

Read:

- twenty-three minutes after _______
- one _______

Another Way Find minutes before the hour.

Look at the time on this clock face.

- What does the hour hand tell you?

____

- What does the minute hand tell you?

____

Now count by fives and ones from the 12 on the clock back to where the minute hand is pointing. Write the missing counting numbers next to the clock.

When a clock shows 31 or more minutes after the hour, you can read the time as a number of minutes *before* the next hour.

Write: 2:43

Read:

- seventeen _______ before three
- two _______

ERROR Alert

Remember that time *after* the hour uses the previous hour, and time *before* the hour uses the next hour.

Name ______________________________

Share and Show MATH BOARD

1. How would you use counting and the minute hand to find the time shown on this clock? Write the time.

____

Write the time. Write one way you can read the time.

2. 

____

____

✓ 3. 

____

____

✓ 4. 

____

____

MATHEMATICAL PRACTICES

Math Talk **Explain** how you know when to stop counting by fives and start counting by ones when counting minutes after an hour.

On Your Own

Write the time. Write one way you can read the time.

5. 

____

____

6. 

____

____

7. 

____

____

Write the time another way.

8. 34 minutes after 5

____

9. 11 minutes before 6

____

10. 22 minutes after 11

____

11. 5 minutes before 12

____

Problem Solving REAL WORLD

Use the clocks for 12–13.

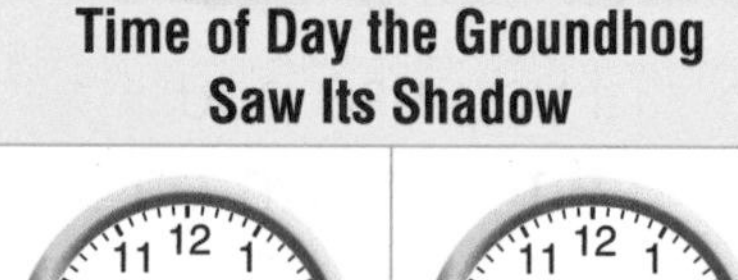

Time of Day the Groundhog Saw Its Shadow	

NY	PA

12. How many minutes later in the day did the groundhog see its shadow in Pennsylvania than in New York?

____

13. **What if** the groundhog in Pennsylvania saw its shadow 5 minutes later? What time would this be?

____

SHOW YOUR WORK

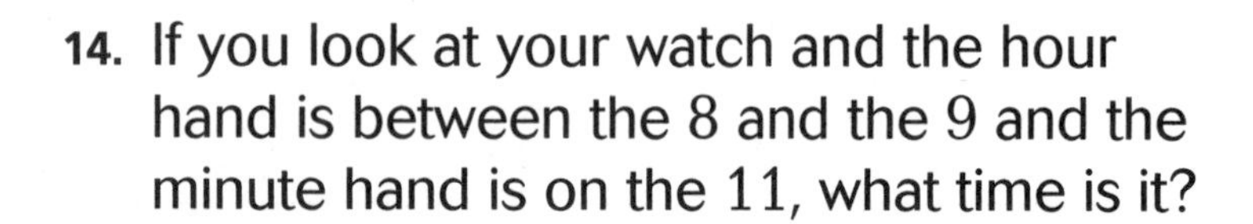

14. If you look at your watch and the hour hand is between the 8 and the 9 and the minute hand is on the 11, what time is it?

____

15. H.O.T. What time is it when the hour hand and the minute hand are both pointing to the same number?

____

16. Write Math Lucy said the time is 4:46 on her digital watch. **Explain** where the hands on the analog clock are pointing when it is 4:46.

____

____

____

____

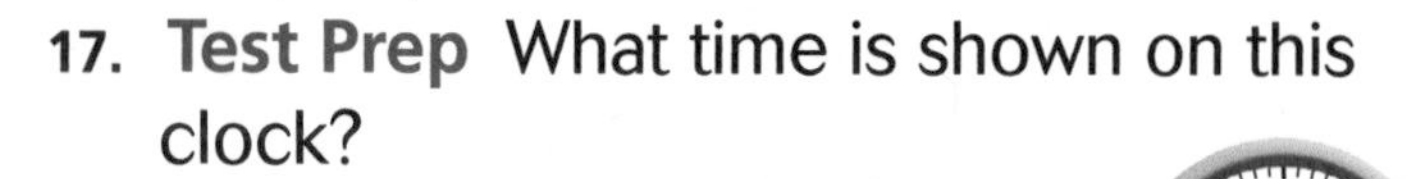

17. **Test Prep** What time is shown on this clock?

Ⓐ 4:23　Ⓒ 9:20

Ⓑ 4:47　Ⓓ 9:23

Name ____________________

A.M. and P.M.

Essential Question How can you tell when to use A.M. and P.M. with time?

UNLOCK the Problem REAL WORLD

Lauren's family is going hiking tomorrow at 7:00. How should Lauren write the time to show that they are going in the morning, not in the evening?

- Circle the helpful information that tells about the hiking time.
- What do you need to find?

____

____

You can use a number line to show the sequence or order of events. It can help you understand the number of hours in a day.

Think: The distance from one mark to the next mark represents one hour.

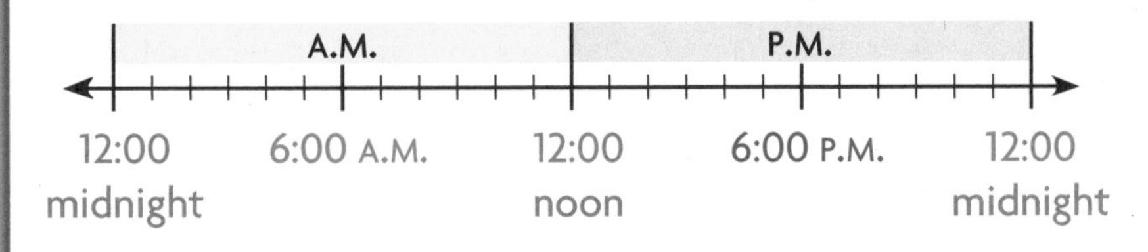

Tell time after midnight.

Midnight is 12:00 at night.

The times after midnight and before noon are written with **A.M.**

7:00 in the morning is written as

7:00 ________

After Midnight and Before Noon

So, Lauren should write the hiking time as 7:00 ________

- Find the mark that shows 7:00 A.M. on the number line above. Circle the mark.

MATHEMATICAL PRACTICES

Math Talk How are the number line on this page and the clock face alike? How are they different?

Tell time after noon.

Callie's family is going for a canoe ride at 3:00 in the afternoon. How should Callie write the time?

Noon is 12:00 in the daytime.

The times after noon and before midnight are written with **P.M.**

3:00 in the afternoon is written as 3:00 ________

After Noon and Before Midnight

So, Callie should write the time as 3:00 ________

Share and Show

1. Name two things you do in the A.M. hours.
   Name two things you do in the P.M. hours.

____

____

Write the time for the activity. Use A.M. or P.M.

2. ride a bicycle

____

3. make a sandwich

____

4. get ready for bed

____

5. This morning Sam woke up at the time shown on this clock. Write the time using A.M. or P.M. ________

MATHEMATICAL PRACTICES

Math Talk Explain how you decide whether to use A.M. or P.M. when you write the time.

Name ____________________

On Your Own

Write the time for the activity. Use A.M. or P.M.

6. eat breakfast

____

7. have science class

____

8. play softball

____

9. go to the store

____

10. leave on a morning airplane flight

____

11. look up at stars

____

Write the time. Use A.M. or P.M.

12. quarter after 9:00 in the morning

____

13. 6 minutes after 7:00 in the morning

____

14. one half hour past midnight

____

15. 18 minutes before noon

____

16. Daylight saving time begins on the second Sunday in March at 2:00 in the morning. Write the time.

Use A.M. or P.M. ____________

17. H.O.T. Write Math From midnight to noon each day, how many times does the minute hand on a clock pass 6? **Explain** how you found your answer.

____

____

____

UNLOCK the Problem REAL WORLD

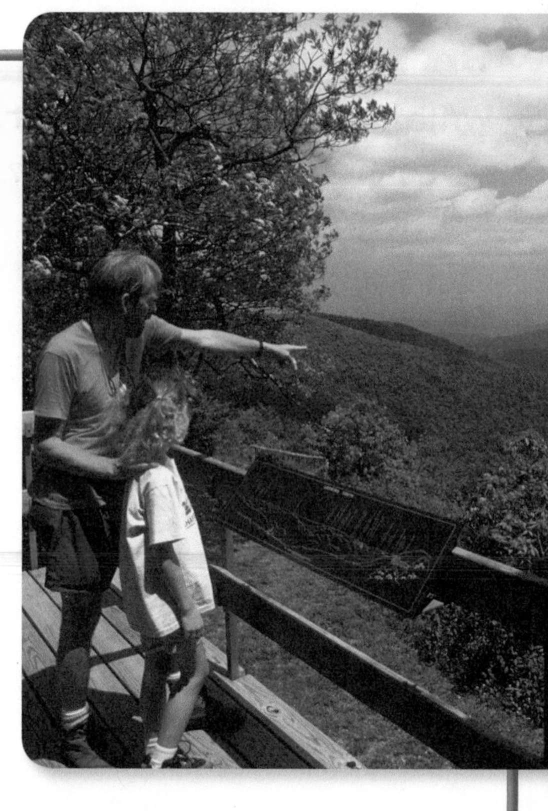

18. Lea and her father arrived at the scenic overlook 15 minutes before noon and left 12 minutes after noon. Using A.M. or P.M., write the time when Lea and her father arrived at the scenic overlook and the time when they left.

a. What do you need to find? ______________

____

____

b. What do you need to find first?

____

c. Show the steps you used to solve the problem.

d. Complete the sentences.

Lea and her father arrived at the overlook _____ minutes __________ noon.

They arrived at ________ ______.M.

Lea and her father left the overlook at _____ minutes __________ noon.

They left at ________ ______.M.

19. Lauren was supposed to meet her family at the parking lot at 10:00 in the morning. Lauren arrived 20 minutes early. At what time did Lauren arrive at the parking lot? Use A.M. or P.M.

____

20. **Test Prep** At which time are most third graders asleep?

Ⓐ 9:23 A.M.

Ⓑ 2:41 P.M.

Ⓒ 7:16 P.M.

Ⓓ 1:38 A.M.

FOR MORE PRACTICE:
Standards Practice Book, pp. P195–P196

Name ______

Measure Time Intervals

Essential Question How can you measure elapsed time in minutes?

UNLOCK the Problem REAL WORLD

Alicia and her family visited the Kennedy Space Center. They watched a movie that began at 4:10 P.M. and ended at 4:53 P.M. How long did the movie last?

- What time did the movie begin?
______
- What time did the movie end?
______
- Underline the question.

To find **elapsed time**, find the amount of time that passes from the start of an activity to the end of the activity.

One Way **Use a number line.**

STEP 1 Find the time on the number line that the movie began.

STEP 2 Count on to the ending time, 4:53. Count on by tens for each 10 minutes. Count on by ones for each minute. Write the times below the number line.

STEP 3 Draw the jumps on the number line to show the minutes from 4:10 to 4:53. Record the minutes. Then add them.

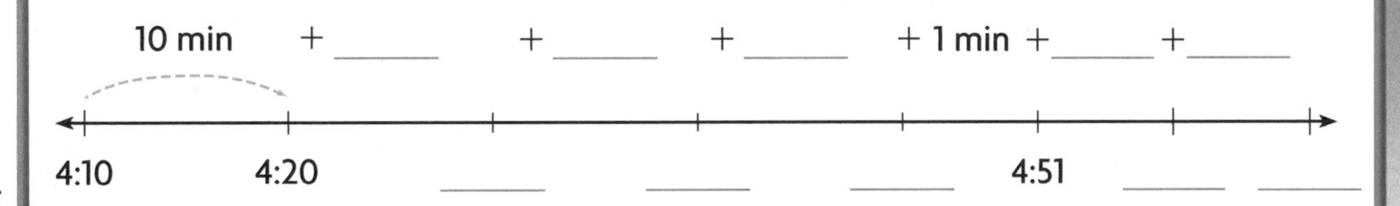

10 + 10 + 10 + 10 + 1 + 1 + 1 = ______

The elapsed time from 4:10 P.M. to 4:53 P.M. is ______ minutes.

So, the movie lasted ______ minutes.

MATHEMATICAL PRACTICES

Math Talk **Describe** another way you can use jumps on the number line to find the elapsed time from 4:10 P.M. to 4:53 P.M.

Other Ways

Start time: 4:10 P.M. End time: 4:53 P.M.

A Use an analog clock.

STEP 1 Find the starting time on the clock.

STEP 2 Count the minutes by counting on by fives and ones to 4:53 P.M. Write the missing counting numbers next to the clock.

So, the elapsed time is _____ minutes.

B Use subtraction.

STEP 1 Write the ending time. Then write the starting time so that the hours and minutes line up.

STEP 2 The hours are the same, so subtract the minutes.

4 : ___ ← end time
− 4 : ___ ← start time
___ ← elapsed time

Try This! **Find the elapsed time in minutes two ways.**

Start time: 10:05 A.M. End time: 10:30 A.M.

A Use a number line.

STEP 1 Find 10:05 on the number line. Count on from 10:05 to 10:30. Draw marks and record the times on the number line. Then draw and label the jumps.

Think: Count on using longer amounts of time that make sense.

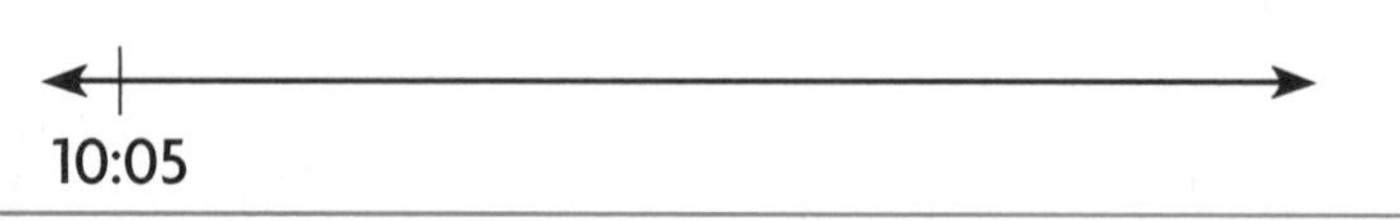

STEP 2 Add to find the total minutes from 10:05 to 10:30.

From 10:05 A.M. to ___________ is _____ minutes.

So, the elapsed time is _____ minutes.

B Use subtraction.

Think: The hours are the same, so subtract the minutes.

10 : 30
− 10 : 05
___

MATHEMATICAL PRACTICES

Math Talk Which method do you prefer to use to find elapsed time? **Explain.**

Name ______________________

Share and Show 

1. Use the number line to find the elapsed time from 1:15 P.M. to 1:40 P.M. __________

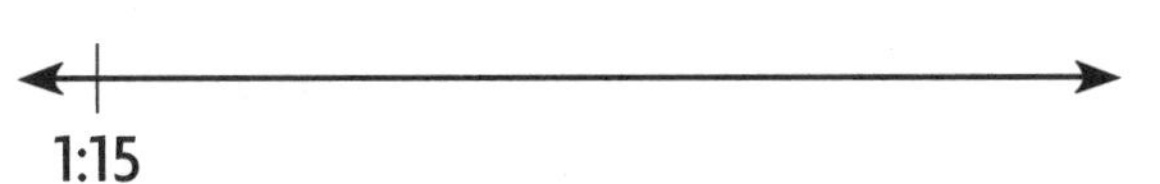

Find the elapsed time.

2. Start: 11:35 A.M. End: 11:54 A.M.

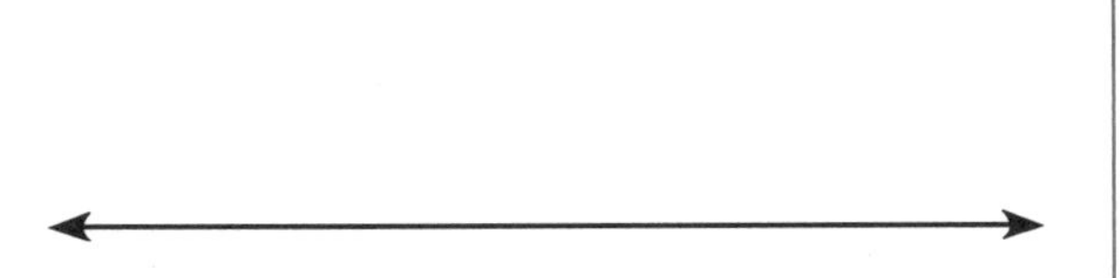

____

3. Start: 4:20 P.M. End: 5:00 P.M.

____

MATHEMATICAL PRACTICES

Math Talk **Explain** how to use a number line to find the elapsed time from 11:10 A.M. until noon.

On Your Own

Find the elapsed time.

4. Start: 8:35 P.M. End: 8:55 P.M.

____

5. Start: 10:10 A.M. End: 10:41 A.M.

____

6. Start: 9:25 A.M. End: 9:43 A.M.

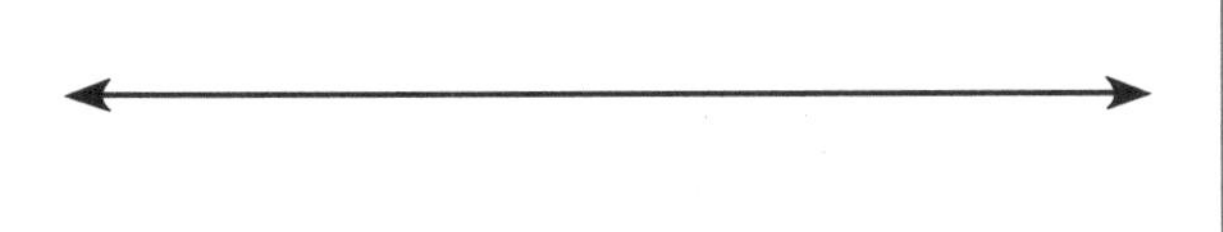

____

7. Start: 2:15 P.M. End: 2:52 P.M.

____

Problem Solving REAL WORLD

8. John started reading his book about outer space at quarter after nine in the morning. He read until quarter to ten in the morning. How long did John read his book?

SHOW YOUR WORK

9. **What's the Question?** Tim and Alicia arrived at the rocket display at 3:40 P.M. Alicia left the display at 3:56 P.M. Tim left at 3:49 P.M. The answer is Alicia.

10. H.O.T. At the space center, Karen bought a model of a shuttle. She started working on the model the next day at 11:13 A.M. She worked until leaving for lunch at 11:51 A.M. After lunch, she worked on the model again from 1:29 P.M. until 1:48 P.M. How long did Karen work on the model?

11. **Write Math** Tim's family toured the Astronaut Hall of Fame from 2:05 P.M. to 2:45 P.M. **Explain** how you know that the tour was more than 30 minutes.

12. **Test Prep** Kira got on the tour bus at 5:15 P.M. She got off the bus at 5:37 P.M. How long was Kira on the bus?

Ⓐ 15 minutes Ⓒ 37 minutes

Ⓑ 22 minutes Ⓓ 52 minutes

FOR MORE PRACTICE: Standards Practice Book, pp. P197–P198

Name ______________________________

Use Time Intervals

Essential Question How can you find a starting time or an ending time when you know the elapsed time?

UNLOCK the Problem REAL WORLD

Al begins working on his oceans project at 1:30 P.M. He spends 42 minutes painting a model of Earth and labeling the oceans. At what time does Al finish working on his project?

- What time is given?

____

- What time do you need to find?

____

One Way Use a number line to find the ending time.

STEP 1 Find the time on the number line when Al started working on the project.

STEP 2 Count forward on the number line to add the elapsed time. Draw and label the jumps to show the minutes.

Think: I can break apart 42 minutes into shorter amounts of time.

STEP 3 Write the times below the number line.

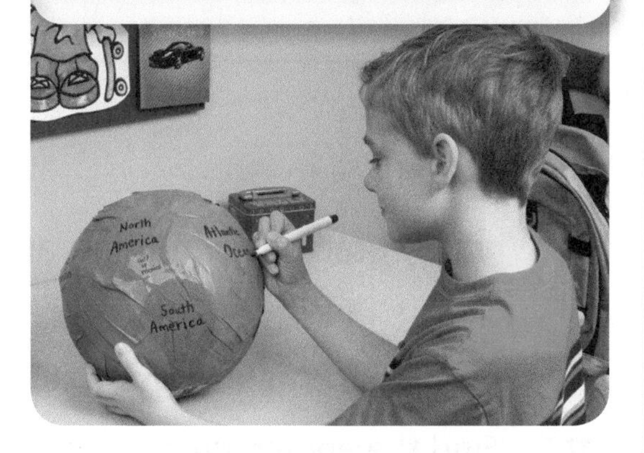

MATHEMATICAL PRACTICES **Math Talk** Explain how you decided what size jumps to make on the number line.

The jumps end at __________

So, Al finishes working on his project at __________

Another Way Use a clock to find the ending time.

STEP 1 Find the starting time on the clock.

STEP 2 Count on by fives and ones for the elapsed time of 42 minutes. Write the missing counting numbers next to the clock.

So, the ending time is __________

Find Starting Times

Whitney went swimming in the ocean for 25 minutes. She finished swimming at 11:15 A.M. At what time did Whitney start swimming?

One Way **Use a number line to find the starting time.**

STEP 1 Find the time on the number line when Whitney finished swimming in the ocean.

STEP 2 Count back on the number line to subtract the elapsed time. Draw and label the jumps to show the minutes.

STEP 3 Write the times below the number line.

11:15 A.M.

MATHEMATICAL PRACTICES

Math Talk **Explain** how the problem on this page is different from the problem on page 401.

You jumped back to ___________

So, Whitney started swimming at ___________

Another Way **Use a clock to find the starting time.**

STEP 1 Find the ending time on the clock.

STEP 2 Count back by fives for the elapsed time of 25 minutes. Write the missing counting numbers next to the clock.

So, the starting time is ___________

Share and Show

1. Use the number line to find the starting time if the elapsed time is 35 minutes. ___________

5:10 P.M.

MATHEMATICAL PRACTICES

Math Talk **Explain** how to find the starting time when you know the ending time and the elapsed time.

Name ____________________

Find the ending time.

2. Starting time: 1:40 P.M.
   Elapsed time: 33 minutes

____

3. Starting time: 9:55 A.M.
   Elapsed time: 27 minutes

____

On Your Own

Find the starting time.

4. Ending time: 3:05 P.M.
   Elapsed time: 40 minutes

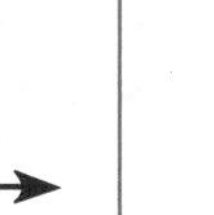

____

5. Ending time: 8:06 A.M.
   Elapsed time: 16 minutes

____

Find the ending time.

6. Starting time: 10:20 A.M.
   Elapsed time: 56 minutes

____

7. Starting time: 5:54 P.M.
   Elapsed time: 15 minutes

____

H.O.T. **Algebra** Complete the table.

	Starting Time	Ending Time	Elapsed Time
8.	3:13 P.M.		65 minutes
9.		12:17 P.M.	35 minutes

Problem Solving REAL WORLD

10. Write Math ▸ Suzi began fishing at 10:30 A.M. and fished until 11:10 A.M. James finished fishing at 11:45 A.M. He fished for the same length of time as Suzi. At what time did James start fishing? **Explain.**

____

____

____

11. **Test Prep** Dante's surfing lesson began at 2:35 P.M. His lesson lasted 45 minutes. At what time did Dante's lesson end?

Ⓐ 1:50 P.M.

Ⓑ 2:45 P.M.

Ⓒ 3:20 P.M.

Ⓓ 3:45 P.M.

Connect to Science

Tides

If you have ever been to the beach, you have seen the water rise and fall along the shore every day. This change in water level is called the tide. Ocean tides are mostly caused by the pull of the moon and the sun's gravity. High tide is when the water is at its highest level. Low tide is when the water is at its lowest level. In most places on Earth, high tide and low tide each occur about twice a day.

Use the table for 12–13.

Tide Times Atlantic City, NJ

	Low Tide	High Tide
Day 1	2:12 A.M.	9:00 A.M.
	2:54 P.M.	9:00 P.M.
Day 2	3:06 A.M.	9:36 A.M.
	3:36 P.M.	9:54 P.M.
Day 3	4:00 A.M.	10:12 A.M.
	4:30 P.M.	10:36 P.M.

12. The first morning, Courtney walked on the beach for 20 minutes. She finished her walk 30 minutes before high tide. At what time did Courtney start her walk?

____

13. The third afternoon, Courtney started collecting shells at low tide. She collected shells for 35 minutes. At what time did Courtney finish collecting shells?

____

FOR MORE PRACTICE:
Standards Practice Book, pp. P199–P200

Name ____________________________

Problem Solving • Time Intervals

Essential Question How can you use the strategy *draw a diagram* to solve problems about time?

UNLOCK the Problem REAL WORLD

Zach and his family are going to New York City. Their airplane leaves at 9:15 A.M. They need to arrive at the airport 60 minutes before their flight. It takes 15 minutes to get to the airport. The family needs 30 minutes to get ready to leave. At what time should Zach's family start getting ready?

Read the Problem

What do I need to find?	What information do I need to use?	How will I use the information?
I need to find what ________ Zach's family should start ________.	the time the ________ leaves; the time the family needs to arrive at the ________; the time it takes to get to the ________; and the time the family needs to ________	I will use a number line to find the answer.

Solve the Problem

- Find 9:15 A.M. on the number line. Draw the jumps to show the time.
- Count back ____ minutes for the time they need to arrive at the airport.
- Count back ____ minutes for the time to get to the airport.
- Count back ____ minutes for the time to get ready.

9:15 A.M.

So, Zach's family should start getting ready at _____ _____.M.

MATHEMATICAL PRACTICES

Math Talk How can you check your answer by starting with the time the family starts getting ready?

Try Another Problem

Bradley gets out of school at 2:45 P.M. It takes him 10 minutes to walk home. Then he spends 10 minutes eating a snack. He spends 8 minutes putting on his soccer uniform. It takes 20 minutes for Bradley's father to drive him to soccer practice. At what time does Bradley arrive at soccer practice?

Read the Problem

What do I need to find?	What information do I need to use?	How will I use the information?

Solve the Problem

Draw a diagram to help you explain your answer.

1. At what time does Bradley arrive at soccer practice? ____________

2. How do you know your answer is reasonable? ____________

MATHEMATICAL PRACTICES

Math Talk Do you need to draw jumps on the number line in the same order as the times in the problem? **Explain.**

Name ______________________________

Share and Show 

Tips

UNLOCK the Problem

- ✓ Circle the question.
- ✓ Underline important facts.
- ✓ Choose a strategy you know.

SHOW YOUR WORK

1. Patty went to the shopping mall at 11:30 A.M. She shopped for 25 minutes. She spent 40 minutes eating lunch. Then she met a friend at a movie. At what time did Patty meet her friend?

First, begin with ________ on the number line.

Then, count forward ______________.

Next, count forward ______________.

Think: I can break apart the times into shorter amounts of time that make sense.

11:30 A.M.

So, Patty met her friend at _____ _____M.

2. H.O.T. **What if** Patty goes to the mall at 11:30 A.M. and meets a friend at a movie at 1:15 P.M.? Patty wants to shop and have 45 minutes for lunch before meeting her friend. How much time can Patty spend shopping?

____

3. When Caleb got home from school, he worked on his science project for 20 minutes. Then he studied for a test for 30 minutes. He finished at 4:35 P.M. At what time did Caleb get home from school?

____

4. Avery got on the bus at 1:10 P.M. The trip took 90 minutes. Then she walked for 32 minutes to get home. At what time did Avery arrive at home?

____

On Your Own

5. Olivia modeled her number with 3 hundreds blocks and 9 tens blocks. Todd modeled his number with 2 hundreds, 6 tens, and 5 ones blocks. Who has the greater number? How much greater?

6. H.O.T. Kyle and Josh have a total of 64 CDs. Kyle has 12 more CDs than Josh. How many CDs does each boy have?

7. Jamal spent 60 minutes using the computer to play games and do research for his report. He spent a half hour of the time playing games and the rest of the time researching his report. How many minutes did Jamal spend researching his report?

8. **Write Math** Miguel played video games each day for a week. On Monday, he scored 83 points. His score went up 5 points each day. On what day did Miguel score 103 points? **Explain** how you found your answer.

9. **Test Prep** When Laura arrived at the library, she spent 40 minutes reading a book. Then she spent 15 minutes reading a magazine. She left the library at 4:15 P.M. At what time did Laura arrive at the library?

Ⓐ 3:20 P.M.

Ⓑ 3:35 P.M.

Ⓒ 4:00 P.M.

Ⓓ 5:10 P.M.

Choose a STRATEGY

- Act It Out
- Draw a Diagram
- Find a Pattern
- Make a Table

SHOW YOUR WORK

FOR MORE PRACTICE:
Standards Practice Book, pp. P201–P202

Name ______________________________

Mid-Chapter Checkpoint

Vocabulary

Choose the best term from the box.

Vocabulary
A.M.
minute
P.M.

1. In one ____________, the minute hand moves from one mark to the next on a clock. (p. 389)
2. The times after noon and before midnight are written with ____________. (p. 394)

Concepts and Skills

Write the time for the activity. Use A.M. or P.M.

3. play ball

____

4. eat breakfast

____

5. do homework

____

6. sleep

____

Find the elapsed time.

7. Start: 10:05 A.M. End: 10:50 A.M.

10:05

____

8. Start: 5:30 P.M.
   End: 5:49 P.M.

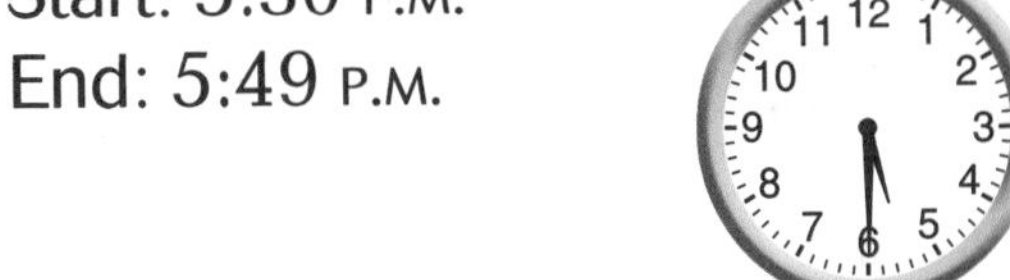

____

Find the starting time or the ending time.

9. Starting time: ____________
   Elapsed time: 50 minutes
   Ending time: 9:05 A.M.

9:05 A.M.

10. Starting time: 2:46 P.M.
    Elapsed time: 15 minutes
    Ending time: ____________

Fill in the bubble for the correct answer choice.

11. Lori started walking to school at 7:45 A.M. She arrived at school 23 minutes later. At what time did Lori arrive at school?

Ⓐ 8:03 A.M.
Ⓑ 8:05 A.M.
Ⓒ 8:08 A.M.
Ⓓ 8:23 A.M.

12. The third-grade art class ends at 3 minutes before 2:00. Which clock shows the time the art class ends?

Ⓒ

Ⓑ

Ⓓ

13. Matt went to his friend's house. He arrived at 5:10 P.M. He left at 5:37 P.M. How long was Matt at his friend's house?

Ⓐ 10 minutes
Ⓑ 27 minutes
Ⓒ 37 minutes
Ⓓ 47 minutes

14. Ruthie's train leaves at 7:30 A.M. She needs to arrive 10 minutes early to buy her ticket. It takes her 20 minutes to get to the train station. At what time should Ruthie leave her house?

Ⓐ 7:00 A.M.
Ⓑ 7:40 A.M.
Ⓒ 8:00 A.M.
Ⓓ 8:30 A.M.

15. At which time are most third graders in school?

Ⓐ 10:12 A.M.
Ⓑ 4:41 P.M.
Ⓒ 7:34 P.M.
Ⓓ 2:03 A.M.

Name ______________________________

Measure Length

Essential Question How can you generate measurement data and show the data on a line plot?

CONNECT You have learned how to measure length to the nearest inch. Sometimes the length of an object is not a whole unit. For example, a paper clip is more than 1 inch but less than 2 inches.

You can measure length to the nearest half inch or fourth inch. The half-inch markings on a ruler divide each inch into two equal parts. The fourth-inch markings divide each inch into four equal parts.

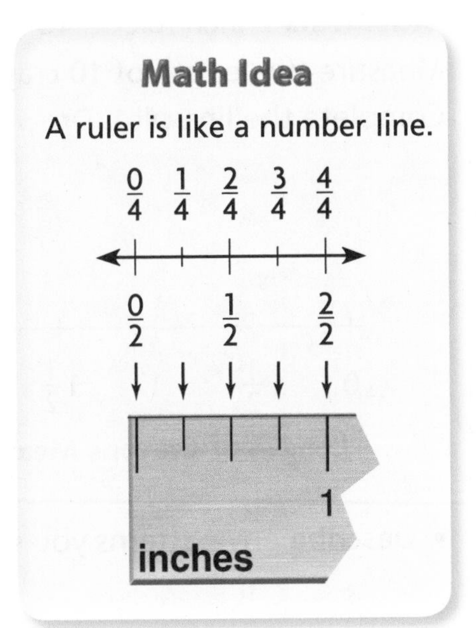

UNLOCK the Problem REAL WORLD

Example 1 **Use a ruler to measure the glue stick to the nearest half inch.**

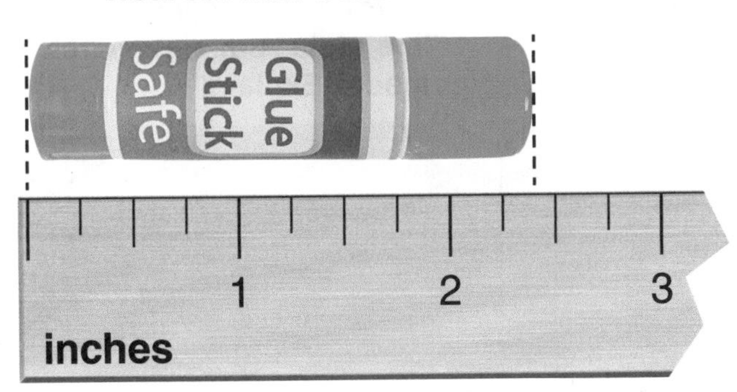

- Line up the left end of the glue stick with the zero mark on the ruler.
- The right end of the glue stick is between the half-inch marks for

  ______ and ______.
- The mark that is closest to the right end of the glue stick is for ______ inches.

So, the length of the glue stick to the nearest half inch is ______ inches.

Example 2 **Use a ruler to measure the paper clip to the nearest fourth inch.**

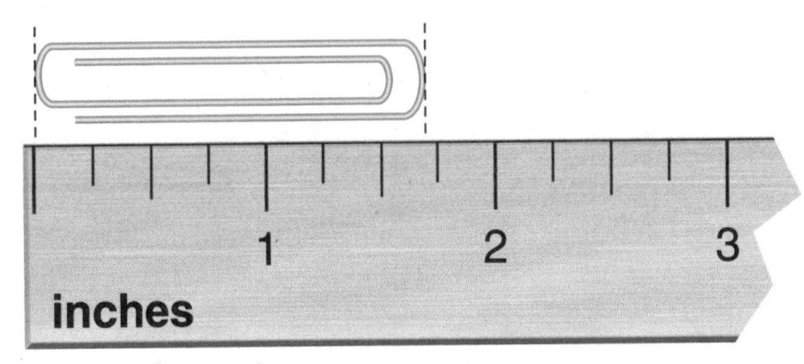

- Line up the left end of the paper clip with the zero mark on the ruler.
- The right end of the paper clip is between the fourth-inch marks for

  ______ and ______.
- The mark that is closest to the right end of the paper clip is for ______ inches.

So, the length of the paper clip to the nearest fourth inch is ______ inches.

Activity Make a line plot to show measurement data.

Materials ■ inch ruler ■ 10 crayons

Measure the length of 10 crayons to the nearest half inch.
Complete the line plot. Draw an ✗ for each length.

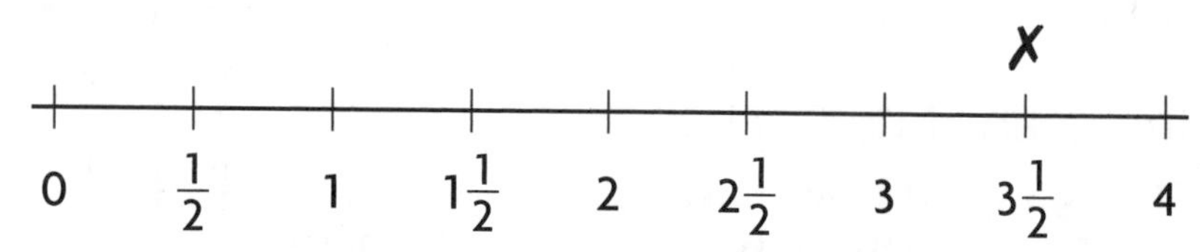

Length of Crayons Measured to the Nearest Half Inch

- **Describe** any patterns you see in your line plot.

____

____

Try This! **Measure the length of your fingers to the nearest fourth inch. Complete the line plot. Draw an ✗ for each length.**

MATHEMATICAL PRACTICES

Math Talk How do you think your line plot compares to line plots your classmates made? **Explain.**

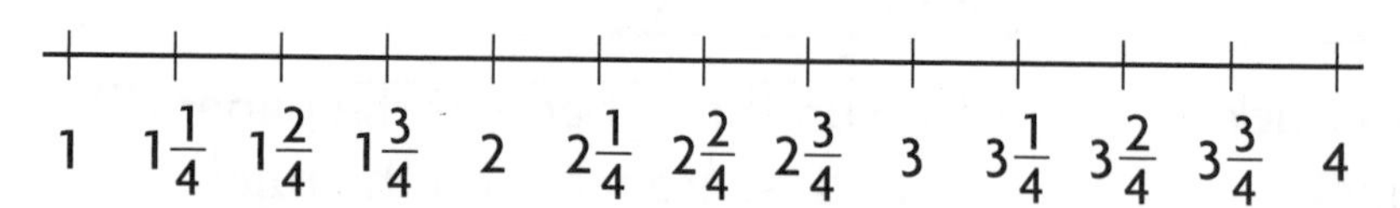

Length of Fingers Measured to the Nearest Fourth Inch

Share and Show

1. Measure the length to the nearest half inch. Is the key closest to $1\frac{1}{2}$ inches, 2 inches, or $2\frac{1}{2}$ inches?

   ______ inches

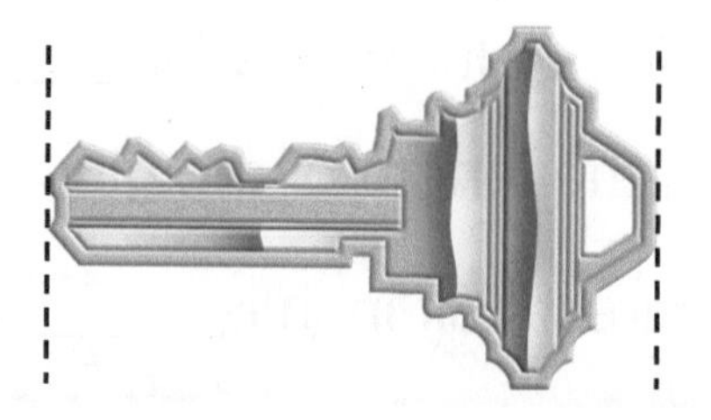

Name ______________________

Measure the length to the nearest half inch.

2. ____ inches

Measure the length to the nearest fourth inch.

3. ____ inches

MATHEMATICAL PRACTICES

Math Talk **Explain** why you would want to measure to the nearest half inch or fourth inch.

On Your Own

Measure the length to the nearest half inch.

4. ____ inches

5. ____ inches

Measure the length to the nearest fourth inch.

6. ____ inches

7. ____ inches

Practice: Copy and Solve **Use the lines for 8–9.**

8. Measure the length of the lines to the nearest half inch and make a line plot.

9. Measure the length of the lines to the nearest fourth inch and make a line plot.

Problem Solving REAL WORLD

SHOW YOUR WORK

10. Tara's hand is $4\frac{3}{4}$ inches long. This length is between which two inch-marks on a ruler?

____

11. H.O.T. **What's the Error?** Joni says that this piece of ribbon is $3\frac{1}{2}$ inches long. Describe her error. How long is the ribbon?

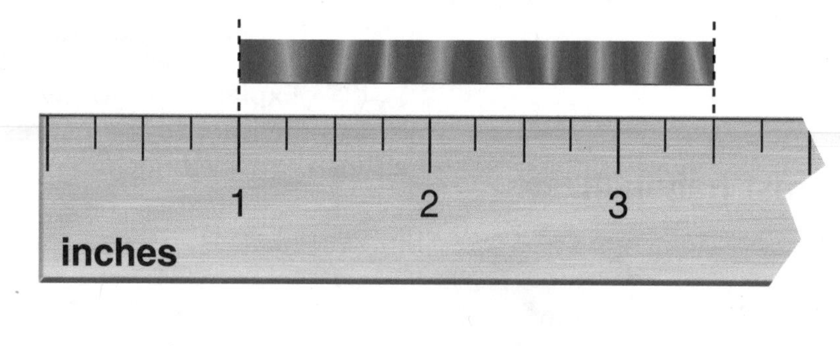

____

____

____

____

12. Find two objects in your desk that are different sizes. Measure the length of each object to the nearest fourth inch. Then use < or > to compare the lengths.

Object _______________ Length _______________

Object _______________ Length _______________

_______________ ◯ _______________

13. **Test Prep** What is the length of the pencil to the nearest half inch?

Ⓐ $3\frac{1}{2}$ inches

Ⓑ 4 inches

Ⓒ $5\frac{1}{2}$ inches

Ⓓ 6 inches

FOR MORE PRACTICE:
Standards Practice Book, pp. P203–P204

Name ______________________

Estimate and Measure Liquid Volume

Essential Question How can you estimate and measure liquid volume in metric units?

REAL WORLD

Liquid volume is the amount of liquid in a container. The **liter (L)** is the basic metric unit for measuring liquid volume.

Activity 1

Materials ■ 1-L beaker ■ 4 containers ■ water ■ tape

STEP 1 Fill a 1-liter beaker with water to the 1-liter mark.

STEP 2 Pour 1 liter of water into a container. Mark the level of the water with a piece of tape. Draw the container below and name the container.

STEP 3 Repeat Steps 1 and 2 with three different-sized containers.

Container 1	Container 2
______________	______________
Container 3	**Container 4**
______________	______________

MATHEMATICAL PRACTICES

Math Talk What can you say about the amount of liquid volume in each container?

1. How much water did you pour into each container? ______________

2. Which containers are mostly full? **Describe** them.

____

3. Which containers are mostly empty? **Describe** them.

____

Compare Liquid Volumes

A full glass holds less than 1 liter.

A water bottle holds about 1 liter.

A fish bowl holds more than 1 liter.

Activity 2 **Materials** ■ 1-L beaker ■ 5 different containers ■ water

STEP 1 Write the containers in order from the one you think will hold the least water to the one you think will hold the most water.

_______________, _______________, _______________,

_______________, _______________

STEP 2 Estimate how much each container will hold. Write *more than 1 liter*, *about 1 liter*, or *less than 1 liter* in the table.

STEP 3 Pour 1 liter of water into one of the containers. Repeat until the container is full. Record the number of liters you poured. Repeat for each container.

Container	Estimate	Number of Liters

STEP 4 Write the containers in order from the least to the greatest liquid volume.

_______________, _______________, _______________,

_______________, _______________

MATHEMATICAL PRACTICES

Math Talk Was the order in Step 1 different than the order in Step 4? **Explain** why they may be different.

Name ____________________

Share and Show 

1. The beaker is filled with water. Is the amount *more than 1 liter, about 1 liter,* or *less than 1 liter*?

____

Estimate how much liquid volume there will be when the container is filled. Write *more than 1 liter, about 1 liter,* or *less than 1 liter.*

2. cup of tea

____

3. kitchen sink

____

4. teapot

____

MATHEMATICAL PRACTICES

Math Talk Explain how you could estimate the liquid volume in a container.

On Your Own

Estimate how much liquid volume there will be when the container is filled. Write *more than 1 liter, about 1 liter,* or *less than 1 liter.*

5. pitcher

____

6. juice box

____

7. punch bowl

____

Use the pictures for 8–10. Ginger pours punch into four bottles that are the same size.

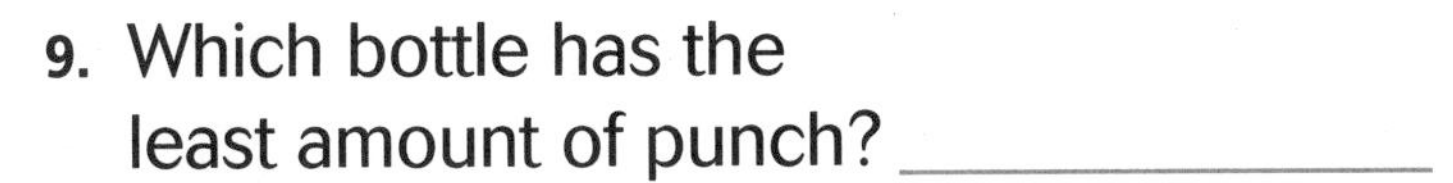

8. Did Ginger pour the same amount into each bottle? ____________________

9. Which bottle has the least amount of punch? ____________________

10. Which bottle has the most punch? ____________________

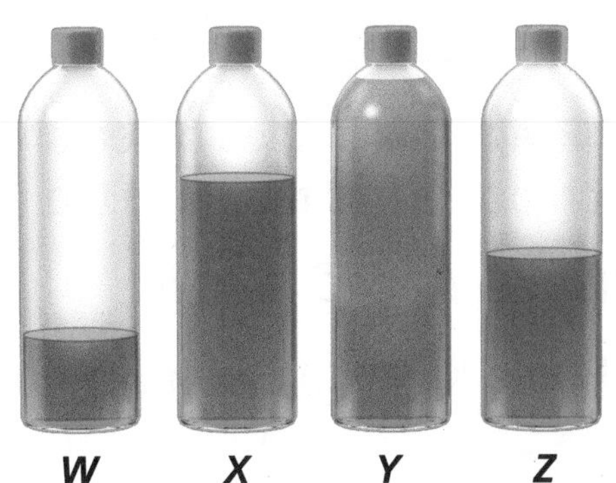

Problem Solving REAL WORLD

H.O.T. **Use the containers for 11–14. Container *A* is full when 1 liter of water is poured into it.**

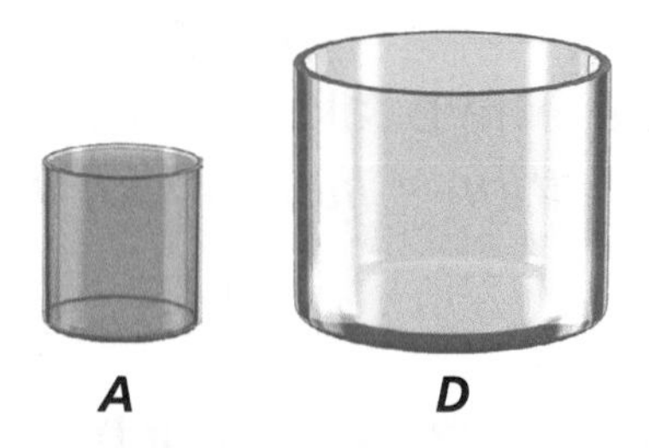

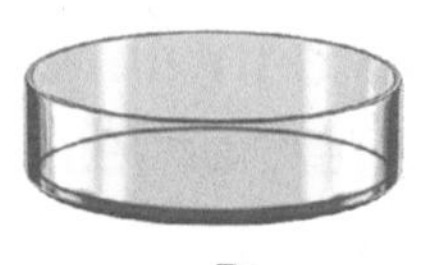

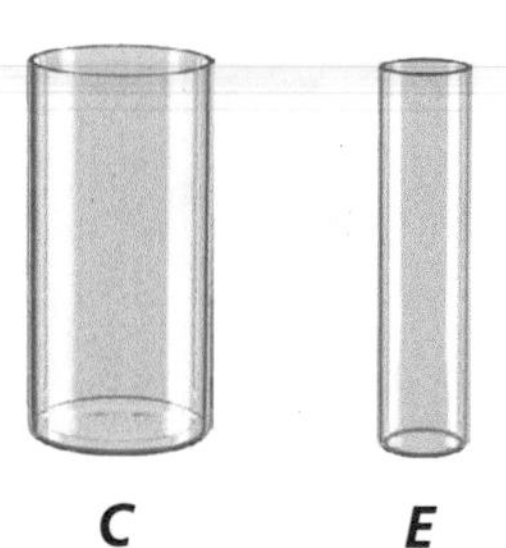

11. Write Math ▸ **What if** you poured 1 liter of water into Container *B*? Describe the way the water fills the container. **Explain** how you know.

____

____

____

12. Estimate how many liters will fill Container *C* and how many liters will fill Container *E*. Which container will hold more water when filled?

____

____

13. Name two containers that will be filled with about the same number of liters of water. **Explain**.

____

____

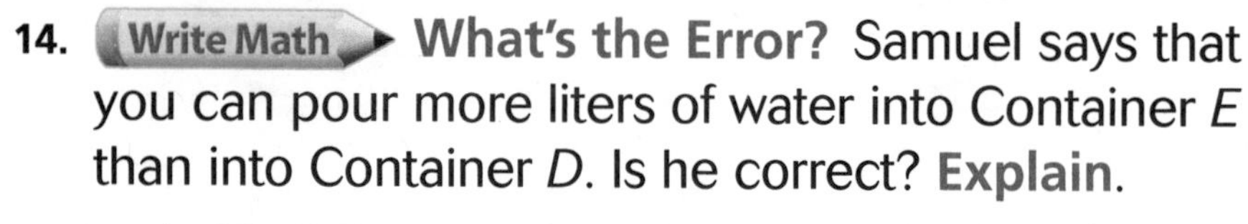

14. Write Math ▸ **What's the Error?** Samuel says that you can pour more liters of water into Container *E* than into Container *D*. Is he correct? **Explain**.

____

____

____

15. **Test Prep** The bottles of tea are all the same size. Which bottle has the most tea in it?

Ⓐ Bottle *J* Ⓒ Bottle *L*

Ⓑ Bottle *K* Ⓓ Bottle *M*

FOR MORE PRACTICE:
Standards Practice Book, pp. P205–P206

Name ______________________________

Estimate and Measure Mass

Essential Question How can you estimate and measure mass in metric units?

UNLOCK the Problem REAL WORLD

Peter has a dollar bill in his pocket. Should Peter measure the mass of the dollar bill in grams or kilograms?

The **gram (g)** is the basic metric unit for measuring **mass**, or the amount of matter in an object. Mass can also be measured by using the metric unit **kilogram (kg)**.

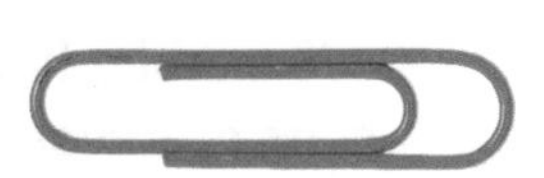

A small paper clip has a mass of about 1 gram.

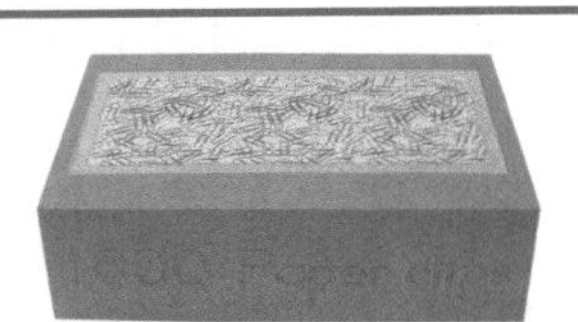

A box of 1,000 paper clips has a mass of about 1 kilogram.

Think: The mass of a dollar bill is closer to the mass of a small paper clip than it is to a box of 1,000 paper clips.

So, Peter should measure the mass of the dollar bill in __________.

Activity 1

Materials ■ pan balance ■ gram and kilogram masses

You can use a pan balance to measure mass.

Do 10 grams have the same mass as 1 kilogram?

- Place 10 gram masses on one side of the balance.
- Place a 1-kilogram mass on the other side of the balance.

Think: If it is balanced, then the objects have the same mass. If it is not balanced, the objects do not have the same mass.

- Complete the picture of the balance above by drawing masses to show your balance.

  The pan balance is ______________.

MATHEMATICAL PRACTICES

Math Talk Which has a greater mass, 10 grams or 1 kilogram? **Explain.**

So, 10 grams and 1 kilogram ______________ the same mass.

Activity 2

Materials ■ pan balance ■ gram and kilogram masses ■ classroom objects

STEP 1 Use the objects in the table. Decide if the object should be measured in grams or kilograms.

STEP 2 Estimate the mass of each object. Record your estimates in the table.

STEP 3 Find the mass of each object to the nearest gram or kilogram. Place the object on one side of the balance. Place gram or kilogram masses on the other side until both sides are balanced.

STEP 4 Add the measures of the gram or kilogram masses. This is the mass of the object. Record the mass in the table.

▲ 189 marbles have a mass of 1 kilogram.

Mass		
Object	**Estimate**	**Mass**
crayon		
stapler		
eraser		
marker		
small notepad		
scissors		

MATHEMATICAL PRACTICES

Math Talk How did your estimates compare with the actual measurements?

- Write the objects in order from greatest mass to least mass.

____________, ____________, ____________,

____________, ____________, ____________

Share and Show

1. Five bananas have a mass of about __________.

Think: The pan balance is balanced, so the objects on both sides have the same mass.

Name ____________________

Choose the unit you would use to measure the mass. Write *gram* or *kilogram*.

2. strawberry

____

3. dog

____

MATHEMATICAL PRACTICES

Math Talk **Explain** how you decided which unit to use to measure mass.

Compare the masses of the objects. Write *is less than, is the same as,* or *is more than.*

4.

The mass of the bowling pin ____________ the mass of the chess piece.

5.

The mass of the erasers ____________ the mass of the clips.

On Your Own

Choose the unit you would use to measure the mass. Write *gram* or *kilogram*.

6. chair

____

7. sunglasses

____

8. watermelon

____

Compare the masses of the objects. Write *is less than, is the same as,* or *is more than.*

9.

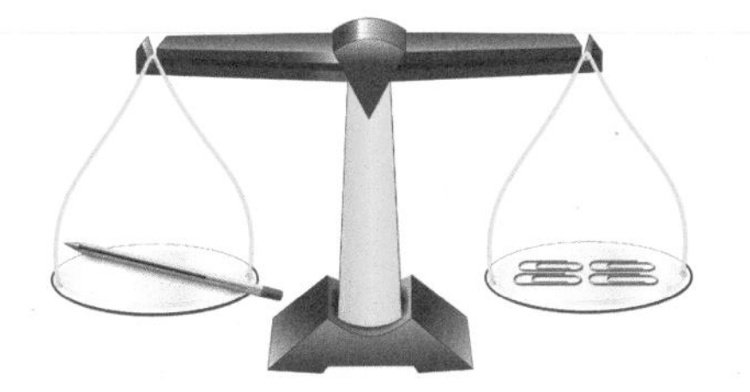

The mass of the pen ____________ the mass of the paper clips.

10.

The mass of the straws ____________ the mass of the blocks.

Problem Solving REAL WORLD

11. Put the sports balls shown at the right in order from greatest mass to least mass.

Golf ball

Table tennis ball

Bowling ball

Baseball

Tennis ball

12. Choose two objects that have about the same mass. Draw a balance with one of these objects on each side.

13. Choose two objects that have different masses. Draw a balance with one of these objects on each side.

14. H.O.T. **Pose a Problem** Write a problem about the objects you chose in Exercise 13. Then solve your problem.

SHOW YOUR WORK

15. Write Math **Sense or Nonsense?** Amber is buying produce at the grocery store. She says that a Fuji apple and a green bell pepper would have the same mass because they are the same size. Does her statement make sense? **Explain.**

16. **Test Prep** Dan wants to find the mass of a large pumpkin. Which unit should he use?

Ⓐ inch Ⓒ kilogram

Ⓑ gram Ⓓ liter

FOR MORE PRACTICE:
Standards Practice Book, pp. P207–P208

Name ____________________

Solve Problems About Liquid Volume and Mass

Essential Question How can you use models to solve liquid volume and mass problems?

UNLOCK the Problem REAL WORLD

A restaurant serves iced tea from a large container that can hold 24 liters. Sadie will fill the container with the pitchers of tea shown below. Will Sadie have tea left over after filling the container?

Example Solve a problem about liquid volume.

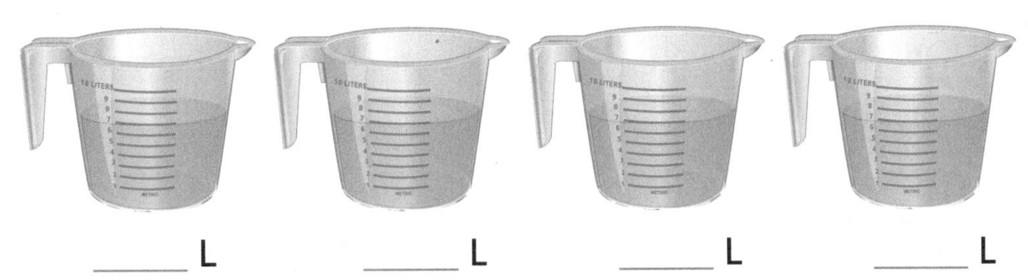

____ L ____ L ____ L ____ L

Since there are ____ equal groups of ____ liters, you can multiply.

____ ◯ ____ = ____

Circle the correct words to complete the sentences.

____ liters is *greater than* / *less than* 24 liters.

So, Sadie *will* / *will not* have tea left over.

Try This! Use a bar model to solve.

Raul's fish tank contains 32 liters of water. He empties it with a bucket that holds 4 liters of water. How many times will Raul have to fill the bucket?

____ times

4 L	- - - - - -	4 L

32 L

____ ◯ ____ = ____

So, Raul will have to fill the bucket ____ times.

Activity Solve a problem about mass.

Materials ■ pan balance ■ glue stick ■ gram masses

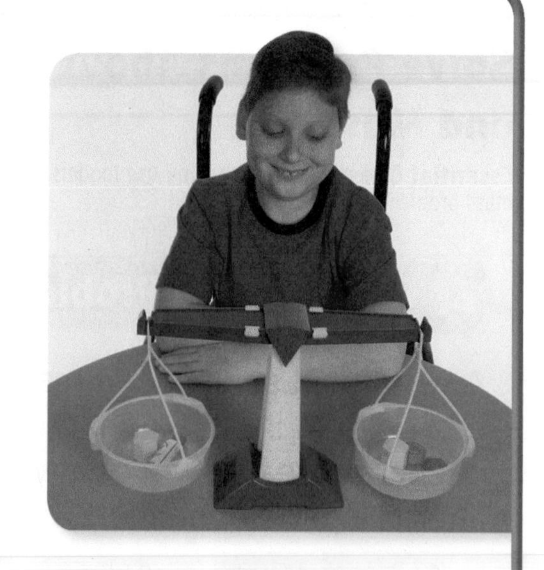

Jeff has a glue stick and a 20-gram mass on one side of a balance and gram masses on the other side. The pan balance is balanced. What is the mass of the glue stick?

STEP 1 Place a glue stick and a 20-gram mass on one side of the balance.

STEP 2 Place gram masses on the other side until the pans are balanced.

STEP 3 To find the mass of the glue stick, remove 20 grams from each side.

Think: I can remove 20 grams from both sides and the pan balance will still be balanced.

STEP 4 Then add the measures of the gram masses on the balance.

The gram masses have a measure of _____ grams.

So, the glue stick has a mass of __________.

MATHEMATICAL PRACTICES

Math Talk What equation can you write to find the mass of the glue stick? **Explain.**

Try This! **Use a bar model to solve.**

A bag of peas has a mass of 432 grams.
A bag of carrots has a mass of 263 grams.
What is the total mass of both bags?

_____ g	_____ g

_____ g

_____ ◯ _____ = _____

So, both bags have a total mass of __________ grams.

Share and Show MATH BOARD

1. Ed's Delivery Service delivered three packages to Ms. Wilson. The packages have masses of 9 kilograms, 12 kilograms, and 5 kilograms. What is the total mass of the three packages? Use the bar model to help you solve.

__ kg	_____ kg	__ kg

_____ kg

____

Name ______________________________

Write an equation and solve the problem.

2. Ariel's recipe calls for 64 grams of apples and 86 grams of oranges. How many more grams of oranges than apples does the recipe call for?

   _____ ◯ _____ = _____ _________

3. Dan's Clams restaurant sold 45 liters of lemonade. If it sold the same amount each hour for 9 hours, how many liters of lemonade did Dan's Clams sell each hour?

   _____ ◯ _____ = _____ _________

MATHEMATICAL PRACTICES

Math Talk **Explain** how you could model Exercise 2.

On Your Own

Write an equation and solve the problem.

4. Sara's box holds 4 kilograms of napkins and 29 kilograms of napkin rings. What is the total mass of the napkins and napkin rings?

   _____ ◯ _____ = _____ _________

5. Josh has 6 buckets for cleaning a restaurant. He fills each bucket with 4 liters of water. How many liters of water are in the buckets?

   _____ ◯ _____ = _____ _________

6. H.O.T. **Write Math** Ellen will pour water into Pitcher *B* until it has 1 more liter of water than Pitcher *A*. How many liters of water will she pour into Pitcher *B*? **Explain** how you found your answer.

____

____

7. **Practice: Copy and Solve** **Use the pictures to write two problems. Then solve your problems.**

UNLOCK the Problem REAL WORLD

8. Ken's Café serves hot cocoa. Each serving has 9 grams of chocolate. How many grams of chocolate are in 8 servings of hot cocoa?

Ⓐ 1 gram
Ⓑ 17 grams
Ⓒ 72 grams
Ⓓ 89 grams

a. What do you need to find? ______________________________

____

b. What operation will you use to find the answer? ______________

c. Draw a diagram to solve the problem.

d. Complete the sentences.

There are _____ servings with _____ grams of chocolate in each.

Since each serving is an __________ group, you can __________.

_____ ◯ _____ = _____

So, there are _____ grams of chocolate in 8 servings of hot cocoa.

e. Fill in the bubble for the correct answer choice above.

9. Alison has a container filled with 12 liters of water. Daniel has a container filled with 16 liters of water. What is the total liquid volume of the containers?

Ⓐ 4 liters
Ⓑ 24 liters
Ⓒ 28 liters
Ⓓ 32 liters

10. A jar holds 21 grams of salad dressing. Sam pours 3 grams of salad dressing on each salad plate. If Sam uses all the dressing, how many salad plates are there?

Ⓐ 24
Ⓑ 18
Ⓒ 8
Ⓓ 7

FOR MORE PRACTICE:
Standards Practice Book, pp. P209–P210

Name ______________________

Chapter Review/Test

▶ Vocabulary

Choose the best term from the box.

Vocabulary
kilograms
liquid volume
liters

1. ______________ is the amount of liquid in a container. (p. 415)
2. Mass can be measured in units such as grams and ______________. (p. 419)

▶ Concepts and Skills

Write the time for the activity. Use A.M. or P.M.

3. eat lunch

____

4. play outside

____

5. watch a sunset

____

6. have math class

____

Estimate how much liquid volume there will be when the container is filled. Write *more than 1 liter*, *about 1 liter*, or *less than 1 liter*.

7. bathtub

____

8. shampoo bottle

____

9. bottle of hand soap

____

Choose the unit you would use to measure the mass. Write *gram* or *kilogram*.

10. headphones

____

11. lamp

____

12. boots

____

Fill in the bubble for the correct answer choice.

13. Todd arrived at the soccer game at sixteen minutes before four. Which is one way to write the time?

Ⓐ 3:00 Ⓒ 4:16

Ⓑ 3:44 Ⓓ 4:44

14. Dora placed a pencil on one side of a balance. What is the mass of the pencil?

Ⓐ 1 gram

Ⓑ 6 grams

Ⓒ 1 kilogram

Ⓓ 6 kilograms

15. Which equation can you use to find the total liquid volume of the containers?

Ⓐ $4 + 3 = 7$ Ⓒ $3 \times 4 = 12$

Ⓑ $10 + 4 = 14$ Ⓓ $3 \times 10 = 30$

16. Megan is using a red crayon to draw a picture.

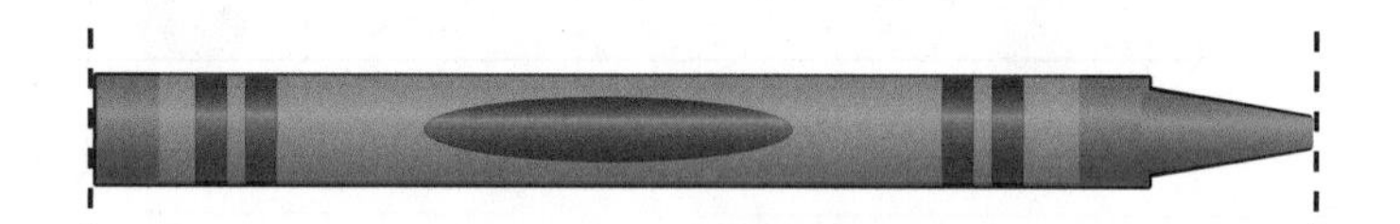

Which is closest to the length of the crayon?

Ⓐ $2\frac{1}{2}$ inches Ⓒ $3\frac{1}{2}$ inches

Ⓑ 3 inches Ⓓ 4 inches

17. Pilar began exercising at 2:15 P.M. She finished at 2:55 P.M. How long did Pilar exercise?

Ⓐ 55 minutes Ⓒ 45 minutes

Ⓑ 50 minutes Ⓓ 40 minutes

Name ______________________________

Fill in the bubble for the correct answer choice.

18. Jared built a model airplane. He started at 10:30 A.M. and finished 55 minutes later. At what time did Jared finish his model airplane?

Ⓐ 10:35 A.M. Ⓒ 11:25 A.M.

Ⓑ 10:55 A.M. Ⓓ 11:35 A.M.

19. The bottles of water are all the same size. Which bottle has the least amount of water?

Ⓐ Bottle *W*

Ⓑ Bottle *X*

Ⓒ Bottle *Y*

Ⓓ Bottle *Z*

W *X* *Y* *Z*

20. Valerie spent 33 minutes at soccer practice. She left practice at 2:15 P.M. At what time did Valerie arrive at soccer practice?

Ⓐ 1:33 P.M. Ⓒ 2:03 P.M.

Ⓑ 1:42 P.M. Ⓓ 2:48 P.M.

21. Alexa is drawing with a blue marker.

Which is closest to the length of the marker?

Ⓐ $4\frac{1}{2}$ inches Ⓒ 4 inches

Ⓑ $4\frac{1}{4}$ inches Ⓓ $3\frac{1}{4}$ inches

22. Allen started his homework at 8:10 P.M. and worked for 45 minutes. Then he phoned a friend and talked for 20 minutes. He went to bed when he finished the call. At what time did Allen go to bed?

Ⓐ 8:30 P.M. Ⓒ 9:15 P.M.

Ⓑ 8:55 P.M. Ⓓ 9:45 P.M.

▶ Constructed Response

23. Anthony's family went out to get dinner. They left at the time shown on the clock. They returned home at 5:52 P.M. How long was Anthony's family gone? **Explain** how you know.

____

____

____

▶ Performance Task

24. Mr. Carter's classroom is collecting data about the length of the students' shoes.

A Andrea's shoe measures $5\frac{1}{2}$ inches. Use a ruler. Draw a line to show the length of her shoe. **Explain** how you drew it.

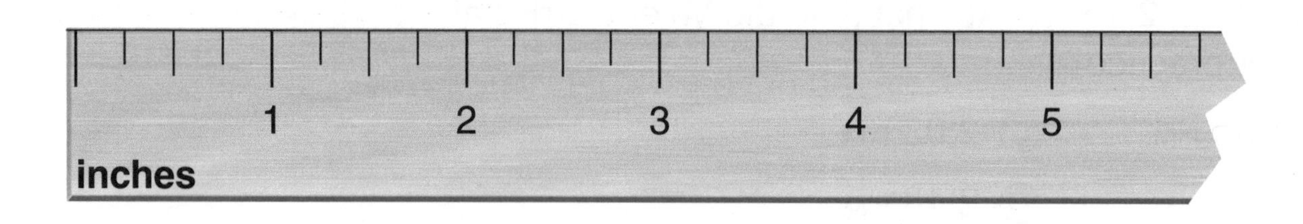

____

____

B Measure the length to the nearest half inch of the shoes of 10 students in your classroom. Complete the line plot to show the length of the shoes. Then write a sentence to **describe** what the line plot shows.

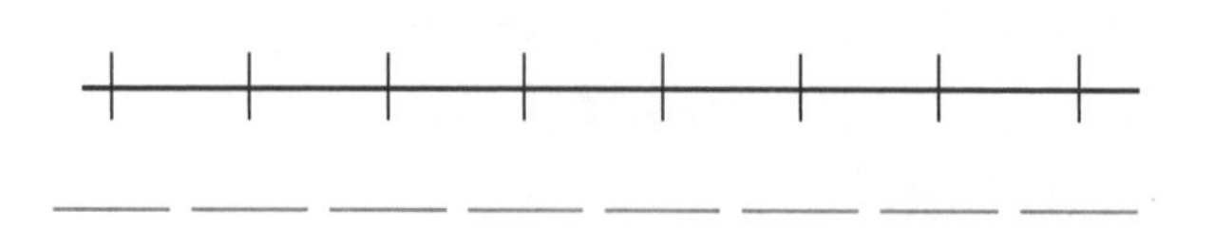

____

Chapter 11

Perimeter and Area

Show What You Know

Check your understanding of important skills.

Name ________________________________

▶ Use Nonstandard Units to Measure Length

Use paper clips to measure the object.

1.  about _____ 

2.  about _____

▶ Add 3 Numbers Write the sum.

3. $2 + 7 + 3 =$ _____

4. $3 + 5 + 2 =$ _____

5. $6 + 1 + 9 =$ _____

▶ Model with Arrays Use the array. Complete.

6.  3 rows of 4

_____ × _____ = _____

7.  4 rows of 2

_____ × _____ = _____

Julia has a picture frame with side lengths of 12 inches and 24 inches. She wants to cut and glue one color of ribbon that will fit exactly around the edge. The green ribbon is 72 inches long. The red ribbon is 48 inches long. Be a Math Detective to find which ribbon she should use to glue around the picture frame.

Vocabulary Builder

▶ Visualize It

Sort the words with a ✓ into the Venn diagram.

Perimeter | Area

Review Words

- addition
- array
- centimeter (cm)
- Distributive Property
- foot (ft)
- inch (in.)
- inverse operations
- ✓ length
- meter (m)
- multiplication
- pattern
- rectangle
- repeated addition
- ✓ unit
- ✓ width

Preview Words

- area
- perimeter
- ✓ square unit (sq un)
- ✓ unit square

▶ Understand Vocabulary

Complete the sentences by using the review and preview words.

1. The distance around a shape is the

   ______________.

2. The ______________ is the measure of the number of unit squares needed to cover a surface.

3. You can count, use ______________, or multiply to find the area of a rectangle.

4. A ______________ is a square with a side length of 1 unit and is used to measure area.

5. The ______________ shows that you can break apart a rectangle into smaller rectangles and add the area of each smaller rectangle to find the total area.

GO Online • eStudent Edition • Multimedia eGlossary

Name ______________________________

Model Perimeter

Essential Question How can you find perimeter?

Investigate

Perimeter is the distance around a shape.

Materials ■ geoboard ■ rubber bands

You can find the perimeter of a rectangle on a geoboard or on dot paper by counting the number of units on each side.

A. Make a rectangle on the geoboard that is 3 units on two sides and 2 units on the other two sides.

B. Draw your rectangle on this dot paper.

C. Write the length next to each side of your rectangle.

D. Add the number of units on each side.

_____ + _____ + _____ + _____ = _____

E. So, the perimeter of the rectangle is _____ units.

- How would the perimeter of the rectangle change if the length of two of the sides was 4 units instead of 3 units?

____

____

Draw Conclusions

1. **Describe** how you would find the perimeter of a rectangle that is 5 units wide and 6 units long.

____

2. H.O.T. A rectangle has two pairs of sides of equal length. **Explain** how you can find the unknown length of two sides when the length of one side is 4 units, and the perimeter is 14 units.

____

____

____

3. **Evaluate** Jill says that finding the perimeter of a shape with all sides of equal length is easier than finding the perimeter of other shapes. Do you agree? **Explain.**

____

____

Make Connections

You can also use grid paper to find the perimeter of shapes by counting the number of units on each side.

Start at the arrow and trace the perimeter. Begin counting with 1. Continue counting each unit around the shape until you have counted each unit.

MATHEMATICAL PRACTICES

Math Talk If a rectangle has a perimeter of 12 units, how many units wide and how many units long could it be? **Explain.**

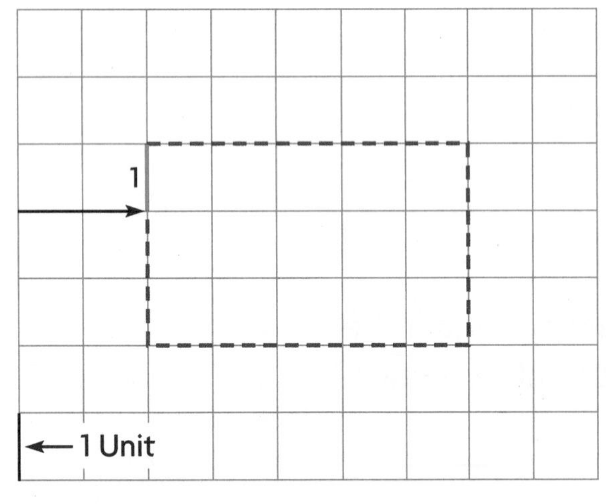

Perimeter = ______ units

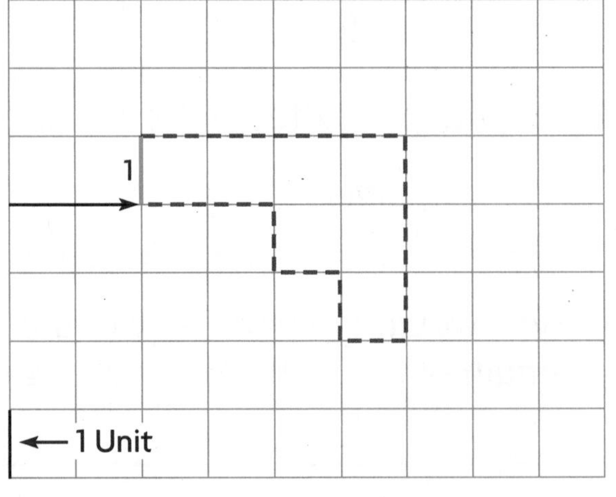

Perimeter = ______ units

Name ______________________________

Share and Show 

Find the perimeter of the shape. Each unit is 1 centimeter.

1.

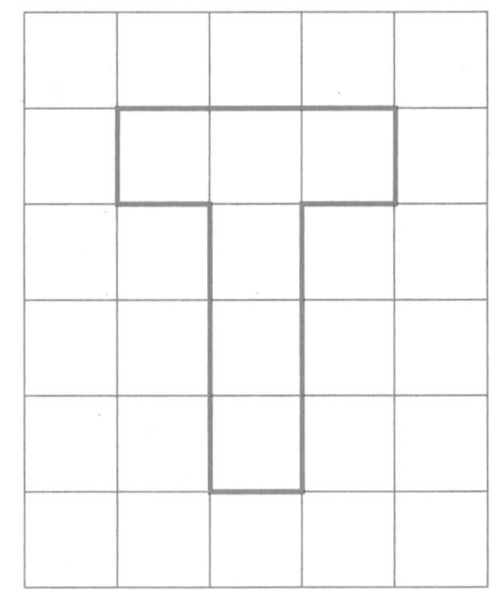

_____ centimeters

 **2.**

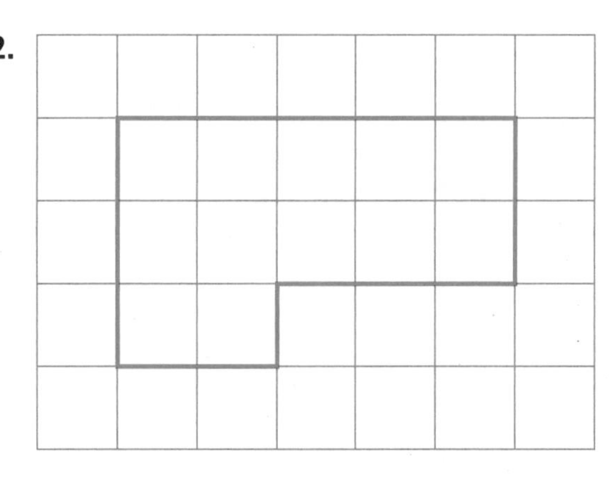

_____ centimeters

3.

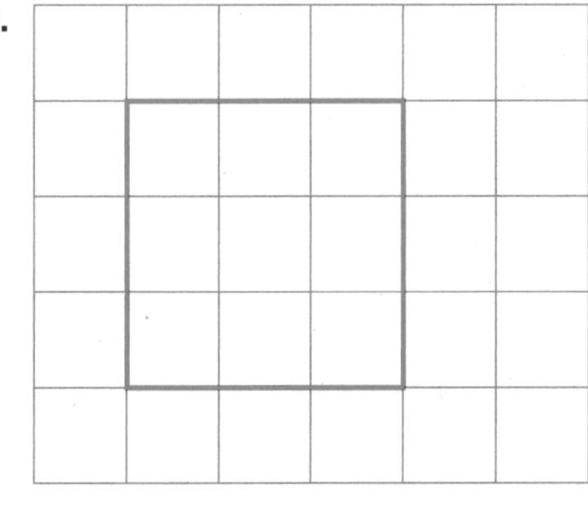

_____ centimeters

4.

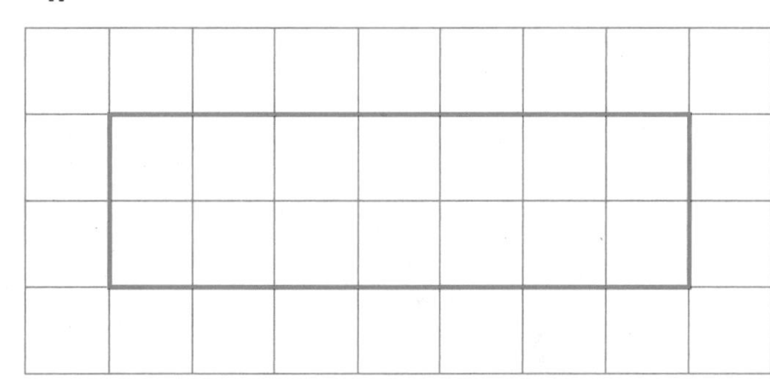

_____ centimeters

Find the perimeter.

5. A shape with four sides that measure 4 centimeters, 6 centimeters, 5 centimeters, and 1 centimeter

_____ centimeters

6. A shape with two sides that measure 10 inches, one side that measures 8 inches, and one side that measures 4 inches

_____ inches

7. H.O.T. Write Math **Explain** how to find the length of each side of a triangle with sides of equal length, and a perimeter of 27 inches.

____

____

Problem Solving

What's the Error?

8. Kevin is solving perimeter problems. He counts the units and says that the perimeter of this shape is 22 units.

Look at Kevin's solution.

Find Kevin's error.

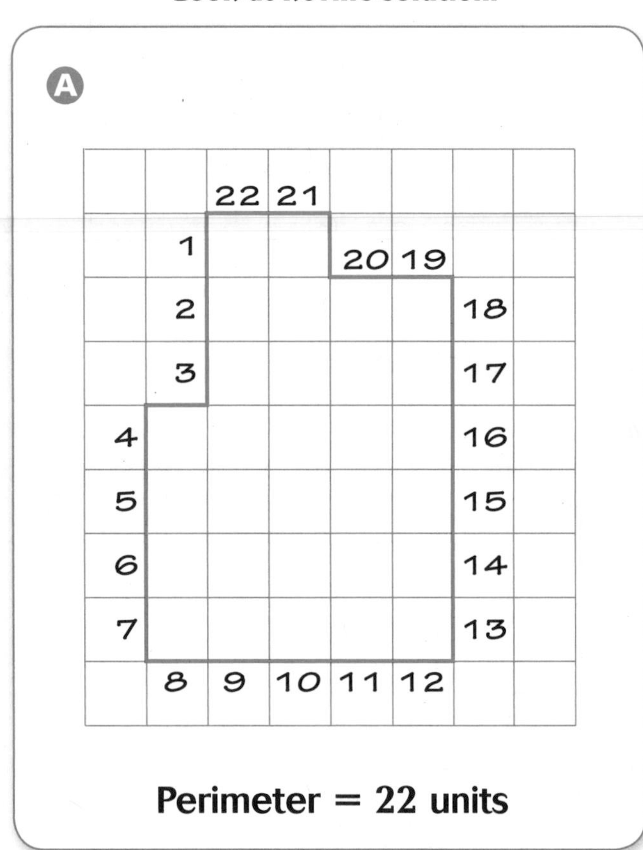

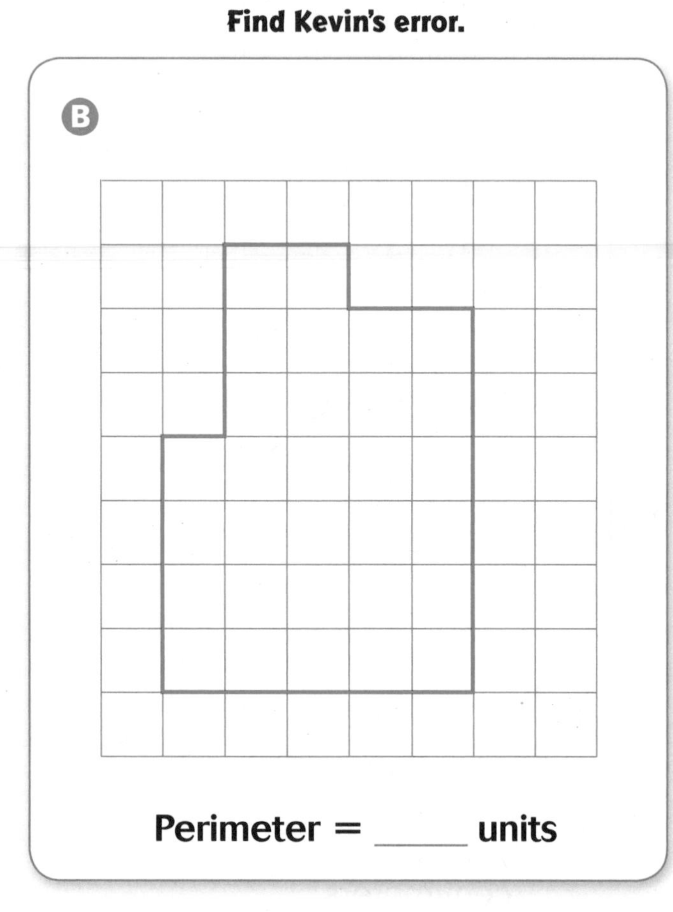

- Use the grid on the right to find the correct perimeter of the shape above.

____

- Describe the error Kevin made.

____

____

____

- Circle the places in the drawing of Kevin's solution where he made an error.

 FOR MORE PRACTICE: Standards Practice Book, pp. P215–P216

Name ____________________

Find Perimeter

Essential Question How can you measure perimeter?

You can estimate and measure perimeter in standard units, such as inches and centimeters.

UNLOCK the Problem REAL WORLD

Find the perimeter of the cover of a notebook.

Activity **Materials** ■ inch ruler

STEP 1 Estimate the perimeter of a notebook in inches. Record your estimate. _____ inches

STEP 2 Use an inch ruler to measure the length of each side of the notebook to the nearest inch.

STEP 3 Record and add the lengths of the sides measured to the nearest inch.

_____ + _____ + _____ + _____ = _____

So, the perimeter of the notebook cover measured to the nearest inch is _____ inches.

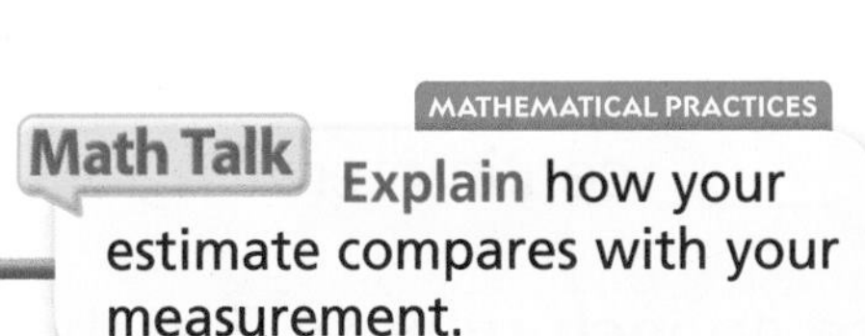

MATHEMATICAL PRACTICES

Math Talk **Explain** how your estimate compares with your measurement.

Try This! **Find the perimeter.**

Use an inch ruler to find the length of each side.

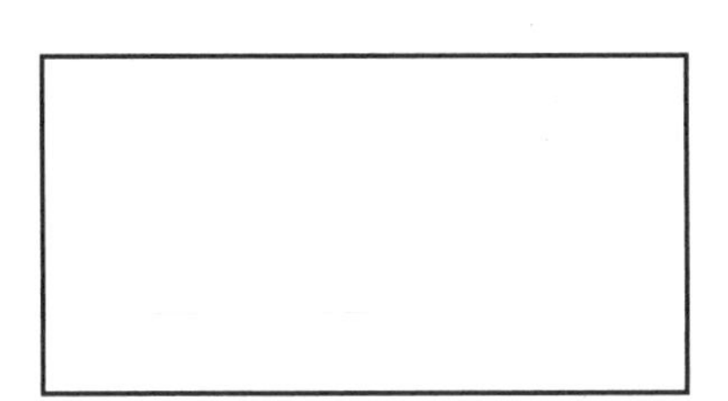

Add the lengths of the sides:

_____ + _____ + _____ + _____ = _____

The perimeter is _____ inches.

Use a centimeter ruler to find the length of each side.

Add the lengths of the sides:

_____ + _____ + _____ + _____ = _____

The perimeter is _____ centimeters.

Share and Show 

MATHEMATICAL PRACTICES

Math Talk Explain how many numbers you add together to find the perimeter of a shape.

1. Find the perimeter of the triangle in inches.

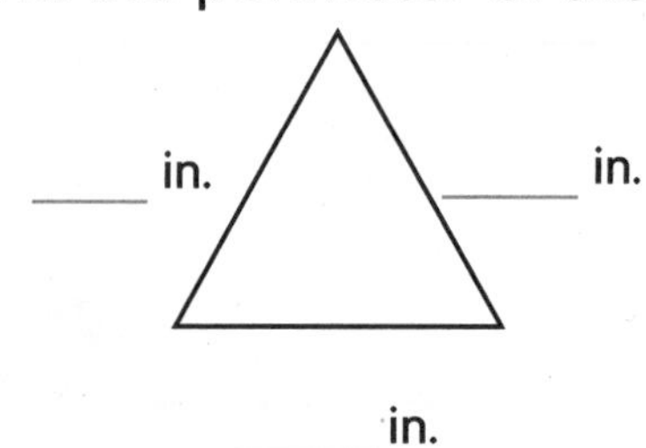

Think: How long is each side?

____ inches

Use a centimeter ruler to find the perimeter.

2.

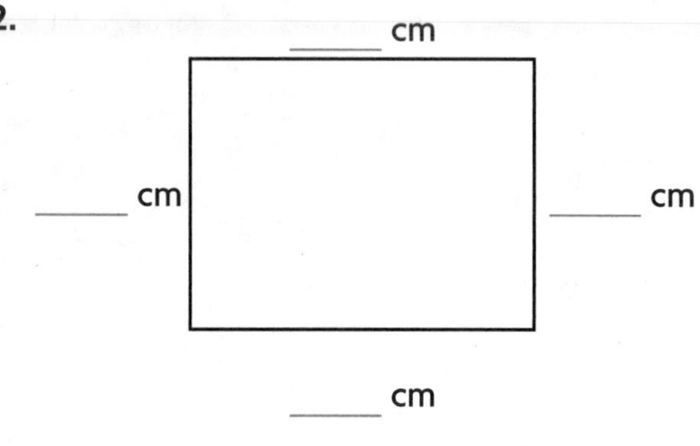

____ centimeters

3.

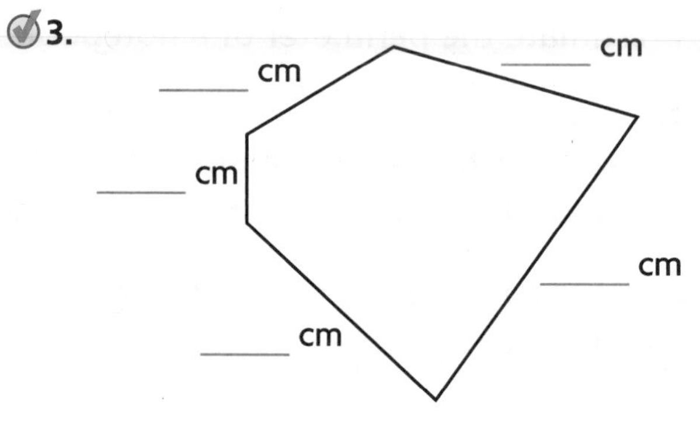

____ centimeters

Use an inch ruler to find the perimeter.

4.

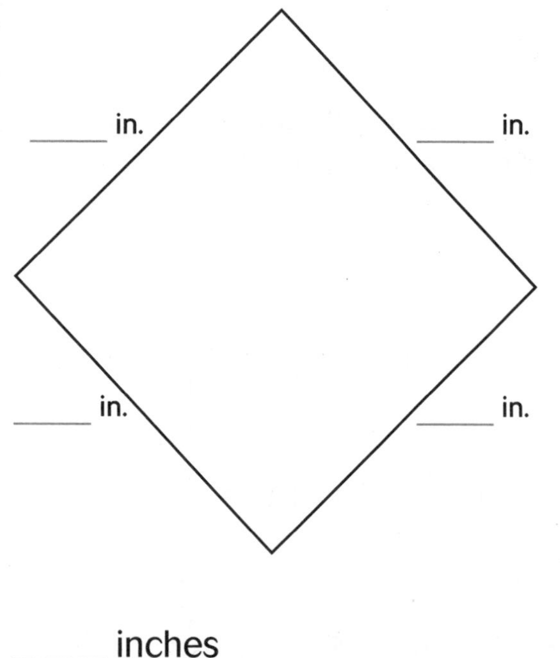

____ inches

5.

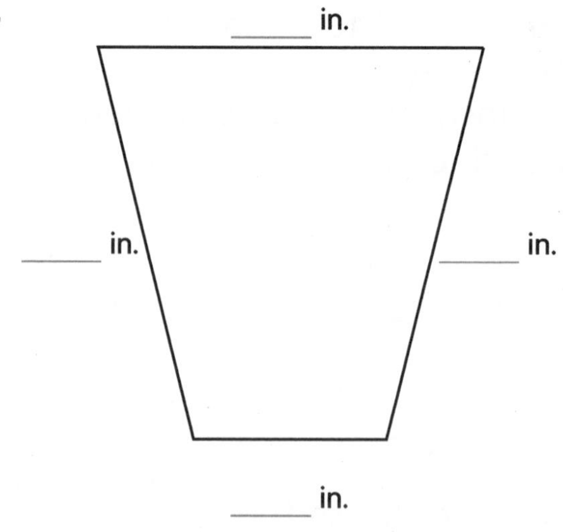

____ inches

Name ______________________

On Your Own

Use a ruler to find the perimeter.

6.

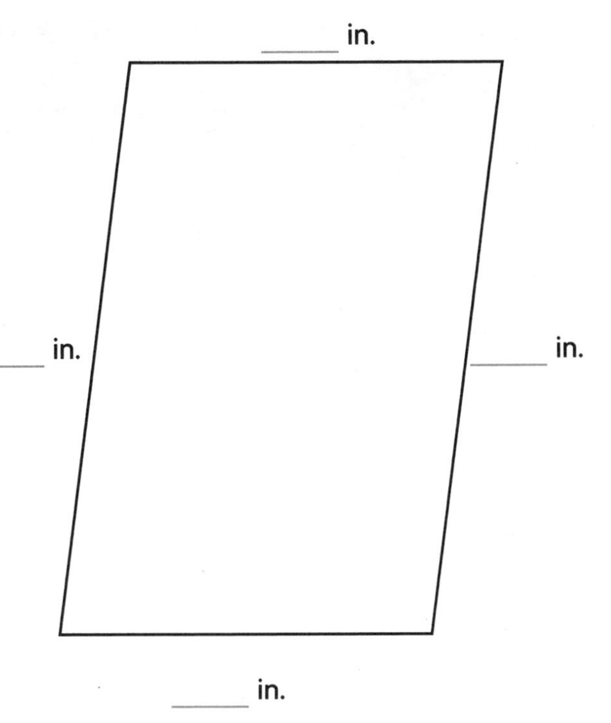

_____ inches

7.

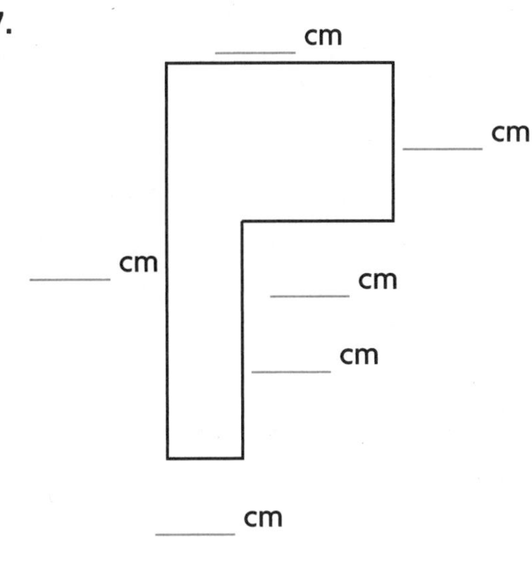

_____ centimeters

8. Use the grid paper to draw a shape that has a perimeter of 24 centimeters. Label the length of each side.

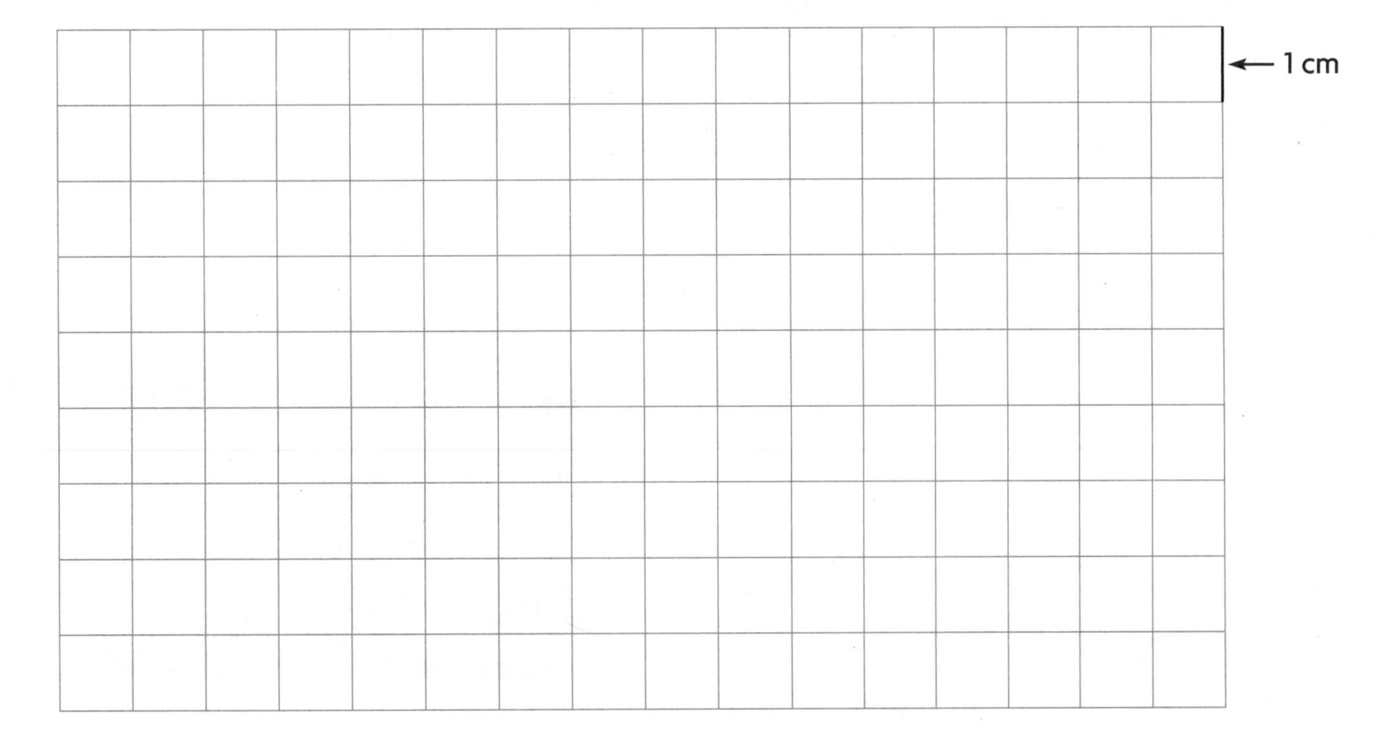

Problem Solving REAL WORLD

Use the photos for 9–10.

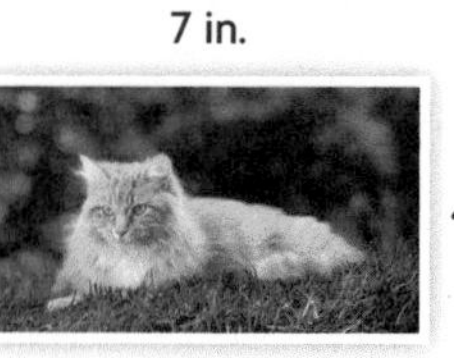

9. Which of the animal photos has a perimeter of 26 inches?

____

10. How much greater is the perimeter of the bird photo than the perimeter of the cat photo?

____

SHOW YOUR WORK

11. H.O.T. Erin is putting a fence around her square garden. Each side of her garden is 3 meters long. The fence costs $5 for each meter. How much will the fence cost?

____

12. H.O.T. Write Math Gary's garden is shaped like a rectangle with two pairs of sides of equal length, and it has a perimeter of 28 feet. **Explain** how to find the lengths of the other sides if one side measures 10 feet.

____

____

____

____

____

13. **Test Prep** Austin's class is making a poster for Earth Day. What is the perimeter of the poster?

Ⓐ 20 feet Ⓒ 24 feet

Ⓑ 21 feet Ⓓ 30 feet

FOR MORE PRACTICE:
Standards Practice Book, pp. P217–P218

Name ______________________

Find Unknown Side Lengths

Essential Question How can you find the unknown length of a side in a plane shape when you know its perimeter?

UNLOCK the Problem REAL WORLD

Chen has 27 feet of fencing to put around his garden. He has already used the lengths of fencing shown. How much fencing does he have left for the last side?

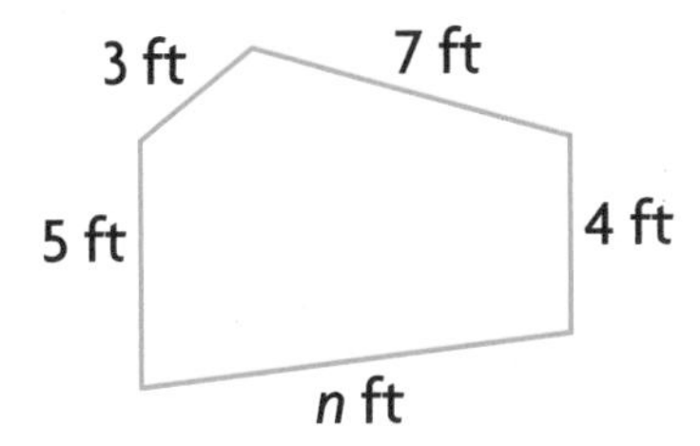

Find the unknown side length.

Write an equation for the perimeter. $5 + 3 +$ ____ $+$ ____ $+ n = 27$

Think: If I knew the value of n, I would add all the side lengths to find the perimeter. $5 + 3 + 7 + 4 + n = 27$

Add the lengths of the sides you know. ____ $+ n = 27$

Think: Addition and subtraction are inverse operations.

Write a related equation. $n = 27 - 19$

So, Chen has ____ feet of fencing left. ____ $= 27 - 19$

Math Idea

A symbol or letter can stand for an unknown side length.

Try This!

The perimeter of the shape is 24 meters. Find the unknown side length.

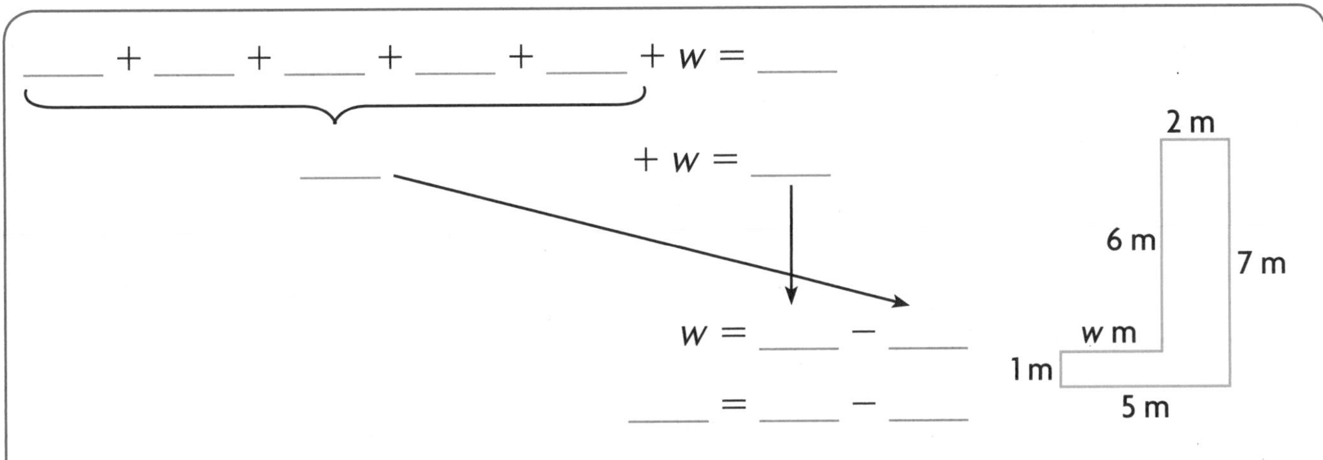

____ + ____ + ____ + ____ + ____ $+ w =$ ____

____ $+ w =$ ____

$w =$ ____ $-$ ____

____ $=$ ____ $-$ ____

So, the unknown side length is ____ meters long.

Example Find unknown side lengths of a rectangle.

Lauren has a rectangular blanket. The perimeter is 28 feet. The width of the blanket is 5 feet. What is the length of the blanket?

Hint: A rectangle has two pairs of opposite sides that are equal in length.

You can predict the length and add to find the perimeter. If the perimeter is 28 feet, then that is the correct length.

Predict	Check	Does it check?
$l = 7$	5 + ____ + 5 + ____ = ____	**Think:** Perimeter is not 28 feet, so the length does not check.
$l = 8$	5 + ____ + 5 + ____ = ____	**Think:** Perimeter is not 28 feet, so the length does not check.
$l = 9$	5 + ____ + 5 + ____ = ____	**Think:** Perimeter is 28 feet, so the length is correct. ✓

So, the length of the blanket is ____ feet.

Try This! Find unknown side lengths of a square.

The square has a perimeter of 20 inches. What is the length of each side of the square?

s in.

s in. *s* in.

s in.

Think: A square has four sides that are equal in length.

You can multiply to find the perimeter.

- Write a multiplication equation for the perimeter. $4 \times s = 20$
- Use a multiplication fact you know to solve. $4 \times$ ____ $= 20$

So, the length of each side of the square is ____ inches.

Name ____________________

Share and Show 

Find the unknown side lengths.

1. Perimeter = 25 centimeters

$9 +$ ____ $+$ ____ $+ n = 25$

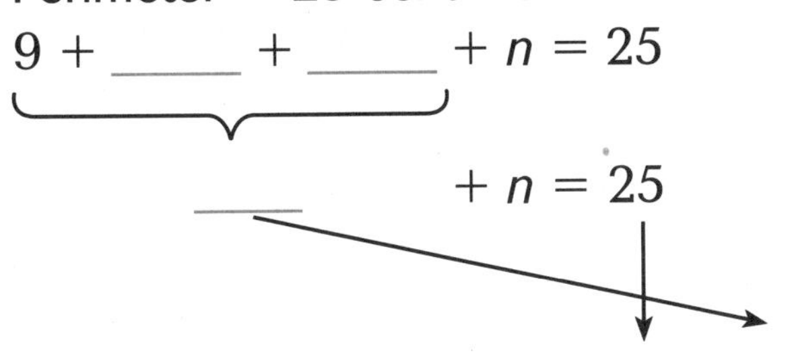

____ $+ n = 25$

____ $=$ ____ $-$ ____

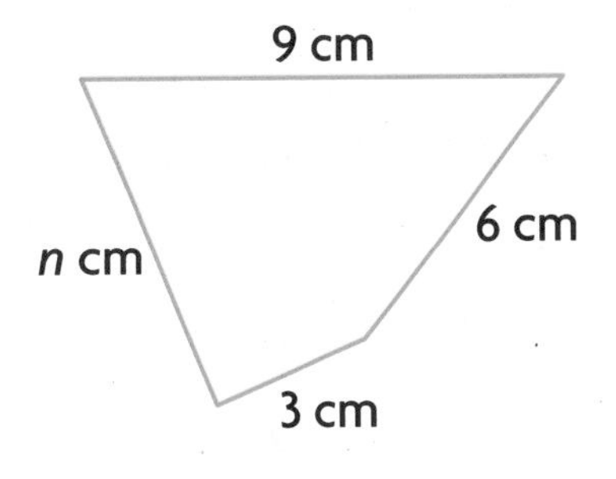

____ centimeters

2. Perimeter = 34 meters

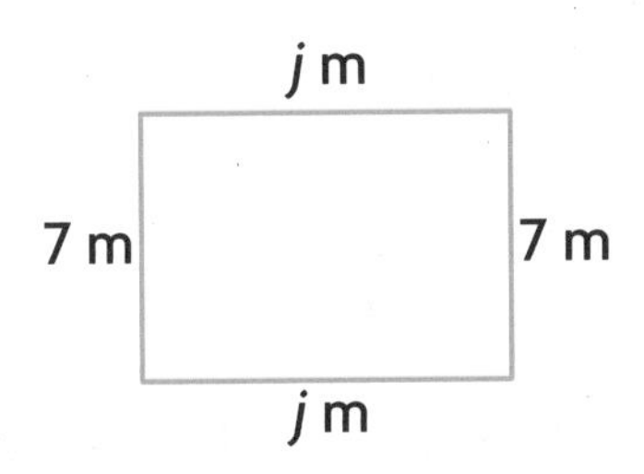

____ meters

3. Perimeter = 12 feet

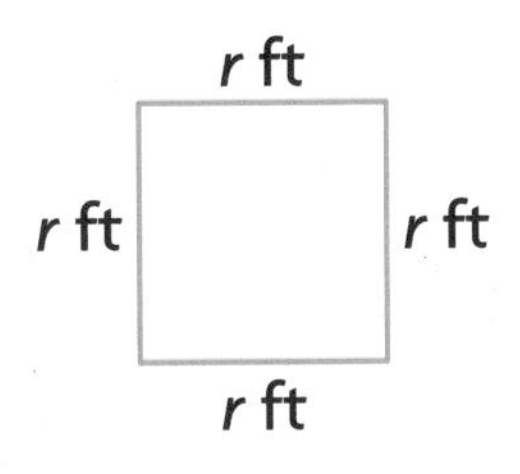

____ feet

MATHEMATICAL PRACTICES

Math Talk Explain how you can use division to find the length of a side of a square.

On Your Own

Find the unknown side lengths.

4. Perimeter = 32 centimeters

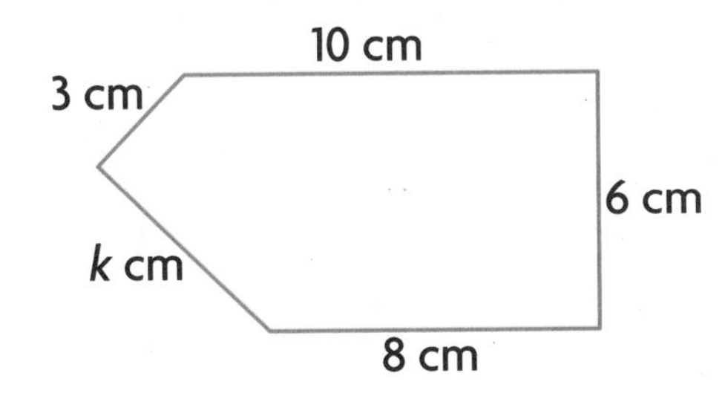

____ centimeters

5. Perimeter = 8 meters

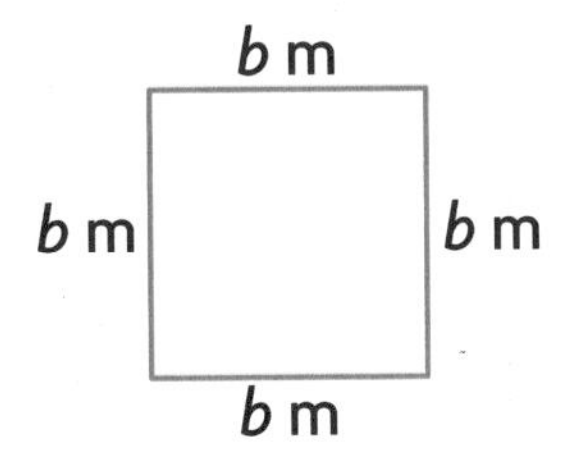

____ meters

6. Perimeter = 16 inches

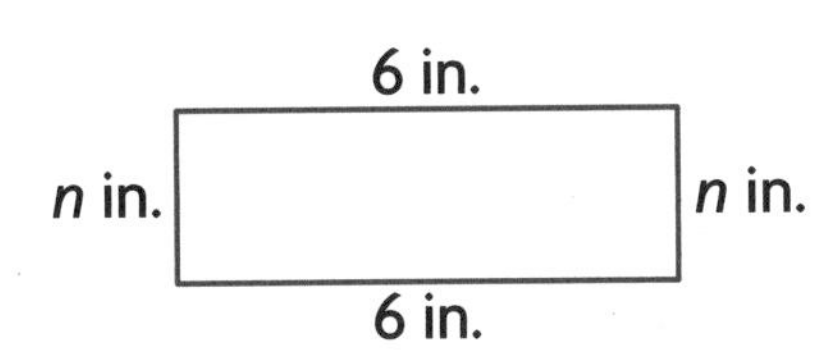

____ inches

7. H.O.T. Perimeter = 42 feet

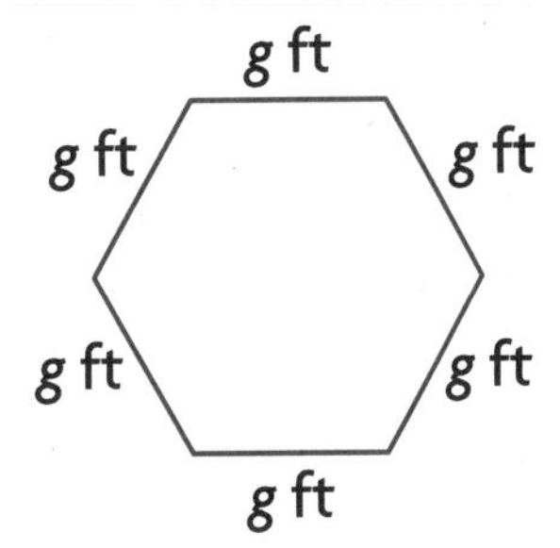

____ feet

UNLOCK the Problem REAL WORLD

8. Latesha wants to make a border with ribbon around a shape she made and sketched at the right. She will use 44 centimeters of ribbon for the border. What is the unknown side length?

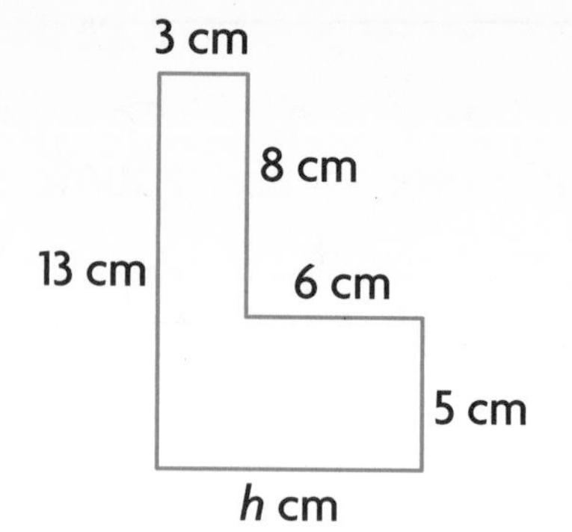

Ⓐ 3 centimeters Ⓒ 9 centimeters

Ⓑ 6 centimeters Ⓓ 13 centimeters

a. What do you need to find? ______________________

b. How will you use what you know about perimeter to help you solve the problem? ______________________

____

c. Write an equation to solve the problem.

d. Complete the sentences.

The perimeter is _____ centimeters.

The sum of the sides I know is _____ centimeters.

A related subtraction equation is _____ = _____ − _____.

So, the unknown side length is _____ centimeters.

e. Fill in the bubble for the correct answer choice above.

9. A rectangle has a perimeter of 34 inches. The left side is 6 inches long. What is the length of the top side?

Ⓐ 6 inches

Ⓑ 11 inches

Ⓒ 28 inches

Ⓓ 40 inches

10. Eleni wants to put up a fence around her square garden. The garden has a perimeter of 28 meters. How long will each side of the fence be?

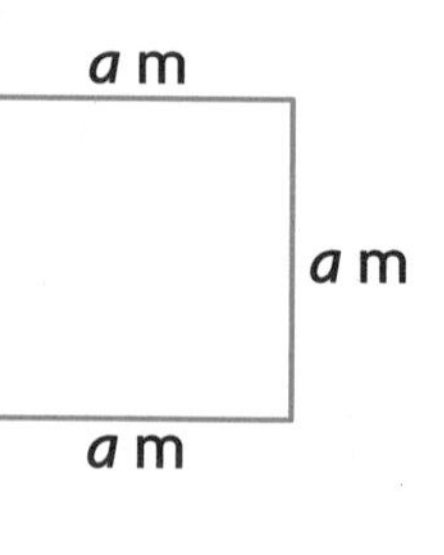

Ⓐ 6 meters Ⓒ 8 meters

Ⓑ 7 meters Ⓓ 14 meters

FOR MORE PRACTICE:
Standards Practice Book, pp. P219–P220

Name ______________________

Understand Area

Essential Question How is finding the area of a shape different from finding the perimeter of a shape?

UNLOCK the Problem REAL WORLD

CONNECT You learned that perimeter is the measure around a shape. It is measured in linear units, or units that are used to measure the distance between two points.

Area is the measure of the number of unit squares needed to cover a flat surface. A **unit square** is a square with a side length of 1 unit. It has an area of 1 **square unit (sq un)**.

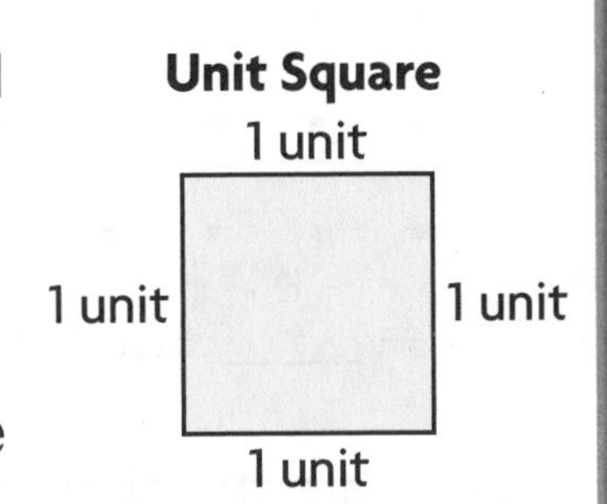

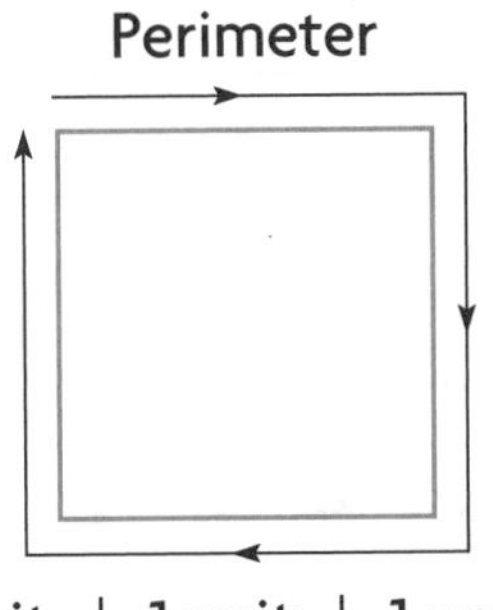

1 unit + 1 unit + 1 unit + 1 unit = 4 units

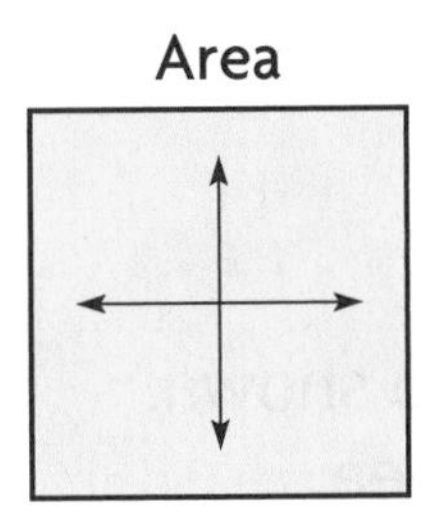

1 square unit

Math Idea

You can count the number of units on each side of a shape to find its perimeter. You can count the number of unit squares inside a shape to find its area in square units.

Activity Materials ■ geoboard ■ rubber bands

A Use your geoboard to form a shape made from 2 unit squares. Record the shape on this dot paper.

What is the area of this shape?

Area = _____ square units

B Change the rubber band so that the shape is made from 3 unit squares. Record the shape on this dot paper.

What is the area of this shape?

Area = _____ square units

MATHEMATICAL PRACTICES

Math Talk For B, did your shape look like your classmate's shape? **Explain.**

Try This! Draw three different shapes that are each made from 4 unit squares. Find the area of the shape.

Shape 1	Shape 2	Shape 3
Area = _____ square units	Area = _____ square units	Area = _____ square units

- How are the shapes the same? How are the shapes different?

____

Share and Show

1. Shade each unit square in the shape shown. Count the unit squares to find the area.

   Area = _____ square units

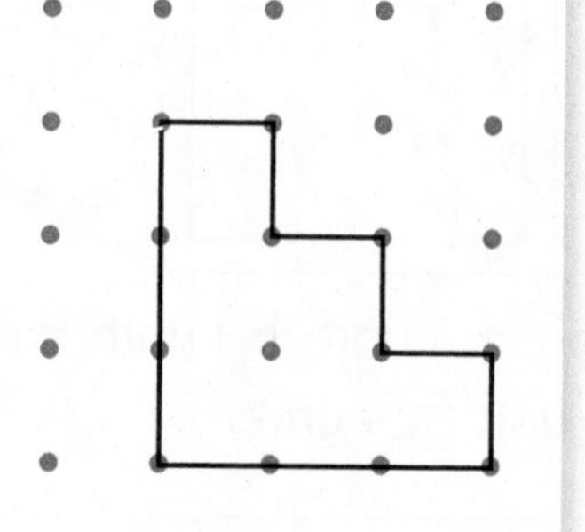

Count to find the area of the shape.

2. 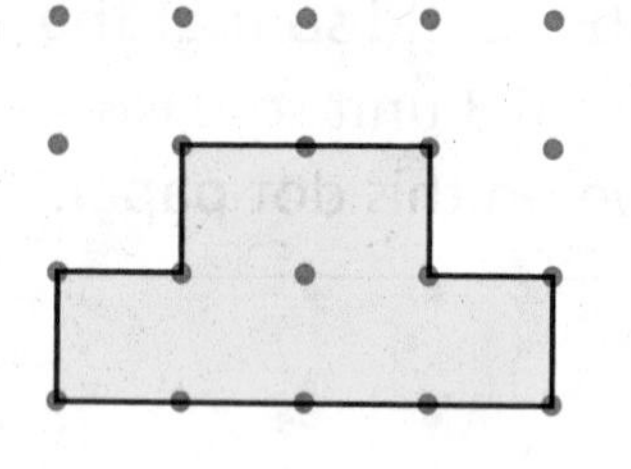

   Area = _____ square units

3. 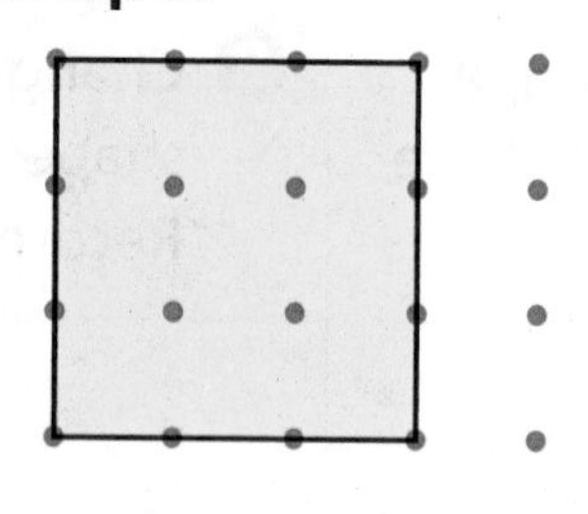

   Area = _____ square units

4. 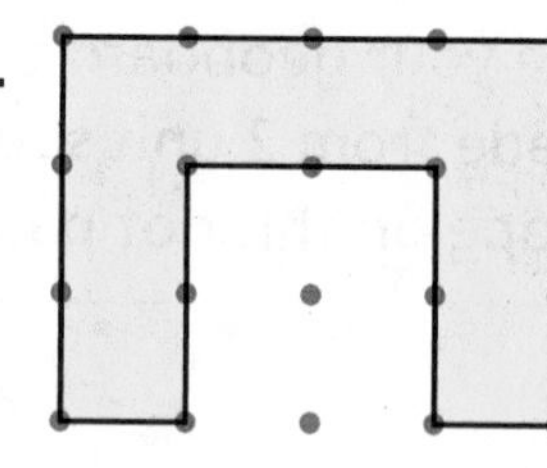

   Area = _____ square units

Write *area* or *perimeter* for the situation.

5. buying a rug for a room

   ____

6. putting a fence around a garden

   ____

MATHEMATICAL PRACTICES

Math Talk What are other situations where you need to find area?

Name ___________________________

On Your Own

Count to find the area of the shape.

7. 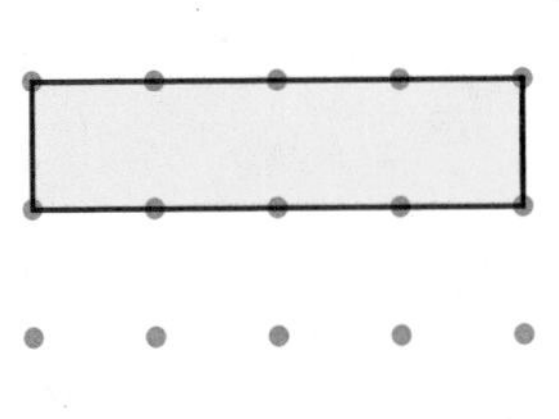

Area = _____ square units

8. 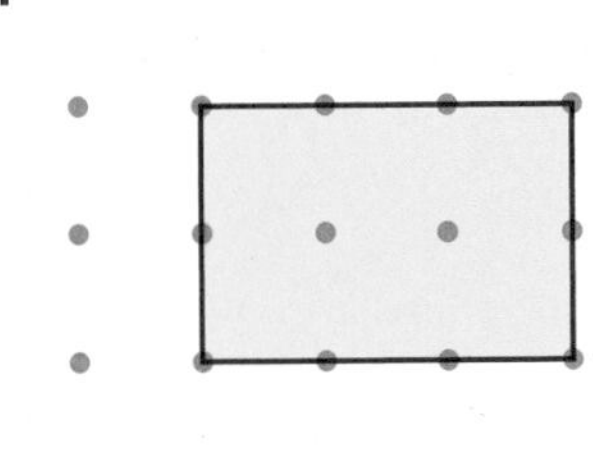

Area = _____ square units

9. 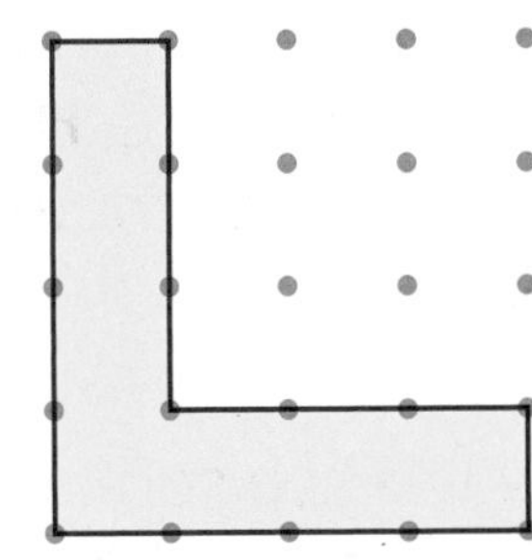

Area = _____ square units

10. 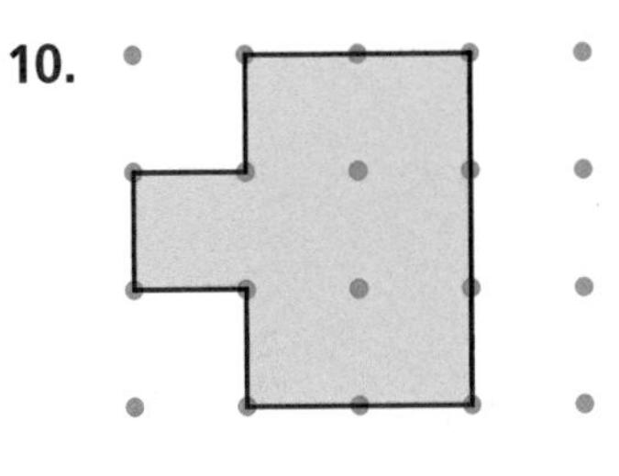

Area = _____ square units

11. 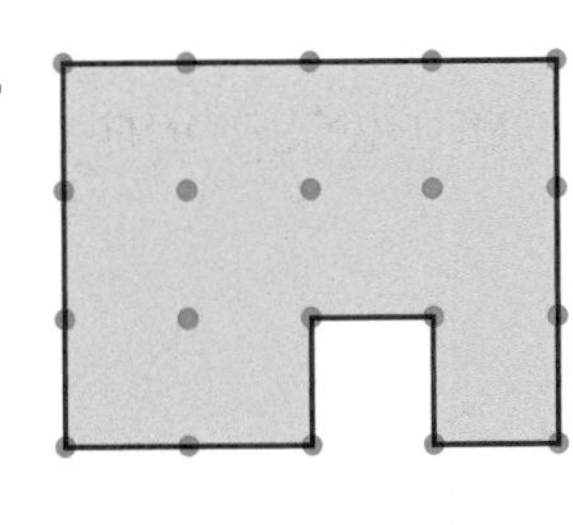

Area = _____ square units

12. 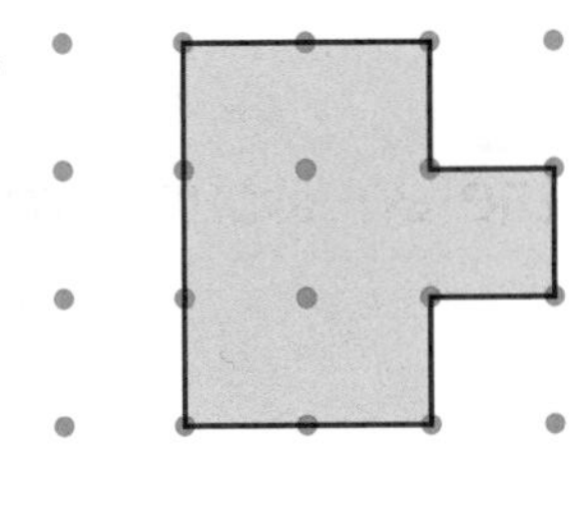

Area = _____ square units

13. 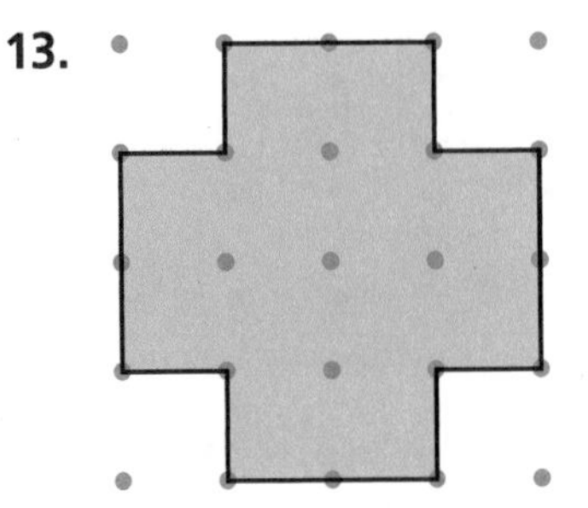

Area = _____ square units

14. 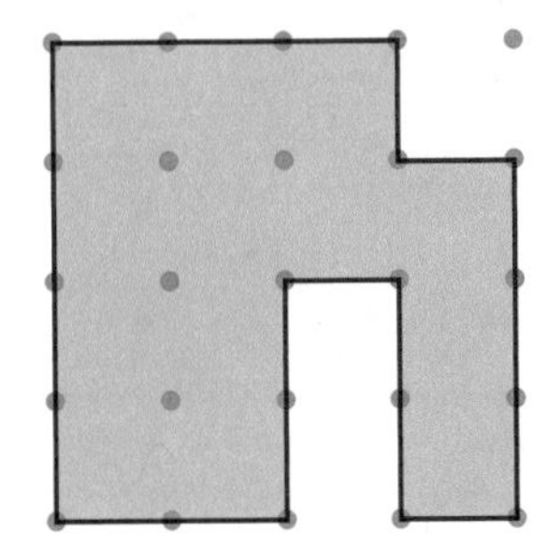

Area = _____ square units

15. 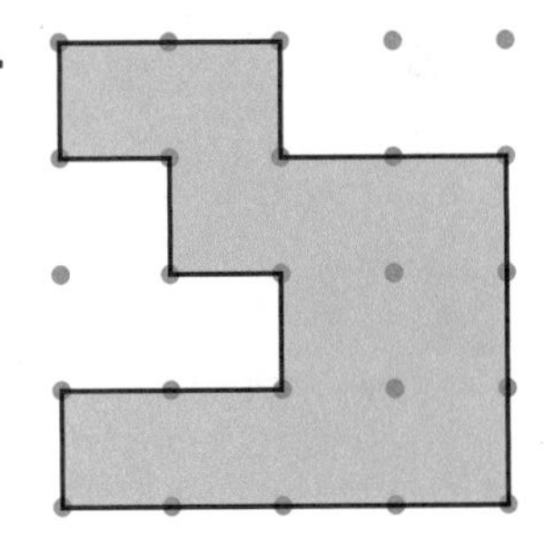

Area = _____ square units

Write *area* or *perimeter* for the situation.

16. painting a wall

____

17. covering a patio with tiles

____

18. putting a wallpaper border around a room

____

19. gluing a ribbon around a picture frame

____

Problem Solving REAL WORLD

Billy is building an enclosure for his small dog, Eli. Use the diagram for 20–22.

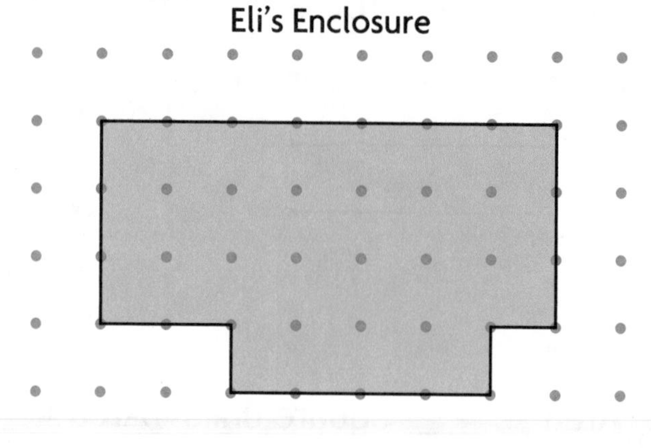

20. Billy will put fencing around the outside of the enclosure. How much fencing does he need for the enclosure?

____

21. Billy will use sod to cover the ground in the enclosure. How much sod does Billy need?

____

22. Write Math ▸ **Explain** how the perimeter and the area of the enclosure are different.

____

____

23. H.O.T. Draw two different shapes, each with an area of 10 square units.

24. **Test Prep** What is the area of the shape at the right?

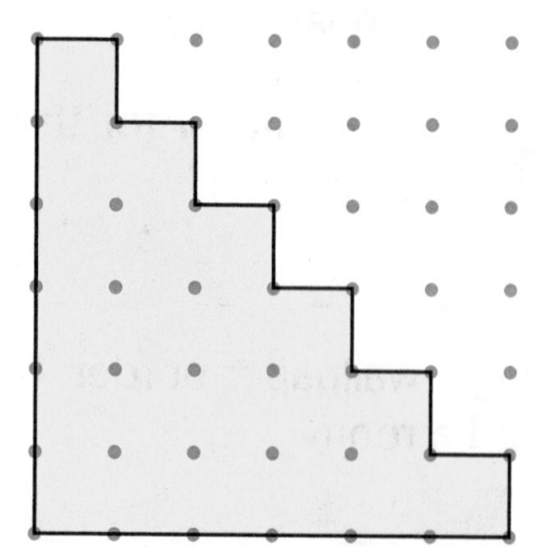

Ⓐ 6 square units

Ⓑ 21 square units

Ⓒ 24 square units

Ⓓ 36 square units

Name ______________________________

Measure Area

Essential Question How can you find the area of a plane shape?

UNLOCK the Problem

Jaime is measuring the area of the rectangles with 1-inch square tiles.

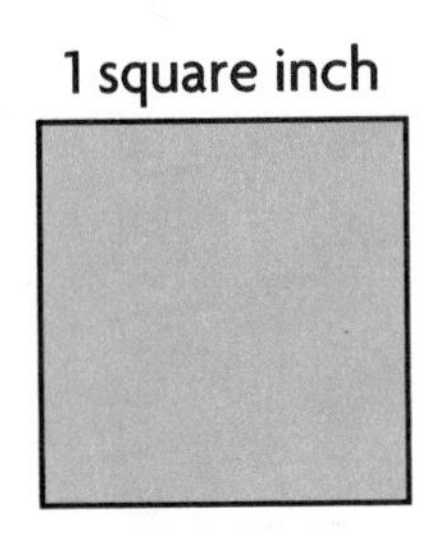

Activity 1 **Materials** ■ 1-inch grid paper ■ scissors

Cut out eight 1-inch squares. Use the dashed lines as guides to place tiles for *A*–*C*.

A Place 4 tiles on Rectangle *A*.

- Are there any gaps? ______
- Are there any overlaps? ______
- Jaime says that the area is 4 square inches. Is Jaime's measurement correct? ______

Rectangle *A*

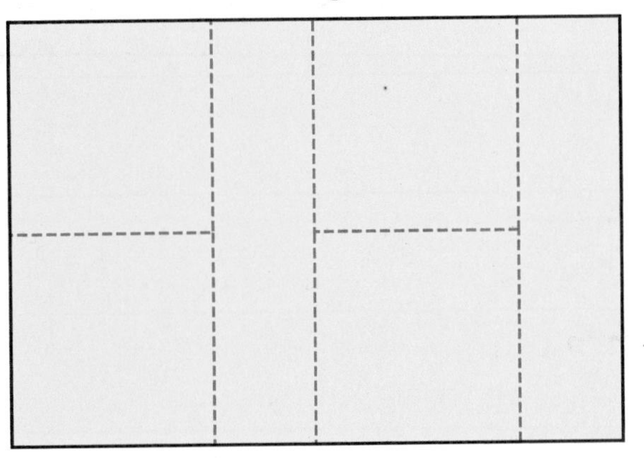

So, when you measure area, there can be no space between the tiles, or no gaps.

B Place 8 tiles on Rectangle *B*.

- Are there any gaps? ______
- Are there any overlaps? ______
- Jaime says that the area is 8 square inches. Is Jaime's measurement correct? ______

Rectangle *B*

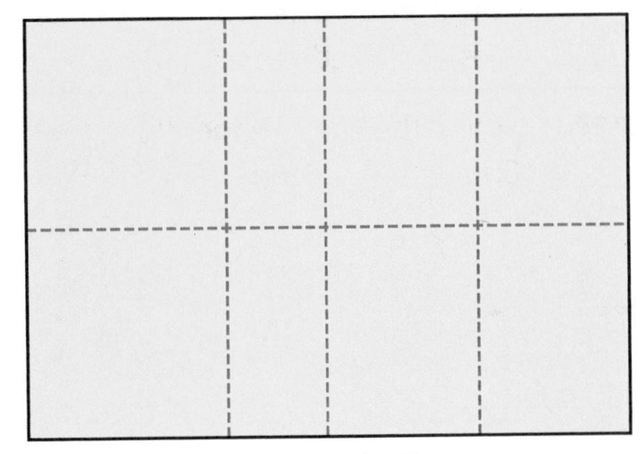

So, when you measure the area, the tiles cannot overlap.

C Place 6 tiles on Rectangle *C*.

- Are there any gaps? ______
- Are there any overlaps? ______
- Jaime says that the area is 6 square inches. Is Jaime's measurement correct? ______

Rectangle *C*

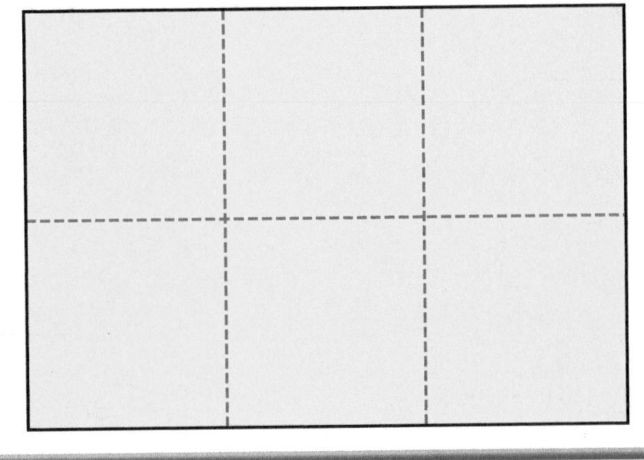

So, the area of the rectangles is

______ square inches.

Activity 2 **Materials** ■ green and blue paper ■ scissors

ERROR Alert

Be sure that there are no gaps or overlaps when you use square tiles to find area.

STEP 1 Estimate the number of blue square tiles it will take to cover the gray shape. _____ blue square tiles

STEP 2 Estimate the number of green tiles it will take to cover the gray shape. _____ green square tiles

STEP 3 Trace the blue square pattern ten times and cut out the squares.

STEP 4 Trace the green square pattern thirty-six times and cut out the squares.

STEP 5 Cover the gray shape with blue square tiles. Count and write the number of blue square tiles you used. Record the area of the shape.
_____ blue square tiles
Area = _____ blue square units

STEP 6 Cover the gray shape with green square tiles. Count and write the number of green square tiles you used. Record the area of the shape.
_____ green square tiles
Area = _____ green square units

MATHEMATICAL PRACTICES

Math Talk **Explain** why the number of green square tiles needed to cover the shape is different than the number of blue square tiles needed.

Try This! Count to find the area of the shape.

☐ is 1 square centimeter.

There are _____ unit squares in the shape.

1
2

So, the area is _____ square centimeters.

Name ________________________________

Share and Show

MATH BOARD

1. Count to find the area of the shape. Each unit square is 1 square centimeter.

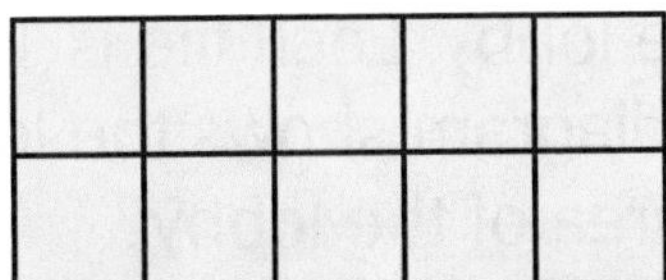

Think: Are there any gaps? Are there any overlaps?

There are _____ unit squares in the shape.

So, the area is _____ square centimeters.

MATHEMATICAL PRACTICES

Math Talk Explain how you can use square centimeters to find the area of the shapes in Exercises 2 and 3.

Count to find the area of the shape. Each unit square is 1 square centimeter.

2. 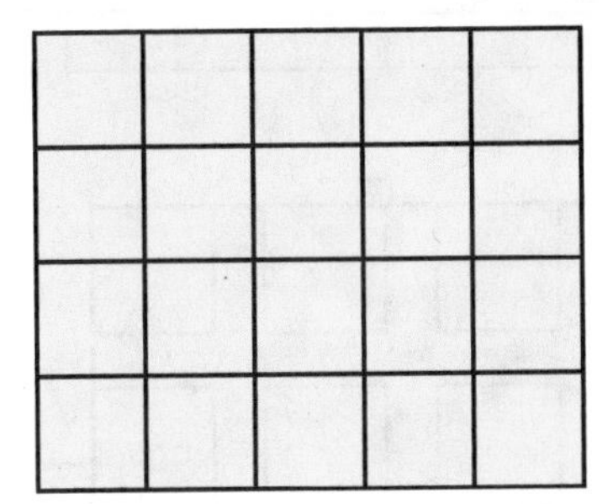

Area = _____ square centimeters

3. 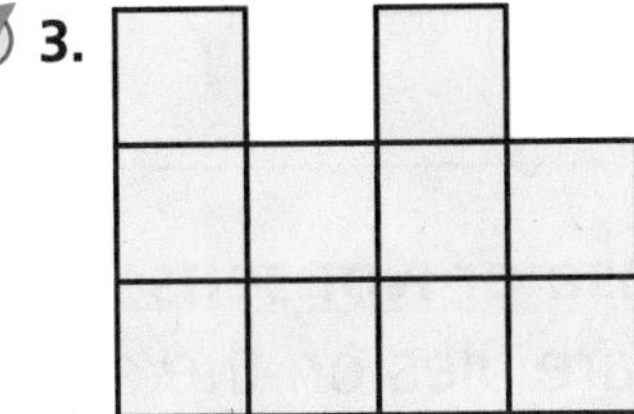

Area = _____ square centimeters

On Your Own

Count to find the area of the shape. Each unit square is 1 square inch.

4. 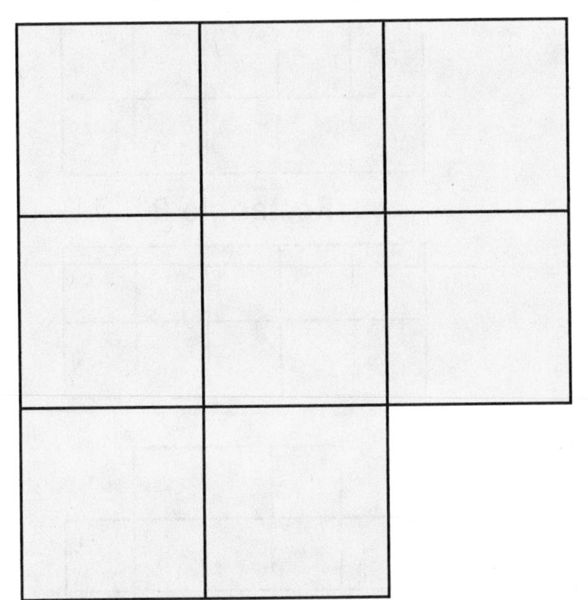

Area = _____ square inches

5. 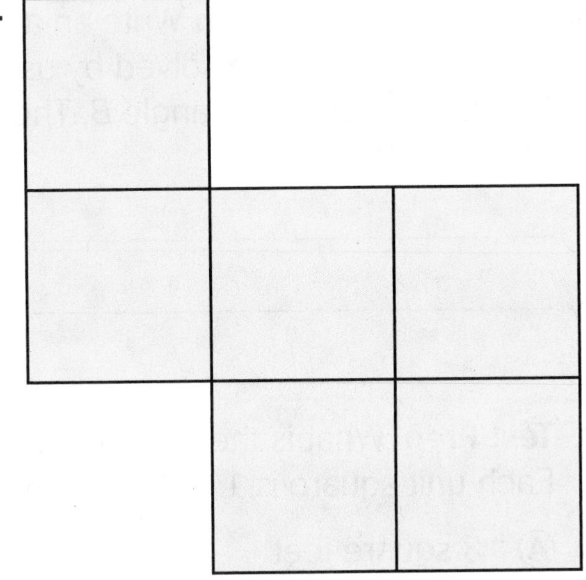

Area = _____ square inches

Problem Solving REAL WORLD

6. Danny is placing tiles on the floor of an office lobby. Each tile is 1 square meter. The diagram shows the lobby. What is the area of the lobby?

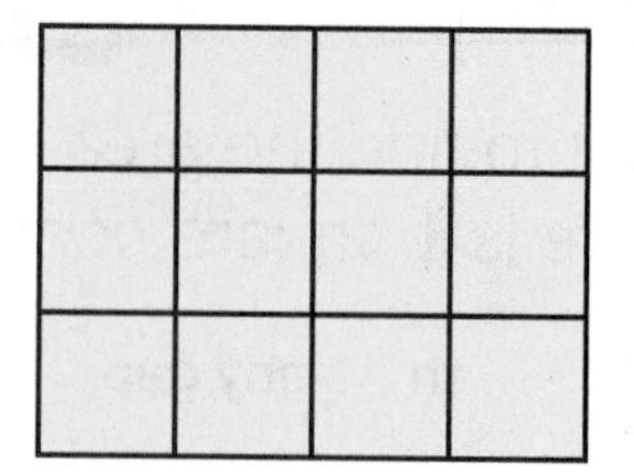

7. Angie is painting a space shuttle mural on a wall. She divides the area into sections that are each one square foot. The diagram shows the mural. What is the area of the mural?

8. Write Math **Sense or Nonsense?** Tom places green square tiles on the shape shown. He says that the shape has an area of 12 square units. Does this make sense? **Explain.**

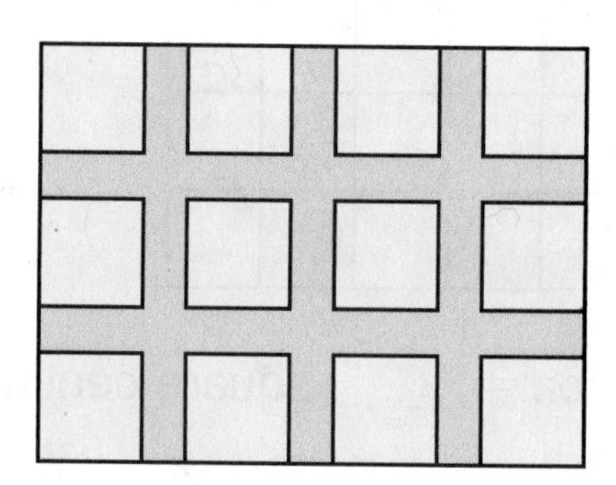

9. H.O.T. **Pose a Problem** Write an area problem that can be solved by using Rectangle *A* and Rectangle *B*. Then solve your problem.

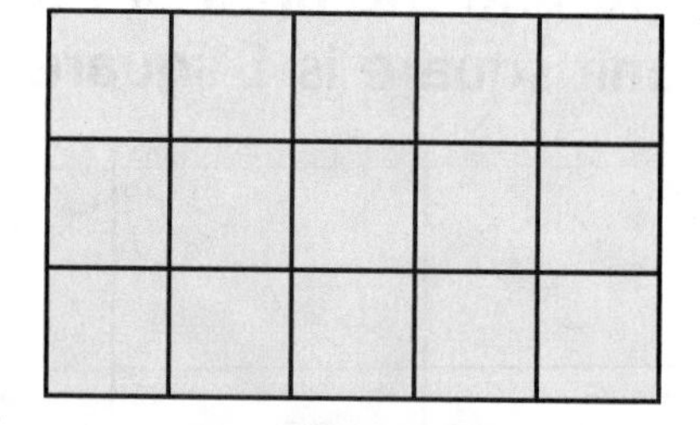

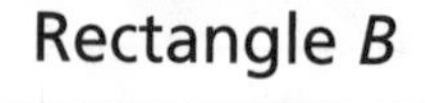

10. **Test Prep** What is the area of the shape? Each unit square is 1 square foot.

    Ⓐ 3 square feet
    Ⓑ 10 square feet
    Ⓒ 13 square feet
    Ⓓ 15 square feet

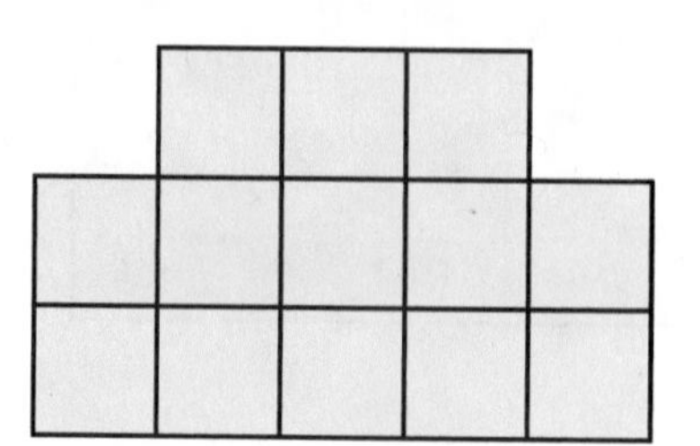

Name ____________________

Use Area Models

Essential Question Why can you multiply to find the area of a rectangle?

Melissa has a garden that is shaped like the rectangle below. Each unit square represents 1 square meter. What is the area of her garden?

- Circle the shape of the garden.

One Way Count unit squares.

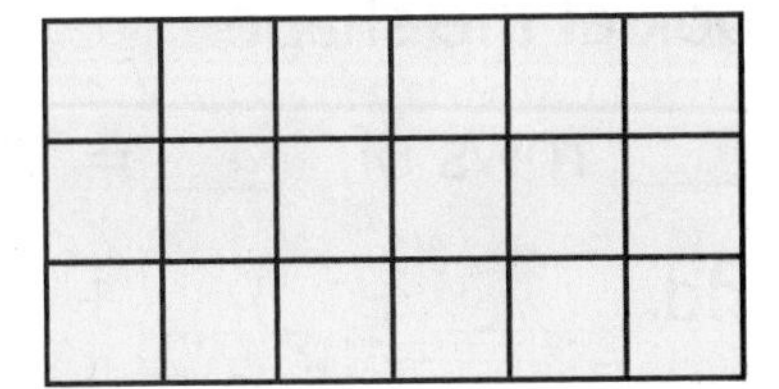

Count the number of unit squares in all.

There are _____ unit squares.

So, the area is _____ square meters.

Other Ways

A Use repeated addition.

Count the number of rows. Count the number of unit squares in each row.

_____ unit squares

_____ unit squares

_____ unit squares

_____ rows of _____ = ■

Write an addition equation.

_____ + _____ + _____ = _____

So, the area is _____ square meters.

B Use multiplication.

Count the number of rows. Count the number of unit squares in each row.

_____ unit squares in each row

_____ rows

_____ rows of _____ = ■

This shape is like an array. How do you find the total number of squares in an array?

____

Write a multiplication equation.

_____ × _____ = _____

So, the area is _____ square meters.

MATHEMATICAL PRACTICES

Math Talk **Explain** when you can use different methods to find the same area.

Try This!

Find the area of the shape.
Each unit square is 1 square foot.

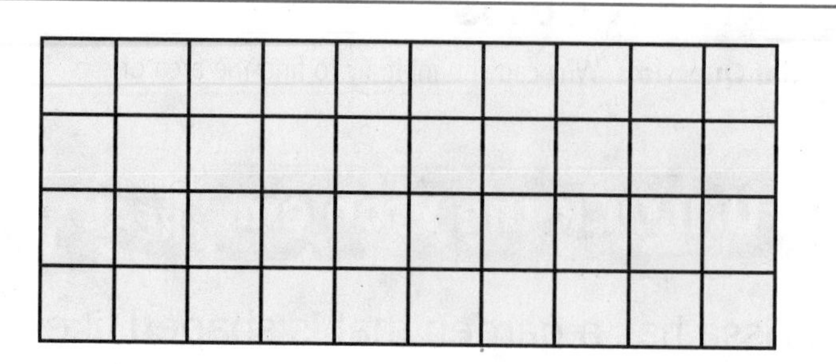

Think: There are 4 rows of 10 unit squares.

_____ × _____ = _____

So, the area is _____ square feet.

Share and Show

1. Look at the shape.

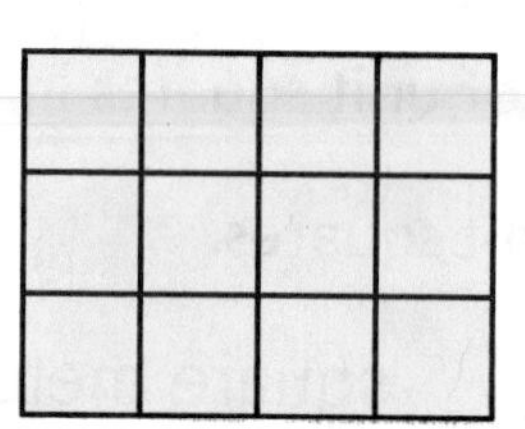

_____ rows of _____ = ■

Add. _____ + _____ + _____ = _____

Multiply. _____ × _____ = _____

What is the area of the shape?

_____ square units

MATHEMATICAL PRACTICES

Math Talk Which method do you prefer using? **Explain.**

Find the area of the shape.
Each unit square is 1 square foot.

2.

____

✓ 3.

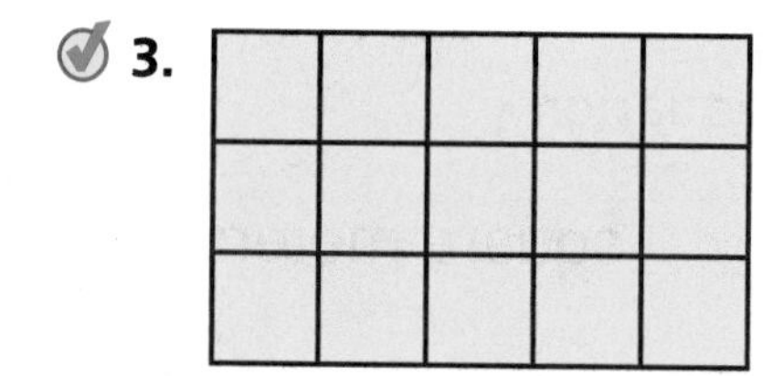

____

Find the area of the shape.
Each unit square is 1 square meter.

4.

____

✓ 5.

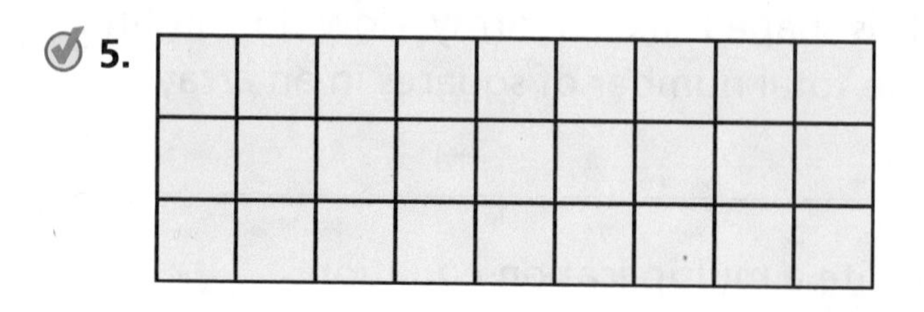

____

Name ______________________

On Your Own

Find the area of the shape.
Each unit square is 1 square foot.

6. 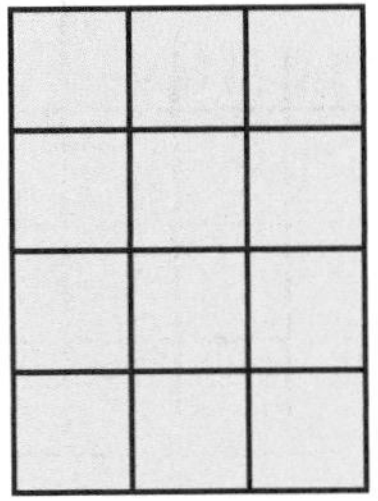

____

7. 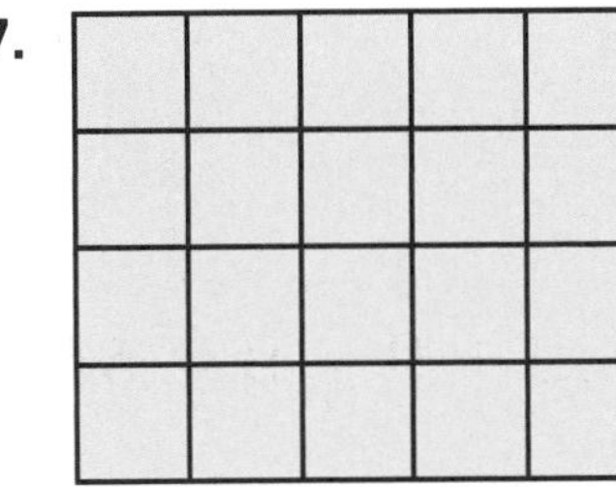

____

Find the area of the shape.
Each unit square is 1 square meter.

8. 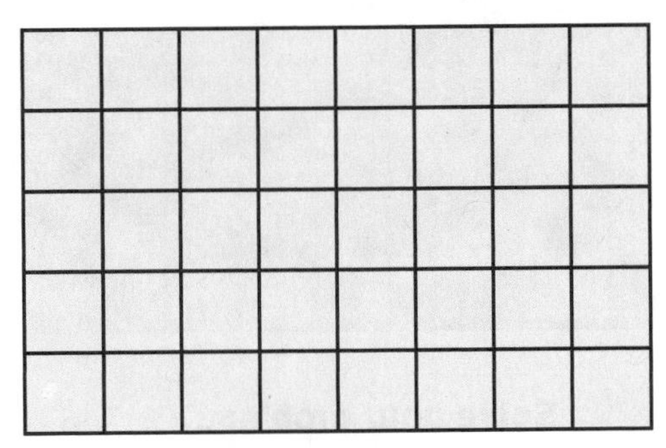

____

9. 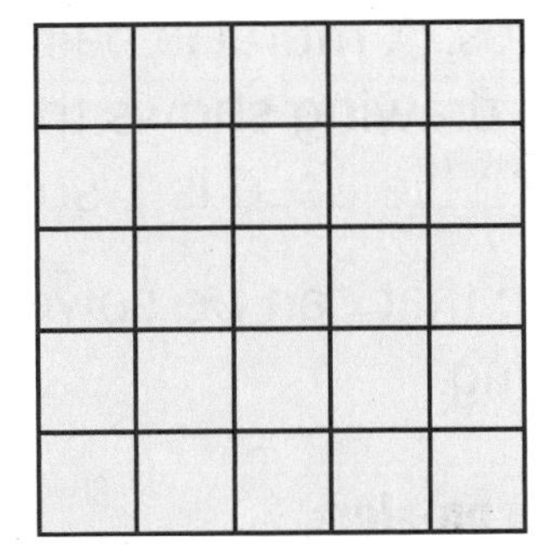

____

10. H.O.T. Draw and shade three rectangles with an area of 24 square units. Then write an addition or multiplication equation for each.

____

____

Problem Solving REAL WORLD

11. Test Prep Heather drew this rectangle. Which multiplication equation can be used to find the area of the rectangle?

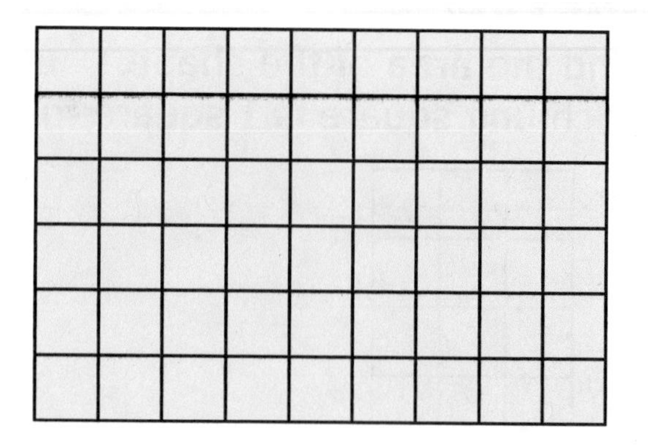

Ⓐ $6 \times 9 = 54$

Ⓑ $9 \times 9 = 81$

Ⓒ $9 + 9 + 9 + 9 + 9 + 9 = 54$

Ⓓ $6 \times 6 = 36$

H.O.T. Pose a Problem

12. Tile Design Company tiled a bathroom wall using square tiles. A mural is painted in the center. The drawing shows the design. The area of each tile used is 1 square foot.

Write a problem that can be solved by using the drawing.

Pose a problem.

Solve your problem.

FOR MORE PRACTICE:
Standards Practice Book, pp. P225–P226

Name ___

Mid-Chapter Checkpoint

▶ Vocabulary

Choose the best term from the box.

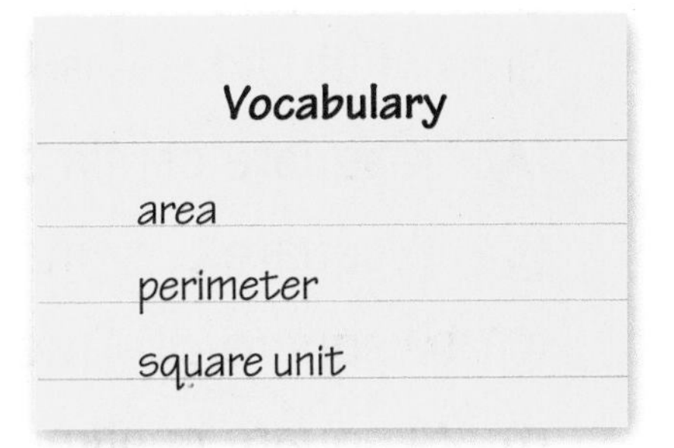

1. The distance around a shape is the ___. (p. 433)

2. The measure of the number of unit squares needed to cover a shape with no gaps or overlaps is the ___. (p. 445)

▶ Concepts and Skills

Find the perimeter of the shape. Each unit is 1 centimeter.

3.

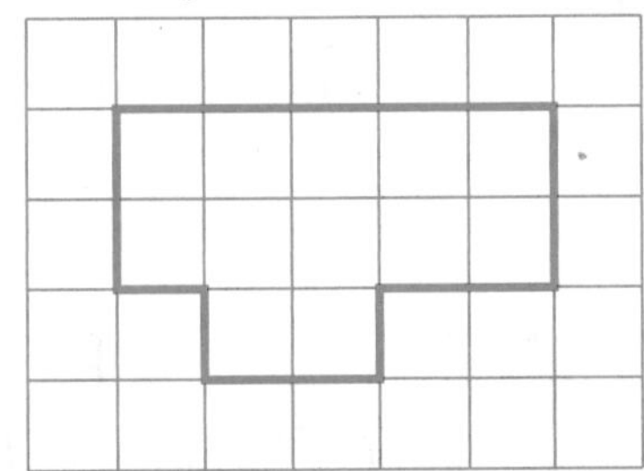

___ centimeters

4.

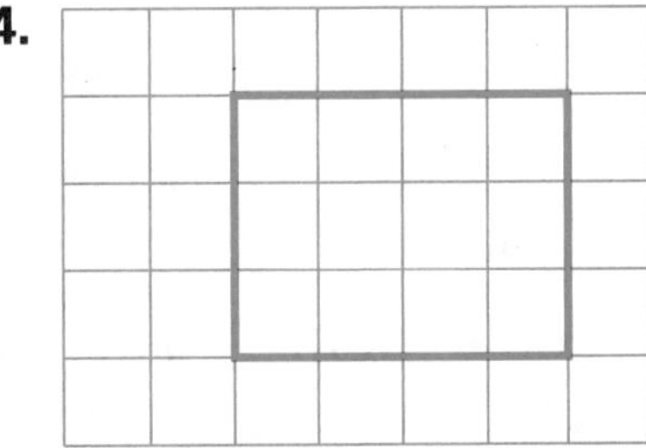

___ centimeters

Find the unknown side lengths.

5. Perimeter = 33 centimeters

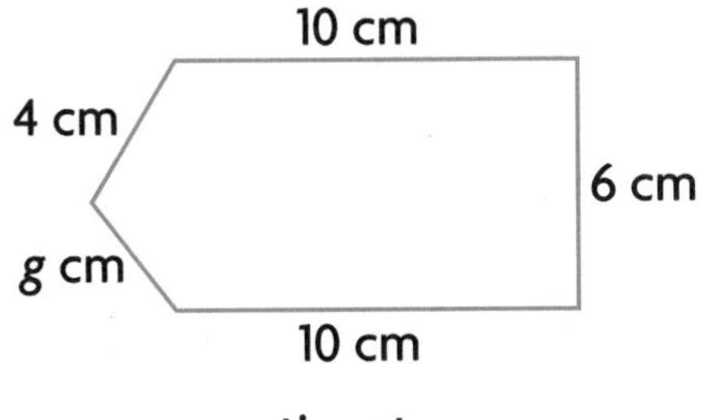

___ centimeters

6. Perimeter = 32 feet

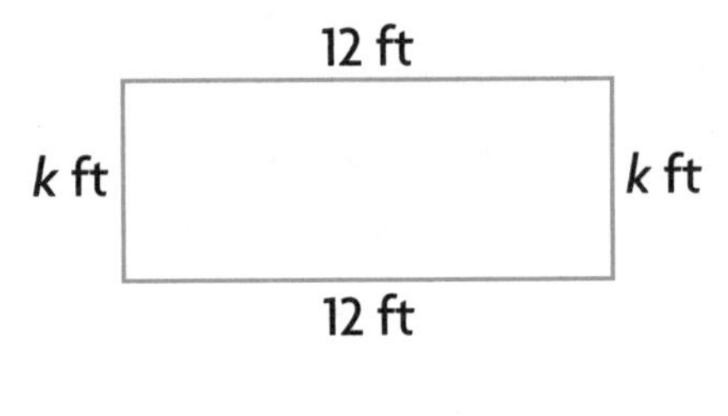

___ feet

Find the area of the shape. Each unit square is 1 square meter.

7.

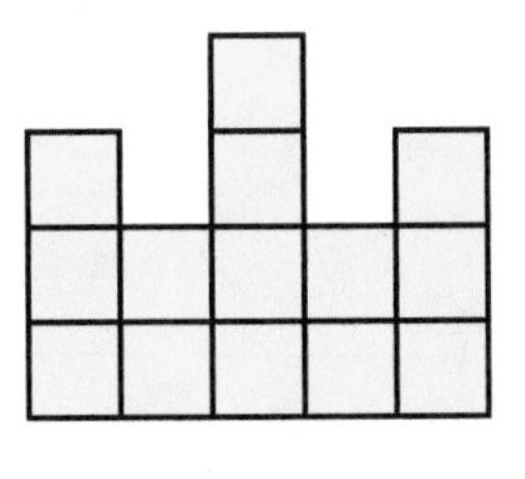

___ square meters

8.

___ square meters

Fill in the bubble for the correct answer choice.

9. Annie is making a lid for her rectangular jewelry box. The jewelry box has side lengths of 6 centimeters and 4 centimeters. What is the area of the lid she is making?

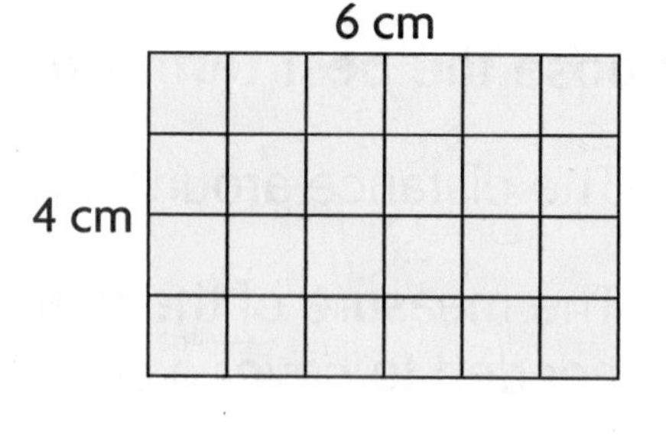

Ⓐ 4 square centimeters

Ⓑ 6 square centimeters

Ⓒ 10 square centimeters

Ⓓ 24 square centimeters

10. Adrienne is decorating a square picture frame. She glued 36 inches of ribbon around the edge of the frame. What is the length of each side of the picture frame?

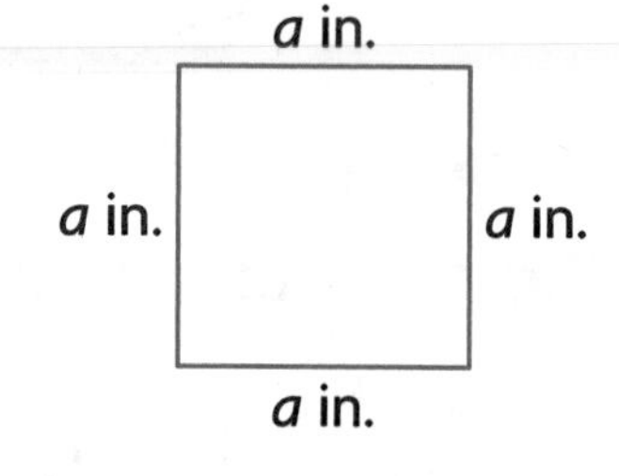

Ⓐ 6 inches　　Ⓒ 30 inches

Ⓑ 9 inches　　Ⓓ 32 inches

11. Margo will sweep a room. A diagram of the floor that she needs to sweep is shown at the right. What is the area of the floor?

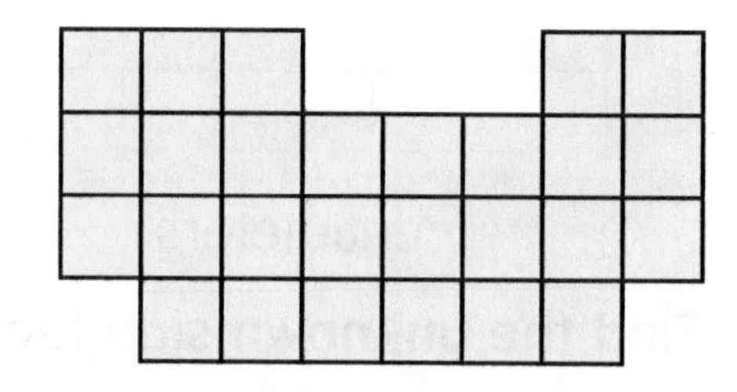

Ⓐ 8 square units　　Ⓒ 27 square units

Ⓑ 21 square units　　Ⓓ 32 square units

12. Jeff is making a poster for a car wash for the Campout Club. What is the perimeter of the poster?

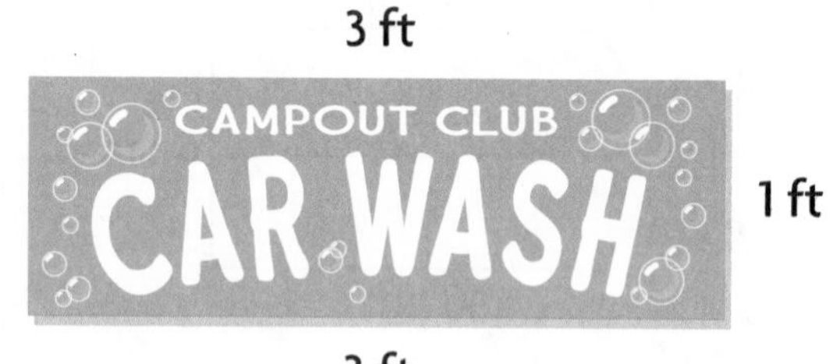

Ⓐ 8 feet　　Ⓒ 4 feet

Ⓑ 6 feet　　Ⓓ 3 feet

13. A rectangle has two side lengths of 8 inches and two side lengths of 10 inches. What is the perimeter of the rectangle?

Ⓐ 18 inches　　Ⓒ 32 inches

Ⓑ 28 inches　　Ⓓ 36 inches

Name ______________________________

Problem Solving • Area of Rectangles

Essential Question How can you use the strategy *find a pattern* to solve area problems?

UNLOCK the Problem REAL WORLD

Mr. Koi wants to build storage buildings, so he drew plans for the buildings. He wants to know how the areas of the buildings are related. How will the areas of Buildings *A* and *B* change? How will the areas of Buildings *C* and *D* change?

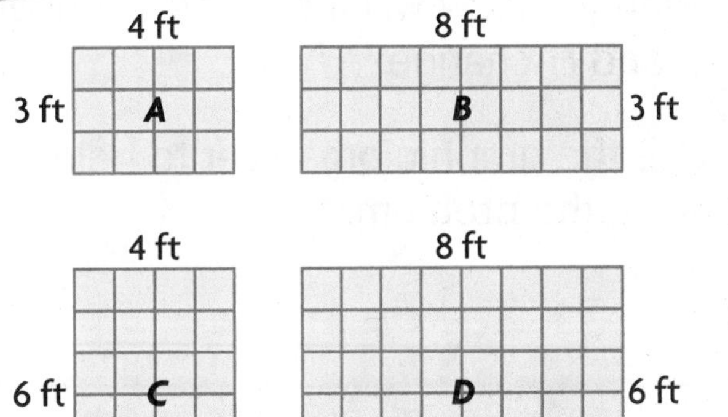

Use the graphic organizer to help you solve the problem.

Read the Problem

What do I need to find?

I need to find how the areas will change from *A* to *B* and from _____ to _____.

What information do I need to use?

I need to use the _____ and _____ of each building to find its area.

How will I use the information?

I will record the areas in a table. Then I will look for a pattern to see how the _____ will change.

Solve the Problem

I will complete the table to find patterns to solve the problem.

	Length	Width	Area		Length	Width	Area
Building *A*	3 ft			Building *C*		4 ft	
Building *B*	3 ft			Building *D*		8 ft	

I see that the lengths will be the same and the widths will be doubled.

The areas will change from _____ to _____ and from _____ to _____.

So, when the lengths are the same and the widths are doubled, the areas will be __________.

Try Another Problem

Mr. Koi is building more storage buildings. He wants to know how the areas of the buildings are related. How will the areas of Buildings *E* and *F* change? How will the areas of Buildings *G* and *H* change?

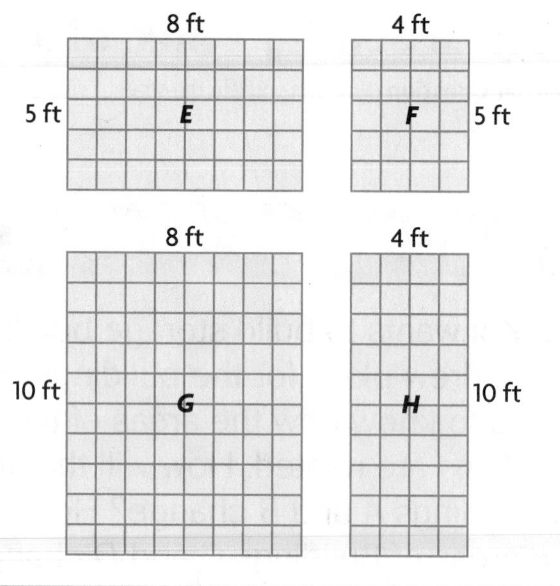

Use the graphic organizer to help you solve the problem.

Read the Problem

What do I need to find?	What information do I need to use?	How will I use the information?

Solve the Problem

I will complete the table to find patterns to solve the problem.

	Length	Width	Area		Length	Width	Area
Building *E*				Building *G*			
Building *F*				Building *H*			

- How did your table help you find a pattern?

____

____

____

MATHEMATICAL PRACTICES

Math Talk What if the length of both sides is doubled? How would the areas change?

Name ______________________________

Share and Show 

Use the table for 1–2.

1. Many pools come in rectangular shapes. How do the areas of the swimming pools change when the widths change?

   First, complete the table by finding the area of each pool.

   Think: I can find the area by multiplying the length and the width.

Swimming Pool Sizes			
Pool	Length (in feet)	Width (in feet)	Area (in square feet)
A	8	20	
B	8	30	
C	8	40	
D	8	50	

   Then, find a pattern of how the lengths change and how the widths change.

   The ________ stays the same.

   The widths ____________________.

   Last, describe a pattern of how the area changes.

   The areas ________ by ____ square feet.

2. H.O.T. **What if** the length of each pool was 16 feet? **Explain** how the areas would change.

   ____

   ____

3. Elizabeth built a sandbox that is 4 feet long and 4 feet wide. She also built a flower garden that is 4 feet long and 6 feet wide. She built a vegetable garden that is 4 feet long and 8 feet wide. How do the areas change?

   ____

   ____

   ____

On Your Own

Choose a STRATEGY

- Act It Out
- Draw a Diagram
- Find a Pattern
- Make a Table

4. Danielle is baking cookies. She makes 30 cookies and sets them in 6 equal rows on the baking tray. How many cookies are in each row? Draw a bar model to find your answer.

____

5. H.O.T. A diagram of Paula's bedroom is at the right. Her bedroom is in the shape of a rectangle. Write the measures for the other sides. What is the perimeter of the room? (Hint: The two pairs of opposite sides are equal lengths.)

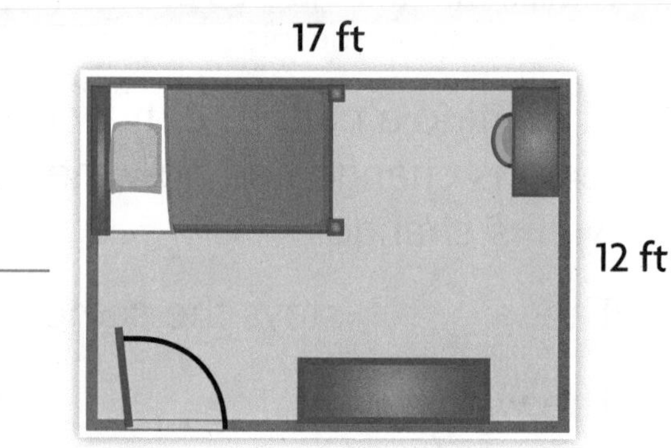

____

6. H.O.T. Jacob has a rectangular garden with an area of 56 square feet. The length of the garden is 8 feet. What is the width of the garden?

____

7. **Test Prep** Heather owns a carpet cleaning company. Her prices are based on the area of the carpet. What is the price of cleaning a room with an area of 200 square feet?

Total Area (in square feet)	50	100	150	200
Price (in dollars)	15	30	45	■

Ⓐ \$30 Ⓒ \$60

Ⓑ \$45 Ⓓ \$245

FOR MORE PRACTICE: Standards Practice Book, pp. P227–P228

Name ____________________

Area of Combined Rectangles

Essential Question How can you break apart a shape to find the area?

UNLOCK the Problem REAL WORLD

Anna's rug has side lengths of 4 feet and 9 feet. What is the area of Anna's rug?

Remember

You can use the Distributive Property to break apart an array.

$3 \times 3 = 3 \times (2 + 1)$

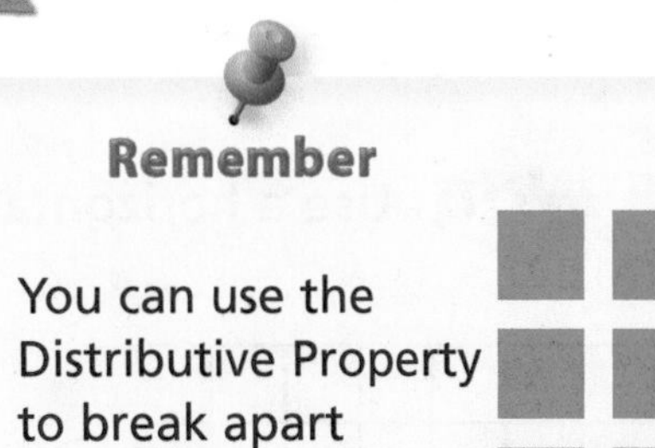

Activity **Materials** ■ square tiles

STEP 1 Use square tiles to model 4×9.

STEP 2 Draw a rectangle on the grid paper to show your model.

STEP 3 Draw a vertical line to break apart the model to make two smaller rectangles.

The side length 9 is broken into _____ plus _____.

STEP 4 Find the area of each of the two smaller rectangles.

Rectangle 1: _____ × _____ = _____

Rectangle 2: _____ × _____ = _____

STEP 5 Add the products to find the total area.

_____ + _____ = _____ square feet

STEP 6 Check your answer by counting the number of square feet.

_____ square feet

So, the area of Anna's rug is _____ square feet.

MATHEMATICAL PRACTICES

Math Talk Did you draw a line in the same place as your classmates? **Explain** why you found the same total area.

CONNECT Using the Distributive Property, you found that you could break apart a rectangle into smaller rectangles, and add the area of each smaller rectangle to find the total area.

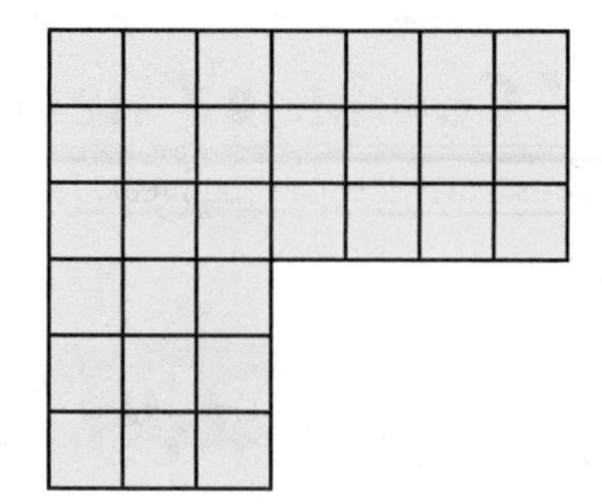

How can you break apart this shape into rectangles to find its area?

One Way Use a horizontal line.

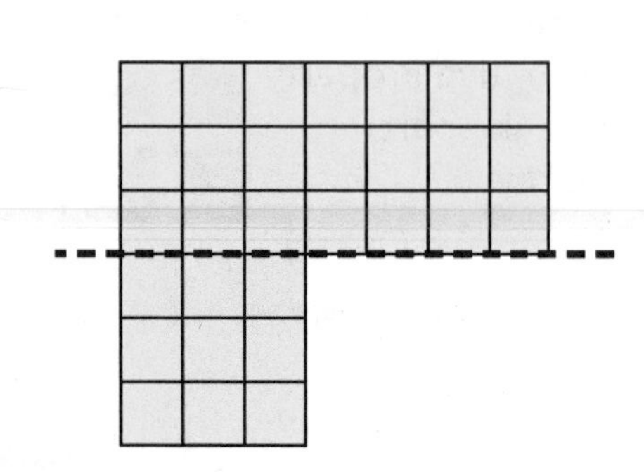

STEP 1 Write a multiplication equation for each rectangle.

Rectangle 1: ____ × ____ = ____

Rectangle 2: ____ × ____ = ____

STEP 2 Add the products to find the total area.

____ + ____ = ____ square units

Another Way Use a vertical line.

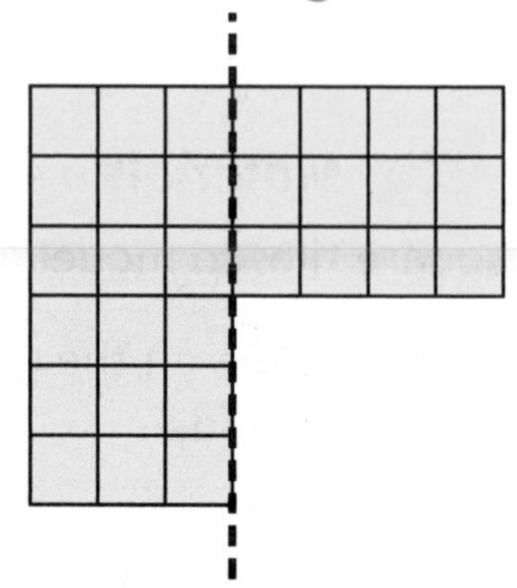

STEP 1 Write a multiplication equation for each rectangle.

Rectangle 1: ____ × ____ = ____

Rectangle 2: ____ × ____ = ____

STEP 2 Add the products to find the total area.

____ + ____ = ____ square units

So, the area is _____ square units.

MATHEMATICAL PRACTICES **Math Talk** Explain how you can check your answer.

Share and Show

1. Draw a line to break apart the shape into rectangles. Find the total area of the shape.

   Think: I can draw vertical or horizontal lines to break apart the shape to make rectangles.

   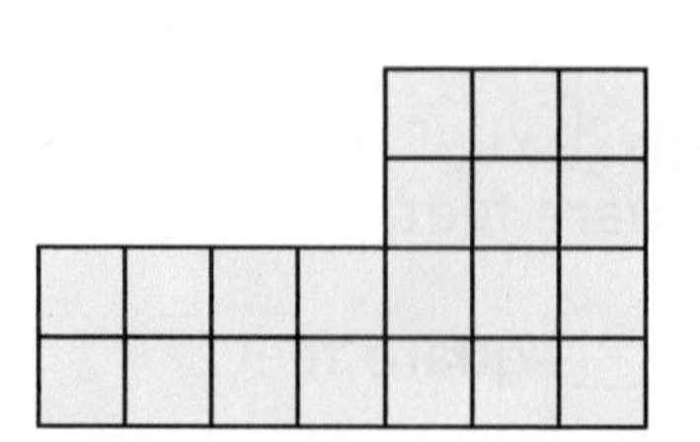

   Rectangle 1: ____ × ____ = ____

   Rectangle 2: ____ × ____ = ____

   ____ + ____ = ____ square units

Name ______________________

Use the Distributive Property to find the area. Show your multiplication and addition equations.

 **2.** 

____

_____ square units

 **3.** 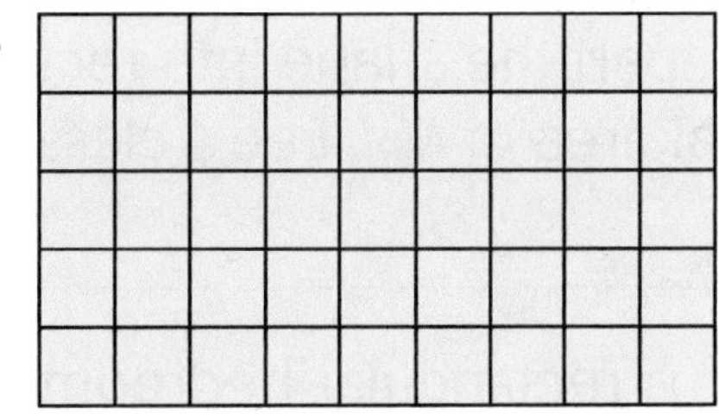

____

_____ square units

On Your Own

Use the Distributive Property to find the area. Show your multiplication and addition equations.

4. 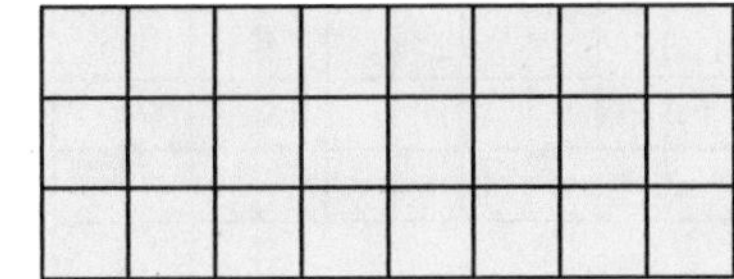

____

_____ square units

5. 

____

_____ square units

Draw a line to break apart the shape into rectangles. Find the area of the shape.

6. 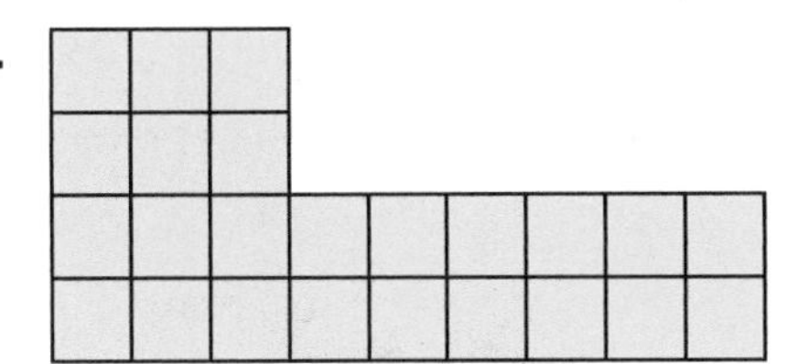

Rectangle 1: ____ × ____ = ____

Rectangle 2: ____ × ____ = ____

____ + ____ = ____ square units

7. H.O.T. 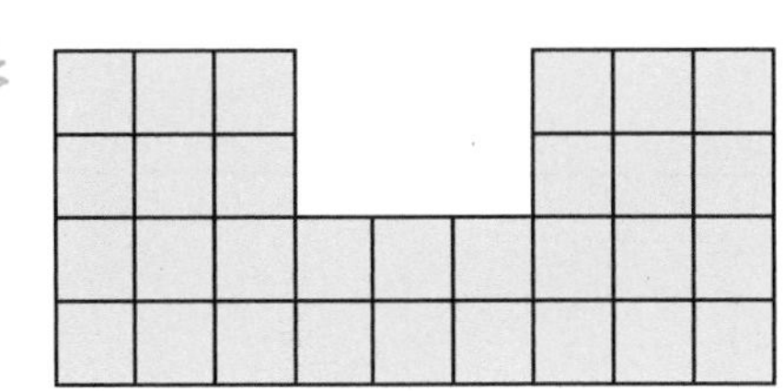

Rectangle 1: ____ × ____ = ____

Rectangle 2: ____ × ____ = ____

Rectangle 3: ____ × ____ = ____

____ + ____ + ____ = ____ square units

Problem Solving REAL WORLD

8. Ms. Lee's classroom is shown at the right. Each unit square is 1 square foot. Draw a line to break apart the shape into rectangles. What is the total area of Ms. Lee's classroom?

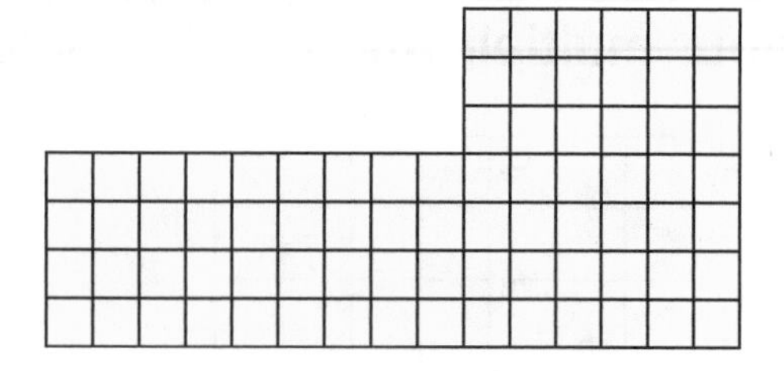

____

9. Jared has a rectangular bedroom with a rectangular closet. Each unit square is 1 square foot. Draw a line to break apart the shape into rectangles. What is the total area of Jared's bedroom?

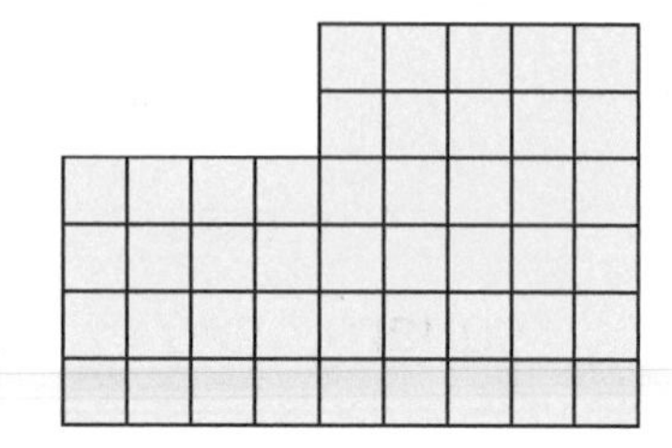

____

10. **Write Math** **Explain** how to break apart the shape to find the area of the shape.

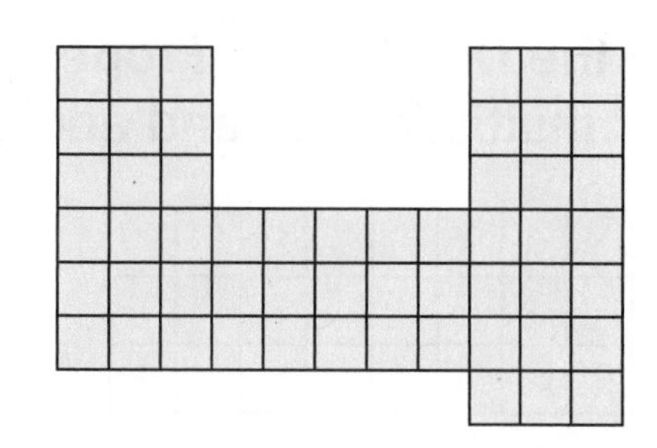

1 unit square = 1 square meter

____

____

____

____

____

11. H.O.T. Use the Distributive Property to find the area of the shape at the right. Write your multiplication and addition equations.

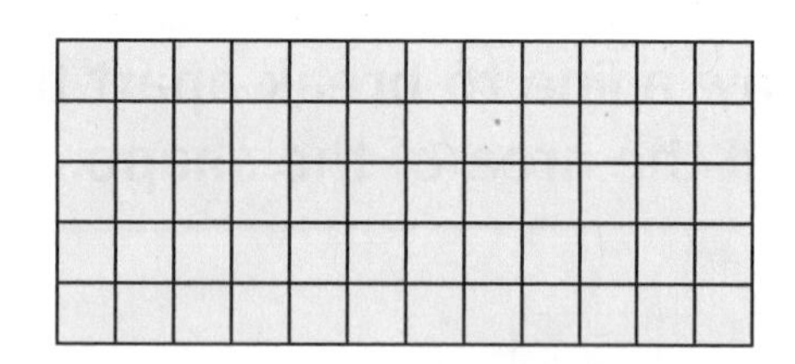

1 unit square = 1 square centimeter

____

____

12. **Test Prep** Pete drew a diagram of his backyard on grid paper. Each unit square is 1 square meter. The area surrounding the patio is grass. How many square meters of grass are in his backyard?

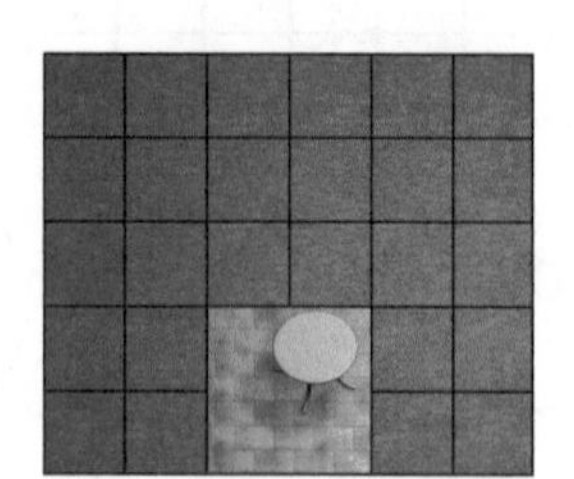

Ⓐ 4 square meters
Ⓑ 18 square meters
Ⓒ 26 square meters
Ⓓ 30 square meters

FOR MORE PRACTICE:
Standards Practice Book, pp. P229–P230

Name ________________________________

Same Perimeter, Different Areas

Essential Question How can you use area to compare rectangles with the same perimeter?

UNLOCK the Problem REAL WORLD

Toby has 12 feet of boards to put around a rectangular sandbox. How long should he make each side so that the area of the sandbox is as large as possible?

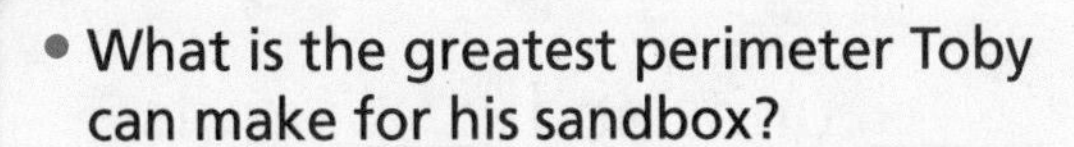

- What is the greatest perimeter Toby can make for his sandbox?

____

Activity

Materials ■ square tiles

Use square tiles to make all the rectangles you can that have a perimeter of 12 units. Draw and label the sandboxes. Then find the area of each.

Sandbox 1

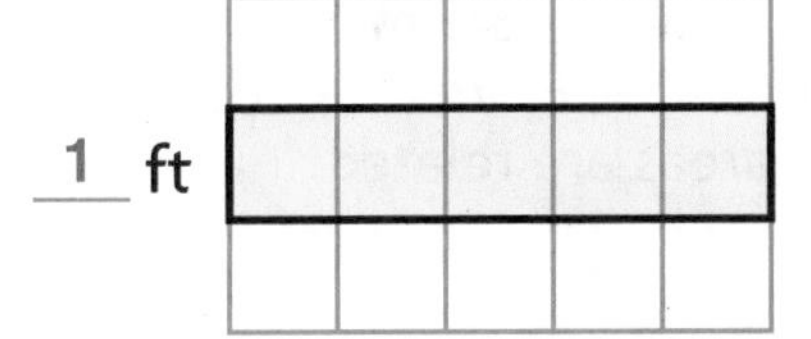

1 ft

5 ft

Sandbox 2

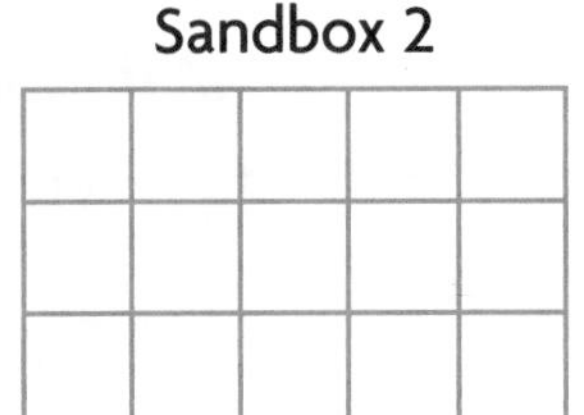

___ ft

___ ft

Sandbox 3

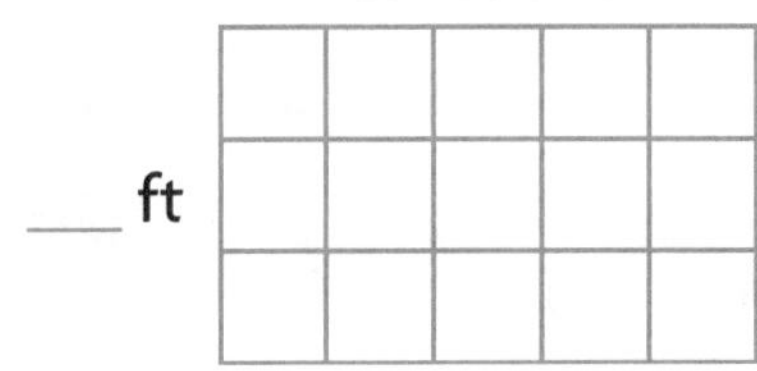

___ ft

___ ft

Find the perimeter and area of each rectangle.

	Perimeter	Area
Sandbox 1	5 + 1 + 5 + 1 = 12 feet	1 × 5 = ___ square feet
Sandbox 2	___ + ___ + ___ + ___ = ___ feet	___ × ___ = ___ square feet
Sandbox 3	___ + ___ + ___ + ___ = ___ feet	___ × ___ = ___ square feet

The area of Sandbox _____ is the greatest.

So, Toby should build a sandbox that is

_____ feet wide and _____ feet long.

MATHEMATICAL PRACTICES

Math Talk How are the sandboxes alike? How are the sandboxes different?

Examples Draw rectangles with the same perimeter and different areas.

A Draw a rectangle that has a perimeter of 20 units and an area of 24 square units.

The sides of the rectangle measure

_____ units and _____ units.

B Draw a rectangle that has a perimeter of 20 units and an area of 25 square units.

The sides of the rectangle measure

_____ units and _____ units.

MATHEMATICAL PRACTICES

Math Talk **Explain** how the perimeters of Example *A* and Example *B* are related. **Explain** how the areas are related.

Share and Show

MATH BOARD

1. The perimeter of the rectangle at the right is

   _____ units. The area is _____ square units.

2. Draw a rectangle that has the same perimeter as the rectangle in Exercise 1 but with a different area.

3. The area of the rectangle in Exercise 2 is

   _____ square units.

4. Which rectangle has the greater area?

   _____

5. If you were given a rectangle with a certain perimeter, how would you draw it so that it has the greatest area?

   _____

MATHEMATICAL PRACTICES

Math Talk **Explain** how you knew what the rectangle for Exercise 5 would look like.

Name ______________________

Find the perimeter and the area. Tell which rectangle has a greater area.

6. 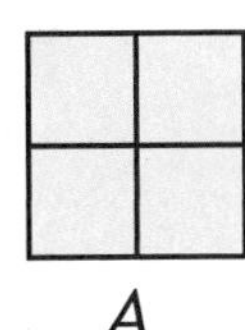

A

B

A: Perimeter = ______; Area = ____________

B: Perimeter = ______; Area = ____________

Rectangle ___ has a greater area.

On Your Own

Find the perimeter and the area. Tell which rectangle has a greater area.

7. 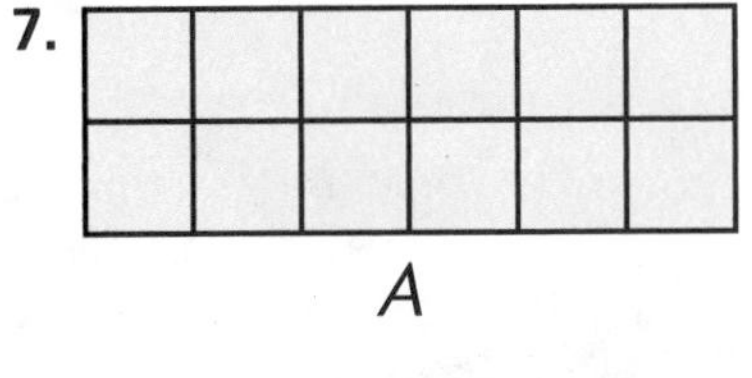

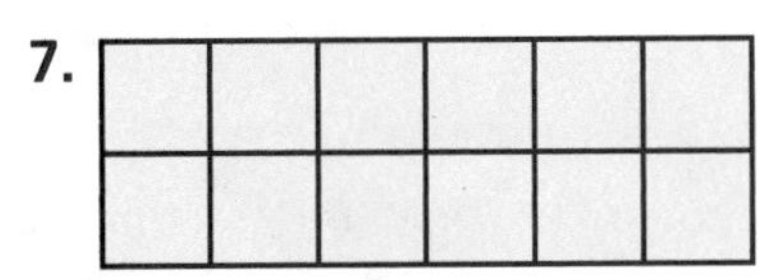

A

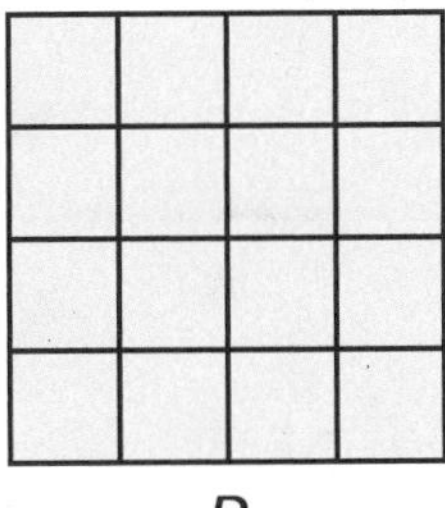

B

A: Perimeter = ______;

Area = ____________

B: Perimeter = ______;

Area = ____________

Rectangle ___ has a greater area.

8. 

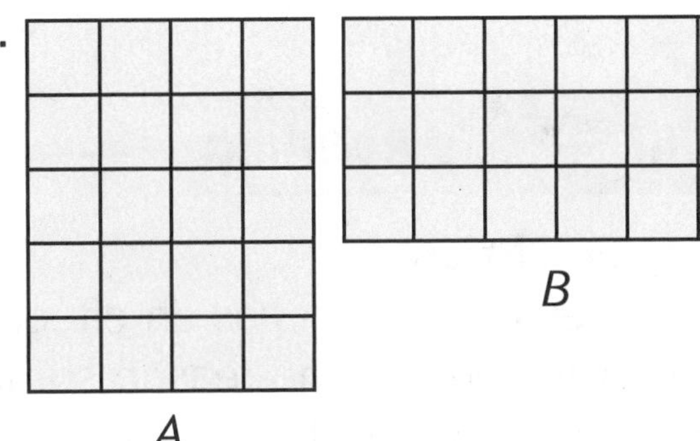

A

B

A: Perimeter = ______;

Area = ____________

B: Perimeter = ______;

Area = ____________

Rectangle ___ has a greater area.

9. H.O.T. Draw a rectangle with the same perimeter as Rectangle *C*, but with a smaller area. What is the area?

Area = ____________

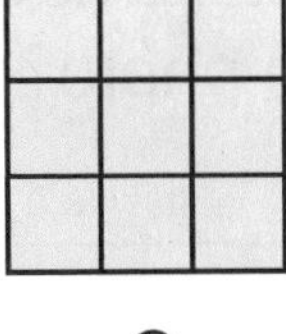

C

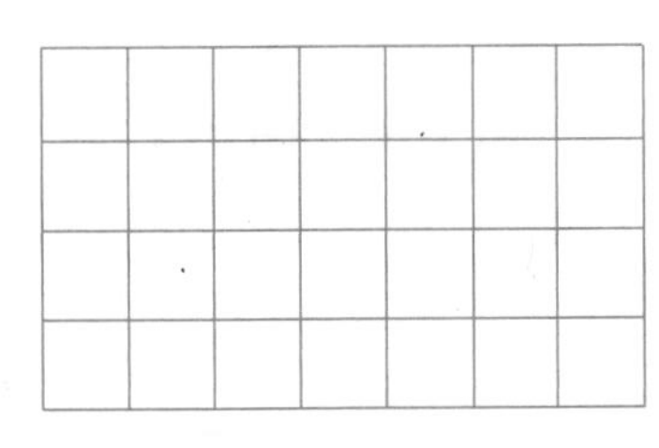

Problem Solving REAL WORLD

10. Write Math What's the Question? Todd's flower garden is 4 feet wide and 8 feet long. The answer is 32 square feet.

____

11. Test Prep Which shape has a perimeter of 20 units and an area of 16 square units?

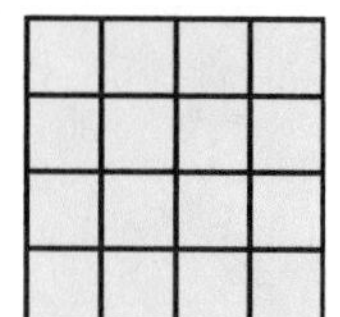

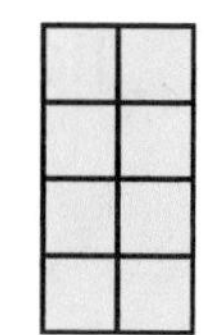

(B)
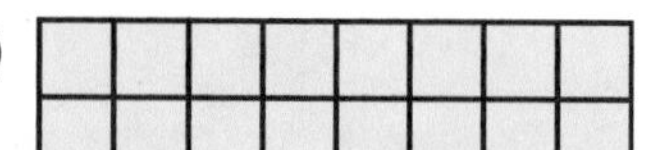

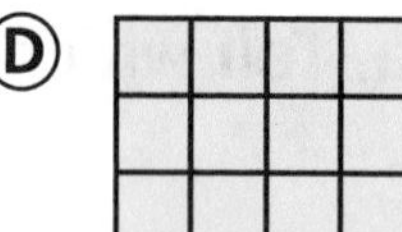

Connect to Reading

Cause and Effect

Sometimes one action has an effect on another action. The *cause* is the reason something happens. The *effect* is the result.

Sam wanted to print a digital photo that is 3 inches wide and 5 inches long. What if Sam accidentally printed a photo that is 4 inches wide and 6 inches long?

Sam can make a table to understand cause and effect.

Cause	Effect
The wrong size photo was printed.	Each side of the photo is a greater length.

Use the information and the strategy to solve the problems.

12. What effect did the mistake have on the perimeter of the photo?

____

____

13. What effect did the mistake have on the area of the photo?

____

____

FOR MORE PRACTICE:
Standards Practice Book, pp. P231–P232

Name ___________________________

Same Area, Different Perimeters

Essential Question How can you use perimeter to compare rectangles with the same area?

UNLOCK the Problem REAL WORLD

Marcy is making a rectangular pen to hold her rabbits. The area of the pen should be 16 square meters with side lengths that are whole numbers. What is the least amount of fencing she needs?

- What does the least amount of fencing represent?

____

Activity **Materials** ■ square tiles

Use 16 square tiles to make rectangles. Make as many different rectangles as you can with 16 tiles. Record the rectangles on the grid, write the multiplication equation for the area shown by the rectangle, and find the perimeter of each rectangle.

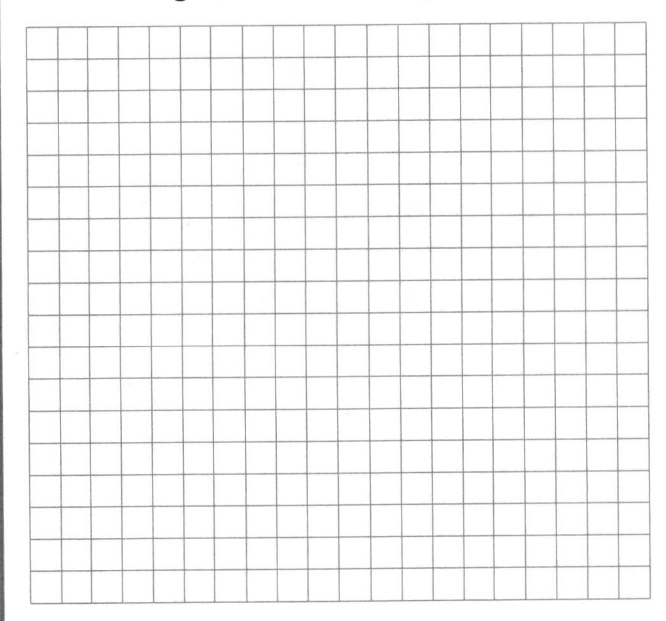

MATHEMATICAL PRACTICES

Math Talk Explain how you found the rectangles.

Area: _____ × _____ = 16 square meters Perimeter: _____ meters

Area: _____ × _____ = 16 square meters Perimeter: _____ meters

Area: _____ × _____ = 16 square meters Perimeter: _____ meters

To use the least amount of fencing, Marcy should make a rectangular pen with side lengths of _____ meters and _____ meters.

So, _____ meters is the least amount of fencing Marcy needs.

Try This!

Draw three rectangles that have an area of 18 square units on the grid. Find the perimeter of each rectangle. Shade the rectangle that has the greatest perimeter.

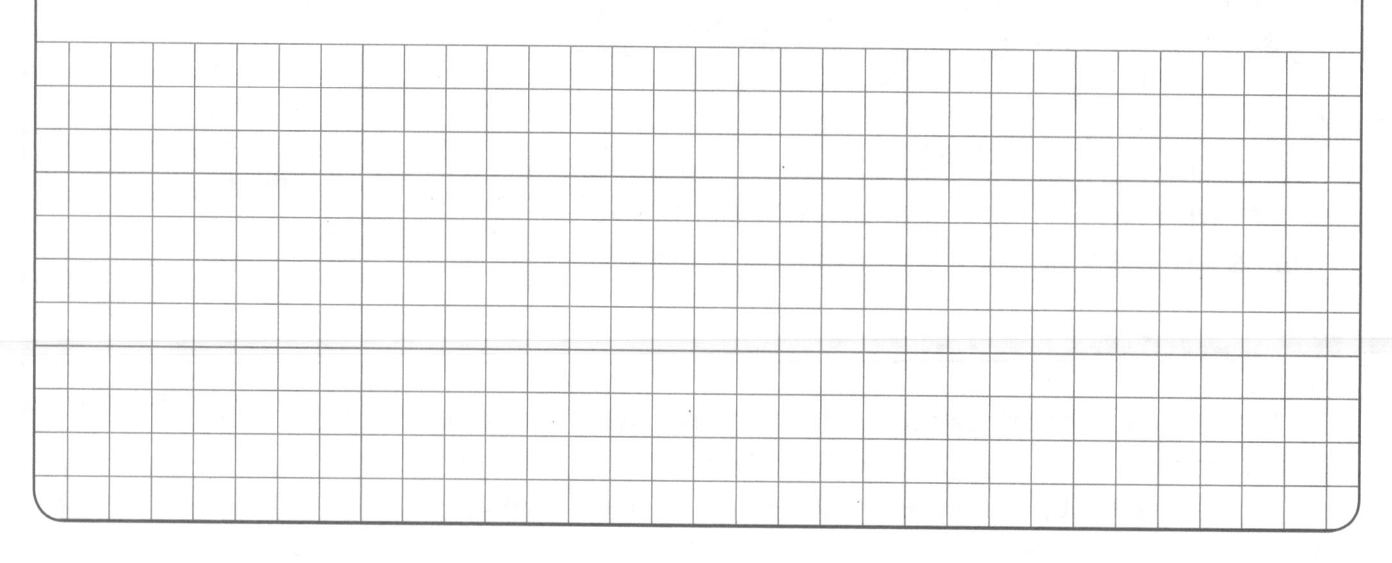

Share and Show

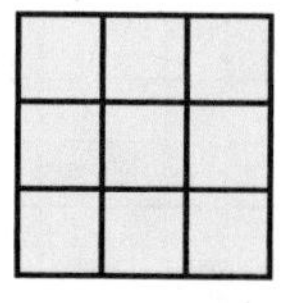

1. The area of the rectangle at the right is _____ square units. The perimeter is _____ units.

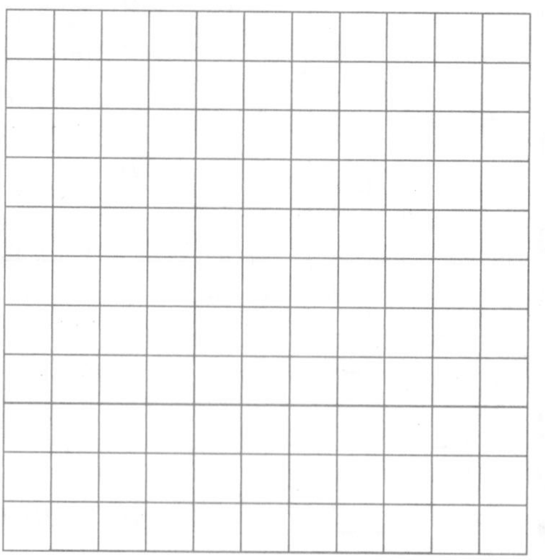

2. Draw a rectangle that has the same area as the rectangle in Exercise 1 but with a different perimeter.

3. The perimeter of the rectangle in Exercise 2 is _____ units.

4. Which rectangle has the greater perimeter?

____

5. If you were given a rectangle with a certain area, how would you draw it so that it had the greatest perimeter?

____

MATHEMATICAL PRACTICES

Math Talk Did you and your classmate draw the same rectangle for Exercise 2? Explain.

Name ______________________

Find the perimeter and the area. Tell which rectangle has a greater perimeter.

6. 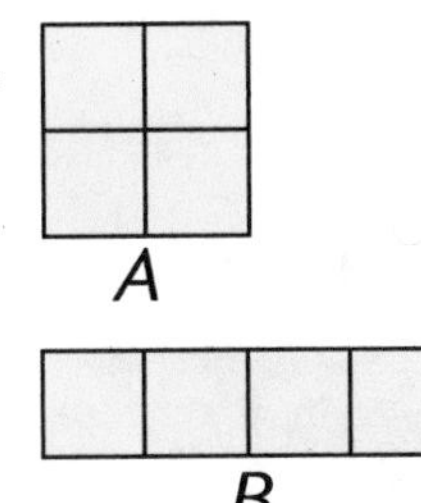

A

B

A: Area = ______________; Perimeter = __________

B: Area = ______________; Perimeter = __________

Rectangle ______ has a greater perimeter.

On Your Own

Find the perimeter and the area. Tell which rectangle has a greater perimeter.

7. 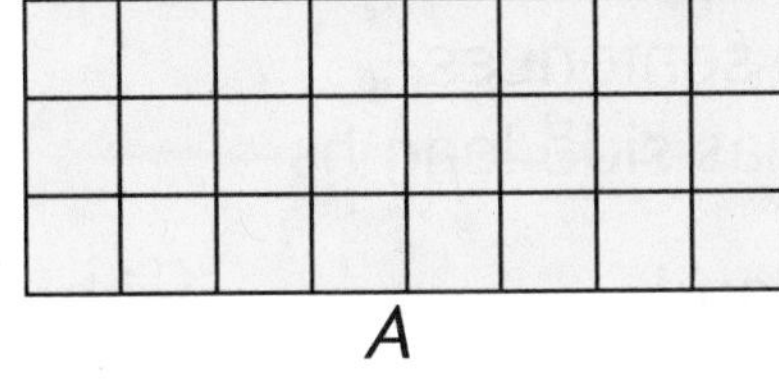

A

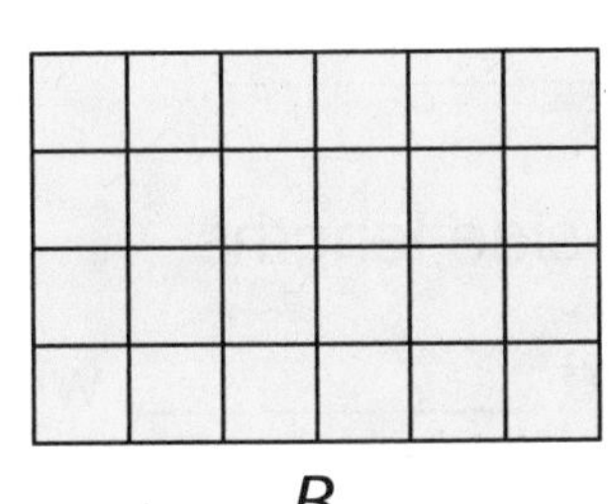

B

A: Area = ______________;

Perimeter = __________

B: Area = ______________;

Perimeter = __________

Rectangle ______ has a greater perimeter.

8. 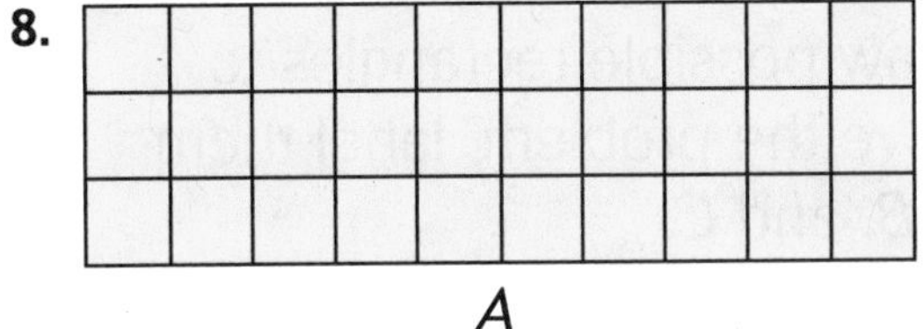

A

B

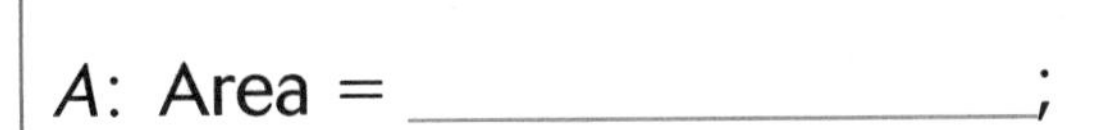

A: Area = ______________;

Perimeter = __________

B: Area = ______________;

Perimeter = __________

Rectangle ______ has a greater perimeter.

9. H.O.T. **Sense or Nonsense?** Dora says that of all the possible rectangles with the same area, the rectangle with the largest perimeter will have two side lengths that are 1 unit. Does her statement make sense? **Explain.**

____

UNLOCK the Problem REAL WORLD

10. Ed has 12 tiles. Each tile is 1 square inch. He will arrange them into a rectangle and glue 1-inch stones around the edge. How can Ed arrange the tiles so that he uses the least number of stones?

a. What do you need to find? ______________________

____

b. How will you use what you know about perimeter to help you solve the problem?

____

c. Draw possible rectangles to solve the problem, label them *A*, *B*, and *C*.

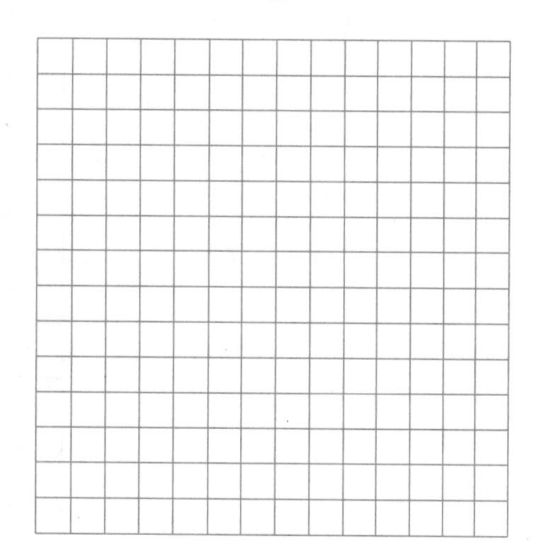

d. Complete the sentences.
Rectangle *A* has side lengths ________ and ________ with a perimeter of ________.

Rectangle *B* has side lengths ________ and ________ with a perimeter of ________.

Rectangle *C* has side lengths ________ and ________ with a perimeter of ________.

So, Ed should arrange the tiles like Rectangle _____.

11. **Test Prep** Which shape has an area of 15 square units and a perimeter of 16 units?

 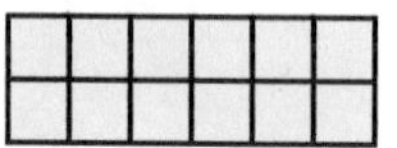

 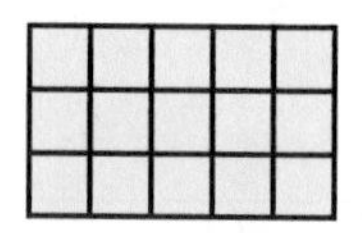

 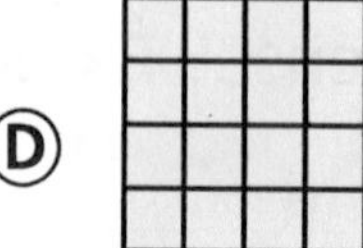

FOR MORE PRACTICE:
Standards Practice Book, pp. P233–P234

Name ___________________________________

Chapter Review/Test

Vocabulary

Choose the best term from the box.

Vocabulary
area
length
perimeter
unit squares

1. You can find the __________ of a shape by adding the lengths of the sides. (p. 433)
2. You can find the __________ of a rectangle by multiplying the number of unit squares in each row by the number of rows. (p. 445)
3. You can count __________ to find the area of a shape. (p. 445)

Concepts and Skills

Draw a line to break apart the shape into rectangles. Find the area of the shape.

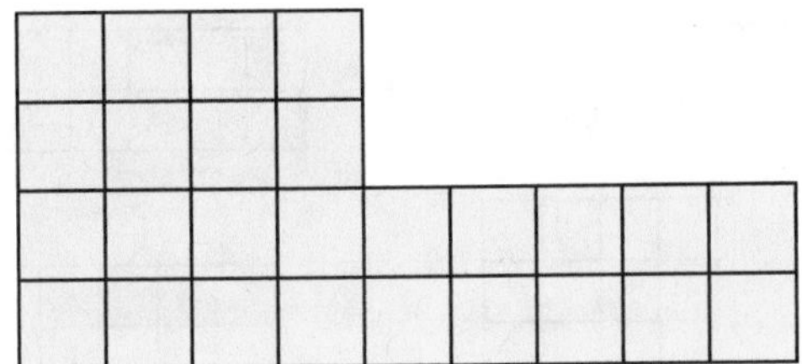

Rectangle 1: ____ × ____ = ____

Rectangle 2: ____ × ____ = ____

____ + ____ = ____ square units

5.

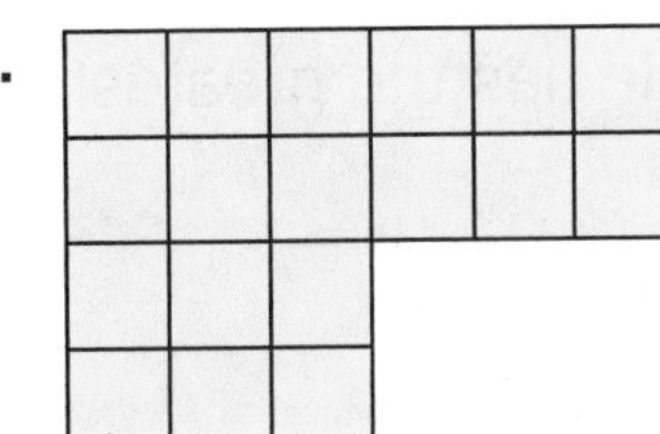

Rectangle 1: ____ × ____ = ____

Rectangle 2: ____ × ____ = ____

____ + ____ = ____ square units

Find the perimeter and the area. Tell which rectangle has a greater area.

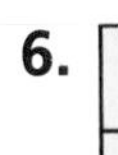

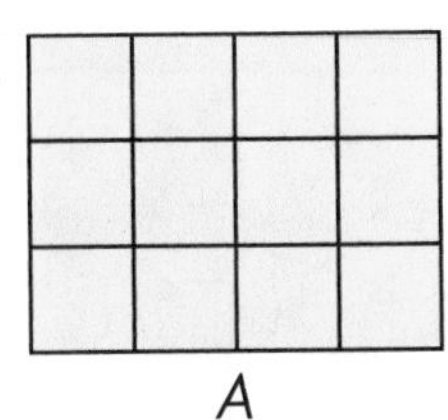

A

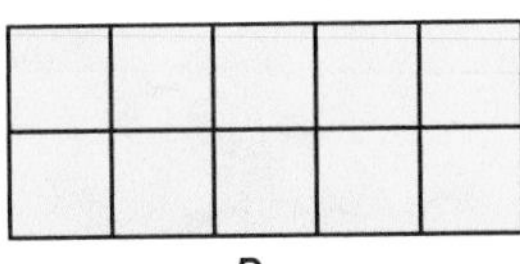

B

A: Perimeter = ______ Area = ____________;

B: Perimeter = ______ Area = ____________;

Rectangle ____ has a greater area.

GO Online Assessment Options Chapter Test

Fill in the bubble for the correct answer choice.

7. Peter drew this rectangle. Which multiplication equation can be used to find the area?

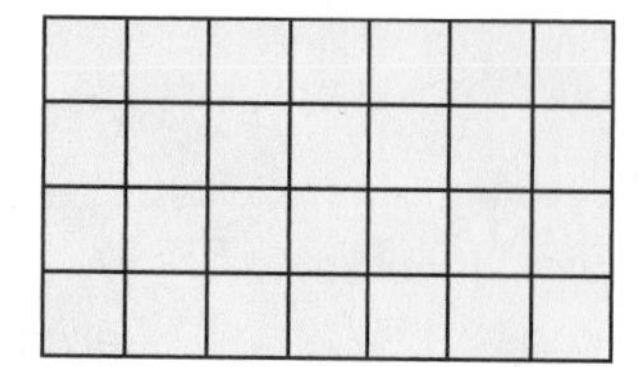

Ⓐ $4 \times 4 = 16$ Ⓒ $4 + 7 + 4 + 7 = 22$

Ⓑ $7 + 7 + 7 + 7 = 28$ Ⓓ $4 \times 7 = 28$

8. Susan cuts a piece of fabric with side lengths of 3 feet and 8 feet. What is the perimeter of the piece of fabric?

Ⓐ 22 feet

Ⓑ 24 feet

Ⓒ 32 feet

Ⓓ 64 feet

9. Which rectangle has the greatest area?

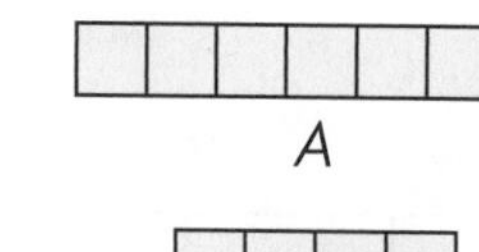

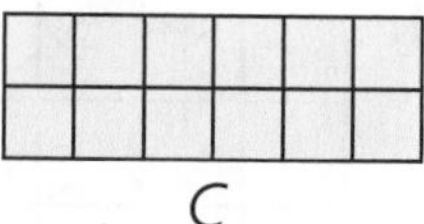

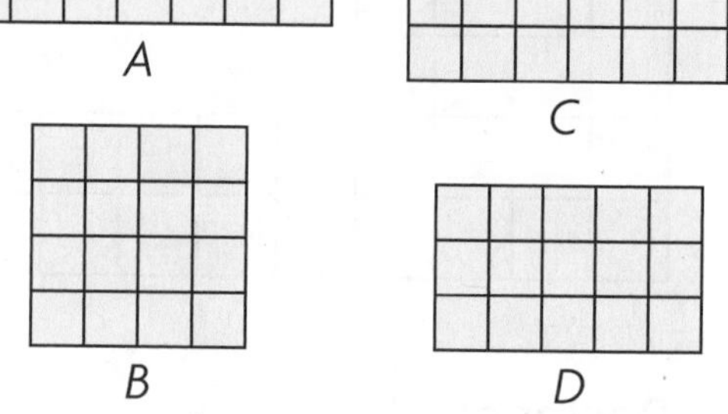

Ⓐ Rectangle *A*

Ⓑ Rectangle *B*

Ⓒ Rectangle *C*

Ⓓ Rectangle *D*

10. Jacob built a toolbox with a perimeter of 70 inches. What is the length of side *m*?

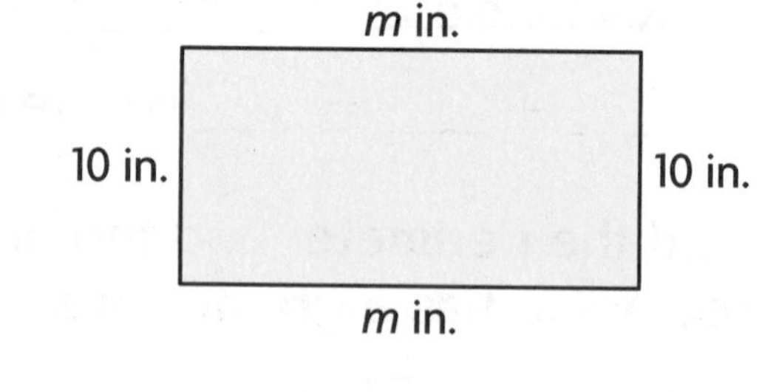

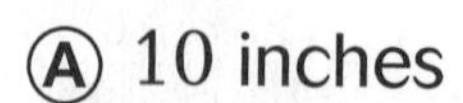

Ⓐ 10 inches

Ⓑ 20 inches

Ⓒ 25 inches

Ⓓ 50 inches

11. What is the area of the shape shown? Each square is 1 square foot.

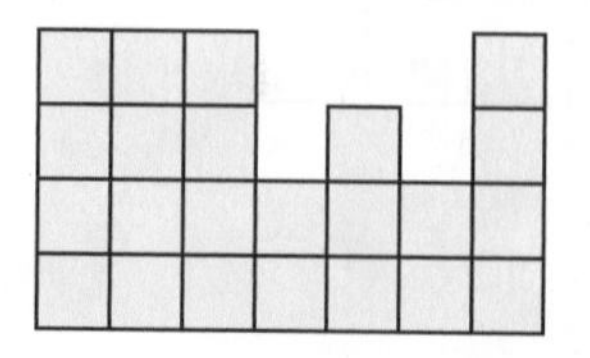

Ⓐ 20 square feet Ⓒ 27 square feet

Ⓑ 23 square feet Ⓓ 28 square feet

Name ________________________________

Fill in the bubble for the correct answer choice.

12. Alfredo used the Distributive Property to find the area of this rectangle. Which set of multiplication and addition equations could he have used?

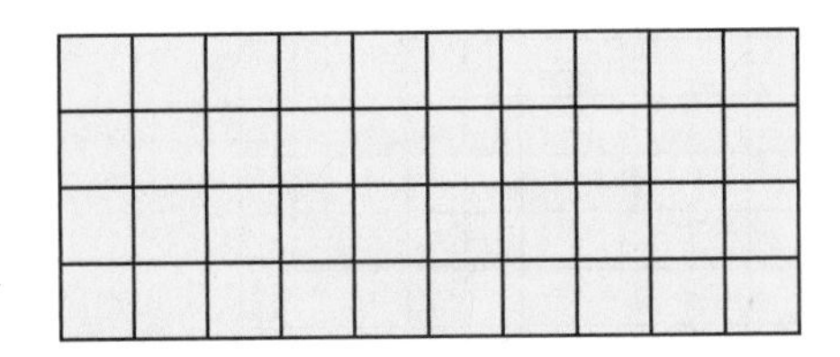

Ⓐ $4 + 5 = 9$; $4 + 5 = 9$; $9 + 9 = 18$

Ⓑ $4 + 5 = 9$; $4 + 5 = 9$; $9 \times 9 = 81$

Ⓒ $4 \times 5 = 20$; $4 \times 5 = 20$; $20 + 20 = 40$

Ⓓ $4 \times 10 = 40$; $4 \times 10 = 40$; $40 + 40 = 80$

13. What is the area of the shape?

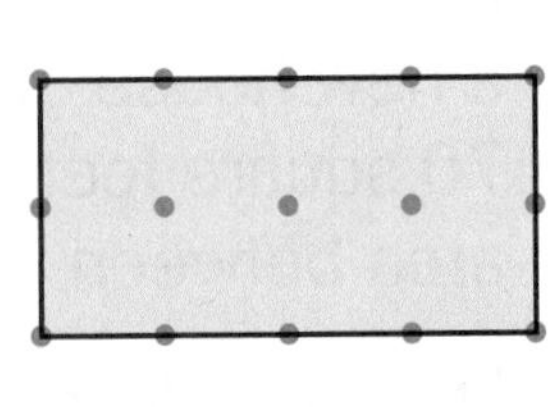

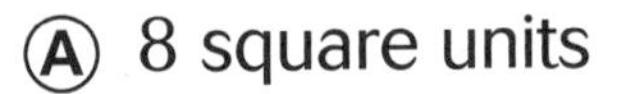

Ⓐ 8 square units

Ⓑ 8 units

Ⓒ 12 square units

Ⓓ 12 units

14. What is the perimeter of the shape? Each unit is 1 centimeter.

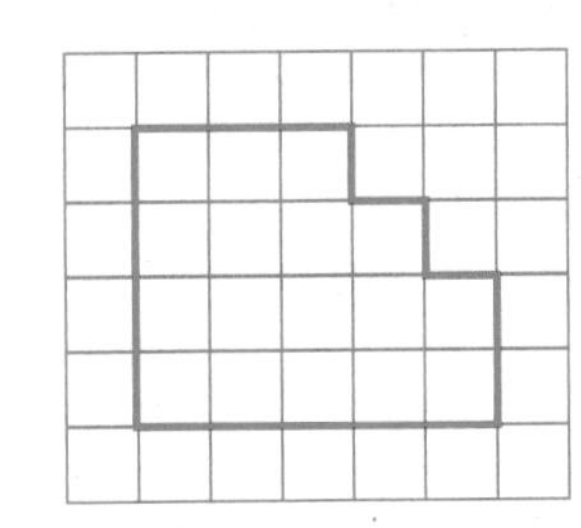

Ⓐ 9 centimeters

Ⓑ 18 centimeters

Ⓒ 20 centimeters

Ⓓ 23 centimeters

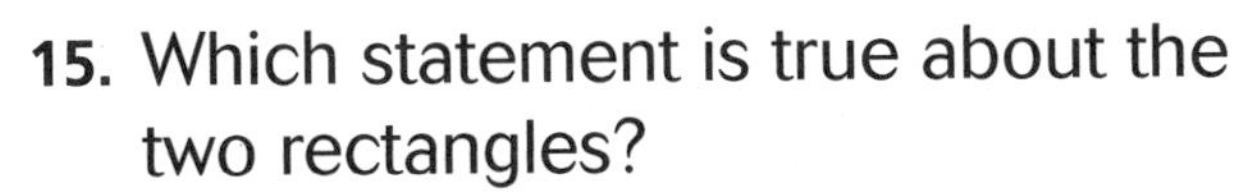

15. Which statement is true about the two rectangles?

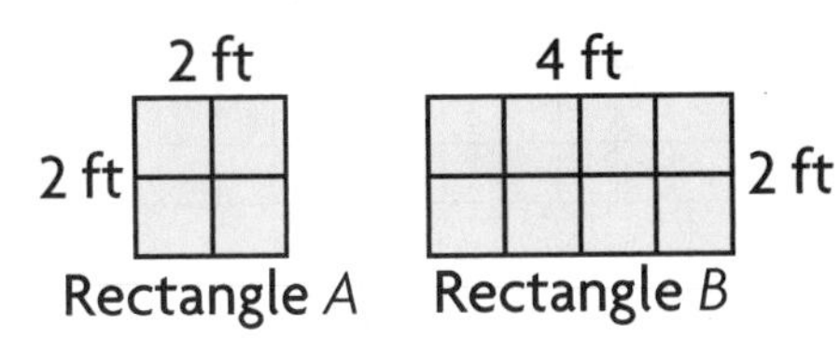

Ⓐ The area of Rectangle *B* is double the area of Rectangle *A*.

Ⓑ The area of Rectangle *A* is double the area of Rectangle *B*.

Ⓒ The area of Rectangle *B* is half of the area of Rectangle *A*.

Ⓓ The area of Rectangle *A* is the same as the area of Rectangle *B*.

▶ Constructed Response

16. Anthony wants to make two gardens, each with an area of 24 square feet. He will buy boards to go around each garden. The garden has side lengths that are whole numbers. Draw two of the possible gardens. Which garden will use fewer boards? **Explain.**

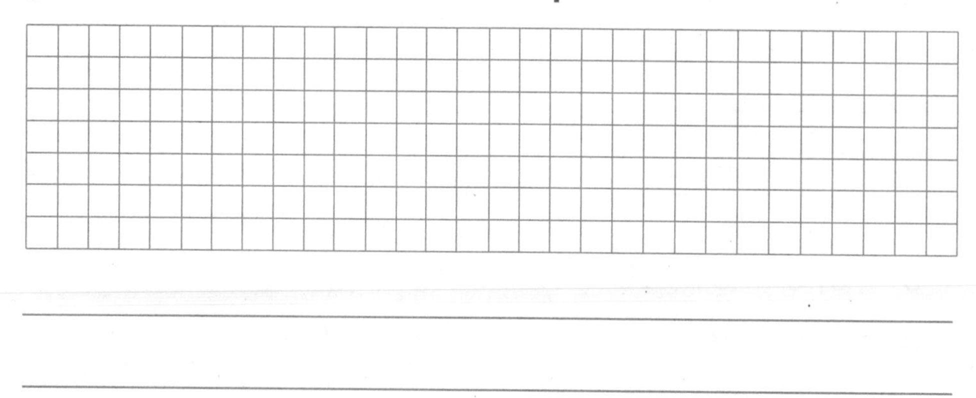

____

____

▶ Performance Task

17. Mr. Wicks designs houses. He is using grid paper to plan a new house design. The kitchen will have an area between 70 square feet and 85 square feet. The pantry will have an area between 4 square feet and 15 square feet.

A Draw and label a diagram to show what Mr. Wicks could design. Find the area of the kitchen. Find the area of the pantry.

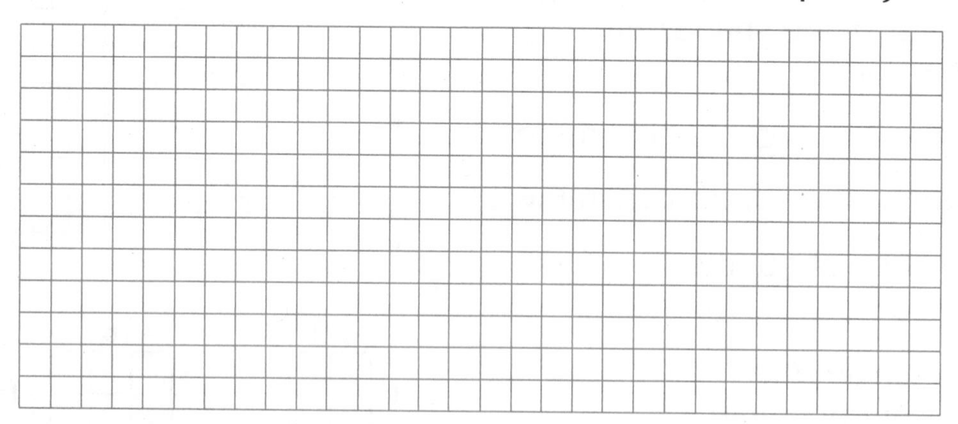

Area of kitchen = _____ square feet

Area of pantry = _____ square feet

B Find the total area of the kitchen and pantry combined.

______ square feet

Geometry

Describing and analyzing two-dimensional shapes

Students at Dommerich Elementary helped design and construct a mosaic to show parts of their community and local plants and animals.

Project

Make a Mosaic

Have you ever worked to put puzzle pieces together to make a picture or design? Pieces of paper can be put together to make a colorful work of art called a mosaic.

Get Started

Materials ■ construction paper ■ glue ■ ruler ■ scissors

Work with a partner to make a paper mosaic. Use the Important Facts to help you.

- Draw a simple pattern on a piece of paper.
- Cut out shapes, such as rectangles, squares, and triangles of the colors you need from construction paper. The shapes should be about 1 inch on each side.
- Glue the shapes into the pattern. Leave a little space between each shape to make the mosaic effect.

Important Facts

- Mosaics is the art of using small pieces of materials, such as tiles or glass, to make a colorful picture or design.
- Mosaic pieces can be small plane shapes, such as rectangles, squares, and triangles.
- Mosaic designs and patterns can be anything from simple flower shapes to common objects found in your home or patterns in nature.

Describe and compare the shapes you used to make your mosaic.

____

____

____

____

____

Completed by ______________________________

Chapter 12

Two-Dimensional Shapes

Show What You Know

Check your understanding of important skills.

Name ______________________________

▶ Plane Shapes

1. Color the triangles blue.

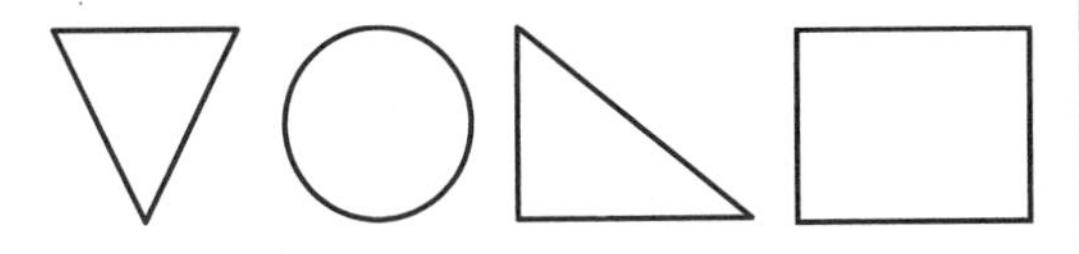

2. Color the rectangles red.

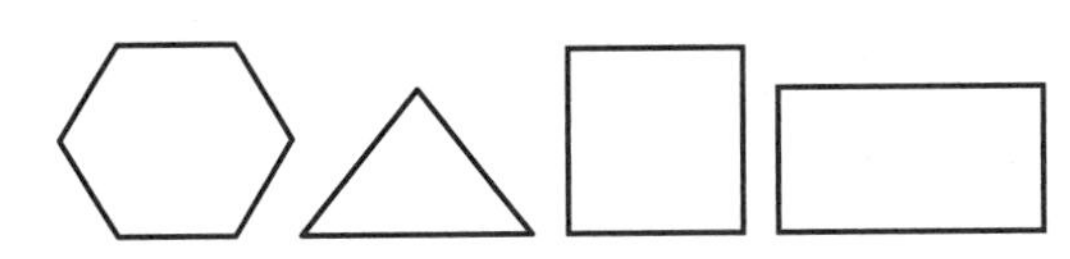

▶ Number of Sides

Write the number of sides.

3. ______ sides

4. ______ sides

5. Circle the shapes that have 4 or more sides.

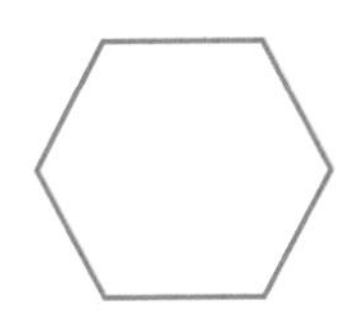  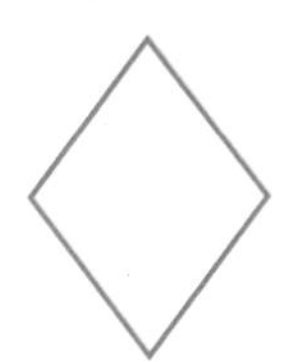  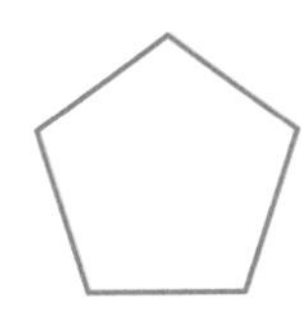 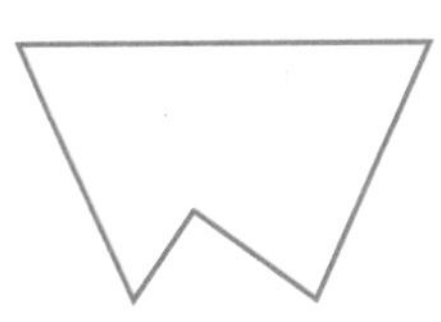

Whitney found this drawing that shows 9 small squares. Be a Math Detective to find larger squares in the drawing. How many squares are there in all? Explain.

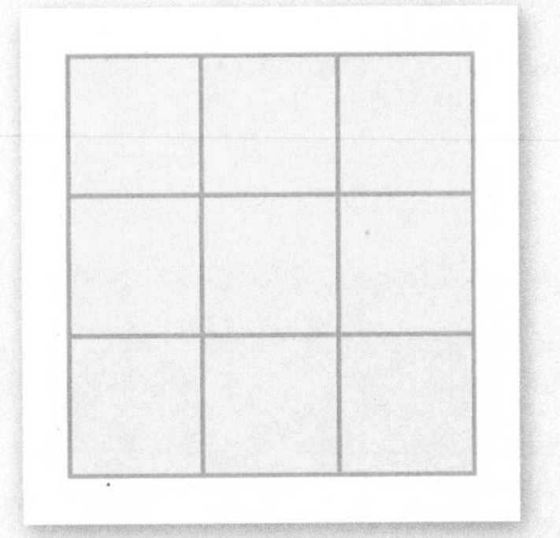

Vocabulary Builder

▶ Visualize It

Complete the tree map by using the words with a ✓.

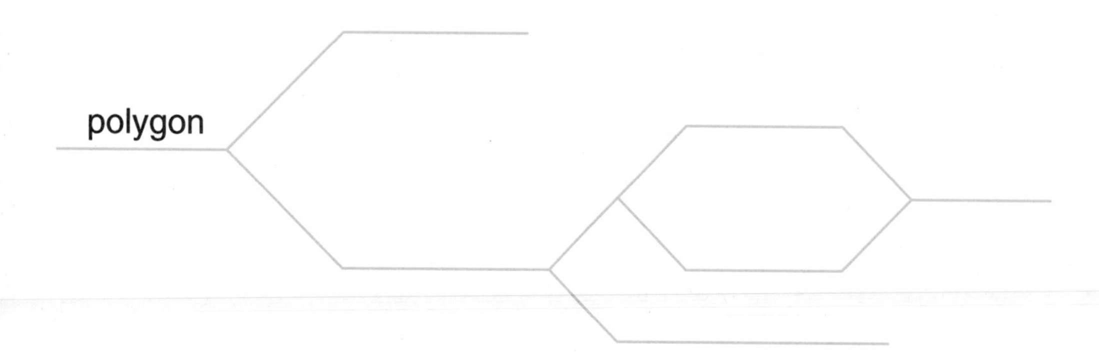

▶ Understand Vocabulary

Draw a line to match the word with its definition.

1. closed shape •	• A part of a line that includes two endpoints and all the points between them
2. line segment •	• A shape formed by two rays that share an endpoint
3. right angle •	• A shape that starts and ends at the same point
4. hexagon •	• An angle that forms a square corner
5. angle •	• A closed plane shape made up of line segments
6. polygon •	• A polygon with 6 sides and 6 angles

Preview Words

- angle
- closed shape
- hexagon
- intersecting lines
- line
- line segment
- open shape
- parallel lines
- perpendicular lines
- point
- polygon
- ✓ quadrilateral
- ray
- ✓ rectangle
- ✓ rhombus
- right angle
- ✓ square
- ✓ trapezoid
- ✓ triangle
- Venn diagram
- vertex

GO Online • eStudent Edition • Multimedia eGlossary

Name ______________________________

Describe Plane Shapes

Essential Question What are some ways to describe two-dimensional shapes?

UNLOCK the Problem REAL WORLD

An architect draws plans for houses, stores, offices, and other buildings. Look at the shapes in the drawing at the right.

A **plane shape** is a shape on a flat surface. It is formed by points that make curved paths, line segments, or both.

point

- is an exact position or location

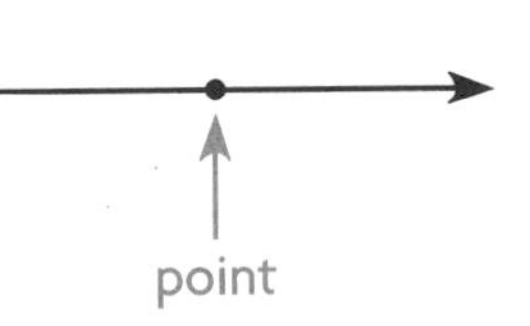

line

- is a straight path
- continues in both directions
- does not end

endpoints

- points that are used to show segments of lines

line segment

- is straight
- is part of a line
- has 2 endpoints

ray

- is straight
- is part of a line
- has 1 endpoint
- continues in one direction

Some plane shapes are made by connecting line segments at their endpoints. One example is a square. Describe a square using math words.

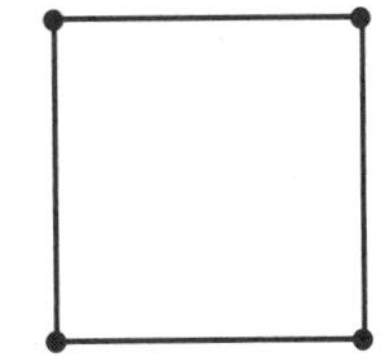

Think: How many line segments and endpoints does a square have?

A square has _____ line segments. The line segments meet only at their ____________.

MATHEMATICAL PRACTICES

Math Talk **Explain** why you cannot measure the length of a line.

Plane shapes have length and width but no thickness, so they are also called **two-dimensional shapes**.

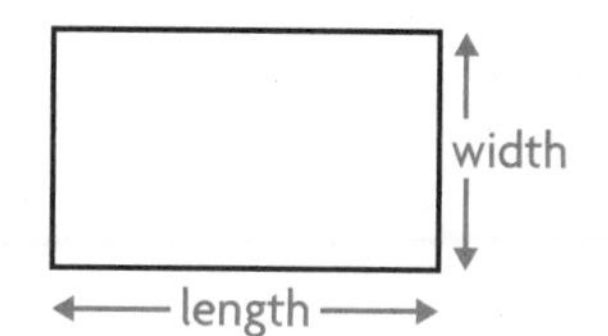

Try This! **Draw plane shapes.**

Plane shapes can be open or closed.

A **closed shape** starts and ends at the same point.

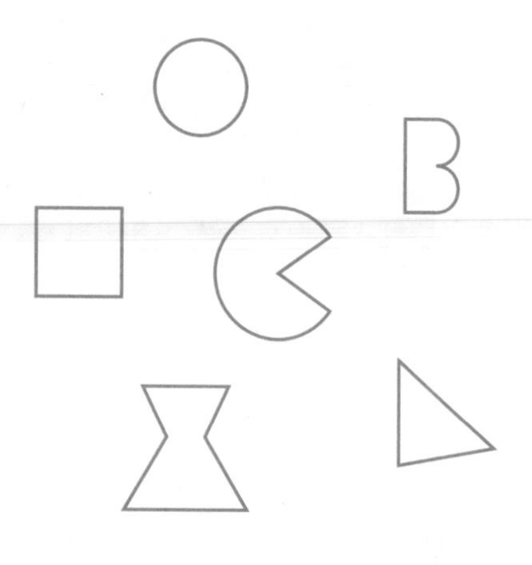

In the space below, draw more examples of closed shapes.

An **open shape** does not start and end at the same point.

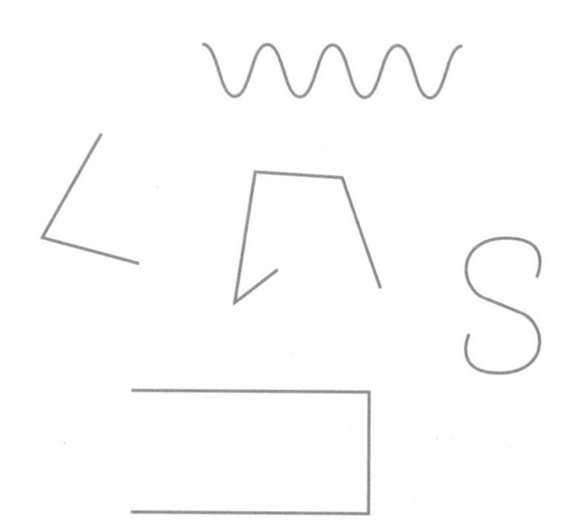

In the space below, draw more examples of open shapes.

MATHEMATICAL PRACTICES

Math Talk **Explain** whether a shape with a curved path must be a closed shape, an open shape, or can be either.

- Is the plane shape at the right a closed shape or an open shape? **Explain** how you know.

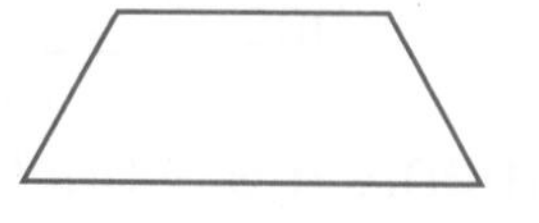

____

____

Name ______________________________

Share and Show 

1. Write how many line segments the shape has. __________

Circle all the words that describe the shape.

2. 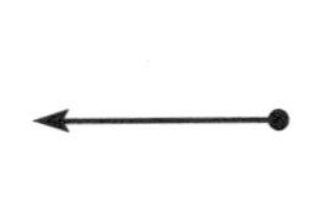

ray

point

3. 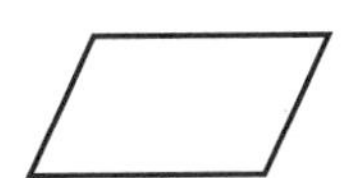

open shape

closed shape

✓4. 

open shape

closed shape

5. 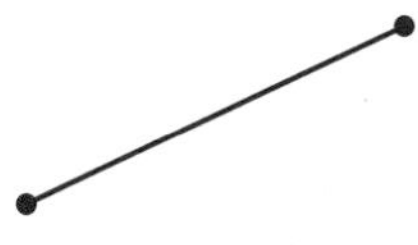

line

line segment

Write whether the shape is *open* or *closed*.

6. 

____

7. 

____

✓8. 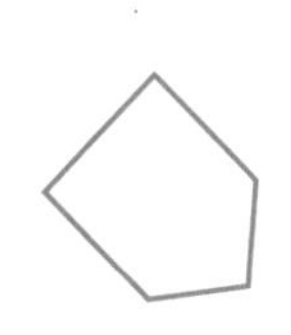

____

9. 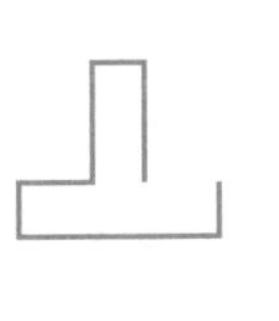

____

MATHEMATICAL PRACTICES

Math Talk Explain how you know the shape in Exercise 9 is an open shape.

On Your Own

Write how many line segments the shape has.

10. 

__________ line segments

11. 

__________ line segments

12. 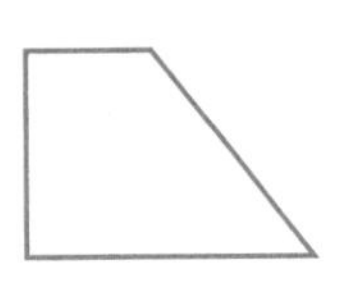

__________ line segments

13. 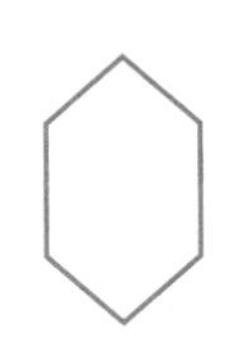

__________ line segments

Write whether the shape is *open* or *closed*.

14. 

____

15. 

____

16. 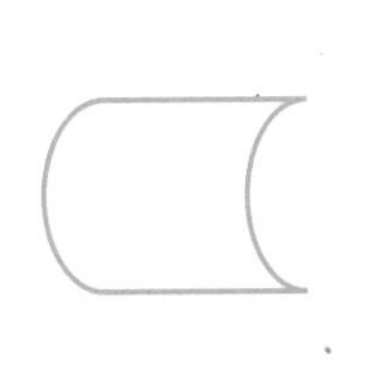

____

17. 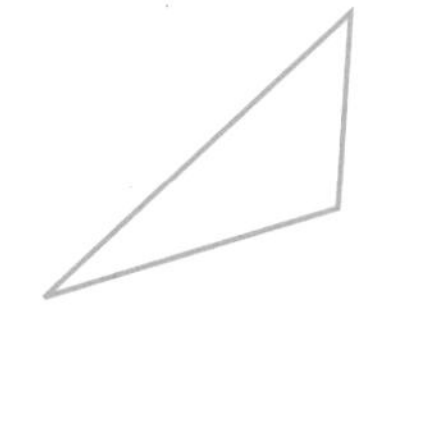

____

Problem Solving

18. **What's the Error?** Brittany says there are two endpoints in the shape shown at the right. Is she correct? **Explain.**

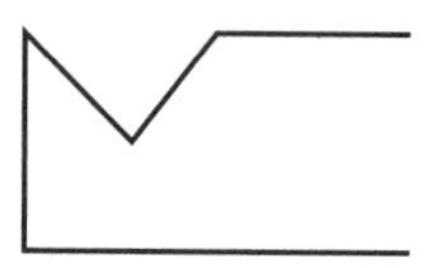

19. Write Math **Explain** how you can make the shape at the right a closed shape. Change the shape so it is a closed shape.

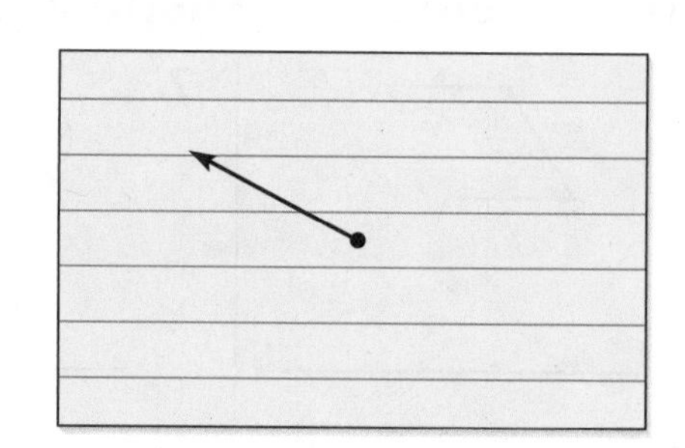

20. Look at Carly's drawing at the right. What did she draw? How is it like a line? How is it different? Change the drawing so that it is a line.

21. H.O.T. Draw a closed shape in the workspace by connecting 5 line segments at their endpoints.

22. **Test Prep** Which is NOT a closed shape?

Ⓐ 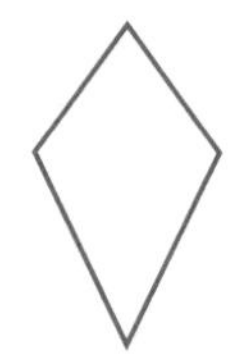

Ⓒ  

Ⓑ 

Ⓓ 

FOR MORE PRACTICE:
Standards Practice Book, pp. P239–P240

Name ______________________________

Describe Angles in Plane Shapes

Essential Question How can you describe angles in plane shapes?

UNLOCK the Problem

An **angle** is formed by two rays that share an endpoint. Plane shapes have angles formed by two line segments that share an endpoint. The shared endpoint is called a **vertex**. The plural of *vertex* is *vertices*.

vertex

Jason drew this shape on dot paper.

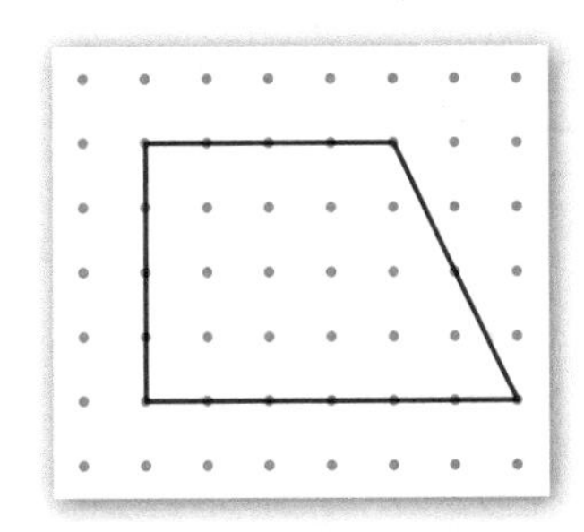

• How many angles are in Jason's shape?

____

Look at the angles in the shape that Jason drew. How can you describe the angles?

Describe angles.

This mark means *right angle*.

A **right angle** is an angle that forms a square corner.

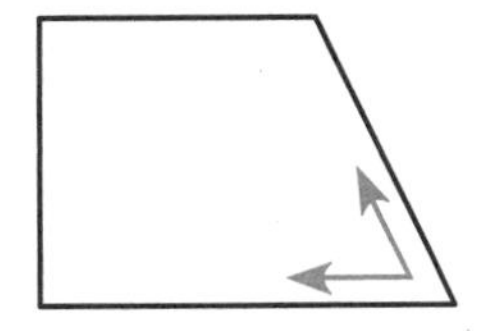

Some angles are less than a right angle.

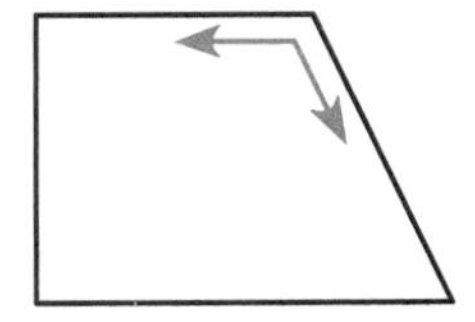

Some angles are greater than a right angle.

Look at Jason's shape.

Two angles are __________ angles, __________ angle is __________ a right angle, and __________ angle is __________ a right angle.

MATHEMATICAL PRACTICES

Math Talk Find examples of each type of angle in your classroom. Describe each angle.

Activity Model angles.

Materials ■ bendable straws ■ scissors ■ paper ■ pencil

- Cut a small slit in the shorter section of a bendable straw. Cut off the shorter section of a second straw and the bendable part. Insert the slit end of the first straw into the second straw.

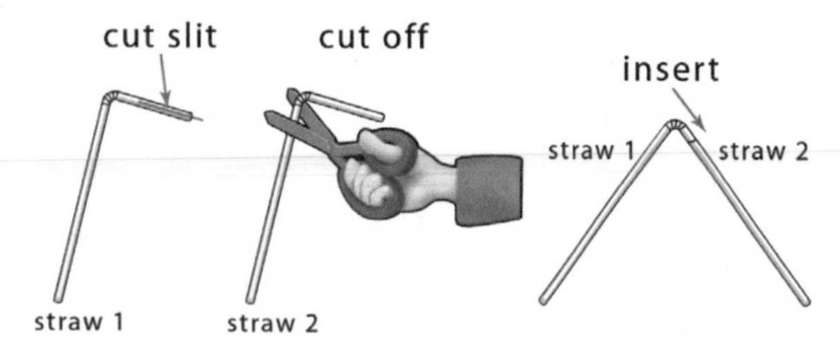

- Make an angle with the straws you put together. Compare the angle you made to a corner of the sheet of paper.
- Open and close the straws to make other types of angles.

In the space below, trace the angles you made with the straws. Label each *right angle, less than a right angle,* or *greater than a right angle.*

Share and Show

1. How many angles are in the triangle at the right?

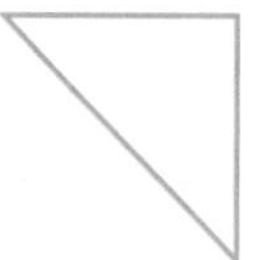

____

MATHEMATICAL PRACTICES

Math Talk Explain how you know an angle is greater than or less than a right angle.

Use the corner of a sheet of paper to tell whether the angle is a *right angle, less than a right angle,* or *greater than a right angle.*

2. 

____

3. 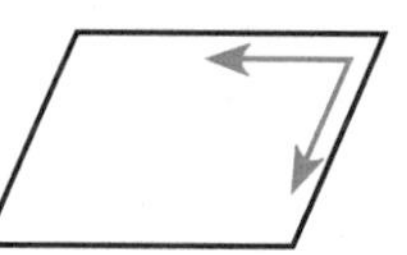

____

4. 

____

Name ______________________________

Write how many of each type of angle the shape has.

5.

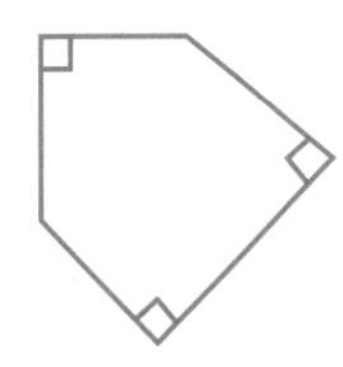

_____ right

_____ less than a right

_____ greater than a right

6.

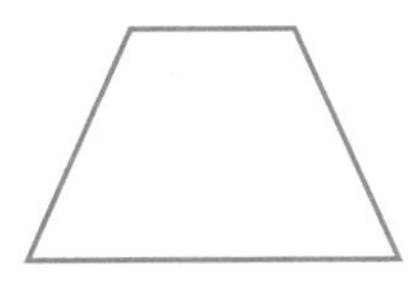

_____ right

_____ less than a right

_____ greater than a right

7.

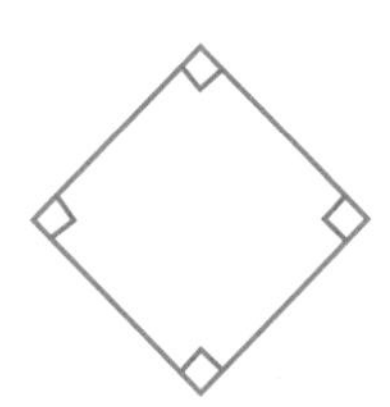

_____ right

_____ less than a right

_____ greater than a right

On Your Own

Use the corner of a sheet of paper to tell whether the angle is a *right angle, less than a right angle,* or *greater than a right angle.*

8.

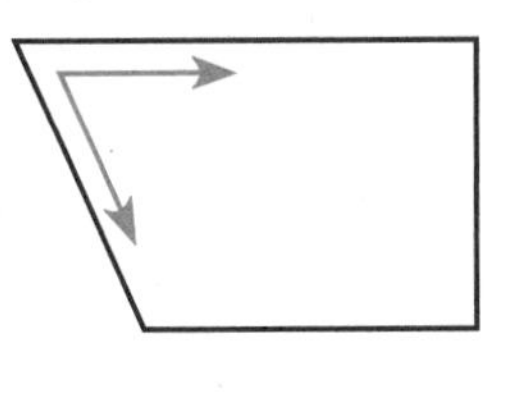

____

9.

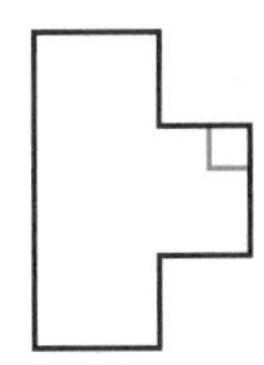

____

10.

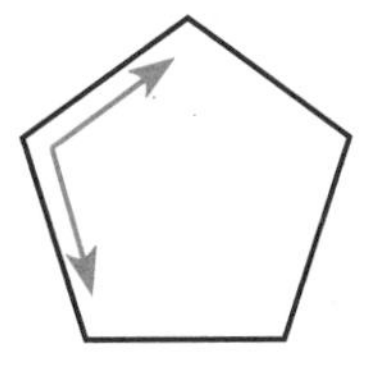

____

Write how many of each type of angle the shape has.

11.

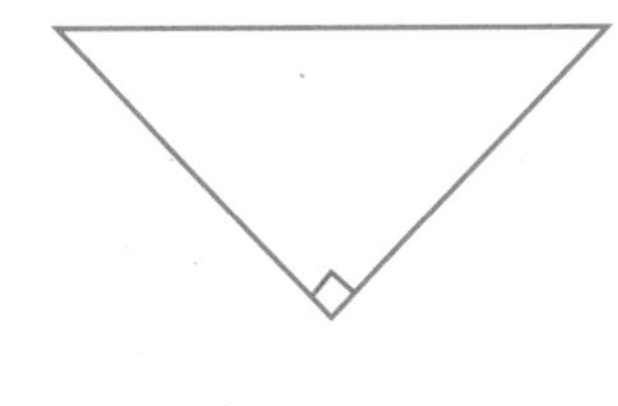

_____ right

_____ less than a right

_____ greater than a right

12.

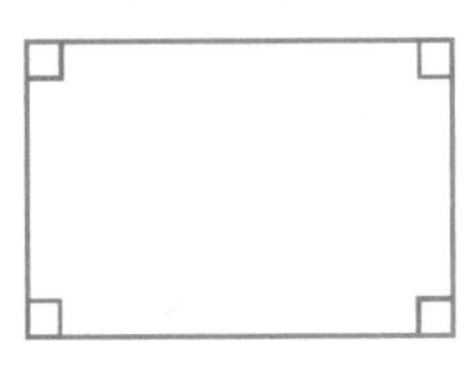

_____ right

_____ less than a right

_____ greater than a right

13.

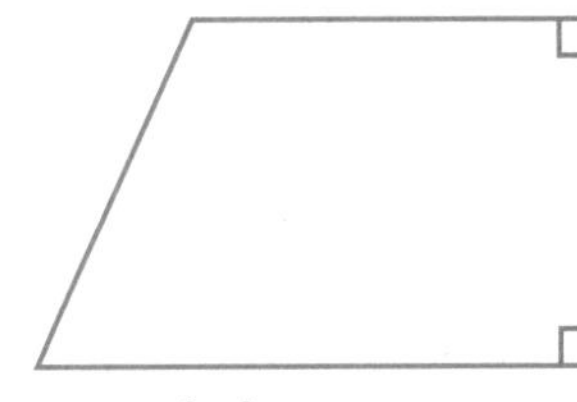

_____ right

_____ less than a right

_____ greater than a right

14. H.O.T. Describe the types of angles formed when you divide a circle into 4 equal parts.

____

UNLOCK the Problem

15. Holly drew a shape that does NOT have a right angle. Which shape did she draw?

Ⓐ  Ⓑ  Ⓒ  Ⓓ 

a. What do you need to know? ______________________________

b. Tell how you might use a sheet of paper to solve the problem.

____

____

c. Shape A has _____ right angle(s), _____ angle(s) greater than a right angle, and _____ angle(s) less than a right angle.

Shape B has _____ right angle(s), _____ angle(s) greater than a right angle, and _____ angle(s) less than a right angle.

Shape C has _____ right angle(s), _____ angle(s) greater than a right angle, and _____ angle(s) less than a right angle.

Shape D has _____ right angle(s), _____ angle(s) greater than a right angle, and _____ angle(s) less than a right angle.

d. Fill in the bubble for the correct answer choice above.

16. How many right angles does the shape have?

Ⓐ 1

Ⓑ 2

Ⓒ 3

Ⓓ 4

17. Which is a true statement about this shape?

Ⓐ There are no right angles.

Ⓑ There are 2 right angles and 4 angles greater than a right angle.

Ⓒ There are 2 right angles and 4 angles less than a right angle.

Ⓓ There are 4 right angles and 2 angles less than a right angle.

FOR MORE PRACTICE:
Standards Practice Book, pp. P241–P242

Name ____________________

Identify Polygons

Essential Question How can you use line segments and angles to make polygons?

CONNECT In earlier lessons, you learned about line segments and angles. In this lesson, you will see how line segments and angles make polygons.

A **polygon** is a closed plane shape that is made up of line segments that meet only at their endpoints. Each line segment in a polygon is a **side**.

Math Idea

All polygons are closed shapes. Not all closed shapes are polygons.

UNLOCK the Problem REAL WORLD

Circle all the words that describe the shape.

A

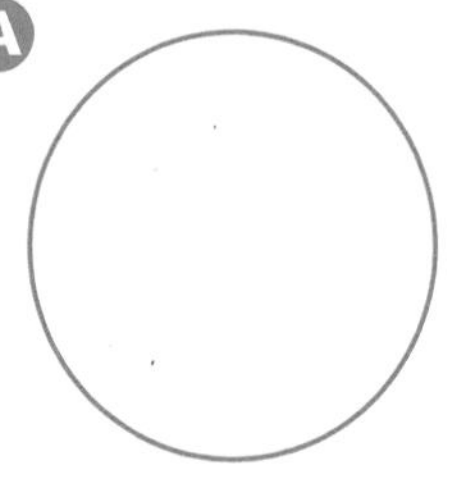

plane shape
open shape
closed shape
curved paths
line segments
polygon

B

plane shape
open shape
closed shape
curved paths
line segments
polygon

C

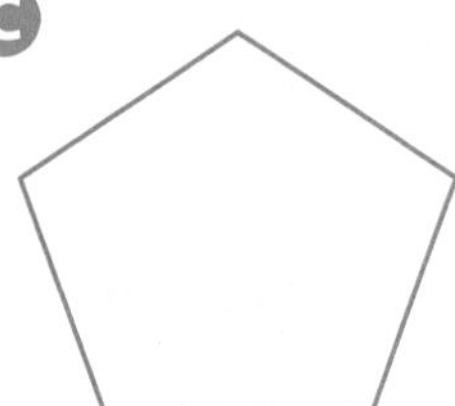

plane shape
open shape
closed shape
curved paths
line segments
polygon

D

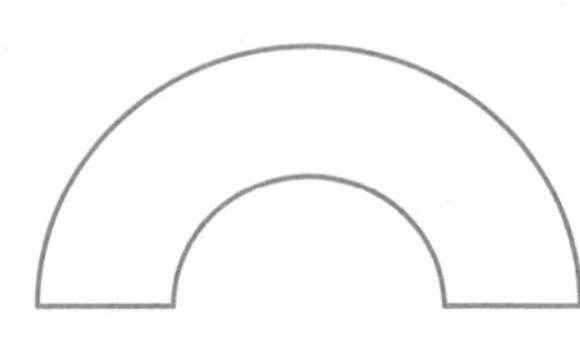

plane shape
open shape
closed shape
curved paths
line segments
polygon

Try This!

Fill in the blanks with *sometimes*, *always*, or *never*.

Polygons are ____________ plane shapes.

Polygons are ____________ closed shapes.

Polygons are ____________ open shapes.

Plane shapes are ____________ polygons.

MATHEMATICAL PRACTICES

Math Talk Explain why not all closed shapes are polygons.

Name Polygons Polygons are named by the number of sides and angles they have.

Some traffic signs are in the shape of polygons. A stop sign is in the shape of which polygon?

angle
side →
STOP

Count the number of sides and angles.

triangle	**quadrilateral**	**pentagon**
3 sides 3 angles	4 sides _______ angles	_______ sides 5 angles
hexagon	**octagon**	**decagon**
_______ sides 6 angles	8 sides _______ angles	_______ sides 10 angles

How many sides does the stop sign have? _____________

How many angles? _____________

So, a stop sign is in the shape of an _____________.

MATHEMATICAL PRACTICES

Math Talk Compare the number of sides and angles. What is a true statement about all polygons?

Share and Show

1. The shape at the right is a polygon. Circle all the words that describe the shape.

plane shape   open shape   closed shape   pentagon

curved paths   line segments   hexagon   quadrilateral

Name ______________________________

Is the shape a polygon? Write *yes* or *no*.

2. 

____

3. 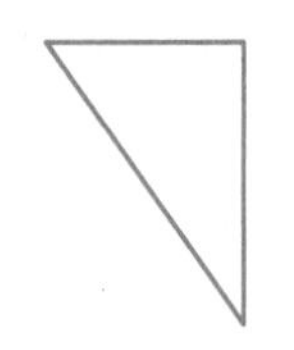

____

✓4. 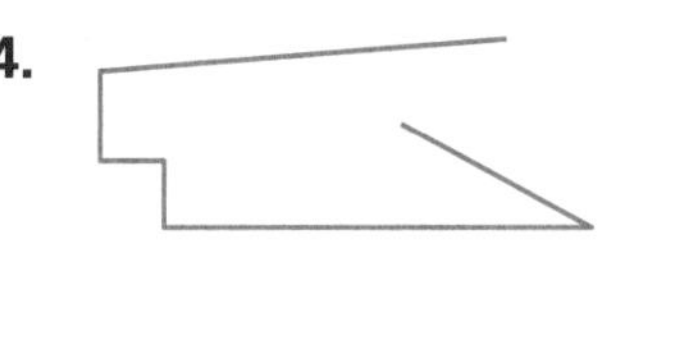

____

MATHEMATICAL PRACTICES

Math Talk **Explain** how you can change the shape in Exercise 4 to make it a polygon.

Write the number of sides and the number of angles. Then name the polygon.

5. 

______ sides

______ angles

____

6. 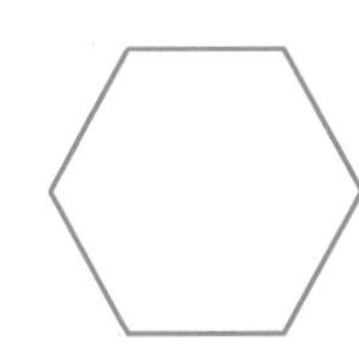

______ sides

______ angles

____

✓7. 

______ sides

______ angles

____

On Your Own

Is the shape a polygon? Write *yes* or *no*.

8. 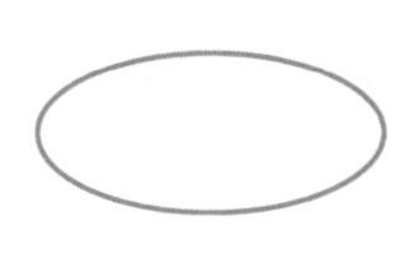

____

9. 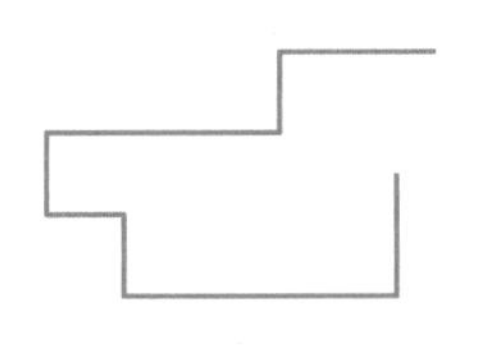

____

10. 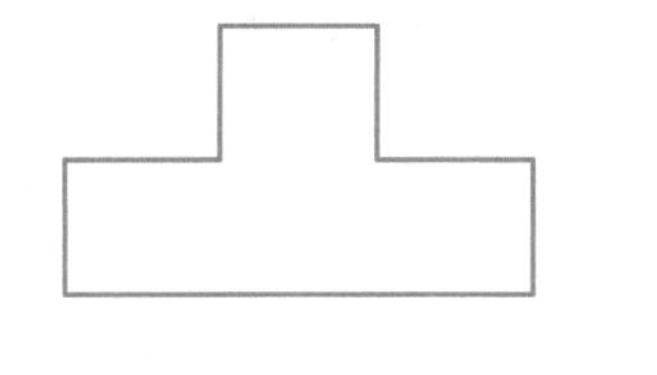

____

Write the number of sides and the number of angles. Then name the polygon.

11. 

______ sides

______ angles

____

12. 

______ sides

______ angles

____

13. 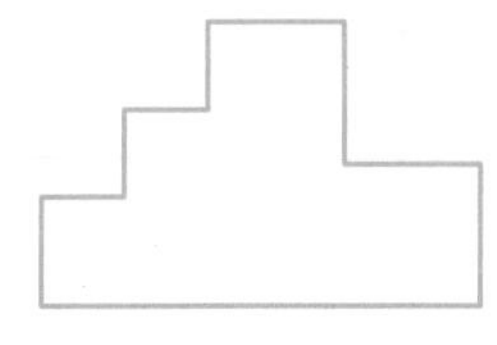

______ sides

______ angles

____

Problem Solving

14. Write Math ▸ **Sense or Nonsense?** Jake said Shapes *A–E* are all polygons. Does this statement make sense? **Explain** your answer.

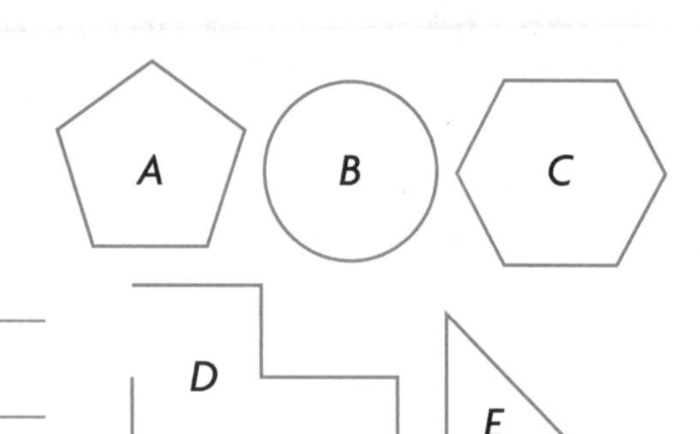

____

____

____

15. **What if** Kim wants to draw a polygon? How can she check her drawing?

____

____

16. I am a closed shape made of 6 line segments. I have 2 angles less than a right angle and no right angles. What shape am I? Draw an example in the workspace.

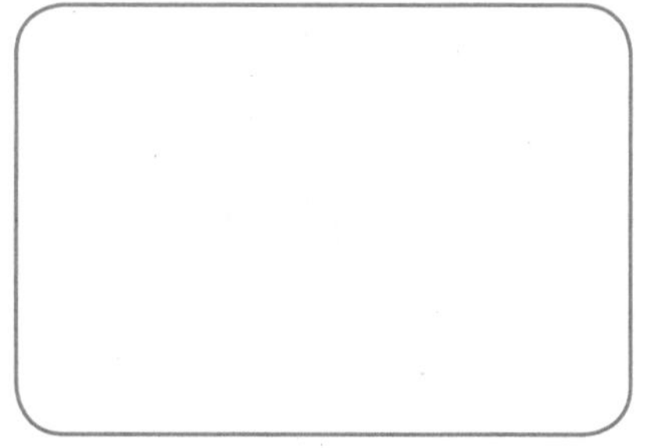

____

17. H.O.T. Is every closed shape a polygon? Use a drawing to help **explain** your answer.

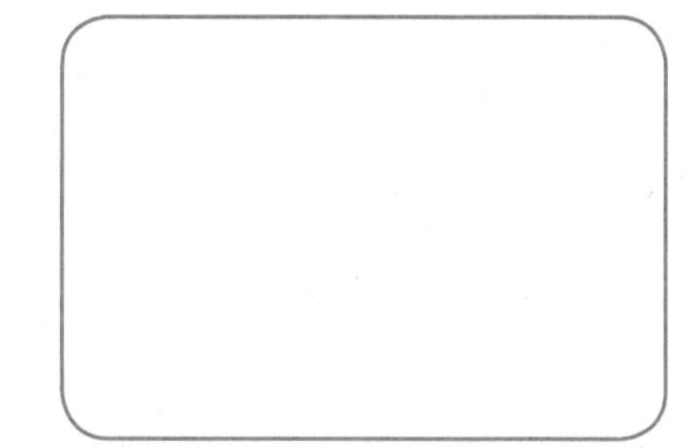

____

____

____

18. **What's the Error?** Eric says that the shape at the right is an octagon. Do you agree or disagree?

Explain. _______________

____

19. **Test Prep** Alicia drew the polygon at the right. What is the name of the polygon she drew?

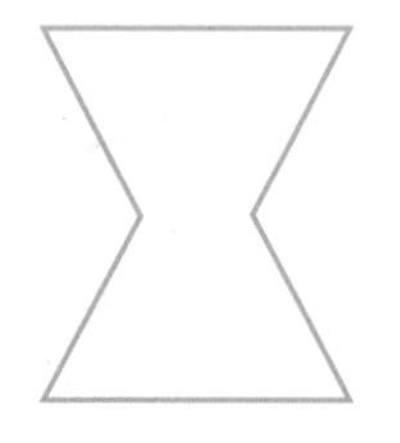

Ⓐ octagon Ⓒ pentagon

Ⓑ hexagon Ⓓ quadrilateral

 FOR MORE PRACTICE: Standards Practice Book, pp. P243–P244

Name ____________________

Describe Sides of Polygons

Essential Question How can you describe line segments that are sides of polygons?

UNLOCK the Problem

Look at the polygon. How many pairs of sides are parallel?

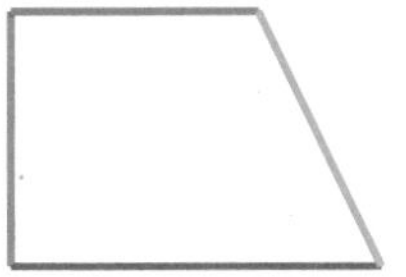

- How do you know the shape is a polygon?

____

____

TYPES OF LINES	TYPES OF LINE SEGMENTS
Lines that cross or meet are **intersecting lines**. Intersecting lines form angles. 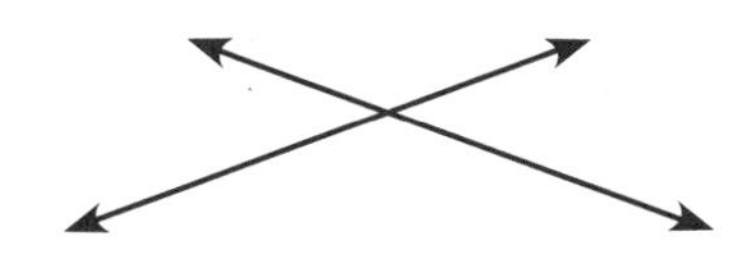	The orange and blue line segments meet and form an angle. So, they are __________.
Intersecting lines that cross or meet to form right angles are **perpendicular lines**. 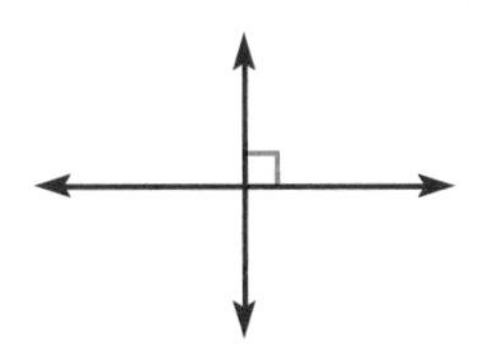	The red and blue line segments meet to form a right angle. So, they are __________.
Lines that appear to never cross or meet and are always the same distance apart are **parallel lines**. They do not form any angles. 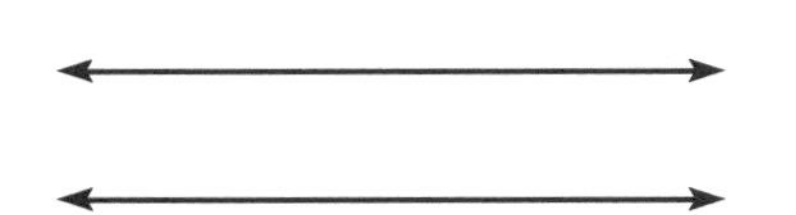	The green and blue line segments would never cross or meet. They are always the same distance apart. So, they appear to be __________.

So, the polygon above has _____ pair of parallel sides.

MATHEMATICAL PRACTICES

Math Talk Why can't parallel lines ever cross?

Try This! Draw a polygon with only 1 pair of parallel sides. Then draw a polygon with 2 pairs of parallel sides. Outline each pair of parallel sides with a different color.

Share and Show 

1. Which sides appear to be parallel?

____

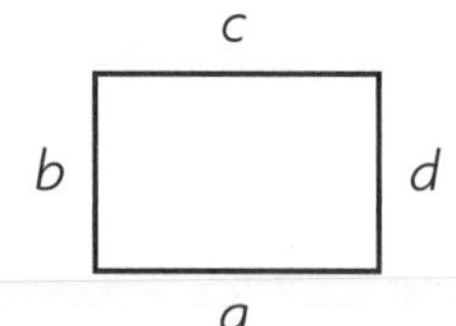

Think: Which pairs of sides appear to be the same distance apart?

Look at the green sides of the polygon. Tell if they appear to be *intersecting*, *perpendicular*, or *parallel*. Write all the words that describe the sides.

2. 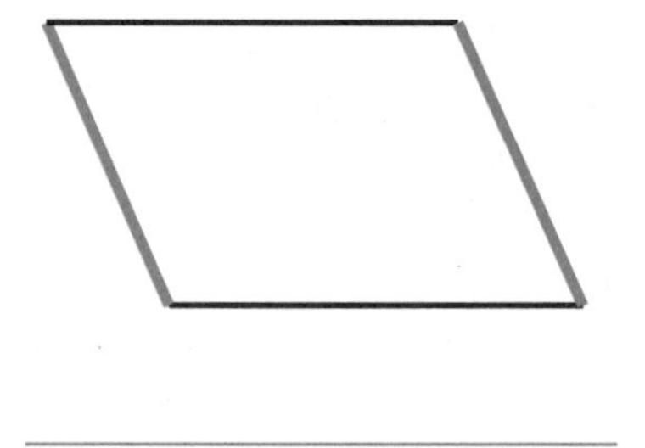

____

____

3. 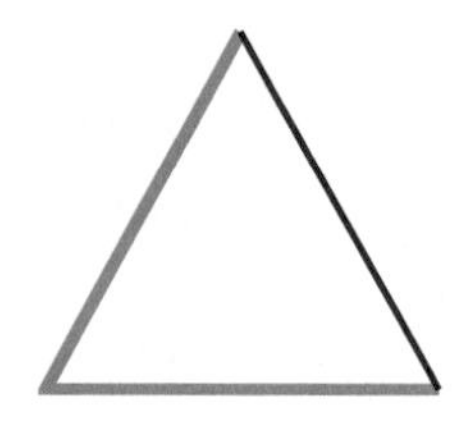

____

____

4. 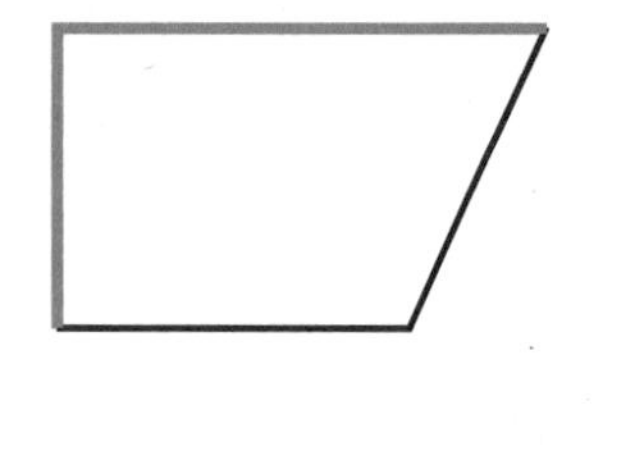

____

____

MATHEMATICAL PRACTICES

Math Talk **Explain** how intersecting and perpendicular lines are alike and how they are different.

On Your Own

Look at the green sides of the polygon. Tell if they appear to be *intersecting*, *perpendicular*, or *parallel*. Write all the words that describe the sides.

5. 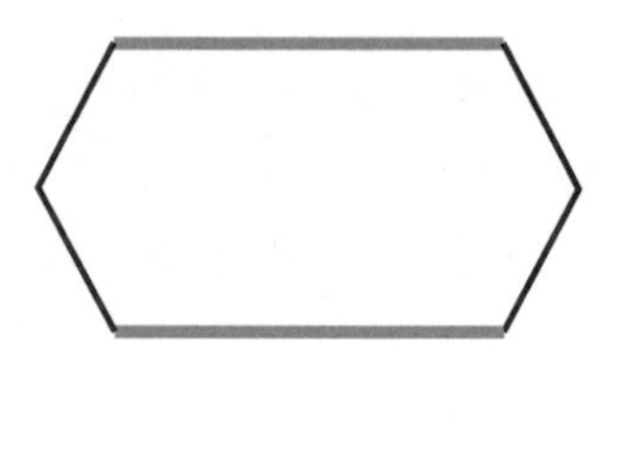

____

____

6. 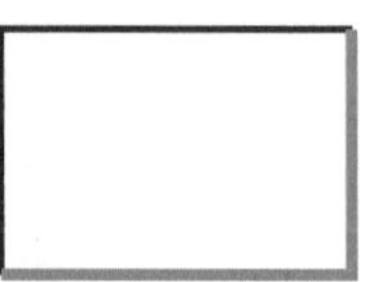

____

____

7. 

____

____

Name ______________________________

Problem Solving

Use pattern blocks *A–E* for 8–11.

Chelsea wants to sort pattern blocks by the types of sides.

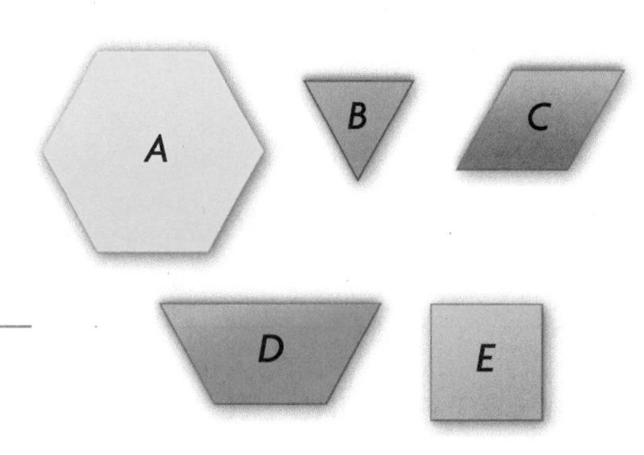

8. Which blocks have intersecting sides?

____

9. Which blocks have parallel sides?

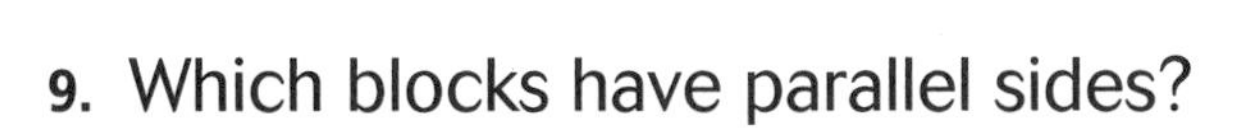

____

10. Which blocks have perpendicular sides?

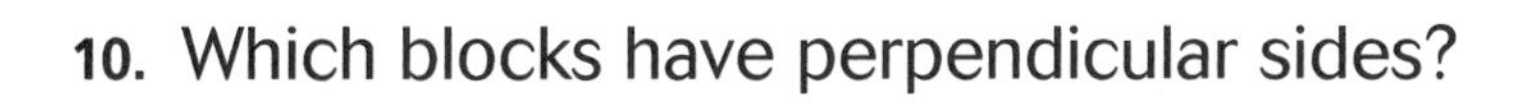

____

11. Which blocks have neither parallel nor perpendicular sides?

____

12. How many pairs of parallel sides are in a square?

____

13. H.O.T. How many pairs of perpendicular line segments are in the box at the right?

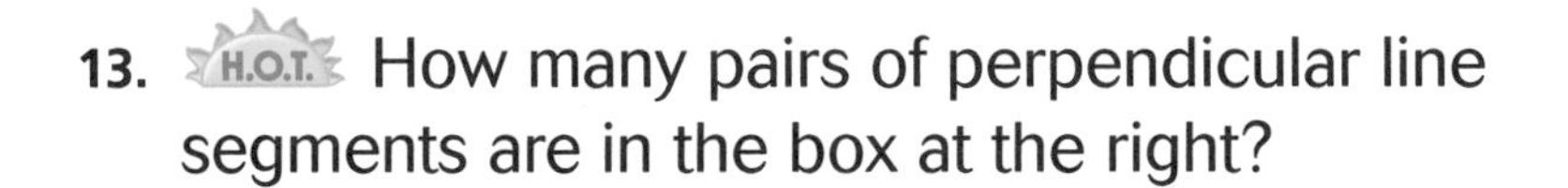

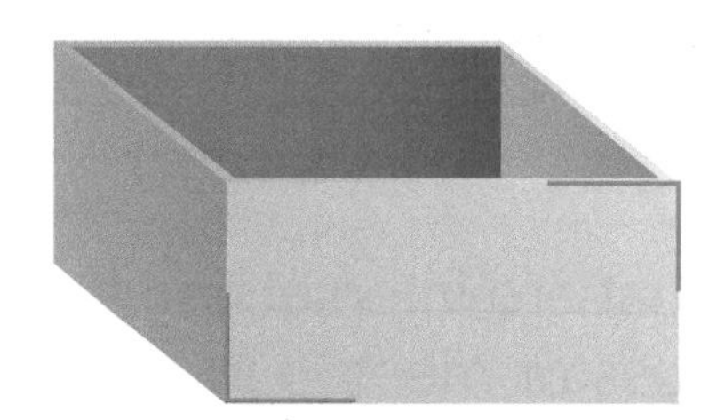

▲ The red line segments show 1 pair of perpendicular line segments.

____

14. H.O.T. Write Math Can the same two lines be parallel, perpendicular, and intersecting at the same time? **Explain** your answer.

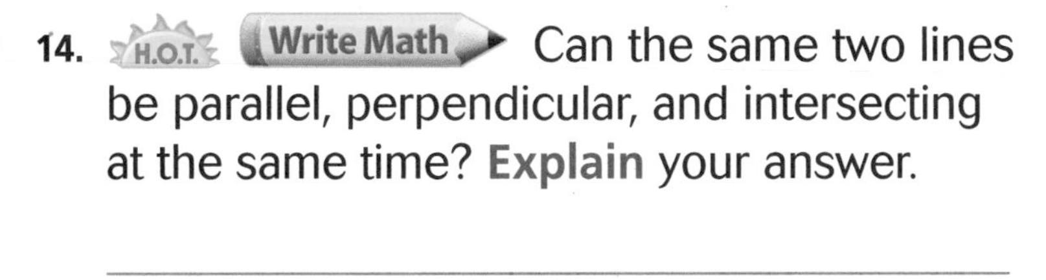

____

____

____

UNLOCK the Problem REAL WORLD

15. I am a pattern block that has 2 fewer sides than a hexagon. I have 2 pairs of parallel sides and 4 right angles. Which shape am I?

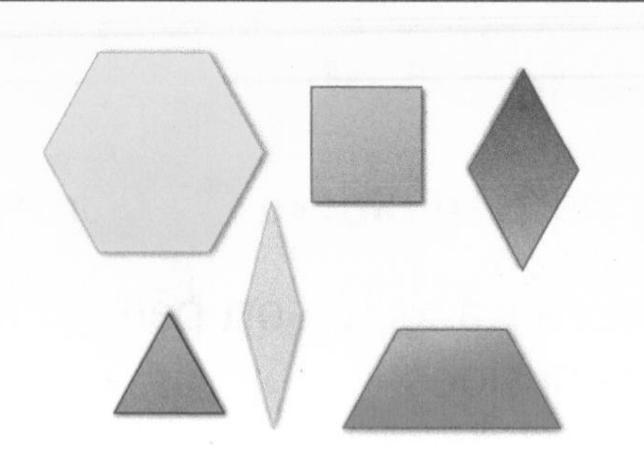

a. What do you need to know? ____________________

____

b. How can you find the answer to the riddle? ____________________

____

c. Write *yes* or *no* in the table to solve the riddle.

		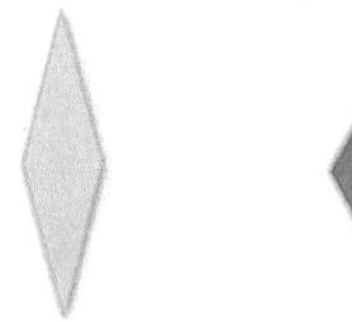			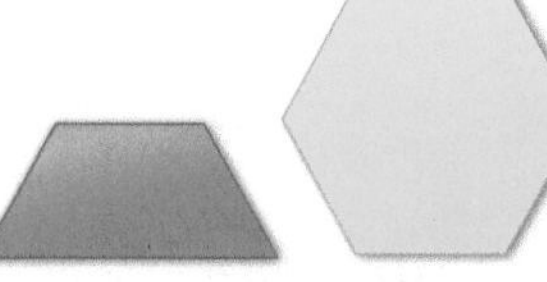	
2 fewer sides than a hexagon						
2 pairs of parallel sides						
4 right angles						

So, the ____________ is the shape.

16. What are intersecting lines that cross or meet to form right angles called?

____

17. **Test Prep** Which shape does NOT have at least one pair of parallel sides?

Ⓒ 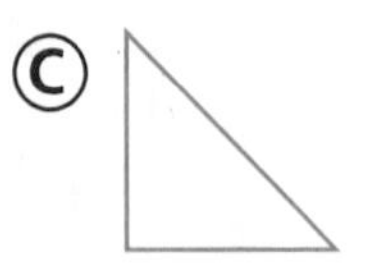

 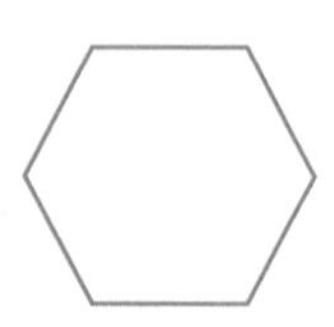

FOR MORE PRACTICE:
Standards Practice Book, pp. P245–P246

Name ___________________________________

Mid-Chapter Checkpoint

▶ Vocabulary

Choose the best term from the box to complete the sentence.

Vocabulary
angle
point
polygon
right angle

1. An _______________ is formed by two rays that share an endpoint. (p. 487)

2. A _______________ is a closed shape made up of line segments. (p. 491)

3. A _______________ forms a square corner. (p. 487)

▶ Concepts and Skills

Use the corner of a sheet of paper to tell whether the angle is a *right angle, less than a right angle,* or *greater than a right angle.*

4. 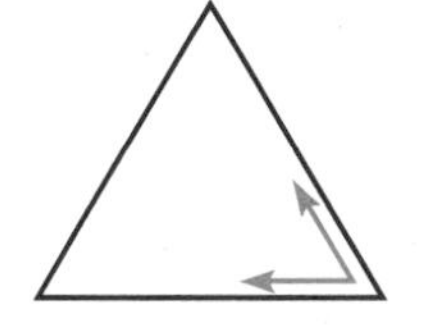

____

5. 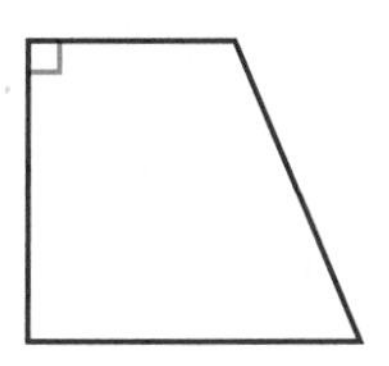

____

6. 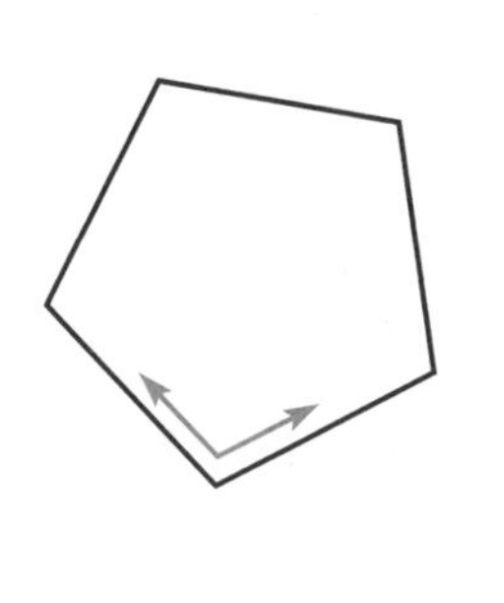

____

Write the number of sides and the number of angles. Then name the polygon.

7. 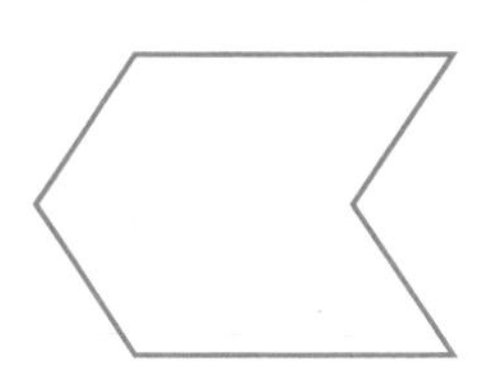

_____ sides

_____ angles

____

8. 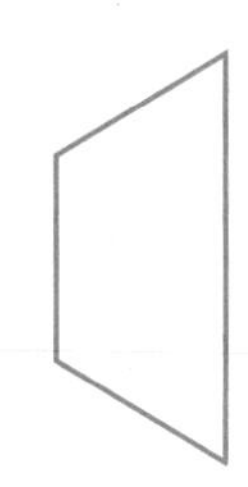

_____ sides

_____ angles

____

9. 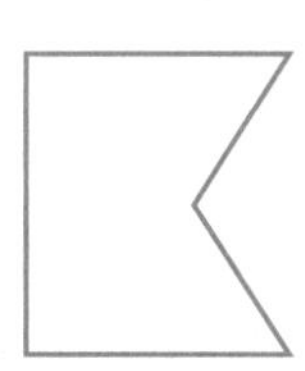

_____ sides

_____ angles

____

Fill in the bubble for the correct answer choice.

10. Anne drew a closed shape. Which of these could be the shape Anne drew?

Ⓐ 

Ⓒ 

Ⓑ 

Ⓓ 

11. This sign tells drivers there is a steep hill ahead. Look at the shape of the sign. How many sides and angles are there?

Ⓐ 4 sides and 4 angles

Ⓑ 5 sides and 5 angles

Ⓒ 6 sides and 6 angles

Ⓓ 8 sides and 8 angles

12. Which shape is NOT a polygon?

Ⓐ 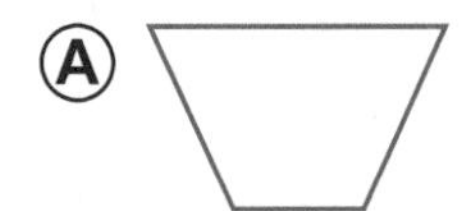

Ⓒ 

Ⓑ 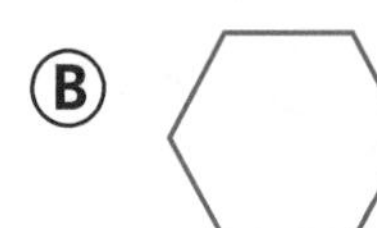

Ⓓ 

13. Sean drew a shape with 2 fewer sides than an octagon. Which shape did he draw?

Ⓐ triangle   Ⓒ hexagon

Ⓑ pentagon   Ⓓ quadrilateral

14. John drew a polygon with two line segments that meet to form a right angle. Which words describe the line segments?

Ⓐ parallel and perpendicular

Ⓑ parallel and intersecting

Ⓒ intersecting and perpendicular

Ⓓ intersecting and curved

Name ______________________________

Classify Quadrilaterals

Essential Question How can you use sides and angles to help you describe quadrilaterals?

UNLOCK the Problem

Quadrilaterals are named by their sides and their angles.

Describe quadrilaterals.

quadrilateral

_______ sides

_______ angles

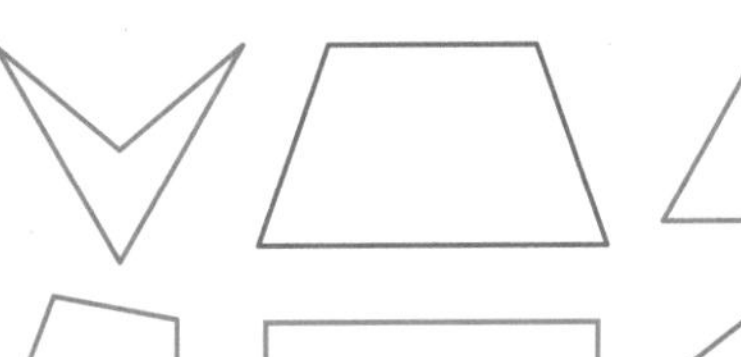

ERROR Alert

Some quadrilaterals cannot be classified as a trapezoid, rectangle, square, or rhombus.

trapezoid

exactly _______ pair of opposite sides that are parallel

lengths of sides could be the same

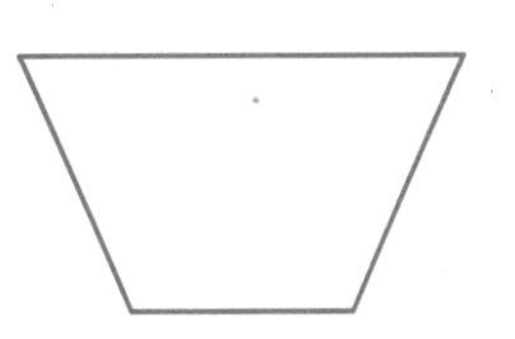

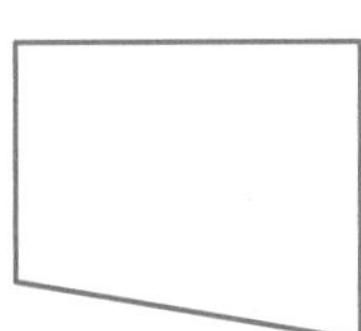

rectangle

_____ pairs of opposite sides that are parallel

_____ pairs of sides that are of equal length

_____ right angles

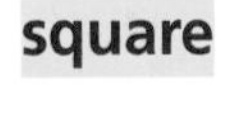

square

_____ pairs of opposite sides that are parallel

_____ sides that are of equal length

_____ right angles

rhombus

_____ pairs of opposite sides that are parallel

_____ sides that are of equal length

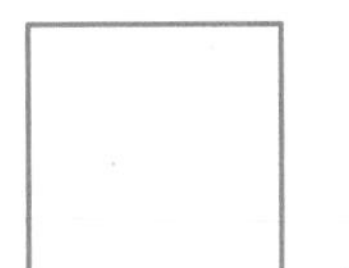

MATHEMATICAL PRACTICES

Math Talk Explain why a square can also be named a rectangle or a rhombus.

Share and Show 

Look at the quadrilateral at the right.

Think: All the angles are right angles.

1. Outline each pair of opposite sides that are parallel with a different color. How many pairs of opposite sides appear to be parallel? __________

2. Look at the parallel sides you colored.

   The sides in each pair are of __________ length.

3. Name the quadrilateral. ________________

Circle all the words that describe the quadrilateral.

4. 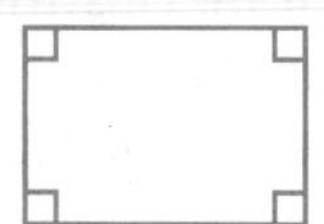

   rectangle

   rhombus

   square

   trapezoid

5. 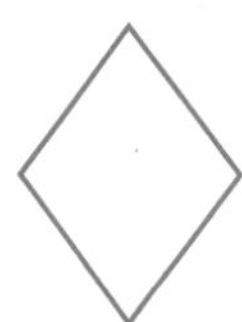

   rhombus

   quadrilateral

   square

   rectangle

6. 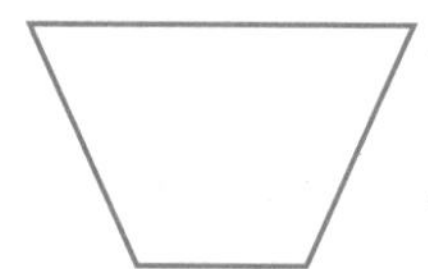

   rectangle

   rhombus

   trapezoid

   quadrilateral

MATHEMATICAL PRACTICES

Math Talk Explain how you can have a rhombus that is not a square.

On Your Own

Circle all the words that describe the quadrilateral.

7. 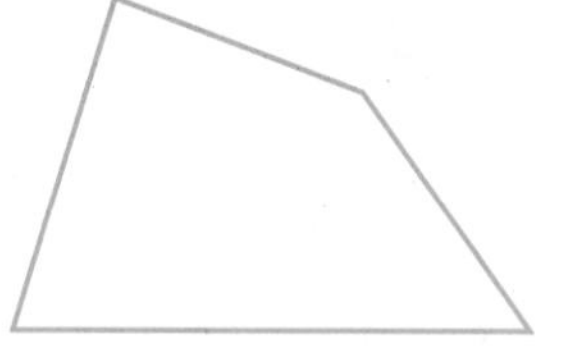

   rectangle

   trapezoid

   quadrilateral

   rhombus

8. 

   rectangle

   rhombus

   trapezoid

   square

9. 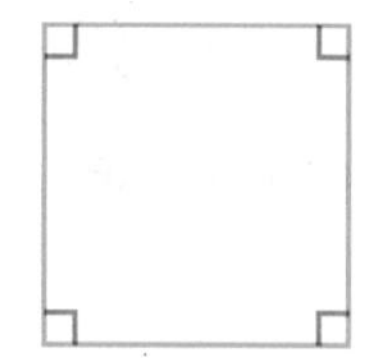

   quadrilateral

   square

   rectangle

   rhombus

Name ______________________________

Problem Solving

Use the quadrilaterals at the right for 10–12.

10. Which quadrilaterals appear to have 4 right angles?

____

11. Which quadrilaterals appear to have 2 pairs of opposite sides that are parallel?

____

12. Which quadrilaterals appear to have no right angles?

____

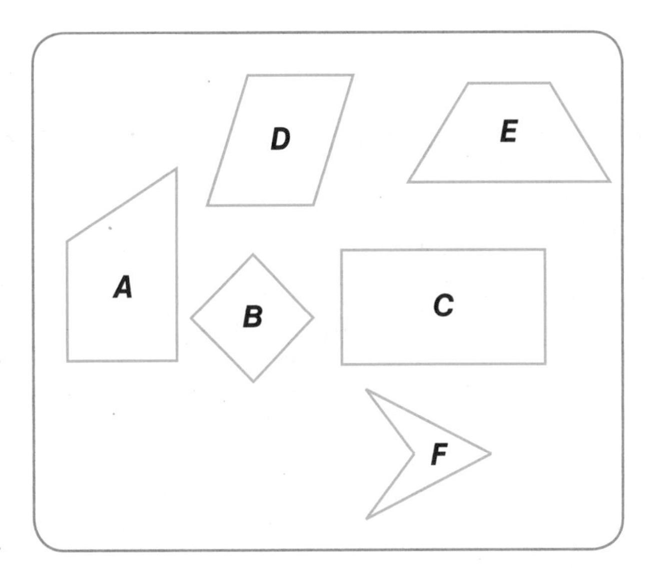

Write *all* or *some* to complete the sentence for 13–18.

13. The opposite sides of ______ rectangles are parallel.

14. ______ sides of a rhombus are the same length.

15. ______ squares are rectangles.

16. ______ rhombuses are squares.

17. ______ quadrilaterals are polygons.

18. ______ polygons are quadrilaterals.

19. **Write Math** Circle the shape at the right that is not a quadrilateral. **Explain** your choice.

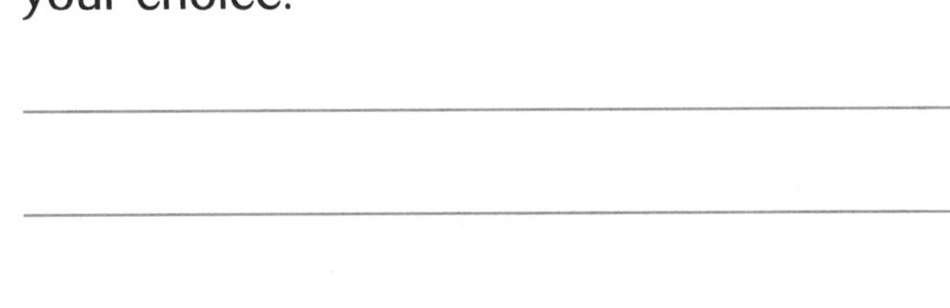

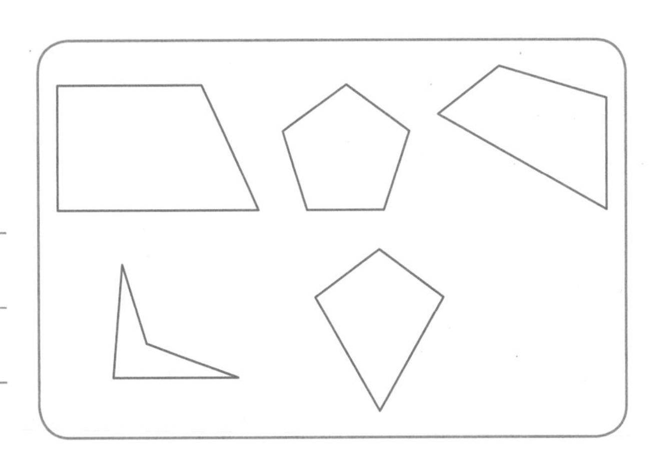

20. H.O.T. I am a polygon that has 4 sides and 4 angles. At least one of my angles is less than a right angle. Circle all the shapes that I could be.

quadrilateral   rectangle   square   rhombus   trapezoid

21. **Test Prep** Rita glued craft sticks together to make this shape. Which best describes the quadrilateral Rita made?

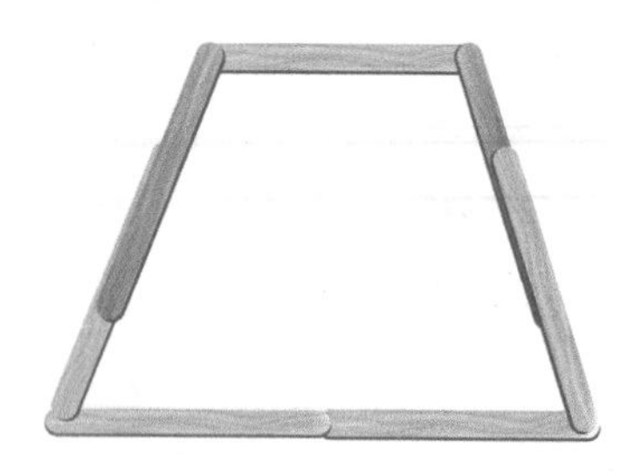

Ⓐ square Ⓑ rectangle Ⓒ rhombus Ⓓ trapezoid

Connect to Reading

Compare and Contrast

When you *compare,* you look for ways that things are alike. When you *contrast,* you look for ways that things are different.

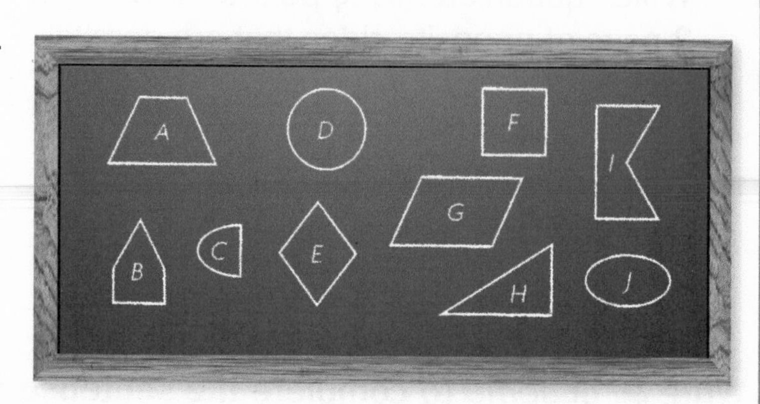

Mr. Briggs drew some shapes on the board. He asked the class to tell how the shapes are alike and how they are different.

Complete the sentences.

- Shapes ______, ______, ______, ______, ______, ______, and ______ are polygons.
- Shapes ______, ______, and ______ are not polygons.
- Shapes ______, ______, ______, and ______ are quadrilaterals.
- Shapes ______, ______, and ______ appear to have only 1 pair of opposite sides that are parallel.
- Shapes ______, ______, and ______ appear to have 2 pairs of opposite sides that are parallel.
- All 4 sides of shapes ______ and ______ appear to be the same length.
- In these polygons, all sides do not appear to be the same length. ______________
- These shapes can be called rhombuses. ______________
- Shapes ______ and ______ are quadrilaterals, but cannot be called rhombuses.
- Shape ______ is a rhombus and can be called a square.

FOR MORE PRACTICE:
Standards Practice Book, pp. P247–P248

Name ______________________

Draw Quadrilaterals

Essential Question How can you draw quadrilaterals?

UNLOCK the Problem REAL WORLD

CONNECT You have learned to classify quadrilaterals by the number of pairs of opposite sides that are parallel, by the number of pairs of sides of equal length, and by the number of right angles.

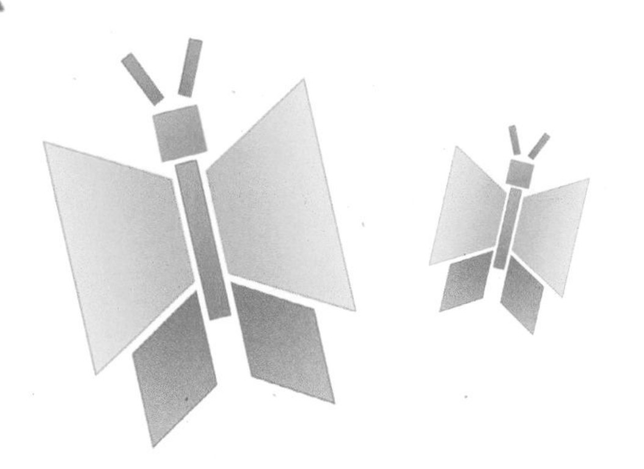

How can you draw quadrilaterals?

Activity 1 Use grid paper to draw quadrilaterals.

Materials ■ ruler

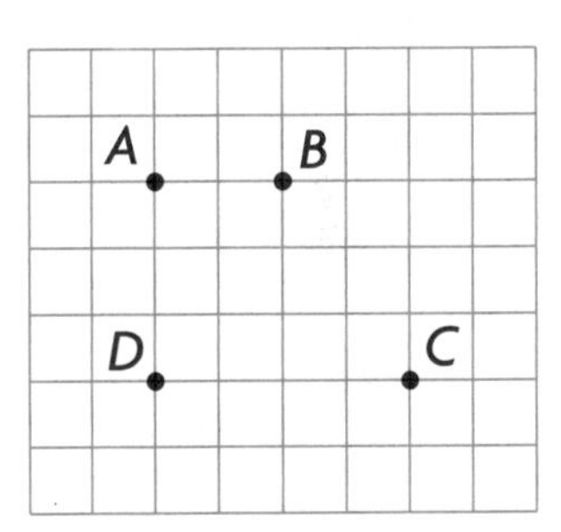

- Use a ruler to draw line segments from points *A* to *B*, from *B* to *C*, from *C* to *D*, and from *D* to *A*.
- Write the name of your quadrilateral.

____

Activity 2 Draw a shape that does not belong.

Materials ■ ruler

A Here are three examples of a quadrilateral. Draw an example of a polygon that is not a quadrilateral.

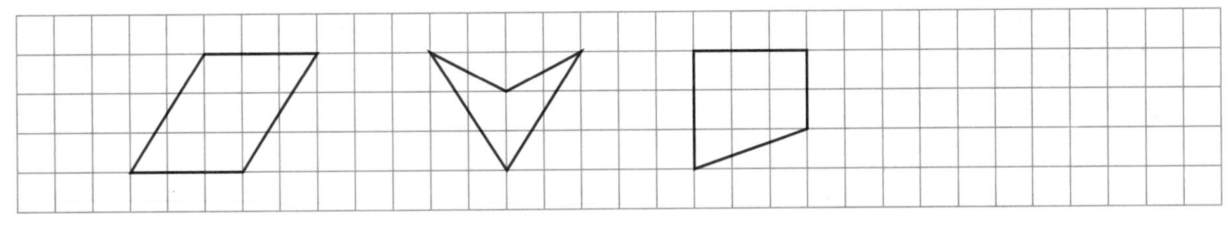

- **Explain** why your polygon is not a quadrilateral.

____

B **Here are three examples of a square. Draw a quadrilateral that is not a square.**

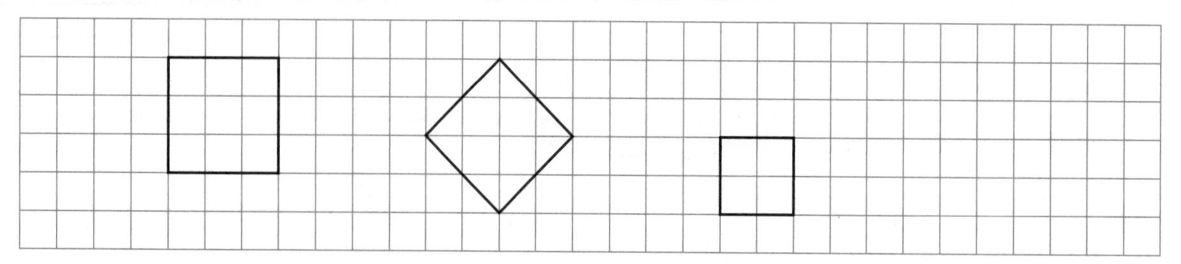

- **Explain** why your quadrilateral is not a square.

____

____

C **Here are three examples of a rectangle. Draw a quadrilateral that is not a rectangle.**

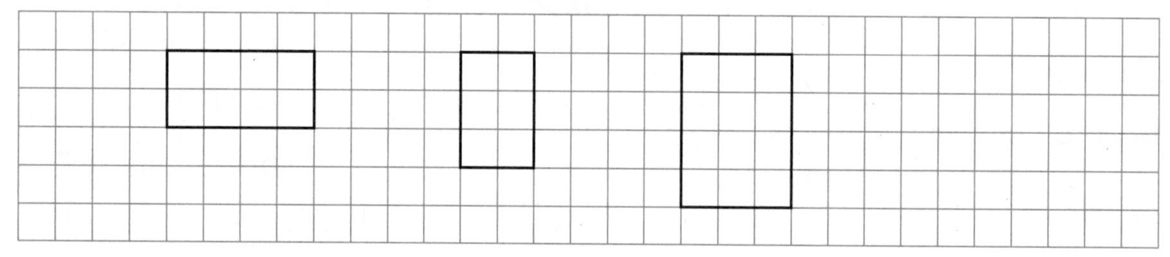

- **Explain** why your quadrilateral is not a rectangle.

____

____

D **Here are three examples of a rhombus. Draw a quadrilateral that is not a rhombus.**

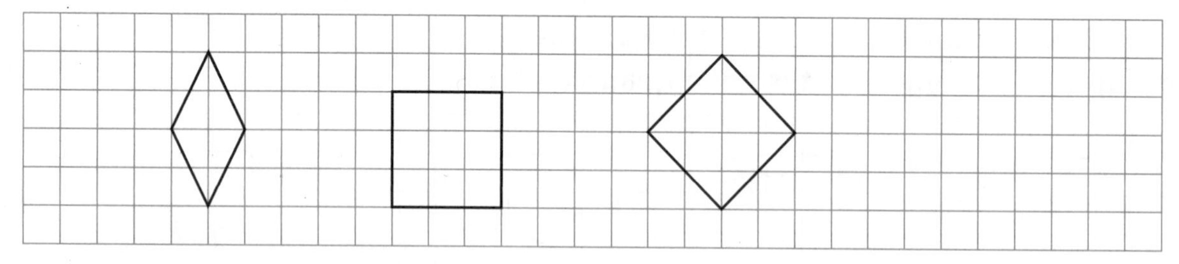

- **Explain** why your quadrilateral is not a rhombus.

____

____

MATHEMATICAL PRACTICES

Math Talk Compare your drawings with your classmates. **Explain** how your drawings are alike and how they are different.

Name ______________________________

Share and Show MATH BOARD

1. Choose four endpoints that connect to make a rectangle.

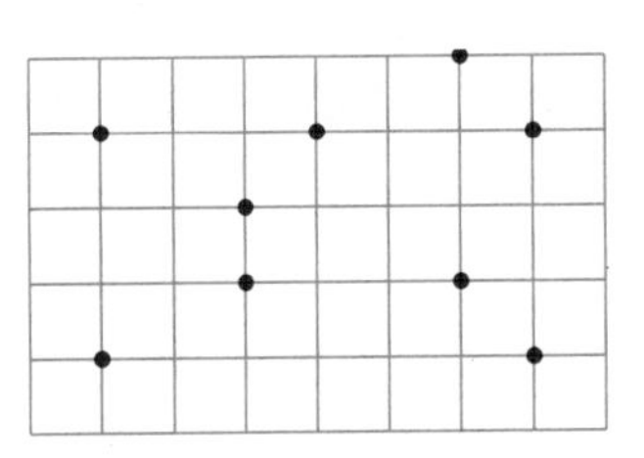

Think: A rectangle has 2 pairs of opposite sides that are parallel, 2 pairs of sides of equal length, and 4 right angles.

Draw a quadrilateral that is described. Name the quadrilateral you drew.

2. 2 pairs of equal sides

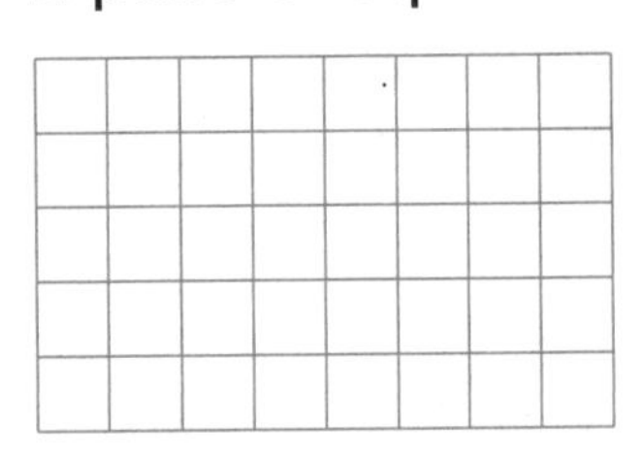

Name __________

3. 4 sides of equal length

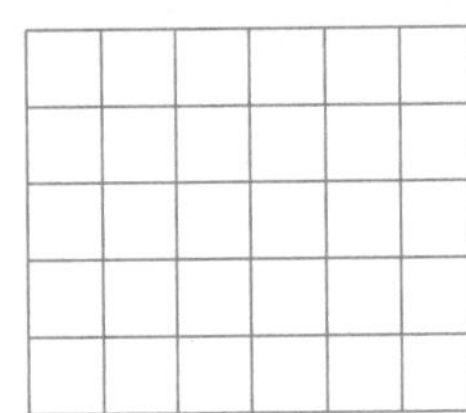

Name __________

MATHEMATICAL PRACTICES

Math Talk Explain one way the quadrilaterals you drew are alike and one way they are different.

On Your Own

Practice: Copy and Solve Use grid paper to draw a quadrilateral that is described. Name the quadrilateral you drew.

4. exactly 1 pair of opposite sides that are parallel

5. 4 right angles

6. 2 pairs of sides of equal length

Draw a quadrilateral that does not belong. Then explain why.

7.

____

8.

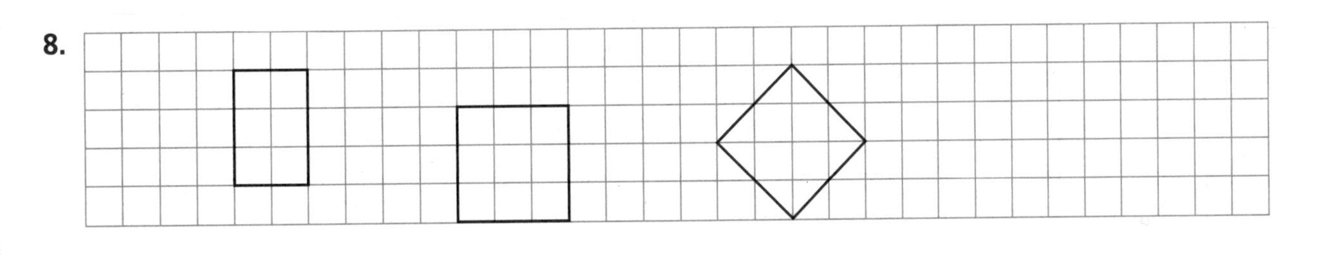

____

Problem Solving REAL WORLD

9. Write Math What's the Error? Jacki drew the shape at the right. She said it is a rectangle because it has 2 pairs of opposite sides that are parallel. **Describe** her error.

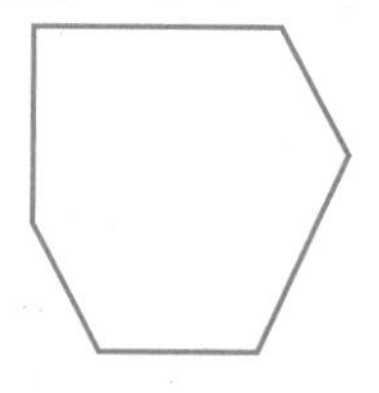

____

____

10. Adam drew three quadrilaterals. One quadrilateral had no pairs of parallel sides, one quadrilateral had 1 pair of opposite sides that are parallel, and the last quadrilateral had 2 pairs of opposite sides that are parallel. Draw the three quadrilaterals that Adam could have drawn. Name the quadrilaterals.

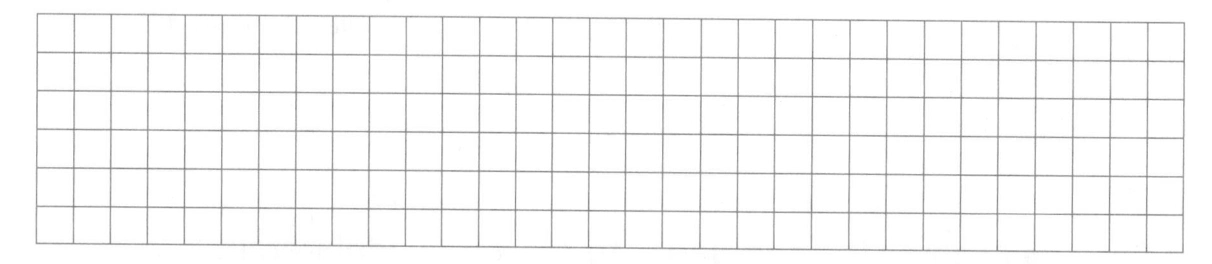

__________ __________ __________

11. H.O.T. Amy has 4 straws of equal length. Name the quadrilaterals that can be made using these 4 straws. __________

Amy cuts one of the straws in half. She uses the two halves and two of the other straws to make a quadrilateral. Name a quadrilateral that can be made using these 4 straws. __________

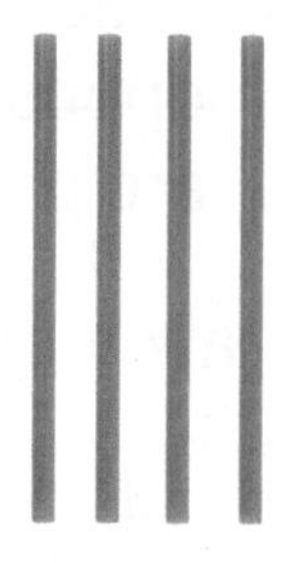

12. **Test Prep** Jordan drew a quadrilateral with 2 pairs of opposite sides that are parallel. Which shape could NOT be the quadrilateral Jordan drew?

Ⓐ 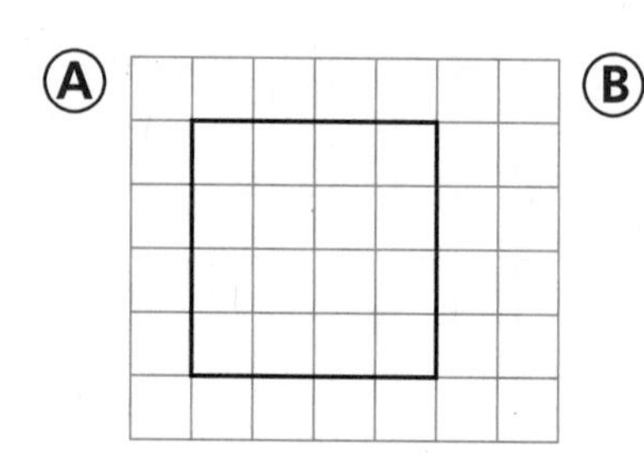

Ⓑ 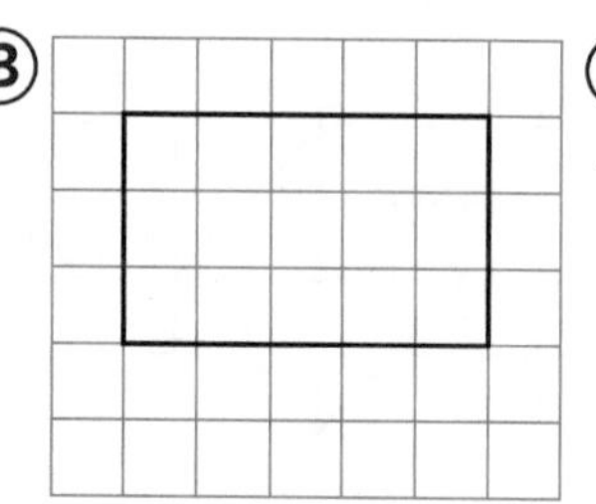

Ⓒ 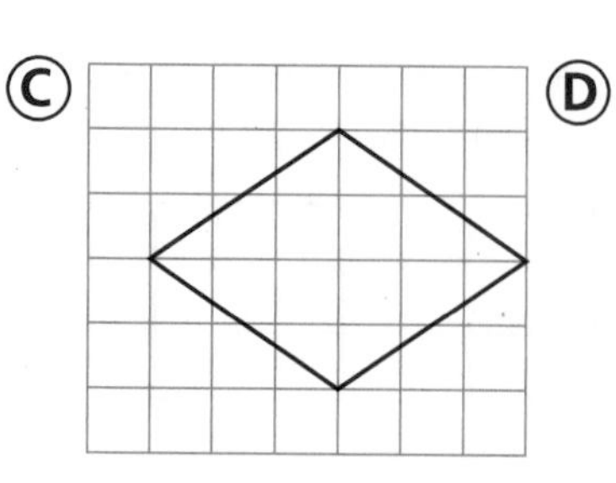

Ⓓ 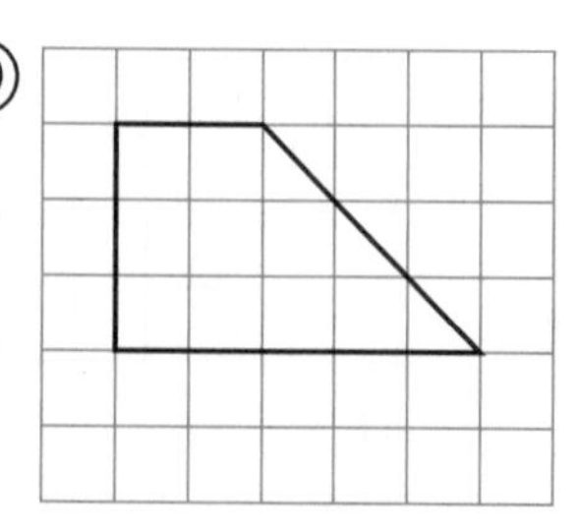

FOR MORE PRACTICE:
Standards Practice Book, pp. P249–P250

Name ______________________________

Describe Triangles

Essential Question How can you use sides and angles to help you describe triangles?

UNLOCK the Problem

How can you use straws of different lengths to make triangles?

Activity **Materials** ■ straws ■ scissors ■ MathBoard

STEP 1 Cut straws into different lengths.

STEP 2 Find straw pieces that you can put together to make a triangle. Draw your triangle on the MathBoard.

STEP 3 Find straw pieces that you cannot put together to make a triangle.

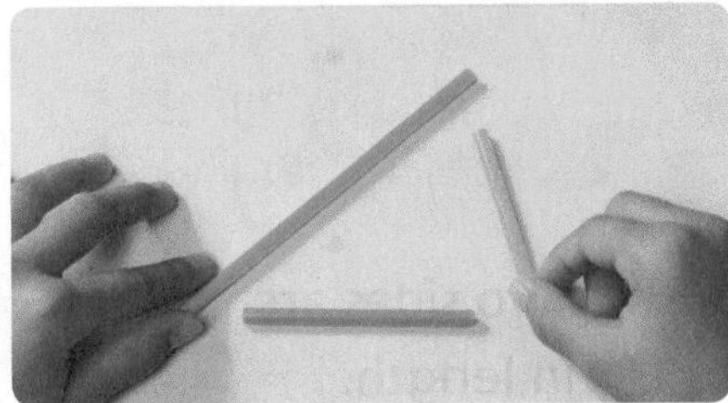

1. Compare the lengths of the sides. **Describe** when you can make a triangle.

____

____

MATHEMATICAL PRACTICES

Math Talk What if you had three straws of equal length? Can you make a triangle? **Explain.**

2. **Describe** when you cannot make a triangle.

____

____

____

3. **Explain** how you can change the straw pieces in Step 3 to make a triangle. ______________________________

____

Ways to Describe Triangles

What are two ways triangles can be described?

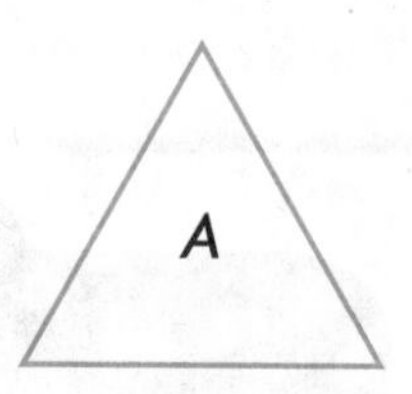
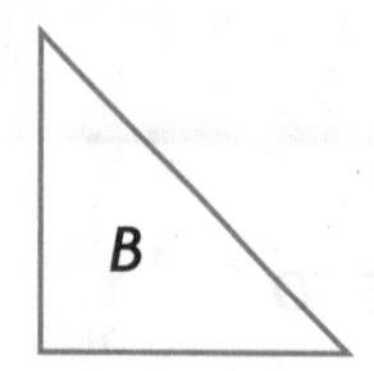
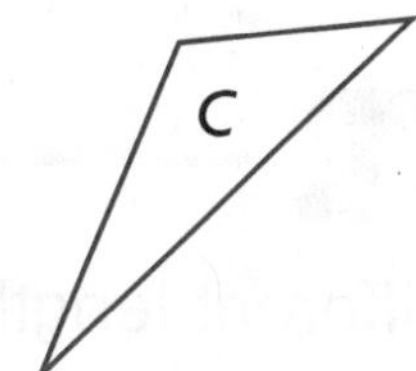
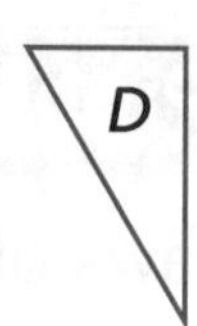
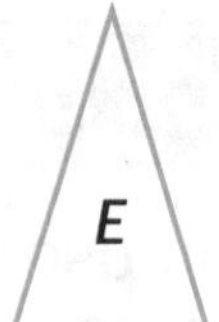

One Way

Triangles can be described by the number of sides that are of equal length.

Draw a line to match the description of the triangle(s).

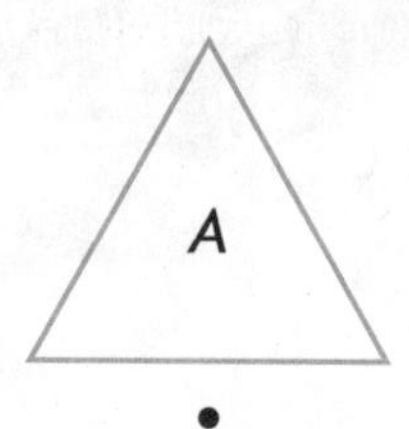
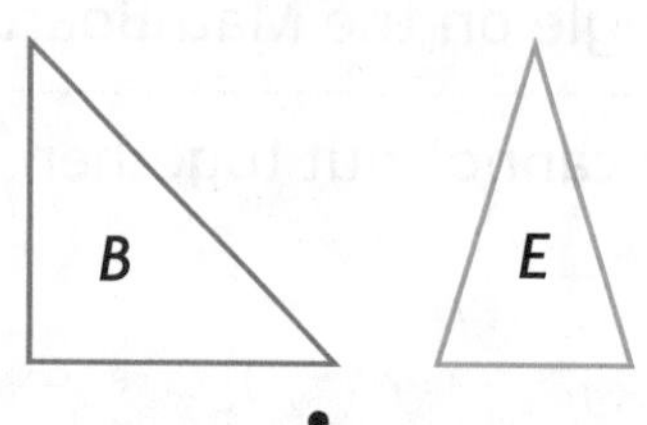
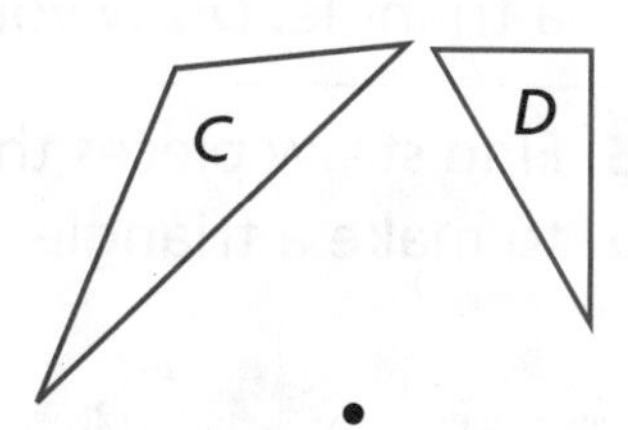

No sides are equal in length.

Two sides are equal in length.

Three sides are equal in length.

Another Way

Triangles can be described by the types of angles they have.

Draw a line to match the description of the triangle(s).

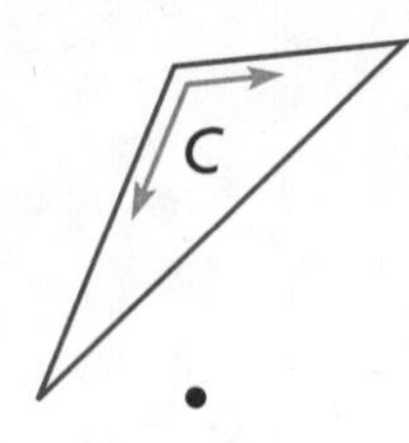
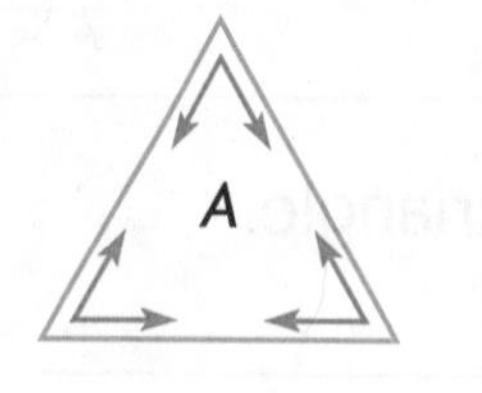

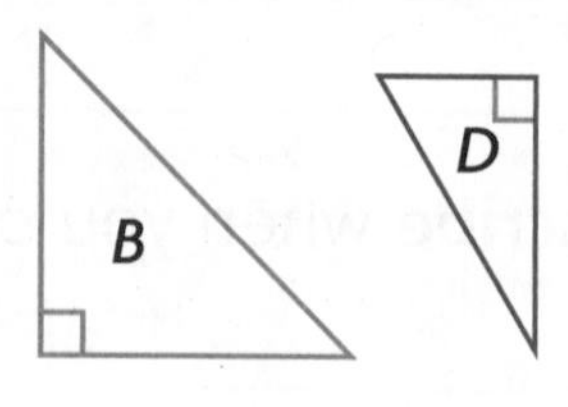

One angle is a right angle.

One angle is greater than a right angle.

Three angles are less than a right angle.

MATHEMATICAL PRACTICES

Math Talk Can a triangle have two right angles? **Explain.**

Name ___________________________

Share and Show 

1. Write the number of sides of equal length the triangle appears to have.

____

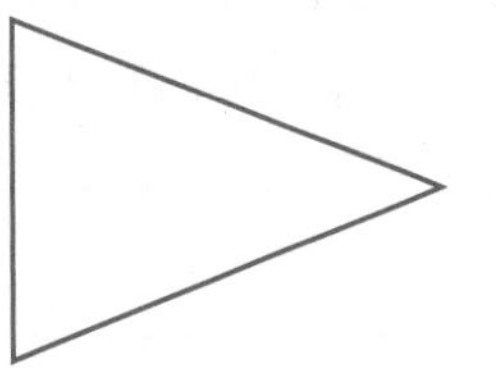

Use the triangles for 2–4. Write *F, G,* or *H*.

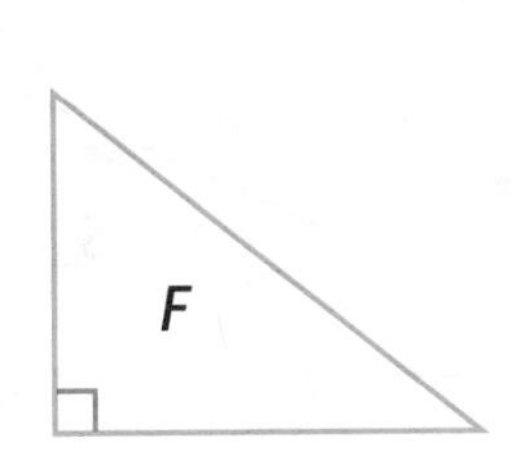

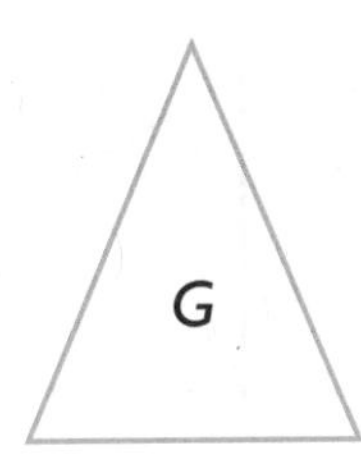

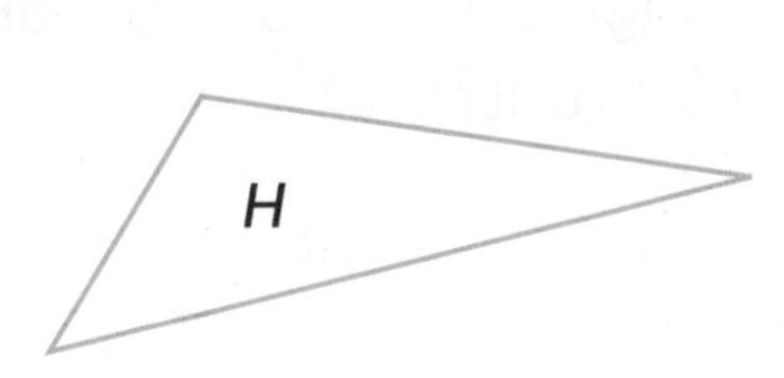

MATHEMATICAL PRACTICES

Math Talk Explain another way you can describe triangle *H*.

2. Triangle ______ has 1 right angle.

3. Triangle ______ has 1 angle greater than a right angle.

4. Triangle ______ has 3 angles less than a right angle.

On Your Own

Use the triangles for 5–7. Write *K, L,* or *M*. Then complete the sentences.

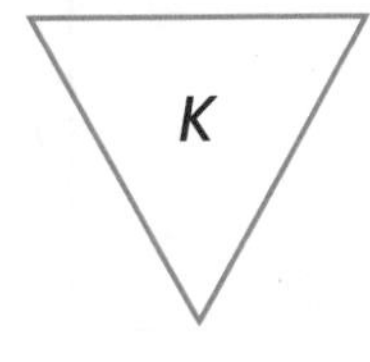

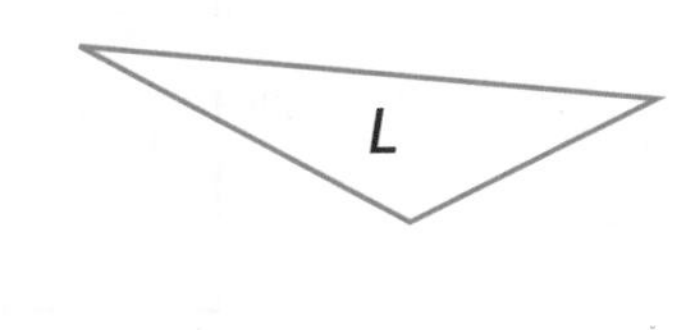

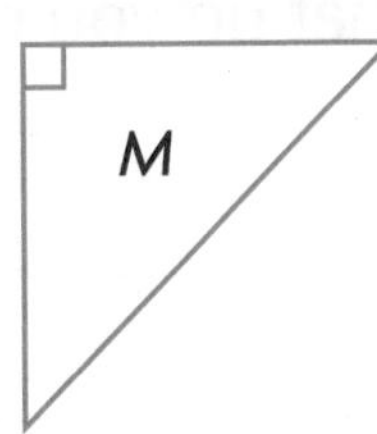

5. Triangle ______ has 1 right angle and appears to have ______ sides of equal length.

6. Triangle ______ has 3 angles less than a right angle and appears to have ______ sides of equal length.

7. Triangle ______ has 1 angle greater than a right angle and appears to have ______ sides of equal length.

Problem Solving REAL WORLD

8. **Sense or Nonsense?** Martin said a triangle can have two sides that are parallel. Does his statement make sense? **Explain.**

____

____

9. Write Math Compare Triangles *R* and *S*. How are they alike? How are they different?

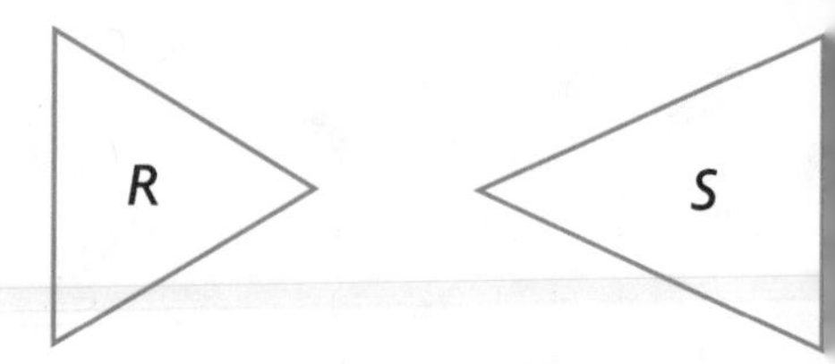

____

____

____

10. **Describe** the shape of the front of the building by its sides and by its angles.

____

____

11. H.O.T. Use a ruler to draw a straight line from one corner of this rectangle to the opposite corner. What shapes did you make? What do you notice about the shapes?

____

____

12. **Test Prep** Savannah made a group with Triangles *L*, *N*, *P*, and *R*. How are these triangles alike?

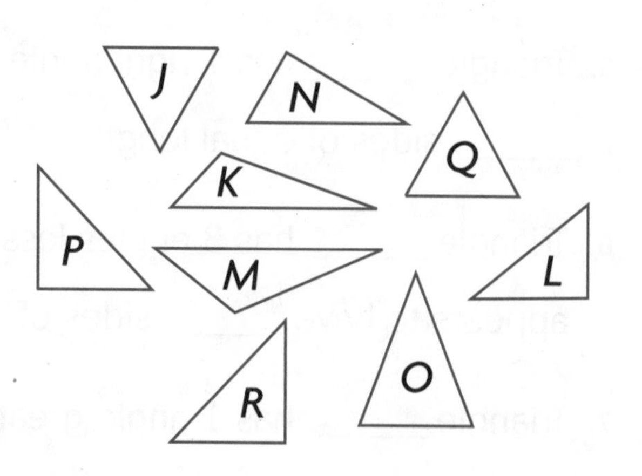

Ⓐ All of their sides are of equal length.

Ⓑ They all have 1 right angle.

Ⓒ They all have 1 angle greater than a right angle.

Ⓓ They all have 3 angles less than a right angle.

Name ______________________________

Problem Solving • Classify Plane Shapes

Essential Question How can you use the strategy *draw a diagram* to classify plane shapes?

UNLOCK the Problem REAL WORLD

A **Venn diagram** shows how sets of things are related. In the Venn diagram at the right, one circle has shapes that are rectangles. Shapes that are rhombuses are in the other circle. The shapes in the section where the circles overlap are both rectangles and rhombuses.

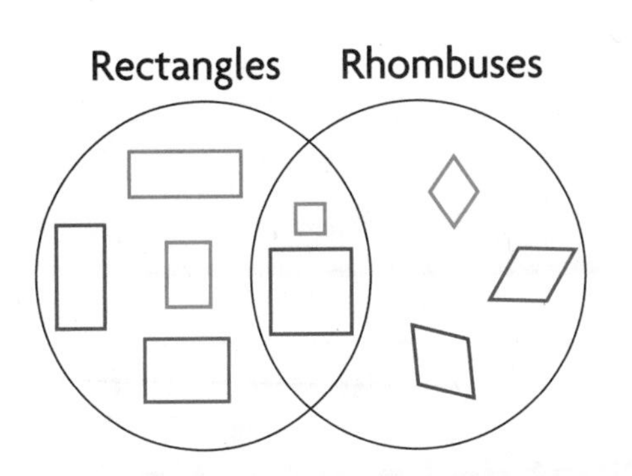

What type of quadrilateral is in both circles?

Read the Problem	Solve the Problem
What do I need to find? ______________ ______________	What is true about all quadrilaterals? ______________ Which quadrilaterals have 2 pairs of opposite sides that are parallel? ______________
What information do I need to use? the circles labeled ________ and ________	Which quadrilaterals have 4 sides of equal length? ______________ Which quadrilaterals have 4 right angles? ______________
How will I use the information? ______________ ______________ ______________	The quadrilaterals in the section where the circles overlap have _____ pairs of opposite sides that are parallel, _____ sides of equal length, and _____ right angles. So, ________ are in both circles.

MATHEMATICAL PRACTICES

Math Talk Does a ⏢ fit in the Venn diagram? **Explain.**

Try Another Problem

The Venn diagram shows the shapes Abbie used to make a picture. Where would the shape shown below be placed in the Venn diagram?

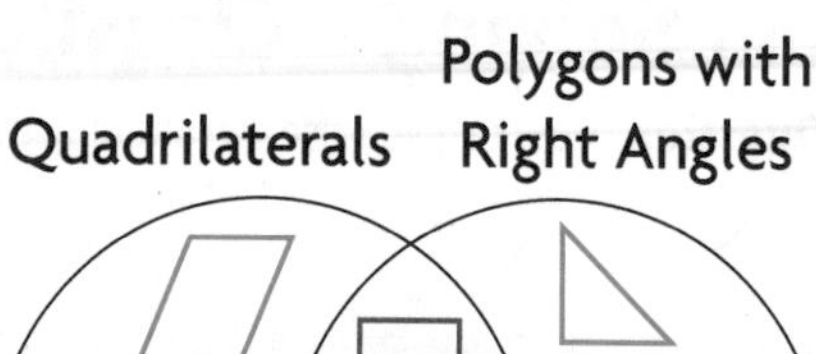

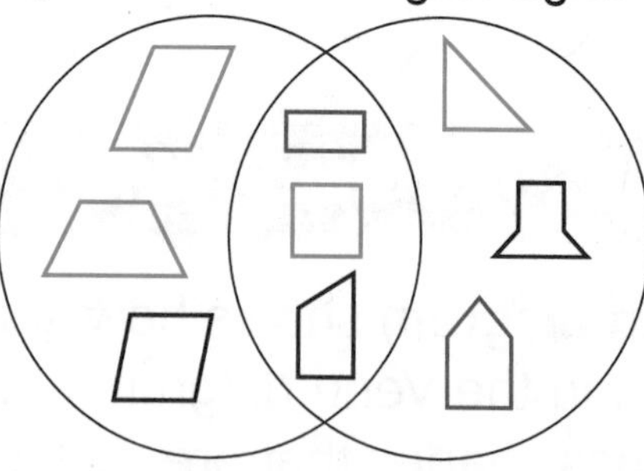

Read the Problem	Solve the Problem
What do I need to find?	**Record the steps you used to solve the problem.**
What information do I need to use?	
How will I use the information?	

1. How many shapes do not have right angles?

 ____

2. How many red shapes have right angles but are not quadrilaterals? __________

3. What is a different way to sort the shapes?

 ____

 ____

MATHEMATICAL PRACTICES

Math Talk What name can be used to describe all the shapes in the Venn diagram? **Explain** how you know.

Name ______________________________

Share and Show 

Use the Venn diagram for 1–3.

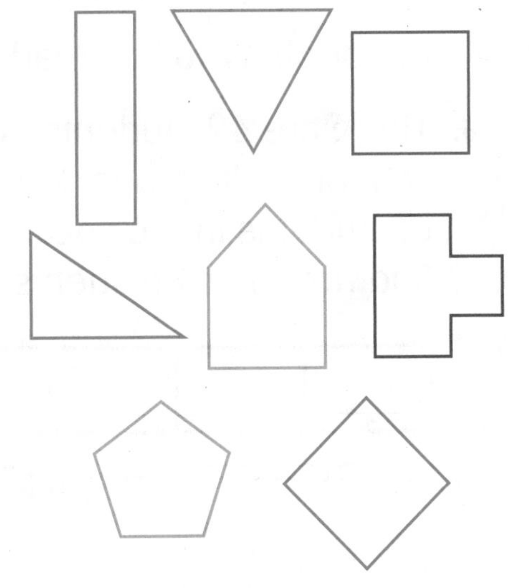

1. Jordan is sorting the shapes at the right in a Venn diagram. Where does the ◇ go?

   First, look at the sides and angles of the polygons.

   Next, draw the polygons in the Venn diagram.

   The shape has _____ sides of equal length

   and _____ right angles.

   So, the shape goes in the

   ____.

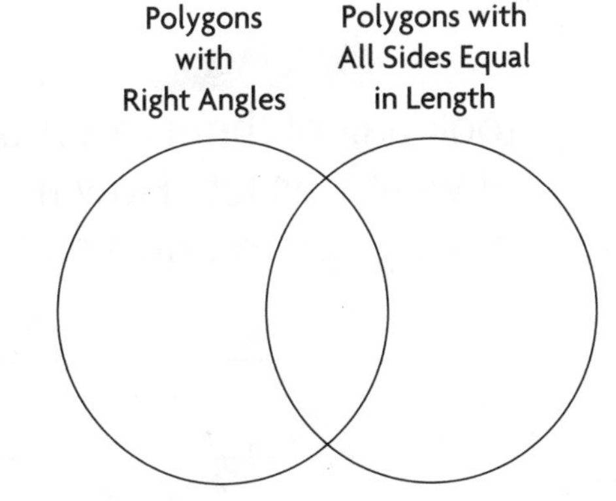

2. H.O.T. **What if** Jordan sorted the shapes by Polygons with Right Angles and Polygons with Angles Less Than a Right Angle? Would the circles still overlap? **Explain.**

   ____

   ____

3. Where would you place a ⏢?

   ____

4. H.O.T. Eva drew the Venn diagram below. What labels could she have used for the diagram?

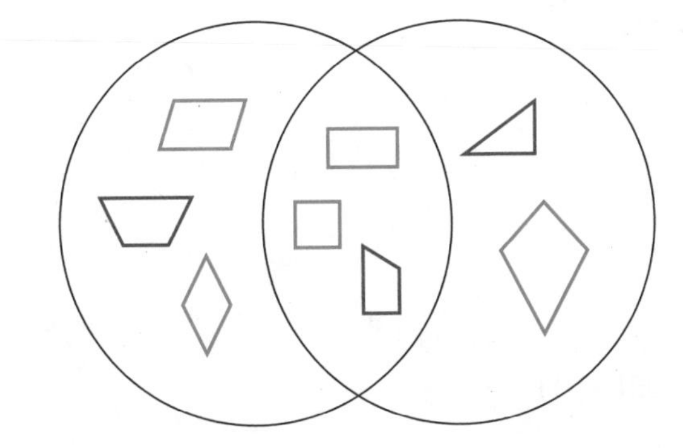

On Your Own

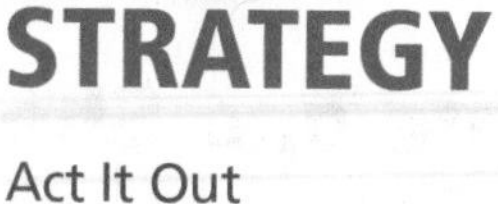

Choose a STRATEGY

- Act It Out
- Draw a Diagram
- Find a Pattern
- Make a Table

5. Ben and Marta are both reading the same book. Ben has read $\frac{1}{3}$ of the book. Marta has read $\frac{1}{4}$ of the book. Who has read more? ________

6. There are 42 students from 6 different classes in the school spelling bee. Each class has the same number of students in the spelling bee. Use the bar model to find how many students are from each class.

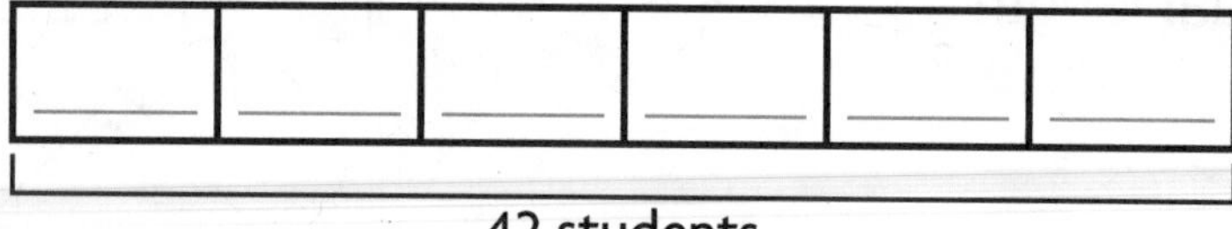

42 students

_____ students ÷ _____ classes = _____ students

7. H.O.T. **Write Math** Cara baked 16 cupcakes. She took half of them to school. She gave Frank one fourth of what was left. How many cupcakes does Cara have now? **Explain** how you know.

____

____

____

8. Ashley is making a quilt with squares of fabric. There are 9 rows with 8 squares in each row. How many squares of fabric are there?

____

9. **Test Prep** What label could describe Circle *A*?

Ⓐ Polygons with Perpendicular Sides

Ⓑ Polygons with 2 Pairs of Opposite Sides That Are Parallel

Ⓒ Polygons with 2 Pairs of Sides of Equal Length

Ⓓ Polygons with All Sides of Equal Length

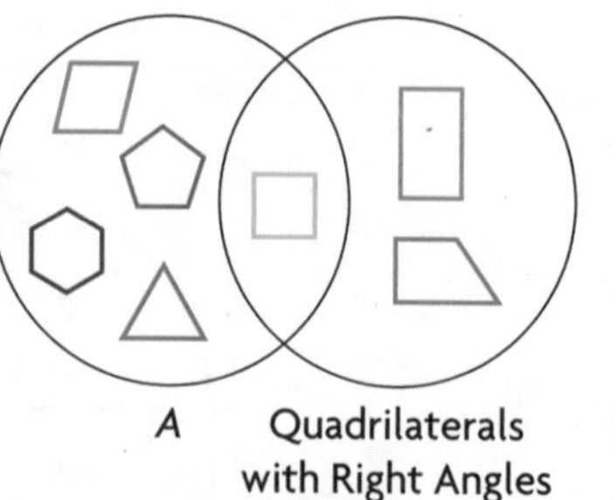

FOR MORE PRACTICE:
Standards Practice Book, pp. P253–P254

Name ______________________________

Relate Shapes, Fractions, and Area

Essential Question How can you divide shapes into parts with equal areas and write the area as a unit fraction of the whole?

Investigate

Materials ■ pattern blocks ■ color pencils ■ ruler

CONNECT You can use what you know about combining and separating plane shapes to explore the relationship between fractions and area.

A. Trace a hexagon pattern block.

B. Divide your hexagon into two parts with equal area.

C. Write the names of the new shapes. __________

D. Write the fraction that names each part of the whole you divided. _____
Each part is $\frac{1}{2}$ of the whole shape's area.

E. Write the fraction that names the whole area. _____

Math Idea

Equal parts of a whole have equal area.

Draw Conclusions

1. **Explain** how you know the two shapes have the same area.

____

____

2. **Predict** what would happen if you divide the hexagon into three shapes with equal area. What fraction names the area of each part of the divided hexagon? What fraction names the whole area?

____

3. H.O.T. **Apply** Show how you can divide the hexagon into four shapes with equal area.

Each part is _____ of the whole shape's area.

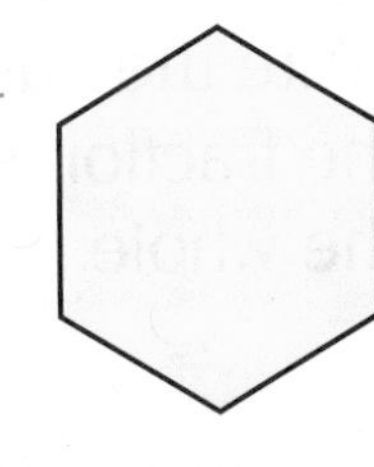

Make Connections

The rectangle at the right is divided into four parts with equal area.

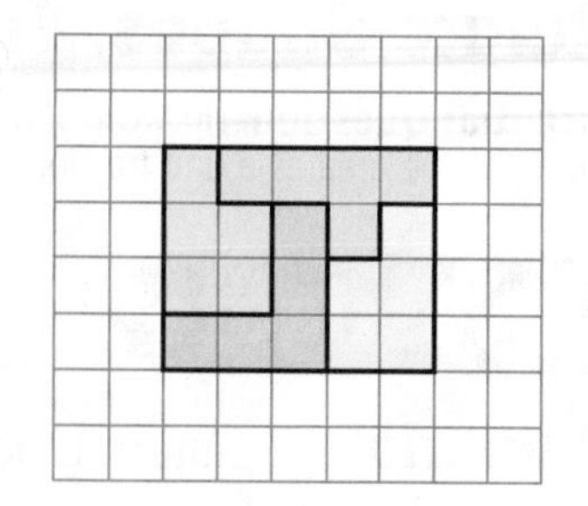

- Write the unit fraction that names each part of the divided whole. ______
- What is the area of each part? __________
- How many $\frac{1}{4}$ parts does it take to make one whole? _____
- Is the shape of each of the $\frac{1}{4}$ parts the same? _____
- Is the area of each of the $\frac{1}{4}$ parts the same? **Explain** how you know.

____

____

Divide the shape into equal parts.

Draw lines to divide the rectangle below into six parts with equal area.

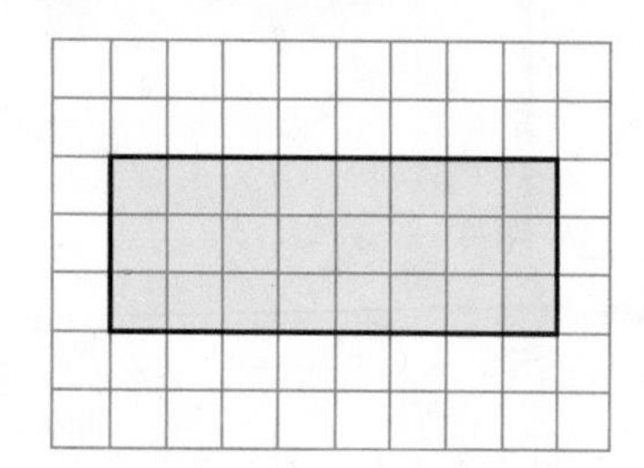

MATHEMATICAL PRACTICES

Math Talk Explain how you know the areas of all the parts are equal.

- Write the fraction that names each part of the divided whole. _____
- Write the area of each part. __________
- Each part is _____ of the whole shape's area.

Share and Show

1. Divide the trapezoid into 3 parts with equal area. Write the names of the new shapes. Then write the fraction that names the area of each part of the whole.

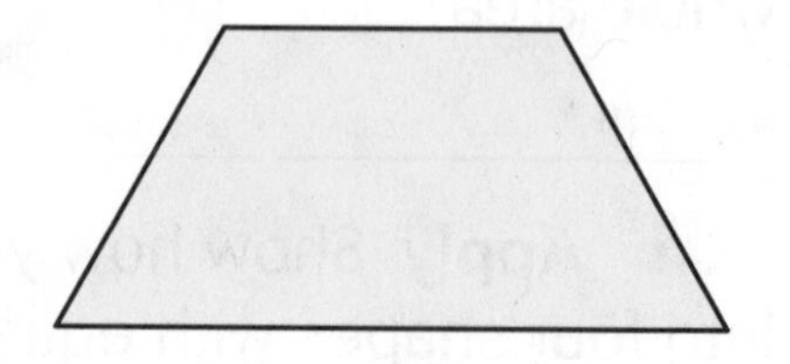

____

Name ______________________

Draw lines to divide the shape into equal parts that show the fraction given.

2.

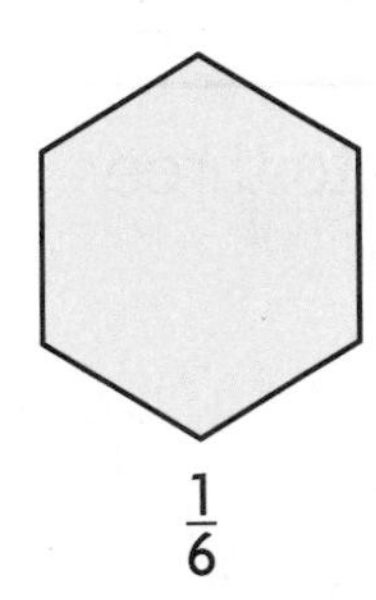

$\frac{1}{6}$

3.

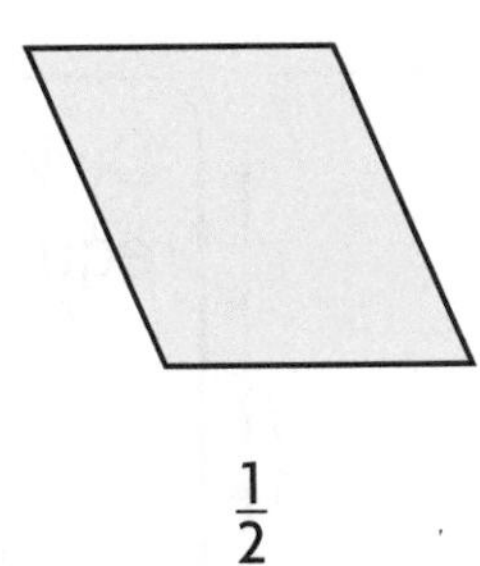

$\frac{1}{2}$

4.

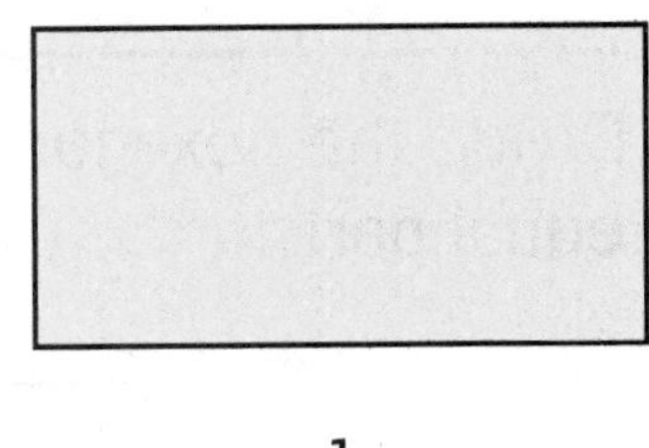

$\frac{1}{8}$

Draw lines to divide the shape into parts with equal area. Write the area of each part as a unit fraction.

5.

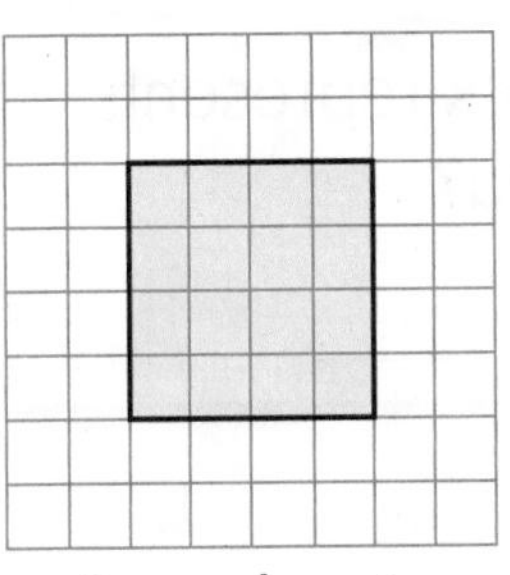

8 equal parts

____

6.

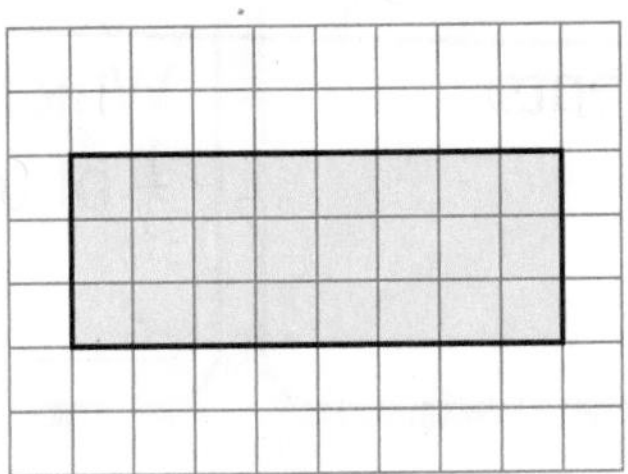

6 equal parts

____

7.

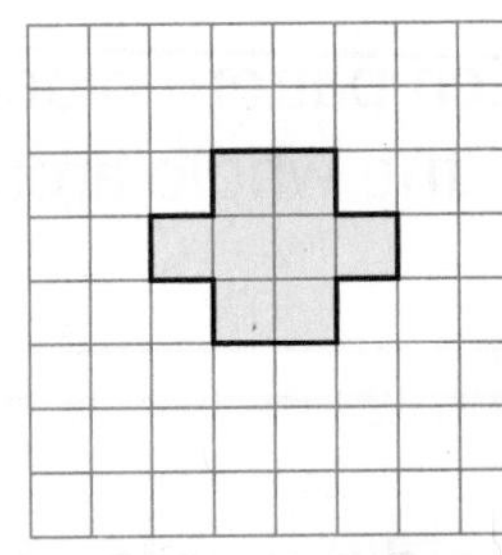

4 equal parts

____

8.

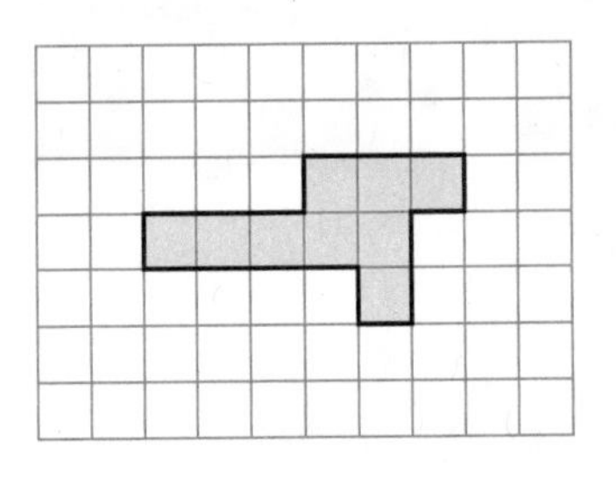

3 equal parts

____

9.

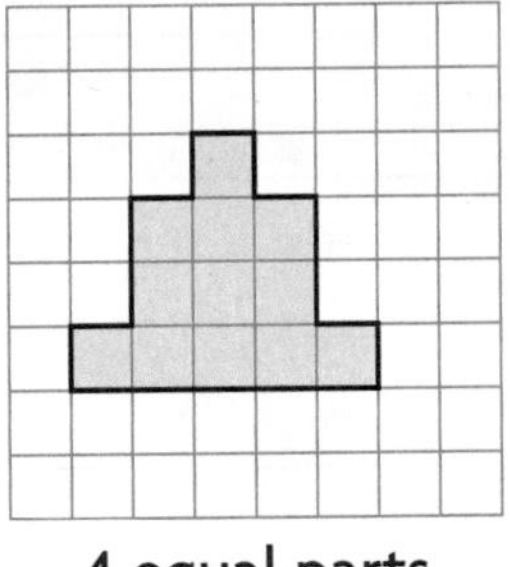

4 equal parts

____

10.

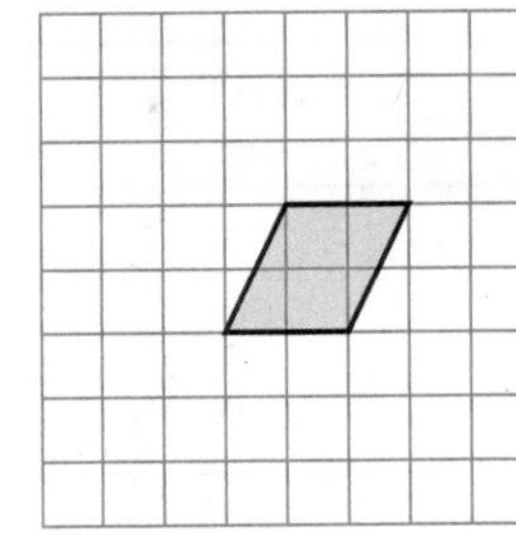

2 equal parts

____

11. H.O.T. **Write Math** If the area of three ◇ is equal to the area of one ⬡, the area of how many ◇ equals four ⬡? **Explain** your answer.

____

Problem Solving REAL WORLD

12. H.O.T. **Sense or Nonsense?**

Divide the hexagon into six equal parts.

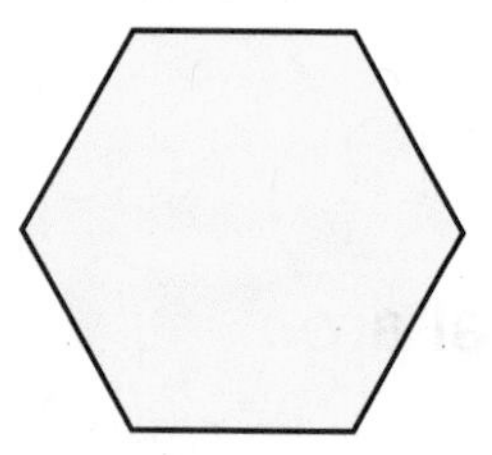

Which pattern block represents $\frac{1}{6}$ of the whole area?

____

Divide the trapezoid into three equal parts.

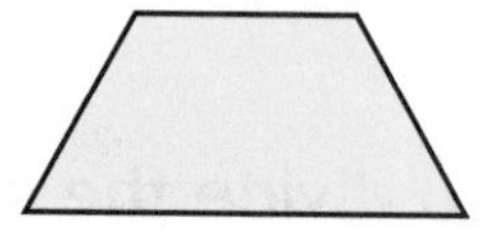

Which pattern block represents $\frac{1}{3}$ of the whole area?

____

Alexis said the area of $\frac{1}{3}$ of the trapezoid is greater than the area of $\frac{1}{6}$ of the hexagon because $\frac{1}{3} > \frac{1}{6}$. Does her statement make sense? **Explain** your answer.

____

____

____

____

- Write a statement that makes sense.

____

____

- **What if** you divide the hexagon into 3 equal parts? Write a sentence that compares the area of each equal part of the hexagon to each equal part of the trapezoid.

____

____

 FOR MORE PRACTICE: Standards Practice Book, pp. P255–P256

Name ___________________________

Chapter Review/Test

Vocabulary

Choose the best term from the box to complete the sentence.

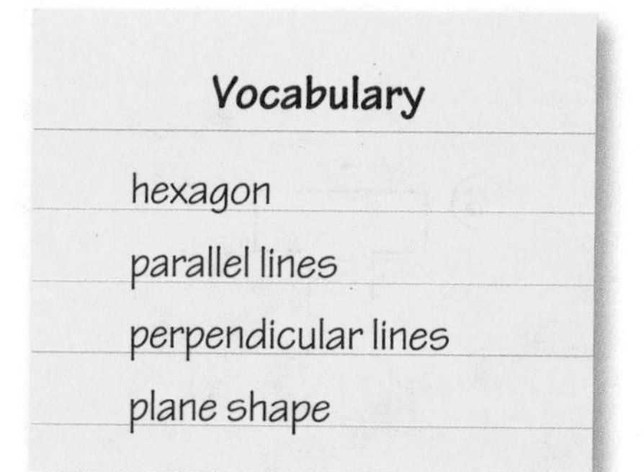

1. A ________________ has 6 sides and 6 angles. (p. 492)

2. A ________________ is formed by points that make curved paths, line segments, or both. (p. 483)

3. ________________ appear to never cross or meet and are always the same distance apart. (p. 495)

Concepts and Skills

Look at the green sides of the polygon. Tell if they appear to be *intersecting, perpendicular,* or *parallel*. Write all the words that describe the sides.

4. 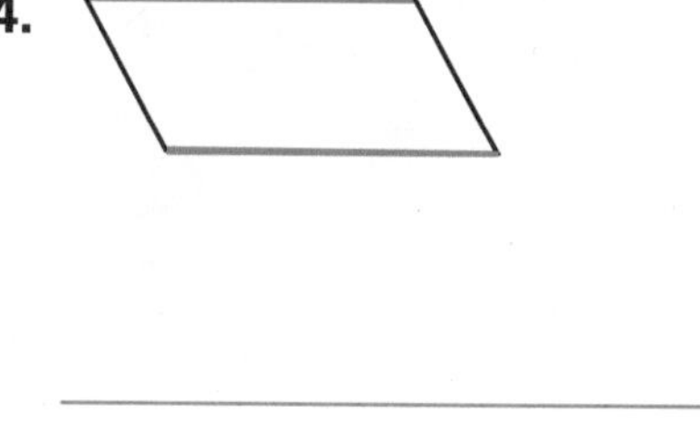

____

____

5. 

____

____

6. 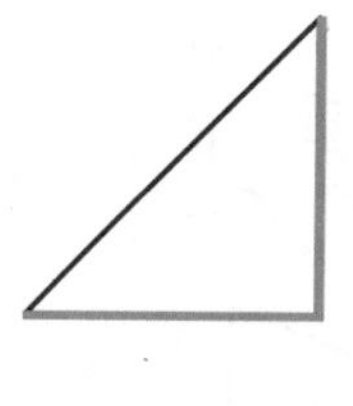

____

____

Circle all the words that describe the quadrilateral.

7. 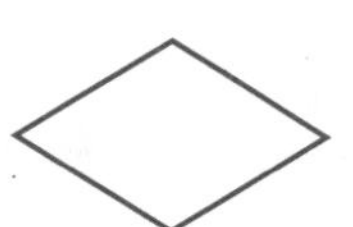

rectangle
rhombus
trapezoid
square

8. 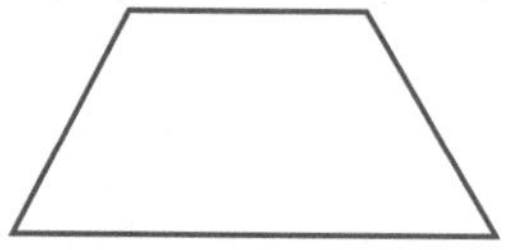

quadrilateral
rhombus
trapezoid
rectangle

9. 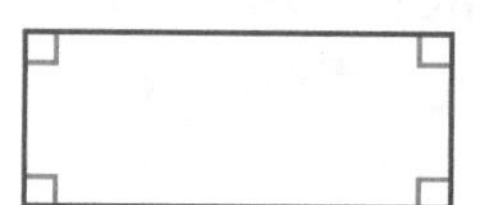

rhombus
rectangle
square
trapezoid

GO Online Assessment Options Chapter Test

Fill in the bubble for the correct answer choice.

10. Which is NOT an example of a closed shape?

Ⓐ

Ⓑ 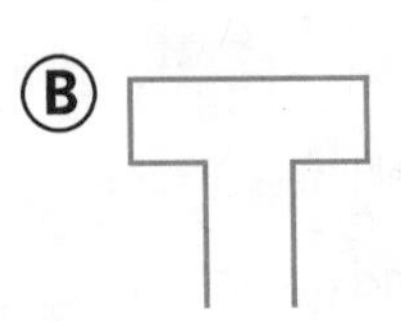

Ⓒ

Ⓓ 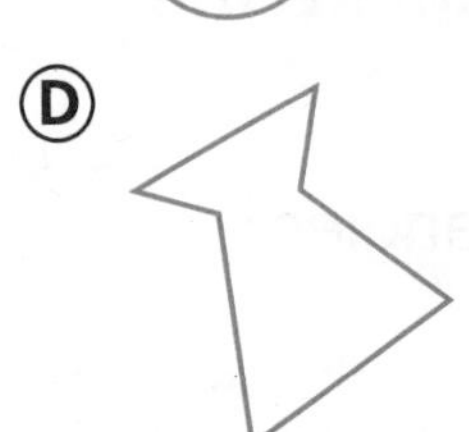

11. Anne drew a line segment. Which of these could be the shape Anne drew?

Ⓐ •

Ⓑ

Ⓒ 

Ⓓ 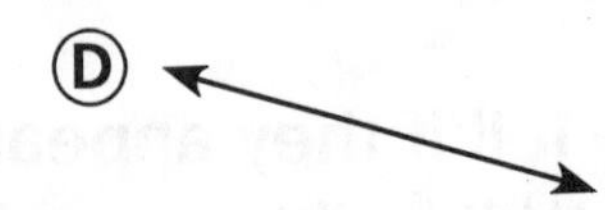

12. Which shape is NOT a polygon?

Ⓐ 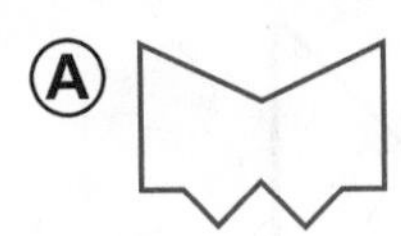

Ⓑ 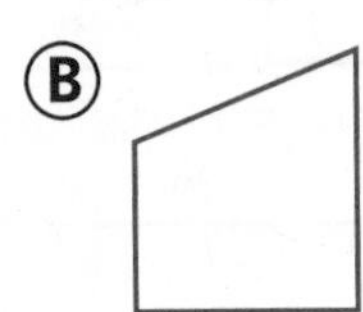

Ⓒ 

Ⓓ 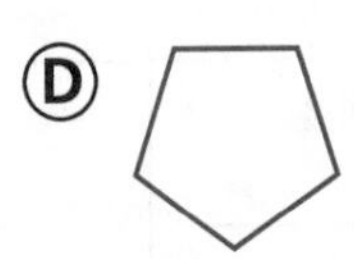

13. Sam divided the shape into parts with equal area. Which fraction names the area of each part of the divided shape?

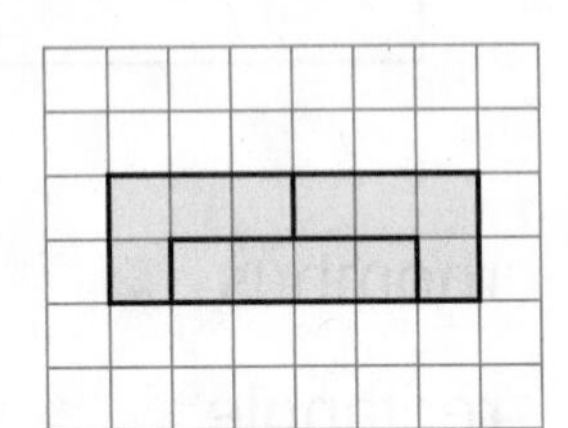

Ⓐ $\frac{1}{2}$

Ⓑ $\frac{1}{3}$

Ⓒ $\frac{1}{4}$

Ⓓ $\frac{1}{6}$

Name ______________________________

Fill in the bubble for the correct answer choice.

14. Philip drew a shape with 2 fewer sides than a hexagon. Which shape did he draw?

Ⓐ triangle Ⓑ pentagon Ⓒ octagon Ⓓ quadrilateral

15. Jennifer drew these polygons. Which shape is NOT a rhombus?

Ⓐ 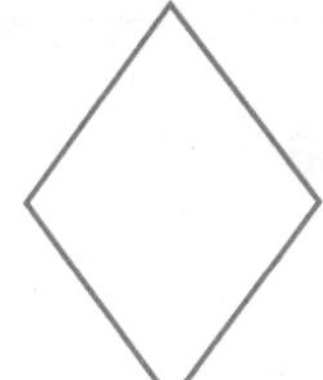

Ⓒ 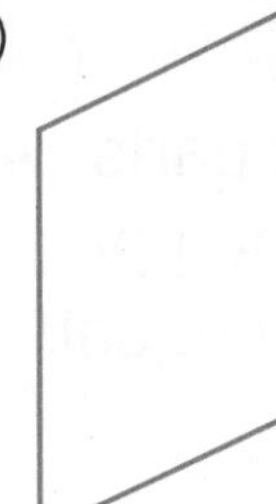

Ⓑ 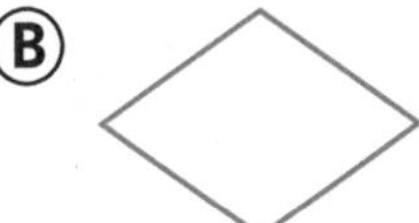

Ⓓ 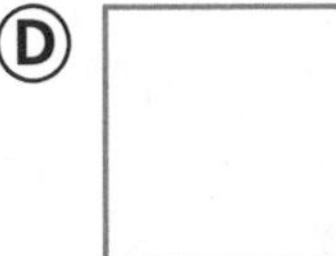

16. Which quadrilateral appears to have only 1 pair of parallel sides?

Ⓐ 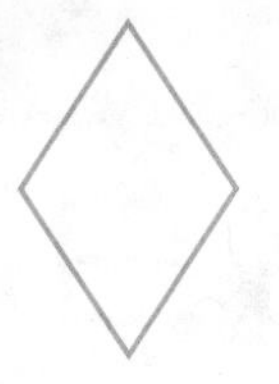

Ⓒ 

Ⓑ 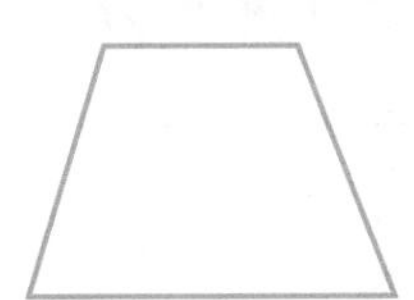

Ⓓ 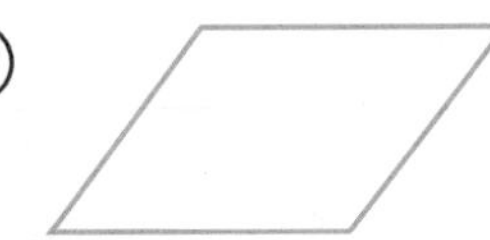

17. Which polygon does NOT have a right angle?

Ⓐ 

Ⓒ 

Ⓑ 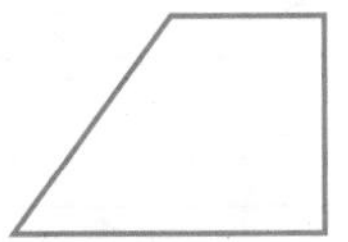

Ⓓ 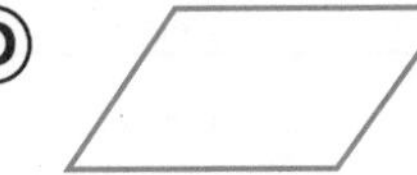

▶ Constructed Response

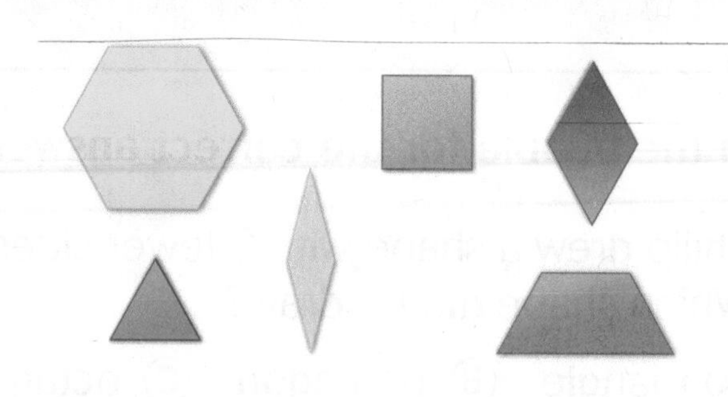

18. Sort the pattern blocks. How many groups did you make? **Explain** how you sorted the shapes.

___

___

___

19. Draw lines to divide the rectangle into four parts with equal area. Write the fraction that names the area of each part. **Explain** how you know the areas of all the parts are equal.

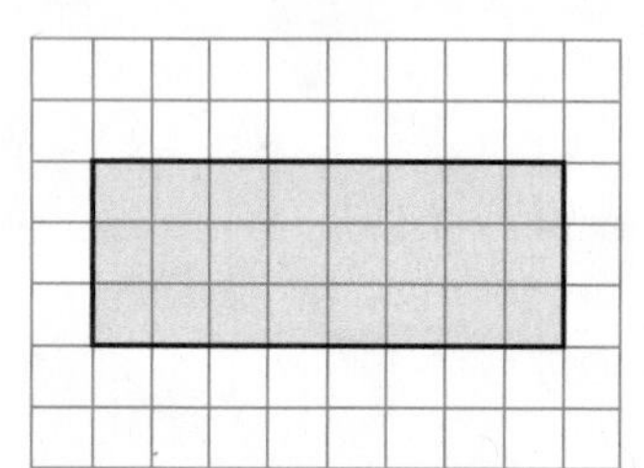

___

▶ Performance Task

20. Jack drew a picture of his house for the art fair. His teacher copied the picture and used it for math class.

A Write about Jack's house by describing four of the shapes you see in it.

___

___

___

B Write more about Jack's house by describing four of the angles you see in it.

___

___

___

Glossary

Pronunciation Key

a	add, map	f	fit, half	n	nice, tin	p	pit, stop	û(r)	burn, term
ā	ace, rate	g	go, log	ng	ring, song	r	run, poor	yo͞o	fuse, few
â(r)	care, air	h	hope, hate	o	odd, hot	s	see, pass	v	vain, eve
ä	palm, father	i	it, give	ō	open, so	sh	sure, rush	w	win, away
b	bat, rub	ī	ice, write	ô	order, jaw	t	talk, sit	y	yet, yearn
ch	check, catch	j	joy, ledge	oi	oil, boy	th	thin, both	z	zest, muse
d	dog, rod	k	cool, take	ou	pout, now	th	this, bathe	zh	vision, pleasure
e	end, pet	l	look, rule	o͝o	took, full	u	up, done		
ē	equal, tree	m	move, seem	o͞o	pool, food	ù	pull, book		

ə the schwa, an unstressed vowel representing the sound spelled *a* in *above*, *e* in *sicken*, *i* in *possible*, *o* in *melon*, *u* in *circus*

Other symbols:
- • separates words into syllables
- ′ indicates stress on a syllable

addend [a′dend] **sumando** Any of the numbers that are added in addition
Examples: 2 + 3 = 5
↑ ↑
addend addend

addition [ə•dish′ən] **suma** The process of finding the total number of items when two or more groups of items are joined; the opposite operation of subtraction

A.M. [ā•em] **a.m.** The time after midnight and before noon (p. 393)

analog clock [an′ə•log kläk] **reloj analógico** A tool for measuring time, in which hands move around a circle to show hours and minutes
Example:

angle [ang′gəl] **ángulo** A shape formed by two rays that share an endpoint (p. 487)
Example:

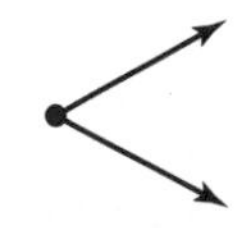

Word History

When the letter *g* is replaced with the letter *k* in the word ***angle***, the word becomes *ankle*. Both words come from the same Latin root, *angulus*, which means "a sharp bend."

area [âr′ē•ə] **área** The measure of the number of unit squares needed to cover a surface (p. 445)
Example:

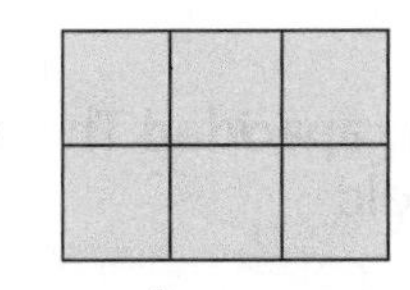

Area = 6 square units

array [ə•rā′] **matriz** A set of objects arranged in rows and columns (p. 115)
Example:

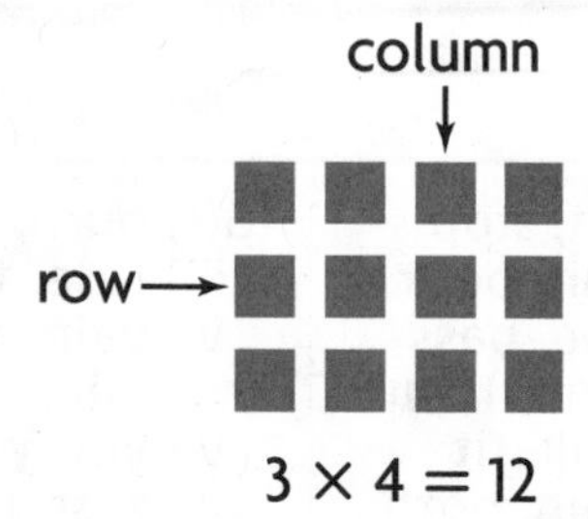

Associative Property of Addition [ə•sō′shē•āt•iv präp′ər•tē əv ə•dish′ən] **propiedad asociativa de la suma** The property that states that you can group addends in different ways and still get the same sum (p. 21)
Example:
4 + (2 + 5) = 11
(4 + 2) + 5 = 11

Associative Property of Multiplication [ə•sō′shē•āt•iv präp′ər•tē əv mul•tə•pli•kā′shən] **propiedad asociativa de la multiplicación** The property that states that when the grouping of factors is changed, the product remains the same (p. 155)
Example:
(3 × 2) × 4 = 24
3 × (2 × 4) = 24

bar graph [bär graf] **gráfica de barras** A graph that uses bars to show data (p. 75)
Example:

capacity [kə•pas′i•tē] **capacidad** The amount a container can hold
Example:
1 liter = 1,000 milliliters

cent sign (¢) [sent sīn] **simbolo de centauo** A symbol that stands for *cent* or *cents*
Example: 53¢

centimeter (cm) [sen′tə•mēt•ər] **centímetro (cm)** A metric unit that is used to measure length or distance
Example:

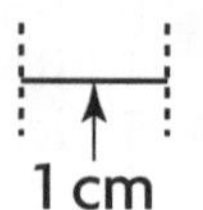

circle [sûr′kəl] **círculo** A round closed plane shape
Example:

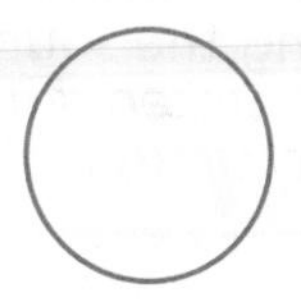

closed shape [klōzd shāp] **figura cerrada** A shape that begins and ends at the same point (p. 484)
Examples:

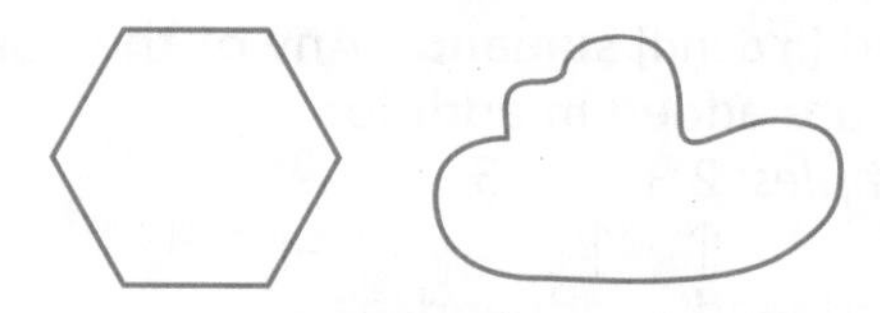

Commutative Property of Addition [kə•myo͞ot′ə•tiv präp′ər•tē əv ə•dish′ən] **propiedad conmutativa de la suma** The property that states that you can add two or more numbers in any order and get the same sum (p. 6)
Example: 6 + 7 = 13
7 + 6 = 13

Commutative Property of Multiplication [kə•myo͞ot′ə•tiv präp′ər•tē əv mul•tə•pli•kā′shən] **propiedad conmutativa de la multiplicación** The property that states that you can multiply two factors in any order and get the same product (p. 120)
Example: 2 × 4 = 8
4 × 2 = 8

compare [kəm•pâr′] **comparar** To describe whether numbers are equal to, less than, or greater than each other

compatible numbers [kəm•pat′ə•bəl num′bərz] **números compatibles** Numbers that are easy to compute with mentally (p. 13)

cone [kōn] **cono** A three-dimensional, pointed shape that has a flat, round base
Example:

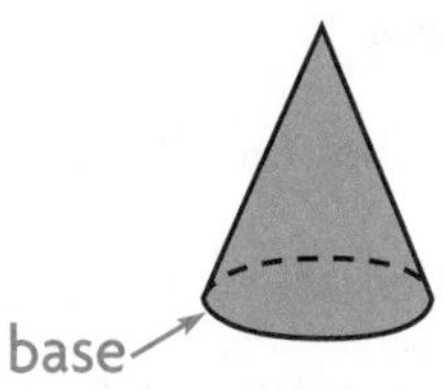

counting number [kount′ing num′bər] **número natural** A whole number that can be used to count a set of objects (1, 2, 3, 4 . . .)

cube [kyōōb] **cubo** A three-dimensional shape with six square faces of the same size
Example:

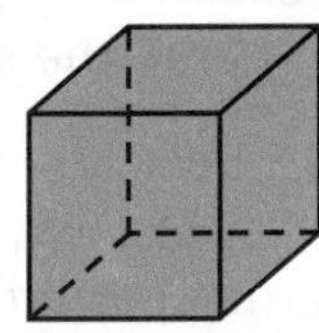

cylinder [sil′ən•dər] **cilindro** A three-dimensional object that is shaped like a can
Example:

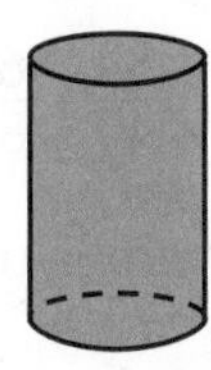

data [dāt′ə] **datos** Information collected about people or things

decagon [dek′ə•gän] **decágono** A polygon with ten sides and ten angles (p. 492)
Example:

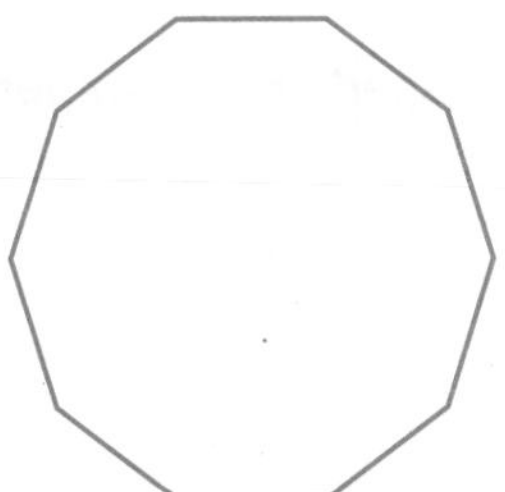

decimal point [des′ə•məl point] **punto decimal** A symbol used to separate dollars from cents in money
Example: \$4.52
↑ decimal point

denominator [dē•näm′ə•nāt•ər] **denominador** The part of a fraction below the line, which tells how many equal parts there are in the whole or in the group (p. 319)
Example: $\frac{3}{4}$ ← denominator

difference [dif′ər•əns] **diferencia** The answer to a subtraction problem
Example: 6 − 4 = 2
↑ difference

digital clock [dij′i•təl kläk] **reloj digital** A clock that shows time to the minute, using digits
Example:

digits [dij′its] **dígitos** The symbols 0, 1, 2, 3, 4, 5, 6, 7, 8, and 9

dime [dīm] **moneda de 10¢** A coin worth 10 cents and with a value equal to that of 10 pennies; 10¢
Example:

Distributive Property [di•strib′yōō•tiv präp′ər•tē] **propiedad distributiva** The property that states that multiplying a sum by a number is the same as multiplying each addend by the number and then adding the products (p. 145)
Example:
$5 \times 8 = 5 \times (4 + 4)$
$5 \times 8 = (5 \times 4) + (5 \times 4)$
$5 \times 8 = 20 + 20$
$5 \times 8 = 40$

divide [də•vīd′] **dividir** To separate into equal groups; the opposite operation of multiplication (p. 213)

dividend [div′ə•dend] **dividendo** The number that is to be divided in a division problem (p. 222)
Example: 35 ÷ 5 = 7
↑ dividend

division [də•vizh′ən] **división** The process of sharing a number of items to find how many groups can be made or how many items will be in a group; the opposite operation of multiplication

divisor [de•vī′zər] **divisor** The number that divides the dividend (p. 222)
Example: $35 \div 5 = 7$
↑ divisor

dollar [däl′ər] **dólar** Paper money worth 100 cents and equal to 100 pennies; $1.00
Example:

edge [ej] **arista** A line segment formed where two faces meet

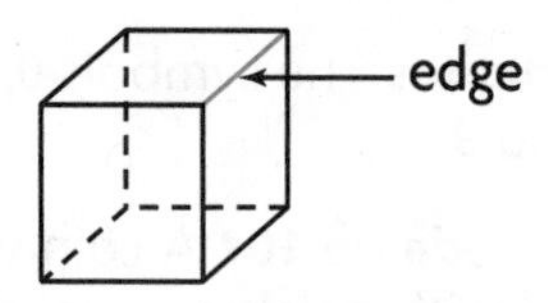

eighths [ātths] **octavos**

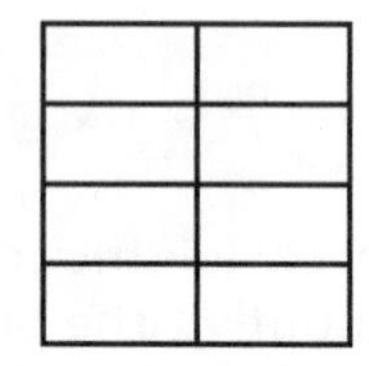

These are eighths (p. 307)

elapsed time [ē•lapst′ tīm] **tiempo transcurrido** The time that passes from the start of an activity to the end of that activity (p. 397)

endpoint [end′point] **extremo** The point at either end of a line segment (p. 483)

equal groups [ē′kwəl gro͞opz] **grupos iguales** Groups that have the same number of objects (p. 97)

equal parts [ē′kwəl pärts] **partes iguales** Parts that are exactly the same size (p. 307)

equal sign (=) [ē′kwəl sīn] **signo de igualdad** A symbol used to show that two numbers have the same value
Example: $384 = 384$

equal to (=) [ē′kwəl to͞o] **igual a** Having the same value
Example: $4 + 4$ is equal to $3 + 5$.

equation [ē•kwā′zhən] **ecuación** A number sentence that uses the equal sign to show that two amounts are equal (p. 185)
Examples:
$3 + 7 = 10$
$4 - 1 = 3$
$6 \times 7 = 42$
$8 \div 2 = 4$

equivalent [ē•kwiv′ə•lənt] **equivalente** Two or more sets that name the same amount

equivalent fractions [ē•kwiv′ə•lənt frak′shənz] **fracciones equivalentes** Two or more fractions that name the same amount (p. 373)
Example:

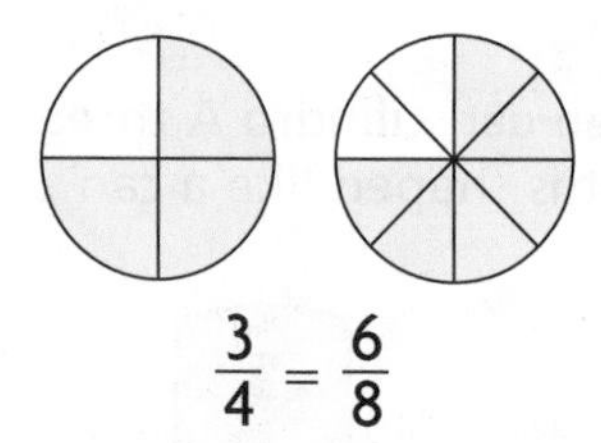

$\frac{3}{4} = \frac{6}{8}$

estimate [es′tə•māt] *verb* **estimar** To find about how many or how much

estimate [es′tə•mit] *noun* **estimación** A number close to an exact amount (p. 13)

even [ē′vən] **par** A whole number that has a 0, 2, 4, 6, or 8 in the ones place

expanded form [ek•span′did fôrm] **forma desarrollada** A way to write numbers by showing the value of each digit
Example: $721 = 700 + 20 + 1$

experiment [ek•sper′ə•mənt] **experimento** A test that is done in order to find out something

face [fās] **cara** A polygon that is a flat surface of a solid shape

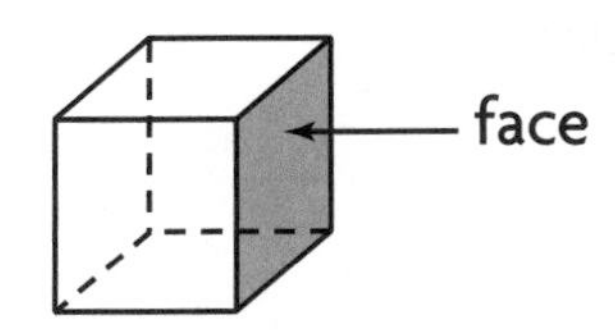

factor [fak′tər] **factor** A number that is multiplied by another number to find a product (p. 102)
Examples: $3 \times 8 = 24$

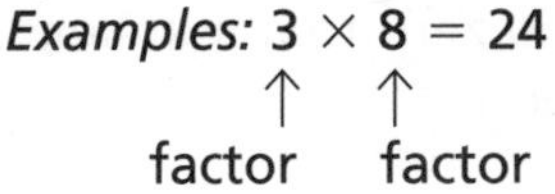

foot (ft) [fo͝ot] **pie** A customary unit used to measure length or distance;
1 foot = 12 inches

fourths [fôrths] **cuartos**

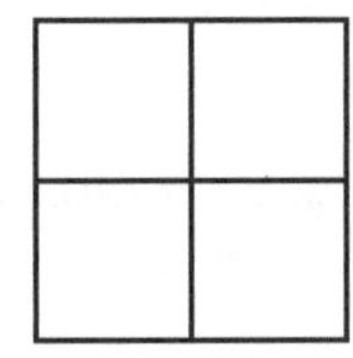

These are fourths (p. 307)

fraction [frak′shən] **fracción** A number that names part of a whole or part of a group (p. 315)
Examples:

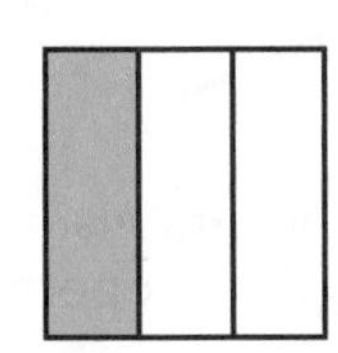

$\frac{1}{3}$

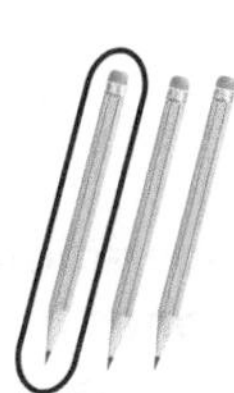

Word History

Often, a ***fraction*** is a part of a whole that is broken into pieces. *Fraction* comes from the Latin word *frangere*, which means "to break."

fraction greater than 1 [frak′shən grāt′ər t͟han wun] **fracción mayor que 1** A number which has a numerator that is greater than its denominator (p. 330)
Examples:

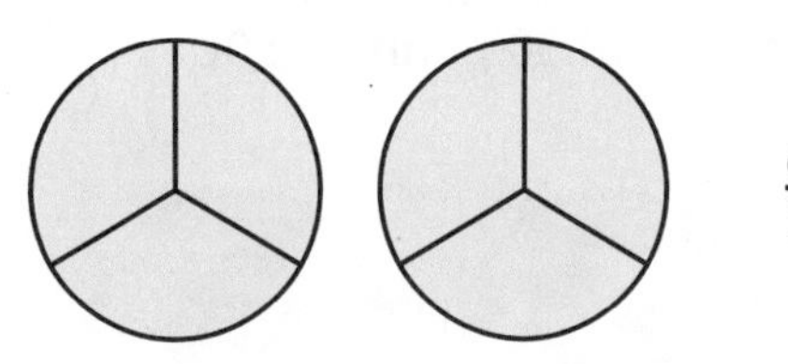

$\frac{6}{3}$ $\frac{2}{1}$

frequency table [frē′kwən•sē tā′bəl] **tabla de frecuencia** A table that uses numbers to record data (p. 61)
Example:

Favorite Color	
Color	**Number**
Blue	10
Green	8
Red	7
Yellow	4

gram (g) [gram] **gramo (g)** A metric unit that is used to measure mass;
1 kilogram = 1,000 grams (p. 419)

greater than (>) [grāt′ər t͟han] **mayor que** A symbol used to compare two numbers when the greater number is given first
Example:
Read $6 > 4$ as "six is greater than four."

Grouping Property of Addition [gro͞op′ing präp′ər•tē əv ə•dish′ən] **propiedad de agrupación de la suma** *See* Associative Property of Addition.

Grouping Property of Multiplication [gro͞op′ing präp′ər•tē əv mul•tə•pli•kā′shən] **propiedad de agrupación de la multiplicación** *See* Associative Property of Multiplication.

half dollar [haf dol′ər] **moneda de 50¢** A coin worth 50 cents and with a value equal to that of 50 pennies; 50¢
Example:

half hour [haf our] **media hora** 30 minutes
Example: Between 4:00 and 4:30 is one half hour.

halves [havz] **mitades**

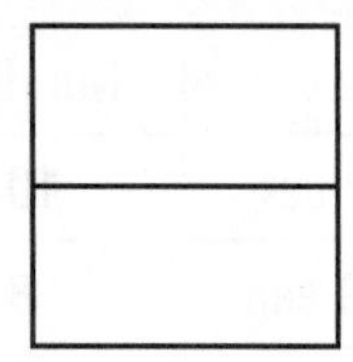

These are halves (p. 307)

hexagon [hek′sə•gän] **hexágono** A polygon with six sides and six angles (p. 492)
Examples:

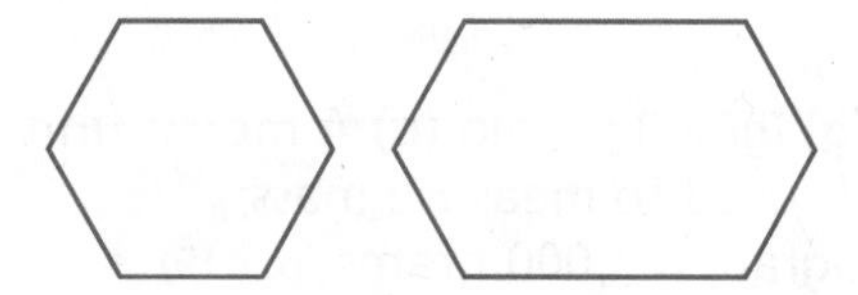

horizontal bar graph [hôr•i•zänt′l bär graf] **gráfica de barras horizontales** A bar graph in which the bars go from left to right (p. 76)
Examples:

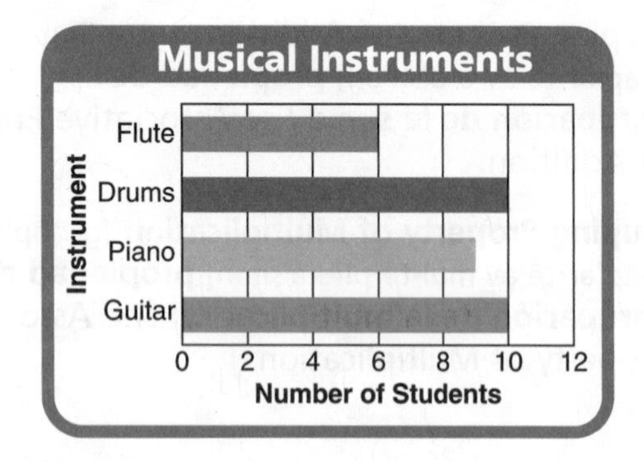

hour (hr) [our] **hora (h)** A unit used to measure time; in one hour, the hour hand on an analog clock moves from one number to the next; 1 hour = 60 minutes

hour hand [our hand] **horario** The short hand on an analog clock

Identity Property of Addition [ī•den′tə•tē präp′ər•tē əv ə•dish′ən] **propiedad de identidad de la suma** The property that states that when you add zero to a number, the result is that number (p. 5)
Example: $24 + 0 = 24$

Identity Property of Multiplication [ī•den′tə•tē präp′ər•tē əv mul•tə•pli•kā′shən] **propiedad de identidad de la multiplicación** The property that states that the product of any number and 1 is that number (p. 124)
Examples: $5 \times 1 = 5$
$1 \times 8 = 8$

inch (in.) [inch] **pulgada (pulg.)** A customary unit used to measure length or distance
Example:

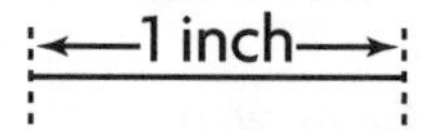

intersecting lines [in•tər•sekt′ing līnz] **líneas secantes** Lines that meet or cross (p. 495)
Example:

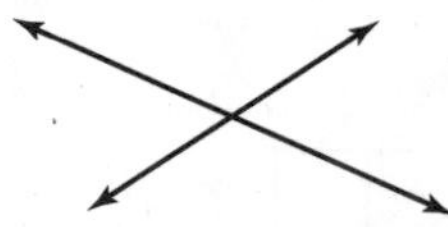

inverse operations [in′vûrs äp•ə•rā′shənz] **operaciones inversas** Opposite operations, or operations that undo one another, such as addition and subtraction or multiplication and division (p. 235)

key [kē] **clave** The part of a map or graph that explains the symbols (p. 65)

kilogram (kg) [kil′ō•gram] **kilogramo (kg)** A metric unit used to measure mass; 1 kilogram = 1,000 grams (p. 419)

length [lengkth] **longitud** The measurement of the distance between two points

less than (<) [les t͟han] **menor que** A symbol used to compare two numbers when the lesser number is given first
Example:
Read $3 < 7$ as "three is less than seven."

line [līn] **línea** A straight path extending in both directions with no endpoints (p. 483)
Example:

Word History

The word *line* comes from *linen*, a thread spun from the fibers of the flax plant. In early times, thread was held tight to mark a straight line between two points.

line plot [līn plät] **diagrama de puntos** A graph that records each piece of data on a number line (p. 87)
Example:

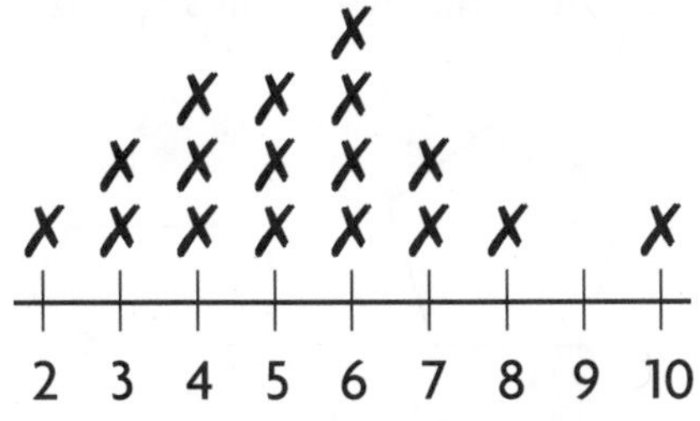

Height of Bean Seedlings to the Nearest Centimeter

line segment [līn seg′mənt] **segmento** A part of a line that includes two points, called endpoints, and all of the points between them (p. 483)
Example:

liquid volume [lik′wid väl′yo͞om] **volumen de un líquido** The amount of liquid in a container (p. 415)

liter (L) [lēt′ər] **litro (L)** A metric unit used to measure capacity and liquid volume;
1 liter = 1,000 milliliters (p. 415)

mass [mas] **masa** The amount of matter in an object (p. 419)

meter (m) [mēt′ər] **metro (m)** A metric unit used to measure length or distance;
1 meter = 100 centimeters

midnight [mid′nīt] **medianoche** 12:00 at night (p. 393)

milliliter (mL) [mil′i•lēt•ər] **mililitro (mL)** A metric unit used to measure capacity and liquid volume

minute (min) [min′it] **minuto (min)** A unit used to measure short amounts of time; in one minute, the minute hand on an analog clock moves from one mark to the next (p. 389)

minute hand [min′it hand] **minutero** The long hand on an analog clock

multiple [mul′tə•pəl] **múltiplo** A number that is the product of two counting numbers (p. 137)
Examples:

6	6	6	6	counting
× 1	× 2	× 3	× 4	← numbers
6	12	18	24	← multiples of 6

multiplication [mul•tə•pli•kā′shən] **multiplicación** The process of finding the total number of items in two or more equal groups; the opposite operation of division

multiply [mul′tə•plī] **multiplicar** To combine equal groups to find how many in all; the opposite operation of division (p. 102)

nickel [nik′əl] **moneda de 5¢** A coin worth 5 cents and with a value equal to that of 5 pennies; 5¢
Example:

noon [no͞on] **mediodía** 12:00 in the day (p. 394)

number line [num′bər līn] **recta numérica** A line on which numbers can be located
Example:

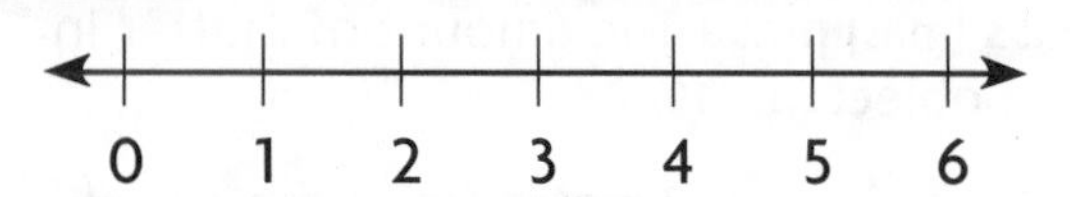

number sentence [num′bər sent′ns] **enunciado numérico** A sentence that includes numbers, operation symbols, and a greater than symbol, a less than symbol, or an equal sign
Example: $5 + 3 = 8$

numerator [nōō′mər•āt•ər] **numerador** The part of a fraction above the line, which tells how many parts are being counted (p. 319)
Example: $\frac{3}{4}$ ← numerator

octagon [äk′tə•gän] **octágono** A polygon with eight sides and eight angles (p. 492)
Examples:

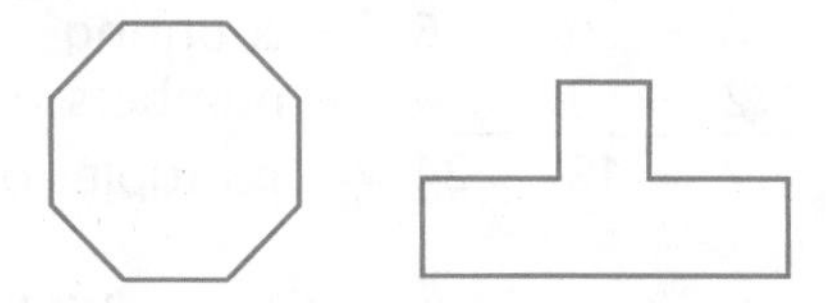

odd [od] **impar** A whole number that has a 1, 3, 5, 7, or 9 in the ones place

open shape [ō′pən shāp] **figura abierta** A shape that does not begin and end at the same point (p. 484)
Examples:

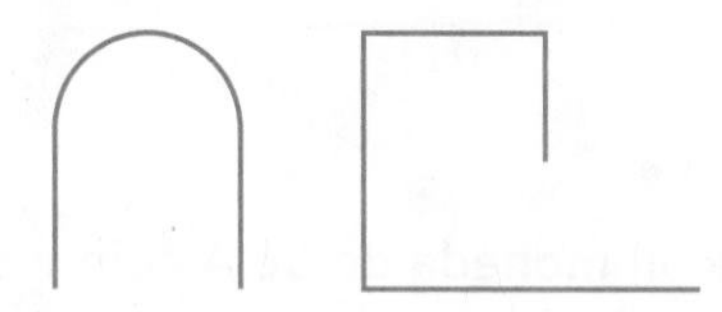

order [ôr′dər] **orden** A particular arrangement or placement of numbers or things, one after another

order of operations [ôr′dər əv äp•ə•rā′shənz] **orden de las operaciones** A special set of rules that gives the order in which calculations are done (p. 296)

Order Property of Addition [ôr′dər präp′ər•tē əv ə•dish′ən] **propiedad de orden de la suma** *See* Commutative Property of Addition.

Order Property of Multiplication [ôr′dər präp′ər•tē əv mul•tə•pli•kā′shən] **propiedad de orden de la multiplicación** *See* Commutative Property of Multiplication.

parallel lines [pâr′ə•lel līnz] **líneas paralelas** Lines in the same plane that never cross and are always the same distance apart (p. 495)
Example:

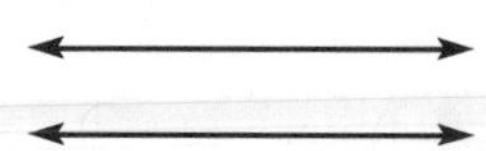

pattern [pat′ərn] **patrón** An ordered set of numbers or objects in which the order helps you predict what will come next (p. 5)
Examples:
2, 4, 6, 8, 10

pentagon [pen′tə•gän] **pentágono** A polygon with five sides and five angles (p. 492)
Examples:

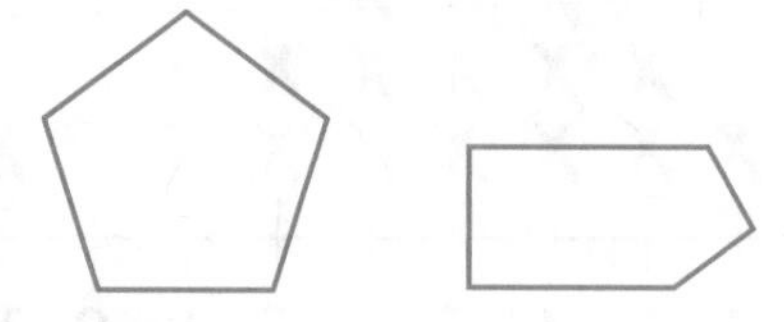

perimeter [pə•rim′ə•tər] **perímetro** The distance around a shape (p. 433)
Example:

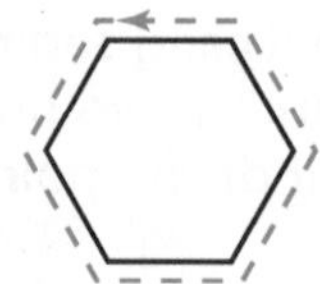

perpendicular lines [pər•pən•dik′yōō•lər līnz] **líneas perpendiculares** Lines that intersect to form right angles (p. 495)
Example:

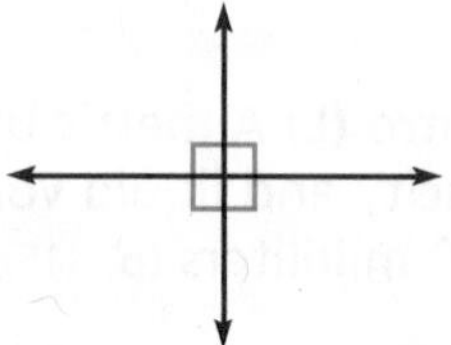

picture graph [pik′chər graf] **gráfica con dibujos** A graph that uses pictures to show and compare information (p. 65)
Example:

How We Get to School	
Walk	❋ ❋ ❋
Ride a Bike	❋ ❋ ❋ ❋
Ride a Bus	❋ ❋ ❋ ❋ ❋ ❋
Ride in a Car	❋ ❋
Key: Each ❋ = 10 students.	

place value [plās val′yo͞o] **valor posicional** The value of each digit in a number, based on the location of the digit

plane [plān] **plano** A flat surface that extends without end in all directions
Example:

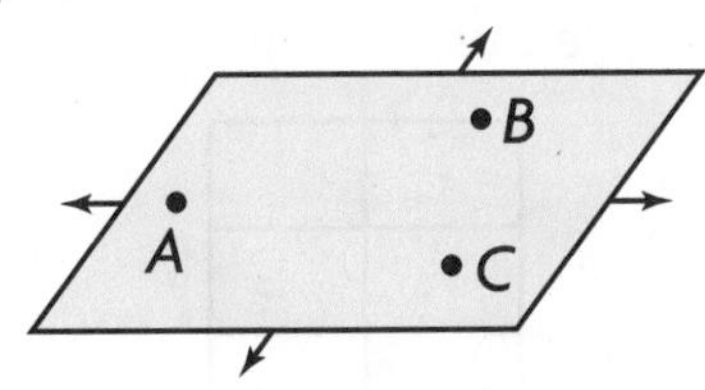

plane shape [plān shāp] **figura plana** A shape in a plane that is formed by curves, line segments, or both (p. 483)
Example:

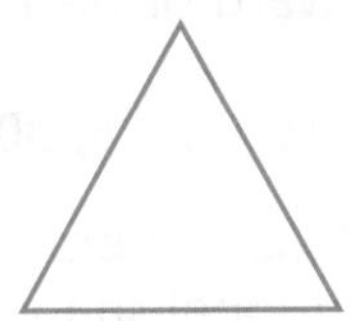

P.M. [pē•em] **p.m.** The time after noon and before midnight (p. 394)

point [point] **punto** An exact position or location (p. 483)

polygon [päl′i•gän] **polígono** A closed plane shape with straight sides that are line segments (p. 491)
Examples:

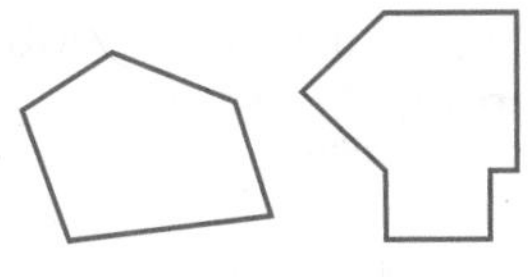

polygons

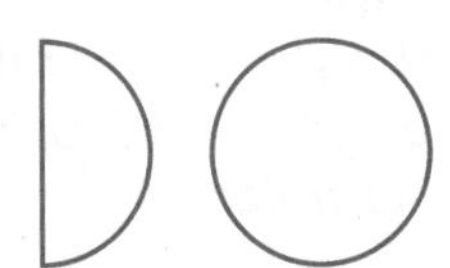

not polygons

Word History

Did you ever think that a ***polygon*** looks like a bunch of knees that are bent? This is how the term got its name. *Poly-* is from the Greek word *polys,* which means "many." The ending *-gon* is from the Greek word *gony,* which means "knee."

product [präd′əkt] **producto** The answer in a multiplication problem (p. 102)
Example: $3 \times 8 = 24$ ← product

quadrilateral [kwäd•ri•lat′ər•əl] **cuadrilátero** A polygon with four sides and four angles (p. 492)
Example:

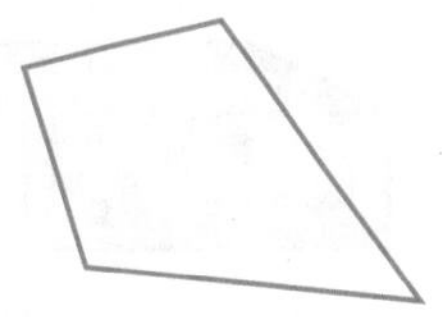

quarter [kwôrt′ər] **moneda de 25¢** A coin worth 25 cents and with a value equal to that of 25 pennies; 25¢
Example:

quarter hour [kwôrt′ər our] **cuarto de hora** 15 minutes
Example: Between 4:00 and 4:15 is one quarter hour.

quotient [kwō′shənt] **cociente** The number, not including the remainder, that results from division (p. 222)
Example: $8 \div 4 = 2$ ← quotient

ray [rā] **semirrecta** A part of a line, with one endpoint, that is straight and continues in one direction (p. 483)
Example:

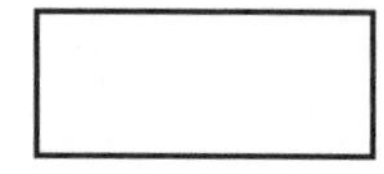

rectangle [rek′tang•gəl] **rectángulo** A quadrilateral with two pairs of parallel sides, two pairs of sides of equal length, and four right angles (p. 501)
Example:

rectangular prism [rek•tang′gyə•lər priz′əm] **prisma rectangular** A three-dimensional shape with six faces that are all rectangles
Example:

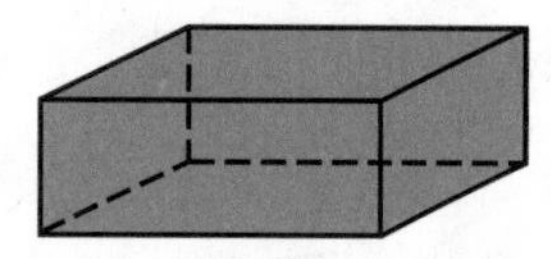

regroup [rē•gro͞op′] **reagrupar** To exchange amounts of equal value to rename a number
Example: 5 + 8 = 13 ones or 1 ten 3 ones

related facts [ri•lāt′id fakts] **operaciones relacionadas** A set of related addition and subtraction, or multiplication and division, number sentences (p. 239)
Examples: $4 \times 7 = 28$ $28 \div 4 = 7$
$7 \times 4 = 28$ $28 \div 7 = 4$

remainder [ri•mān′dər] **residuo** The amount left over when a number cannot be divided evenly

results [ri•zults′] **resultados** The answers from a survey

rhombus [räm′bəs] **rombo** A quadrilateral with two pairs of parallel sides and four sides of equal length (p. 501)
Example:

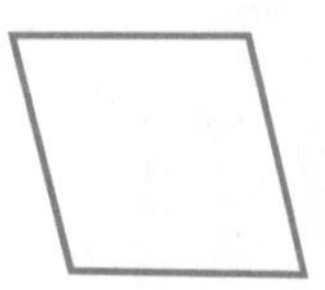

right angle [rīt ang′gəl] **ángulo recto** An angle that forms a square corner (p. 487)
Example:

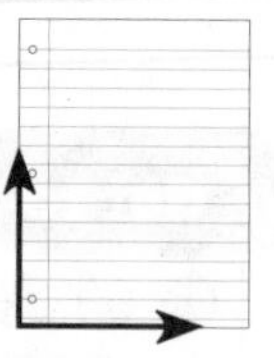

round [round] **redondear** To replace a number with another number that tells about how many or how much (p. 9)

S

scale [skāl] **escala** The numbers placed at fixed distances on a graph to help label the graph (p. 75)

side [sīd] **lado** A straight line segment in a polygon (p. 491)

sixths [siksths] **sextos**

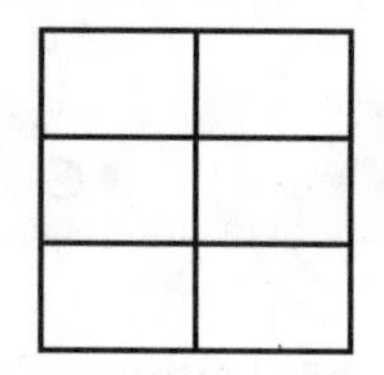

These are sixths (p. 307)

skip count [skip kount] **contar salteado** A pattern of counting forward or backward (p. 83)
Example: 5, 10, 15, 20, 25, 30, . . .

solid shape [sä′lid shāp] **cuerpo geométrico** *See* three-dimensional shape.

sphere [sfir] **esfera** A three-dimensional shape that has the shape of a round ball
Example:

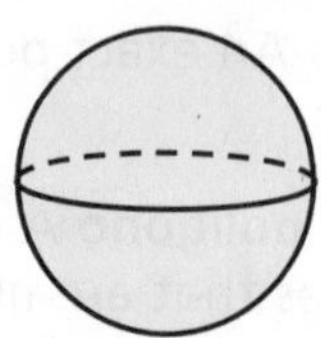

square [skwâr] **cuadrado** A quadrilateral with two pairs of parallel sides, four sides of equal length, and four right angles (p. 501)
Example:

square unit [skwâr yo͞o′nit] **unidad cuadrada** A unit used to measure area such as square foot, square meter, and so on (p. 445)

standard form [stan′dərd fôrm] **forma normal** A way to write numbers by using the digits 0–9, with each digit having a place value
Example: 345 ← standard form

subtraction [səb•trak′shən] **resta** The process of finding how many are left when a number of items are taken away from a group of items; the process of finding the difference when two groups are compared; the opposite operation of addition

sum [sum] **suma o total** The answer to an addition problem
Example: $6 + 4 = 10$ ← sum

survey [sûr′vā] **encuesta** A method of gathering information

tally table [tal′ē tā′bəl] **tabla de conteo** A table that uses tally marks to record data
Example:

Favorite Sport	
Sport	Tally
Soccer	𝍸 III
Baseball	III
Football	𝍸
Basketball	𝍸 I

thirds [thûrdz] **tercios**

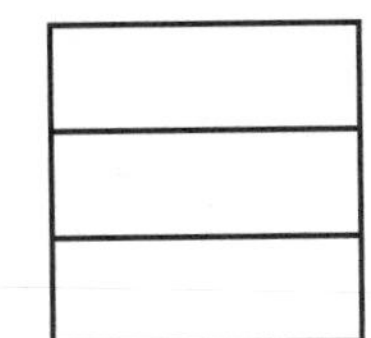

These are thirds (p. 307)

three-dimensional shape [thrē də•men′shə•nəl shāp] **figura tridimensional** A shape that has length, width, and height
Example:

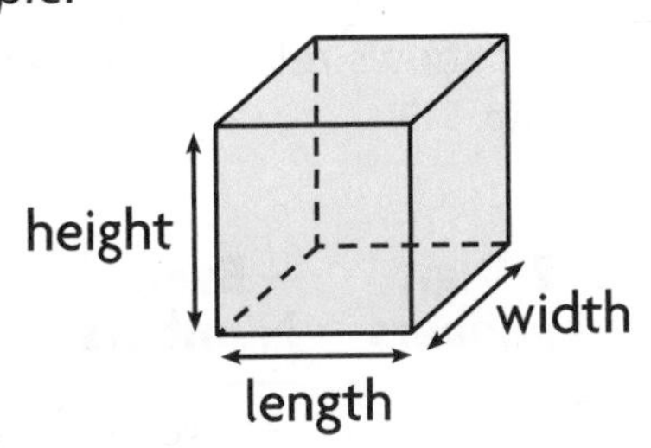

time line [tīm līn] **línea cronológica** A drawing that shows when and in what order events took place

trapezoid [trap′i•zoid] **trapecio** A quadrilateral with exactly one pair of parallel sides (p. 501)
Example:

triangle [trī′ang•gəl] **triángulo** A polygon with three sides and three angles (p. 492)
Examples:

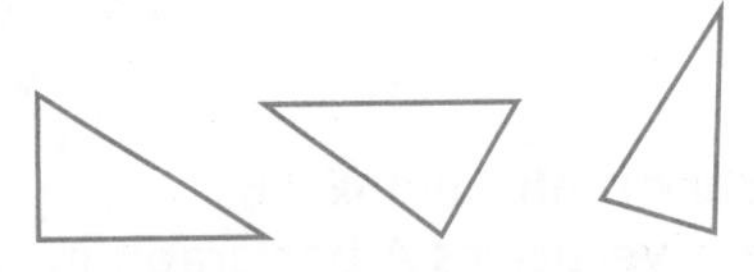

two-dimensional shape [to͞o də•men′shə•nəl shāp] **figura bidimensional** A shape that has only length and width (p. 484)
Example:

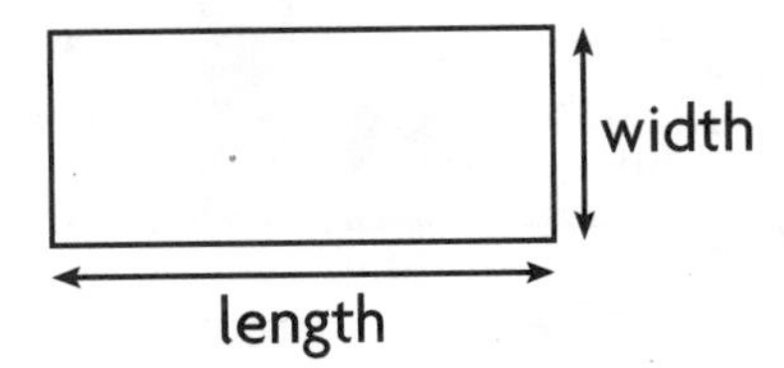

unit fraction [yo͞o′nit frak′shən] **fracción unitaria** A fraction that has 1 as its top number, or numerator (p. 315)
Examples: $\frac{1}{2}$ $\frac{1}{3}$ $\frac{1}{4}$

unit square [yo͞o′nit skwâr] **caudrado de una unidad** A square with a side length of 1 unit, used to measure area (p. 445)

Venn diagram [ven dī′ə•gram] **diagrama de Venn** A diagram that shows relationships among sets of things (p. 513)
Example:

2-Digit Numbers **Even Numbers**

35, 17, 29 | 12, 10 | 8, 6, 4

vertex [vûr′teks] **vértice** The point at which two rays of an angle or two (or more) line segments meet in a plane shape or where three or more edges meet in a solid shape (p. 487)
Examples:

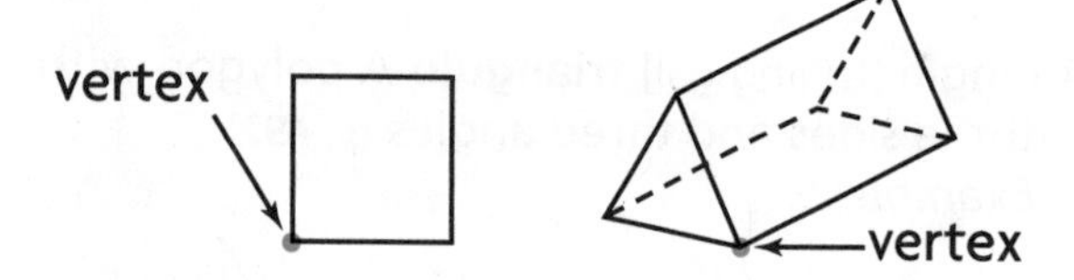

vertical bar graph [vûr′ti•kəl bär graf] **gráfica de barras verticales** A bar graph in which the bars go up from bottom to top (p. 76)

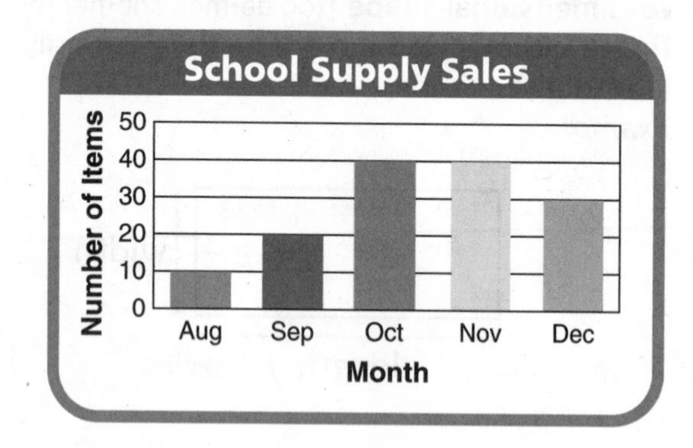

whole [hōl] **entero** All of the parts of a shape or group (p. 307)
Example:

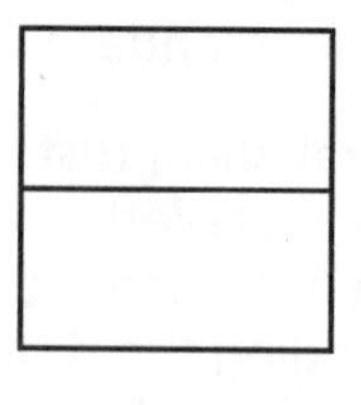

$\frac{2}{2} = 1$

This is one whole.

whole number [hōl num′bər] **número entero** One of the numbers 0, 1, 2, 3, 4, . . . The set of whole numbers goes on without end

word form [wûrd fôrm] **en palabras** A way to write numbers by using words
Example: The word form of 212 is two hundred twelve.

Zero Property of Multiplication [zē′rō präp′ər•tē əv mul•tə•pli•kā′shən] **propiedad del cero de la multiplicación** The property that states that the product of zero and any number is zero (p. 124)
Example: $0 \times 6 = 0$

Table of Measures

METRIC	CUSTOMARY
Length	
1 centimeter (cm) = 10 millimeters (mm)	
1 decimeter (dm) = 10 centimeters (cm)	1 foot (ft) = 12 inches (in.)
1 meter (m) = 100 centimeters	1 yard (yd) = 3 feet, or 36 inches
1 meter (m) = 10 decimeters	1 mile (mi) = 1,760 yards, or 5,280 feet
1 kilometer (km) = 1,000 meters	
Capacity and Liquid Volume	
1 liter (L) = 1,000 milliliters (mL)	1 pint (pt) = 2 cups (c)
	1 quart (qt) = 2 pints
	1 gallon (gal) = 4 quarts
Mass/Weight	
1 kilogram (kg) = 1,000 grams (g)	1 pound (lb) = 16 ounces (oz)

TIME	
1 minute (min) = 60 seconds (sec)	1 year (yr) = 12 months (mo), or about 52 weeks
1 hour (hr) = 60 minutes	1 year = 365 days
1 day = 24 hours	1 leap year = 366 days
1 week (wk) = 7 days	1 decade = 10 years
	1 century = 100 years

MONEY
1 penny = 1 cent (¢)
1 nickel = 5 cents
1 dime = 10 cents
1 quarter = 25 cents
1 half dollar = 50 cents
1 dollar ($) = 100 cents

SYMBOLS
< is less than
> is greater than
= is equal to

Photo Credits

KEY: (t) top, (b) bottom, (l) left, (r) right, (c) center, (bg) background, (fg) foreground, (i) inset

Cover: (Sky) con Tanasiuk/Design Pics/Corbis; (Grass) PBNJ Productions/Corbis; (Frog) Natural Selection/Bill Byrne/ Design Pics/Corbis; (Moth) Michael Crowley/Getty Images.

Table of Contents: v Getty Images; vii Gallo Images/Terra/Corbis; viii (bg) Comstock Images/Jupiterimages Corporation; (bg) Corbis; xi Corbis; Joel Sartore/National Geographic/Getty Images.

Critical Area: 9 Colleen Cahill/Alamy Images; 12 Peter Barritt/Alamy Images; 16 image source/Alamy Images; 20 Creatas/Jupiterimages/Getty Images; 21 photolink/Getty Images; 24 Brand X Pictures/Getty Images; 25 Michael Halberstadt/Alamy; 35 (cr) Stephen Frink Collection/Alamy; 39 Echo/Getty Images; 42 (br) Picture Press/Alamy; (cr) Peter Arnold, Inc./Alamy; (tr) Marilyn Shenton/Alamy Images; 51 (b) Jose Luis Pelaez Inc; 61 Keith Ovregaard/ Getty Images; 66 PhotoDisc/Getty Images; 75 Blend Images/Alamy; 80 (cr) Jeff Greenberg/Alamy; 86 Clouds over Kalahari Gemsbok National Park: digital vision/Getty Images; 88 Comstock/Getty Images; 95 David Frazier/Corbis; 105 Roman Milert/Alamy Images; 108 Dave King/Dorling Kindersley/Getty Images; 111 Ben Molyneux People/Alamy Images; 115 (cr) PhotoDisc/Getty Images; 137 C Squared Studios/Getty Images; 144 Index Stock/Alamy; 149 Geostock/ PhotoDisc/Getty Images; 162 Sonny Senser; 163 David Aubrey/Photo Researchers, Inc.; 181 Sonya Farrell/Getty Images; 182 (b) Goodshoot/Corbis; 191 Image Source/Getty Images; 192 Visions of America, LLC/Alamy Images; 198 Comstock/ Getty Images; 199 Siede Preis/Getty Images; 202 George Doyle/Getty Images; 207 D. Hurst/ Alamy; 209 (tr) Tim Krieger/ Brand X Pictures/PictureQuest; 213 Donna Ikenberry/Art Directors/Alamy; 217 Helene Rogers-Commercial/Alamy; 220 Kitch Bain/Alamy Images; 221 (cr) Juniors Bildarchiv/Alamy; 224 (tr) Lawrence Manning/Corbis; 235 Randy Faris/ Corbis; 238 PhotoDisc/Getty Images; 246 Georgette Douwma/Getty Images; 251 (t) Radius Images/Alamy; 253 Visuals Unlimited/Corbis; 256 (tr) Joe McDonald/Corbis; (tr) PhotoDisc/Getty Images; 257 Pete Ryan/National Geographic/ Getty Images; 261 Peter Cade/Getty Images; 264 Adams Picture Library t/a apl/Alamy Images; 269 Todd Muskopf/ Alamy; 272 (t) James Randklev/Getty Images; 287 Francisco Martinez/Alamy; 291 (t) George Doyle/Getty Images; 292 C Squared Studios/Getty Images; 298 Jack Hollingsworth/Getty Images.

Critical Area: 304 (bl) Weronica Ankarörn; (tr) Richard T. Nowitz/Corbis; 305 PhotoDisc/ Getty Images; 311 PhotoDisc/ Getty Images; 326 (tr) Don Mason/Getty Images; 329 (cr) Stockbyte/Getty Images; 334 (tr) Eric Camden; 341 (tr) Getty Images; 351 (tr) The Photolibrary Wales/Alamy Images; 356 (cr) Deborah Waters - Studio/Alamy Images; 366 (tr) Dorling Kindersley/Getty Images; 372 (tr) Vince Streano/Corbis.

Critical Area: 385 Mischa Photo Ltd/Getty Images; 387 (br) Ken Welsh/Alamy; 394 (c) Ariel Skelley/Getty Images; 396 Andre Jenny/Alamy Images; 397 Scott Andrews/Science Faction/Corbis; 402 Rubberball/Alamy Images; thenakedsnail/Getty Images; 405 PhotoDisc/Getty Images; 413 Stockbyte/Getty Images; 416 (2 images) PhotoDisc/ Getty Images; 417 C Squared Studios/Getty Images; Nikreates/Alamy Images; Stockbyte/Getty Images; 421 GK Hart/ Vikki Hart/Getty Images; Ryan McVay/Getty Images; comstock/Getty Images; artville/Getty Images; 422 (2 images) Stockbyte/Getty Images; comstock/Getty Images; photonic 3/Alamy Images; 423 Ted Foxx/Alamy Images; 426 D. Hurst/ Alamy Images; 442 D. Hurst/Alamy Images; 463 (cr) Sdbphoto carpets/Alamy Images; 470 (cr) Elizabeth Whiting & Associates/Alamy; 471 (tr) somos/veer/Getty Images; 476 jupitermedia/Alamy Images.

Critical Area: 512 (cr) Mark Bond/Alamy Images.

All other photos Houghton Mifflin Harcourt libraries and photographers; Guy Jarvis, Weronica Ankarorn, Eric Camden, Don Couch, Doug Dukane, Ken Kinzie, April Riehm, and Steve Williams.